A MODERN INTRODUCTION TO DYNAMICAL SYSTEMS

A Modern Introduction to Dynamical Systems

RICHARD J. BROWN

OXFORD
UNIVERSITY PRESS

OXFORD
UNIVERSITY PRESS

Great Clarendon Street, Oxford, OX2 6DP,
United Kingdom

Oxford University Press is a department of the University of Oxford.
It furthers the University's objective of excellence in research, scholarship,
and education by publishing worldwide. Oxford is a registered trade mark of
Oxford University Press in the UK and in certain other countries

Published in the United States of America by Oxford University Press
198 Madison Avenue, New York, NY 10016, United States of America

British Library Cataloguing in Publication Data
Data available

Library of Congress Control Number: 2017960673

ISBN 978-0-19-874328-6 (hbk.)
978-0-19-874327-9 (pbk.)

Printed and bound by
CPI Group (UK) Ltd, Croydon, CR0 4YY

This work is dedicated, in loving memory, to my father-in-law
Angelo Detragiache

Preface

The following text comprises the content of a course I have designed and have been running at Johns Hopkins University since the Spring of 2007. It is a senior undergraduate (read: 400-level) analysis-based course in the basic tools, techniques, theory, and development of what is sometimes called the modern theory of dynamical systems. The modern theory, as best as I can define it, focuses on the study and structure of dynamical systems as little more than the study of the properties of one-parameter groups of transformations on a topological space, and what these transformations say about the properties of either the space or the acting group. It is a pure mathematical endeavor in that we study the material simply for the structure inherent in the constructions, and not for any particular application or outside influence. It is understood that many of the topics comprising this theory have natural, beautiful, and important applications, some of which actually dictate the need for the analysis. But the true motivation for the study is little more than the fact that it is beautiful, rich in structure and nuance, and relevant to many tangential areas of mathematics—and that it is there.

When I originally pitched this course to the faculty here at Hopkins, there was no course like it in our department. We have a well-developed engineering school, filled with exceptionally bright and ambitious students, which, along with strong natural and social science programs, provide a ready and eager audience for a course on the pure mathematical study of the theory behind what makes a mathematical model and why we study them. We have a sister department here at Homewood, the Applied Mathematics and Statistics Department, which also offers a course in dynamical systems. However, their course seems to focus on the nature and study of particular models that arise often in other classes, and then to mine those models for relevant information to better understand them. But as a student of the field, I understood that a course on the very nature of using functions as models and then studying their properties in terms of the dynamical information inherent in them was currently missing from our collective curriculum. Hence the birth of this course.

In my personal and humble opinion, it continues to be difficult to find a good text that satisfies all of the properties I think would be possessed by the perfect text for a course such as this one: (1) a focus on the pure mathematical theory of the abstract dynamical system, (2) advanced enough that the course can utilize the relevant topological, analytical, and algebraic nature of the topic without requiring so much prerequisite knowledge as to limit enrollment to just mathematicians, (3) rich enough to develop a good strong story to tell that provides a solid foundation for later individual study, and (4) basic enough that students in the natural sciences and engineering can access the entirety of the content given only the basic foundational material of vector calculus, linear algebra, and differential equations. It is a tall order, this is understood. However, it can be accomplished, I believe, and this text is my attempt at accomplishment.

The original text I chose for the course is *A First Course in Dynamics*, by Boris Hasselblatt and Anatole Katok (Cambridge University Press, 2003), a wonderfully designed storyline from two transformational mathematicians in the field. I saw the line of development they took, from the notion of simple dynamics to the more complicated, as proper and intuitive. I think their focus on using the properties of functions and those of the spaces they are acting upon to develop material is the correct one for this "modern" approach. And their reuse of particular examples over and over again as the story progresses is a strong point. However, in the years I have been teaching and revising the course, I have found myself adding material, redesigning the focus and the schedule, and building in a different storyline; all of this diverging from the text. Encouraged by my students and my general thrill at the field, I decided to create my version of a text. This book is that version.

What the reader will find in this text is my view of the basic foundational ideas that comprise a first (and one-semester) course in the modern theory of dynamical systems. It is geared toward the upper-level undergraduate student studying mathematics, engineering, or the natural and social sciences, with a strong emphasis in learning the theory the way a mathematician would want to teach it. This is a proof-based course. However, when I teach it, I do understand that some of my students do not have experience in writing mathematics in general and proofs in particular. Hence I use the content of the course as a way to also introduce these students to the development of ideas instead of just calculation. It is my hope that these students, upon finishing this course, will begin to look at the models and analysis they see in their other applied classes with an eye to the nature of the model and not just to its mechanics. They are studying to be scholars in their chosen field. Their ability to really "see" the mathematical structure of their tools will be necessary for them to contribute to their field.

This course (this text) is designed to be accessible to a student who has had a good foundational course in the following:

- vector calculus, at least up to the topics of surface integration and the "big three" theorems of Green, Stokes, and Gauss;
- linear algebra, through linear transformations, kernels and images, eigenspaces, orthonormal bases, and symmetric matrices; and
- differential equations, with general first- and second-order equations, linear systems theory, nonlinear analysis, existence and uniqueness of first-order solutions, and the like.

While I make it clear in my class that analysis and algebra are not necessary prerequisites, this course cannot run without a solid knowledge of the convergence of general sequences in a space, the properties that make a set a topological space, and the workings of a group. Hence in the text we introduce these ideas as needed, sometimes through development and sometimes simply through introduction and use. I have found that most of these advanced topics are readily used and workable for students even if they are not fully explored within the confines of a university course. Certainly, having sat through courses in advanced algebra and analysis will be beneficial, but I believe they are not necessary. This text, like all proper endeavors in mathematics, should be seen as a work in progress. The storyline, similar to that of Hasselblatt and Katok, begins with basic definitions of just what is a dynamical system. Once the idea of the dynamical content of a function or differential equation has been established, we take the reader through a number of topics and examples, starting with the notion of simple dynamical systems and moving to the more complicated, all the while developing the language and tools to allow the study to continue. Where possible and where this is illustrative, we bring in applications to base our mathematical study in a more general context, and to provide the reader with examples of the contributing influence the sciences have had on the general theory. We pepper the sections with exercises to broaden the scope of the topic in current discussion, and to extend the theory into areas thought to be of tangential interest to the reader. And we end the text at a place where the course I teach ends, on a notion of dynamical complexity, topological entropy, which is an active area of research. It is my hope that this last topic can serve as a landing on which to begin a more individualized, higher-level study, allowing the reader to further their scholarly endeavor now that the basics have been established.

To all who read this text, I hope that this work contributes well to the learning of high-level mathematics in general and dynamical systems in particular by both students of mathematics and students whose study requires mathematical prowess.

Acknowledgments

I am grateful to many in the mathematics community for facilitating and contributing to this work, both here at Johns Hopkins and beyond. In particular, when I entered the Mathematics Department at JHU in 2005, I came in with an expressed interest in designing and running a course of this nature. The faculty of the Mathematics Department here at Hopkins were very happy to accommodate, and I thank them for this opportunity. This course runs once a year, and others have run the course at times, using my course notes as a guide. These instructors, Qiao Zhang, now at Texas Christian University, Mihai Tohaneanu, now at the University of Kentucky, and Hang Xu, currently a J. J. Sylvester Assistant Professor here at Hopkins, were all instrumental in the development of these notes into the current text. In particular, Mihai was very helpful in prioritizing some of the topics contained herein. And Hang was tireless in his line-by-line edits of a late version of the entire manuscript and a careful scrutiny of exercises. I thank each of them greatly for their help.

Also, I would like to thank my students over these last few years, both for their patience in dealing with early versions of these notes, riddled with both typographical and mathematical errors, and for their dedication, persistent interest, and sometimes excitement about the content of this course. About half of the total students of this course over the last ten years have been math majors. And this participant mix of math and physical science and engineering majors has only added to the value of this work. Their diligence and effort in their quest to understand has helped immensely in creating a more comprehensive and organized storyline. To each of my students, I am especially grateful.

I would also like to thank two people at Oxford University Press for their help in bringing this project to fruition. Keith Mansfield was an early contact at Oxford, and his advice and encouragement were invaluable to me in turning a simple and decent set of lecture notes into a comprehensive text. He was a wonderful source of knowledge about publishing and became a good friend. And Daniel Taber, who took over after Keith's retirement and was the main driver in the final stages of this work. With their leadership and guidance, this work has remained a pleasure throughout.

And lastly, I would like to express my deep gratitude and love to my family, my wife Enrica Detragiache and my two children, Isabelle and Julian. They endured the stresses of a seemingly never-ending project with patience and grace, including extended work sessions on weekends and vacations, and the less-than-great moods that arose during the more trying periods. They are truly the treasure of my life.

Contents

1 What Is a Dynamical System?

1.1 Definitions

As a mathematical discipline, the study of dynamical systems most likely originated at the end of the nineteenth century through the work of Henri Poincaré [44] in his study of celestial mechanics. Once the equations describing the movement of the planets around the sun (describing the interplay between objects of mass in a gravitational field) are formulated, so once the mathematical model is constructed, looking for solutions as a means to describe the planets' motions and make predictions of positions in time is the next step. But when finding solutions to sets of equations is seemingly too complicated or impossible, one is left with studying the mathematical structure of the model to somehow and creatively narrow down the possible solution functions. This view of studying the nature and structure of the equations in a mathematical model for clues as to the nature and structure of its solutions is the general idea behind the techniques and theory of what we now call dynamical systems. Being only 100+ years old, the mathematical concept of a dynamical system is a relatively new idea. And since it really is a focused study of the nature and properties of functions of a single (usually), real (usually) independent variable, one is tempted to classify Dynamical Systems as a subdiscipline of what mathematicians call real analysis. However, one can say that Dynamical Systems draws its theory and techniques from many areas of mathematics, from analysis to geometry and topology, and into algebra. One might call mathematical areas like geometry, topology, and dynamics second-generation mathematics, since they tend to bridge other more pure areas in their theories. But as the study of what it actually means to model phenomena via functions and equations, Dynamical Systems is sometimes called the mathematical study of any mathematical concept that evolves over time. So, as a means to define this concept more precisely, we begin with arguably a most general and yet least helpful statement:

Definition 1.1 A dynamical system *is a mathematical formalization for any* fixed rule *that describes the dependence of the position of a point in some* ambient space *on a* parameter.

- The *parameter* here, usually referred to as "time" due to its reference to applications in the sciences, takes values in the real numbers. Usually, these values come in two varieties:

  (1) discrete (think of the natural numbers $\mathbb{N}$ or the integers $\mathbb{Z}$), or

  (2) continuous (defined by some single interval in $\mathbb{R}$).

  The parameter can sometimes take values in much more general spaces, for instance, subsets of $\mathbb{C}$, $\mathbb{R}^n$, the quaternions, or indeed any set with the structure of an algebraic group. However, classically speaking, a dynamical system really involves a parameter that takes values only in a subset of $\mathbb{R}$. We will hold to this convention.

- The *ambient space* has a "state" to it in the sense that all of its points have a marked position that can change as one varies the parameter. Roughly, every point has a position relative to the other points, and a complete set of *generalized* coordinates on the space often provide this well-defined notion of position. The term "generalized" comes from classical mechanics, referring to coordinates that define configurations of a system relative to a reference configuration. See Section 6.2.1 for a better context. Fixing the coordinate system and allowing the parameter to vary, one can create a functional relationship between the points at one value of the parameter and those at another parameter value. In general, this notion of relative point positions in a space and functional relationships on that space involves the notion of a topology on a set. A topology gives a set the mathematical property of a space; it endows the elements of a set with a notion of nearness to each other and allows for functions on a set to have properties like continuity, differentiability, and such. We will expound more on this later in Section 3.1. We call this ambient space the *state space*: it is the set of all possible states a dynamical system can be in at any parameter value (at any moment of time.)

- The *fixed rule* is usually a recipe for going from one state to the next in the ordering specified by the parameter. For discrete dynamical systems, it is often given as a function. The function, from the state space (the domain) to itself (the codomain), takes each point to its next state. The future states of a point are found by applying the same function to the state space over and over again. This is the process of iteration. If possible, the past states of a point can be found by applying the inverse of the function, or sometimes by choosing an element from the set of elements that the function takes to the current state. This defines the dynamical system recursively via the function. In continuous systems, where it is more involved to define what the successor to a parameter value may be, the continuous movement of points in a space may be defined by a ordinary differential equation (ODE), in a role equivalent to that of the function in a discrete system in that it describes implicitly the method of going from one state to the next, although now defined only infinitesimally. The solution to an ODE (or system of ODEs) would be a function whose domain contains the points of the state space and the parameter space (the product of the state and parameter space is sometimes called the *trajectory space*) and taking values back in the state space (the codomain). Often, this latter function is called the *evolution* of the system, providing a way of going from any particular state to any other state reachable from that initial state via a value of the parameter. As we will see, such a function can be shown to exist, and its properties can often be studied, but in general, it will NOT be known *a priori*, or even knowable *a posteriori*.

Remark 1.2 *It is common in this area of mathematics that the terms fixed rule and evolution are used more or less interchangeably, with both referring to the same objects without distinction. In this book, we will differentiate the two as described above. Namely, the fixed rule will remain the recursively defined recipe for movement within a dynamical system, and the evolution will be reserved for the functional form of the movement of points. Thus, the ODE is simply the fixed rule, while the general solution, if it can be found, is the evolution, for example.*

While this idea of a dynamical system is far too general to be very useful, it is instructive. Before creating a more constructive definition, let's look at some classical examples:

1.1.1 Ordinary Differential Equations (ODEs)

Given the first-order, autonomous (vector) ODE,

(1.1.1)
$$\dot{\mathbf{x}} = \begin{bmatrix} \dot{x}_1 \\ \dot{x}_2 \\ \vdots \\ \dot{x}_n \end{bmatrix} = \begin{bmatrix} f_1(x_1, x_2, \ldots, x_n) \\ f_2(x_1, x_2, \ldots, x_n) \\ \vdots \\ f_n(x_1, x_2, \ldots, x_n) \end{bmatrix} = \mathbf{f}(\mathbf{x}),$$

a solution, if it exists, is a vector of functions $\mathbf{x}(t) = [x_1(t) \; x_2(t) \; \cdots \; x_n(t)]^T$ satisfying the equation $\dot{\mathbf{x}}(t) = \mathbf{f}(\mathbf{x}(t))$, and parameterized by a real variable $t \in \mathbb{R}$ where the common domain of the coordinate functions is some subinterval of $\mathbb{R}$.

Recall that one fairly general form for first-order systems of ODEs is $\dot{\mathbf{x}} = \mathbf{f}(t, \mathbf{x})$, with the independent variable t explicitly represented on the right-hand side. An ODE (or system) is called *autonomous* if t is not explicitly represented, and *nonautonomous* when t is explicit in the ODE. For example, $\dot{x} = \frac{1}{4}tx = f(x, t)$ is nonautonomous, while $\dot{x} = \frac{1}{4}x = f(x)$ is autonomous. We sometimes also call autonomous systems *time-invariant* when the independent variable actually does represent time. The important property of an autonomous ODE is that the laws of motion at any point in time are the same as at any other point in time; the laws of motion are invariant under translations in time.

Exercise 1 Show that if $\mathbf{x}(t)$ is a solution to Equation 1.1.1 defined in an interval $(a, b) \subset \mathbb{R}$, then, for all $c \in \mathbb{R}$, $\mathbf{x}(t - c)$ is a solution to the ODE defined on $(a + c, b + c) \subset \mathbb{R}$.

Indeed, for a first-order autonomous system like Equation 1.1.1, the function on the right-hand side represents a vector field in the state space that does not change in time. This means that going from one state to the next will be the same no matter when the

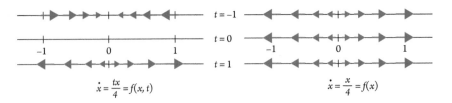

Figure 1 Vector fields on $\mathbb{R}$ under nonautonomous (left) and autonomous (right) ODEs.

motion starts. With time explicit in the function $\mathbf{f}(t,\mathbf{x})$, the vector field would be changing as time progresses. See Figure 1. This time invariance of the vector field of an autonomous ODE is a kind of symmetry that allows one to define an entire 1-parameter family of solutions to the ODE from just one solution.

Of course, it is always possible to render a nonautonomous vector ODE into an autonomous one simply by creating a new state variable equal to the independent time variable, rendering it a dependent variable, and creating a new independent variable as the new time. This increases the size of the state vectors by one, and introduces a new first-order ODE into the system. However, there are two points to consider: (1) the added ODE in the system is just the time derivative set to 1, and (2) there is really no advantage to this rendering in terms of solvability. Indeed, for the system

$$\dot{x}_1 = f_1(x_1,x_2,\ldots,x_n,t),$$
$$\dot{x}_2 = f_2(x_1,x_2,\ldots,x_n,t),$$
$$\vdots$$
$$\dot{x}_n = f_n(x_1,x_2,\ldots,x_n,t),$$

one could simply create a new state variable $x_{n+1} = t$, so that the new autonomous system is now

$$\dot{x}_1 = f_1(x_1,x_2,\ldots,x_n,x_{n+1}),$$
$$\dot{x}_2 = f_2(x_1,x_2,\ldots,x_n,x_{n+1}),$$
$$\vdots$$
$$\dot{x}_n = f_n(x_1,x_2,\ldots,x_n,x_{n+1}),$$
$$\dot{x}_{n+1} = 1.$$

This "trick" may seem straightforward and quite advantageous, as it takes a time-dependent vector field on (a subset of) Euclidean space and associates with it a static (time-invariant) vector field on a one dimension greater space. You did this implicitly when you generated a slope field or graphed solutions to $\dot{x} = f(t,x)$ in the tx-plane, but generated a slope field and graphed solutions to $\dot{x} = f(x,y), \dot{y} = g(x,y)$ in the xy-plane as t-parameterized curves. Where we graph visual representations of our solutions to ODEs is a choice we employ for clarity and understanding, and graphing a t-parameterized solution to a 1-dimensional ODE in $\mathbb{R}$ is difficult to see clearly. Note that this technique of "autonomizing" an ODE system is not so different from rendering an nth-order ODE into a system of first-order ODEs. Creating new dependent variables to untangle an ODE or to expose additional structure or symmetries is a common tactic in gaining insight. However, there may also be pitfalls to this approach. For one, a nonautonomous linear ODE or system of ODEs can easily result in a converted autonomous nonlinear ODE system, losing any advantage of having a linear system. For example, the example $\dot{x} = \frac{1}{4}xt$ is linear, though nonautonomous. Converting this to a autonomous system of two coupled ODEs creates a nonlinear system. Also, one standard way to study nonlinear systems of ODEs is to locate and classify equilibria. However, an autonomous system that was converted from a nonautonomous system will have no equilibria. Nonetheless, the fact that nonautonomous systems can be converted to autonomous ones does render autonomous systems of ODEs as an important and robust, broad category of dynamical systems to study.

Here, for our autonomous system in Equation 1.1.1, we have the following:

- The ODE itself is the fixed rule, describing the infinitesimal way to go from one state to the next by an infinitesimal change in the value of the parameter t. Solving the ODE means finding the unknown function $\mathbf{x}(t)$, at least up to a set of constants determined by some initial state of the system. The inclusion of initial data provides this initial state of the variables of the system, making the system an initial value problem (IVP). A solution to an IVP, $\mathbf{x}(t)$, for valid values of t, provides the various "other" states that the system can reach (either forward or backward in time) as compared with the initial state. Collecting up all the functions $\mathbf{x}(t)$ for all valid sets of initial data (basically, finding the expression that writes the constants of integration of the general solution to the ODE in terms of the initial data variables), into one big function is the evolution.

- This type of dynamical system is called *continuous*, since the parameter t will take values in some domain (an interval with non-empty interior) in $\mathbb{R}$. A dynamical system like this arising from an ODE is also called a *flow*, since the various IVP solutions in phase space look like the flow lines of a fluid in phase space flowing along the slope field (the vector field defined by the ODE).

- In this particular example, the state space is the n-dimensional space parameterized by the n dependent variables that comprise the vector $\mathbf{x}(t)$. Usually, these coordinate functions are simply (subsets of) $\mathbb{R}$, so that the state space is (a subset of) $\mathbb{R}^n$. But there is no restriction that coordinates be rectilinear and no restriction that the state space be Euclidean. In fact, flows on spheres and other non-Euclidean spaces are very interesting to study. Regardless of the properties of the state space, solutions live in it as parameterized curves. These solution curves are often called *trajectories* or *orbits*, although we will make a distinction between these once we actually study these objects. We also call this state space the *phase space*.

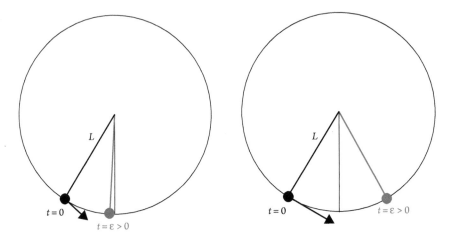

Figure 2 Different velocities render different future configurations.

Remark 1.3 *One should be careful about not confusing a state space, the space of all possible states of a system, with the* configuration space *of, say, a physical system governed by Newton's Second Law of Motion. For example, the set of all positions of a pendulum at any moment of time is simply the circle. This would be the configuration space, the space of all possible configurations. But without knowing the velocity of the pendulum at any particular configuration, one cannot predict future configurations of the system (see Figure 2). The state space, in the case of the pendulum, involves both the position of the pendulum and its velocity (see, for instance, Section 6.2.) For a standard ODE system like the general one above, the state space, phase space, and configuration space all coincide. We will elaborate more on this later.*

So, autonomous first-order ODEs are examples of continuous dynamical systems (actually, differentiable dynamical systems). Solving the ODE (finding the vector of functions $\mathbf{x}(t)$) means finding the rule that stipulates any state of a point in some other parameter value given the state of the point at a starting state. But, as we will soon see, when thinking of ODEs as dynamical systems, we have a different perspective on what we are looking for in solving the ODE.

1.1.2 Maps

Given any set X and a function $f : X \to X$ from X to itself, one can form a dynamical system by simply applying the function over and over (iteratively) to X. When the set has a topology on it (a mathematically precise notion of an "open subset," allowing us to talk about the positions of points in relation to each other), we can then discuss whether the function f is continuous or not. When X has a topology, it is called a space, and a continuous function $f : X \to X$ is called a *map* (we will discuss this in more detail in Section 3.1).

- We will always assume that the sets we specify in our examples are topological spaces, but will detail the topology only as needed. Mostly our state spaces will exist as subsets of real space $\mathbb{R}^n$. Here, one such notion of the nearness of points will result from a precise definition of a distance between points given by a metric. In this context, there should be little confusion. Here the *state space* is X, with the positions of its points given by coordinates on X (defined by the topology).

- The fixed rule is the map f, which is also sometimes called a *cascade*.

- In a purely formal way, f defines the evolution (recursively) by composition with itself. Indeed, for $x \in X$, define $x_0 = x$ and $x_1 = f(x_0)$. Then

$$x_2 = f(x_1) = f(f(x_0)) = f^2(x_0),$$

and for all $n \in \mathbb{N}$ (the natural numbers),

$$x_n = f(x_{n-1}) = f(f(x_{n-2})) = \overbrace{f(f(\cdots f(f(x_0))\cdots))}^{n \text{ times}} = f^n(x_0).$$

- Maps are examples of *discrete dynamical systems*. Some examples of discrete dynamical systems that you may have heard of include discretized ODEs, including difference equations and time-t maps. Also, fractal constructions like Julia sets (Section 7.5) are defined via a discrete dynamical system. One may also say that the famous Mandelbrot set arising from maps of the complex plane to itself is formed via a discrete dynamical system. However, precisely speaking, the Mandelbrot set is actually the parameter space of a dynamical system, recording particular information about an entire family of parameterized maps. Some objects that are not *a priori* considered to be constructed by dynamical systems include fractals like Sierpinski's carpet, Cantor sets, and sequences like Fibonacci's Rabbits (given by a second-order recursion and defined in Section 1.1.5). However, as we shall see, these sets are constructible via dynamical systems and hence play a crucial role in the discipline. Again, we will get to these.

Besides these more classical ideas of a dynamical system, there are much more abstract notions of a dynamical system:

1.1.3 Symbolic Dynamics

Given a set of arbitrary symbols $\mathcal{M} = \{A, B, C, \ldots\}$, consider the "space" of all bi-infinite (i.e., infinite on both sides) sequences of these symbols

$$\Omega_\mathcal{M} = \left\{ (\ldots, x_{-2}, x_{-1}, x_0, x_1, x_2, \ldots) \mid i \in \mathbb{Z}, x_i \in \mathcal{M} \right\}.$$

One can consider $\Omega_\mathcal{M}$ as the space of all functions from $\mathbb{Z}$ to $\mathcal{M}$: each function is just an assignment of a letter in $\mathcal{M}$ to each integer in $\mathbb{Z}$. Now let $f : \Omega_\mathcal{M} \to \Omega_\mathcal{M}$ be the *shift* map: on each sequence, it simply takes the subscript $i \mapsto i+1$; each sequence goes to another sequence that looks like a shift of the original one.

Remark 1.4 *We can always consider this (very large) set of infinite sequences as a space once we give it a topology. This would involve defining open subsets for this set, and we can do this through ϵ-balls by defining a notion of distance between sequences (via a metric). For those who know analysis, what would be a good metric for this set to make it a space using the metric topology? We will define one when we discuss this space in detail in Section 7.1.5. For now, simply think of this example as something to think about.*

This discrete dynamical system is sometimes used as a new dynamical system to study the properties of an old dynamical system whose properties were hard to study. Other times, it is used as a model space for a whole class of dynamical systems that behave similarly. We will revisit these ideas in Section 7.1.4 on inverse limit spaces and in Section 7.1.5.

Sometimes, in a time-dependent system, the actual dynamical system will need to be constructed before it can be studied.

1.1.4 Billiards

Consider two mass-m point-beads moving at constant (possibly different) speeds along a finite-length wire, with perfectly elastic collisions both with each other and with the walls. Recall that this means that the total kinetic energy and the total momentum of the constituent parts in each collision are preserved, so that while energy may be transferred between the point-beads in a collision, no energy is absorbed, either by a wall or another bead. Note that this also means that, while the velocity of a bead may reverse direction at times (after a collision with a wall) or switch with the velocity of the other bead (when the beads collide), there are only a few distinct velocities taken by the beads in the system. In this manner, velocity is not really a variable in this system. So, the state space is still only the set of all positions of the beads. The velocities do play a role in movement around the state space, however.

As an exercise (Exercise 2 below), parameterize the wire from wall to wall as a closed, bounded interval in $\mathbb{R}$. What does the state space look like in this case, then? Taking the position of each bead as a coordinate, the state space is just a triangle in the plane. Work this out. What are the vertices of this triangle? Does it accurately describe all of the states of the system? Are the edges of the triangle part of the state space? Are the vertices? And once you correctly describe the state space, what will motion look like in it as the beads move along the wire at their designated velocities? In other words, how does the dynamical system evolve?

We will revisit this model in detail in Section 5.3.2 as an early example of a type of dynamical system called a *billiard*.

Exercise 2 One way to view the state space, the set of all states of the two point-beads, is to simply view each bead's position as a coordinate along the wire (in a closed subset of $\mathbb{R}$). Then the state of the system at a moment in time can be viewed as a point in the

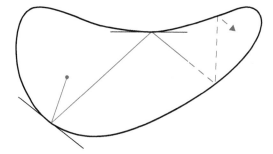

Figure 3 A non-convex billiard.

plane. Parameterize the wire by the interval $[0, 1]$. Then construct the state space as a closed subset of $\mathbb{R}^2$. For given bead velocities v_1 and v_2, describe the motion in the state space by drawing in a representative trajectory. Do this for the following data:

(a) $v_1 = 0, v_2 \neq 0$;

(b) v_1/v_2 a rational number;

(c) v_1/v_2 an irrational number.

By looking at trajectories that get close to a corner, can you describe what happens to a trajectory that intersects a corner directly?

Now consider a single point-ball moving at a constant velocity inside a closed, bounded region of $\mathbb{R}^2$, where the boundary is smooth and collisions with the boundary are specular (mirror-like, meaning that the angle of incidence is equal to the angle of reflection). See Figure 3 for an example of a region that is not convex (see Definition 2.40, if needed). Some questions to ponder: (1) How does the shape of the region affect the types of paths the ball can traverse? (2) Are there closed paths (corresponding to periodic trajectories)? (3) Is there a dense path (one that eventually gets arbitrarily close to any particular point in the region)?

There is a method to study the type of dynamical system called a billiard by creating a discrete dynamical system to record movement by collecting only essential information. In this discrete dynamical system, regardless of the shape of the region, the state space is a cylinder. Can you see how such a state space could be constructed in this situation? Can you identify how the coordinates of a cylinder would relate to this dynamical system? If so, what would the evolution look like?

1.1.5 Higher-Order Recursions

Maps as dynamical systems are examples of first-order recursions, since for $f : X \rightarrow X$, $x_n = f(x_{n-1})$ and each element of a sequence $\{x_n\}_{n \in \mathbb{N}}$ depends only on the previous element. The famous Rabbits of Leonardo of Pisa is a beautiful example of a type of growth that is not exponential, but something called *asymptotically exponential*. We will explore

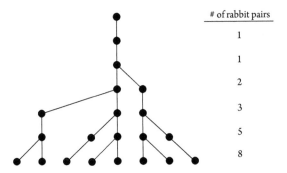

	# of rabbit pairs
	1
	1
	2
	3
	5
	8

Figure 4 A generational chart for Fibonacci's Rabbits.

this more later when we discuss linear maps of the plane, in particular in Section 4.3.5 on fixed points that are saddles. Here, though, we give a brief description.

Place a newborn pair of breeding rabbits in a closed environment. Rabbits of this species produce another pair of rabbits each month after they become fertile (and they never die, nor do they experience menopause). Each new pair of rabbits (again, neglect the incest, gender, and DNA issues) becomes fertile after a month and starts producing each month, starting in the second month. A general idea of how the population changes over the months is given in Figure 4. How many rabbits are there after a year? After 10 years?

Month	a_n	j_n	b_n	Total pairs
1	0	0	1	1
2	0	1	0	1
3	1	0	1	2
4	1	1	1	3
5	2	1	2	5
6	3	2	3	8
7	5	3	5	13

Given the chart in months, we see a way to fashion an expression governing the number of pairs at the end of any given month: Start with r_n, the number of pairs of rabbits in the nth month. Rabbits here will come in three types: Adults a_n, juveniles j_n, and newborns b_n, so that $r_n = a_n + j_n + b_n$. Looking at the chart, we can see that there are constraints on these numbers:

(1) The number of newborns at the $(n+1)$th stage equals the number of adults at the nth stage plus the number of juveniles at the nth stage, so that

$$b_{n+1} = a_n + j_n.$$

(2) This is also precisely equal to the number of adults at the $(n+1)$th stage, so that

$$a_{n+1} = a_n + j_n.$$

(3) Finally, the number of juveniles at the $(n+1)$th stage is just the number of newborns at the nth stage, so that

$$j_{n+1} = b_n.$$

Thus, we have

$$r_n = a_n + j_n + b_n = (a_{n-1} + j_{n-1}) + b_{n-1} + (a_{n-1} + j_{n-1}).$$

And since, in the last set of parentheses, we have $a_{n-1} = a_{n-2} + j_{n-2}$ and $j_{n-1} = b_{n-2}$, we can substitute these to get

$$r_n = a_n + j_n + b_n = (a_{n-1} + j_{n-1}) + b_{n-1} + (a_{n-1} + j_{n-1})$$
$$= a_{n-1} + j_{n-1} + b_{n-1} + a_{n-2} + j_{n-2} + b_{n-2} = r_{n-1} + r_{n-2}.$$

Hence the pattern is ruled by a second-order recursion $r_n = r_{n-1} + r_{n-2}$ with initial data $r_0 = r_1 = 1$. Being a second-order recursion, we cannot go to the next state from a current state without also knowing the previous state. This is an example of a model that is not a dynamical system as stated. We can make it one (by the same method that one would use to turn a higher-order ODE into a first-order system, that is), but we will need a bit more structure, which we will introduce in time.

Remark 1.5 *Leonardo of Pisa, better known as Leonardo Bonacci, or Leonardo Fibonacci (the name is a short form of "filius Bonacci," meaning "son of Bonacci"), or most commonly just Fibonacci, was a thirteenth-century mathematician from the Republic of Pisa, now a part of Italy. He is credited as one of the most talented Western mathematicians of his age, and was a strong early advocate for the use of the Hindu–Arabic number system in Europe, primarily through his composition Liber Abaci, the Book of Abacus or the Book of Calculation. It is not known precisely when he died or even where, but for the mathematical tourist, there is a statue (Figure 5) of Fibonacci in the Camposanto, the cemetery on the Piazza dei Miracoli, close to the famous Tower.*

Exercise 3 A population of opossums, a North American marsupial common to the southeastern part of the United States, breed and evolves, more or less, according to the following stipulations: Given an starting population equal in males and females, each opossum pair will produce four offspring each summer, two of each gender. The young will pair up and produce the following summer. Each opossum will live for two years and die during the third winter. Construct a second-order recursion that will describe the population size of the species. (Note: we will return to this problem in Exercise 157, when discussing first-order vector recursions and linear maps of the plane.)

Figure 5 Leonardo Fibonacci.

Now, with this general idea of what a dynamical system actually is, along with numerous examples, we give a much more accurate and useful definition of a dynamical system:

Definition 1.6 *A dynamical system is a triple* (S, T, Φ)*, where* S *is the state space (or phase space),* T *is the parameter space, and*

$$\Phi : (T \times S) \longrightarrow S$$

is the evolution.

Some notes:

- In the previous discussion, the fixed rule was a map or an ODE that would only define recursively what the evolution would be. In this definition, Φ defines the entire system, mapping where each point $s \in S$ goes for each parameter value $\tau \in T$. It is the functional form of the fixed rule, unraveling the recursion and allowing one to go from a starting point to any point reachable by that point given a value of the parameter.

- In ODEs, Φ plays the role of the *general solution*, as a 1-parameter family of solutions (literally a 1-parameter family of transformations of phase space): In this general solution, one knows for <u>any</u> specified starting value where it will be for <u>any</u> valid parameter value, all in one function of two variables.

  This is the concept of a flow, which you are familiar with from your study of ordinary differential equations: Let $\dot{x} = \mathbf{f}(\mathbf{x})$, $\mathbf{x}(0) = \mathbf{x}_0 \in \mathbb{R}^n$ be an IVP, with a C^1 vector field $\mathbf{f}(\mathbf{x})$. For $I \subset \mathbb{R}$ an interval containing 0, define a continuous map $\varphi : I \times \mathbb{R}^n \to \mathbb{R}^n$ that satisfies the following:

  - $\forall t \in I, \varphi^t = \varphi(t, \cdot) : \mathbb{R}^n \to \mathbb{R}^n$ is a transformation of $\mathbb{R}^n$ (for a given choice of t, this is a map that we will define later as the "time-t map" of the IVP.)
  - $\forall s, t \in I$, where $s + t \in I$, one has

  $$\varphi^s \circ \varphi^t(\mathbf{x}) = \varphi^{s+t}(\mathbf{x}).$$

Hence, flows are the evolutions of ODE systems, if they can be found.

Example 1.7 In the Malthusian growth model $\dot{x} = kx$, with $k \in \mathbb{R}$, and $x(t) \geq 0$ a population, the general solution is given by $x(t) = x_0 e^{kt}$, for $x_0 \in \mathbb{R}_0^+ = [0, \infty)$, the unspecified initial value at $t = 0$. (The notation $\mathbb{R}_0^+$ defines the strictly positive real numbers $\mathbb{R}^+$ together with the value 0: the nonnegative real numbers.) Really, the model works for $x_0 \in \mathbb{R}$, but if the model represents population growth, then initial populations should only be nonnegative, right? Here, $S = \mathbb{R}_0^+$, $T = \mathbb{R}$, and $\Phi(t, s) = se^{kt}$.

Example 1.8 Let $\dot{x} = -x^2 t$, $x(0) = x_0 > 0$. Using the technique commonly referred to as separation of variables, we can integrate to find an expression for the general solution as $x(t) = (t^2/2 + C)^{-1}$. And since $x_0 = C^{-1}$ (you should definitely do these calculations explicitly!), we get

$$\Phi(t, x_0) = \left(\frac{1}{\frac{t^2}{2} + \frac{1}{x_0}} \right)^{-1} = \frac{2x_0}{x_0 t^2 + 2}.$$

Here, we are given $\mathcal{S} = \mathbb{R}^+$, and we can choose $\mathcal{T} = \mathbb{R}$. Question: Do you see any issues with allowing $x_0 < 0$? Let $x_0 = -2$, and describe the particular solution on the interval $t \in (0, 2)$.

Exercise 4 Integrate to find the general solution above for the IVP $\dot{x} = -x^2 t$, $x(0) = x_0 > 0$.

- In discrete dynamics, for a map $f : X \to X$, we would need a single expression to write $\Phi(n, x) = f^n(x)$. This is not always easy or doable, as it would involve finding a functional form for a recursive relation. Try doing this with f a general polynomial of degree more than 1.

Example 1.9 Let $f : \mathbb{R} \to \mathbb{R}$ be defined by $f(x) = rx$, for $r \in \mathbb{R}_+$. Then $\Phi(n, x) = r^n x$.

Example 1.10 For Leonardo of Pisa's (Fibonacci) Rabbits, we will have to use the recursion to calculate every month's population to get to the 10-year mark. However, if we could find a *functional* form for the recursion, giving population in terms of month, we could than simply plug in $12 \cdot 10 = 120$ months to calculate the population after 10 years. It is possible to redefine the model as a true dynamical system, and this dynamical system does have such a functional form, as well shall see in Section 4.3.5. The latter functional form will be the evolution Φ in Definition 1.6.

Exercise 5 Find a closed-form expression for the evolution of $f(x) = rx + a$, in the case where $-1 < r < 1$ and a are constants. Also determine the unique point where $f(x) = x$ in this case.

Exercise 6 For $g(x) = x^2 + 1$, write out the first four iterates $g^i(x)$, $i = 1, 2, 3, 4$. Then look for a pattern with which to write out the nth iterate, $g^n(x)$. Do you see the difficulty? Now if your only interest was to know what the convergence properties of $\{g^n(x)\}_{n \in \mathbb{N}}$ were for arbitrary starting values x_0, what can you assert? And can you prove your assertions?.

In general, finding Φ (in essence, solving the dynamical system) is very difficult, if not impossible, and certainly often impractical and/or tedious. However, it is often the case that the purpose of studying a dynamical system is not to actually solve it. Rather, it is to gain insight into the structure of its solutions. Really, we are trying to make *qualitative* statements about the system rather than quantitative ones. Think about what you did when studying nonlinear systems of first-order ODEs in any standard undergraduate course in differential equations. Think about what you did when studying autonomous first-order ODEs.

Before embarking on a more systematic exploration of dynamical systems, here is another less rigorous definition of a dynamical system:

Definition 1.11 *Dynamical Systems as a field of study attempts to understand the structure of a changing mathematical system by identifying and analyzing the things that do not change.*

There are many ways to identify and classify this notion of an unchanging quantity amidst a changing system. But the general idea is that if a quantity within a system does not change while the system as a whole is evolving, then that quantity holds a special status as a *symmetry*. Identifying symmetries can allow one to possibly locate and identify solutions to an ODE. Or one can use a symmetry to create a new system, simpler than the previous, where the symmetry has been factored out, reducing either the number of variables and/or the size of the system.

More specifically, here are some of the more common notions:

- **Invariance:** First integrals: Sometimes a quantity, defined as a function on all or part of the phase space, is constant along the solution curves of the system. If one could create a new coordinate system of phase space where one coordinate is the value of the first integral, then the solution curves correspond to constant values of this coordinate. The coordinate becomes useless to the system, and it can be discarded. The new system then has fewer degrees of freedom than the original. Phase space volume: In a conservative vector field, as we will see, if we take a small ball of points of a certain volume and then flow along the solution curves to the vector field, the ball of points will typically bend and stretch in very complicated ways. But it will remain an open set, and its total volume will remain the same. This is *phase volume preservation*, and it says a lot about the behavior and types of solution curves.

- **Symmetry:** Periodicity: Sometimes individual solution curves (or sets of them) are closed, and solutions retrace their steps over certain intervals of time or number of iterates. If the entire system behaves like this, the direction of the flow contains limited information about the solution curves of the system. One can in a sense factor out the periodicity, revealing more about the remaining directions of the state space. Or even near an isolated singular periodic solution, one can discretize the system at the period of the periodic orbit. This discretized system has a lower order, or number of variables, than the original.

- **Asymptotics:** In certain autonomous ODEs (systems where the time is not explicitly expressed in the system), one can start at any moment in time and the evolution depends only on the starting time. In systems like these, the long-term behavior of solutions may be more important than where they are in any particular moment in time. In a sense, one studies the asymptotics of the system, instead of attempting to solve it. Special solutions like equilibria and limit cycles are easy to find, and their properties become important elements of the analysis.

Example 1.12 Recall that a first-order differential equation that can be written in either of the forms

$$M(x,y)\,dx + N(x,y)\,dy = 0 \quad \text{or} \quad M(x,y) + N(x,y)\frac{dy}{dx} = 0$$

is called *exact* if both M and N have continuous derivatives and $M_y = \partial M/\partial y = \partial N/\partial x = N_x$. When an ODE is exact, we then know that there exists a function $\phi(x,y)$, where $\partial\phi/\partial x = M$ and $\partial\phi/\partial y = N$. Indeed, given a twice-differentiable function $\phi(x,y)$ defined on a domain in the plane, its level sets are equations $\phi(x,y) = C$, for C a real constant. Each level set defines y implicitly as a function of x. Thinking of y as tied to x implicitly, so that $y = y(x)$, we can differentiate the equation $\phi(x,y) = \phi(x,y(x)) = C$ with respect to x (remember that if two expressions of x are declared to be equal, then their derivatives will be equal also) and we get

$$\frac{d}{dx}\phi(x,y(x)) = \frac{\partial\phi}{\partial x} + \frac{\partial\phi}{\partial y}\frac{dy}{dx} = 0.$$

This last equation will match the original ODE, in the form on the right, precisely when the above two properties $\partial\phi/\partial x = M$ and $\partial\phi/\partial y = N$ hold. The interpretation is then as follows: The solutions to the ODE correspond to the level sets of the function ϕ. We can say that that solutions to the ODE "are forced to live" on the level sets of ϕ. Thus, we can write the general solution set (at least implicitly) as $\phi(x,y) = C$, again a 1-parameter family of solutions. Here ϕ is called a *first integral* of the flow given by the ODE, a concept we will define precisely in Chapter 6.

Exercise 7 Solve the differential equation $12 - 3x^2 + (4 - 2y)dy/dx = 0$ and express the general solution in terms of the initial condition $y(x_0) = y_0$. This is your function $\phi(x,y)$.

Example 1.13 Newton–Raphson: Finding a root of a (twice-differentiable) function $f : \mathbb{R} \to \mathbb{R}$ leads to a discrete dynamical system $x_n = g(x_{n-1})$, where

$$g(x) = x - \frac{f(x)}{f'(x)}.$$

One here does not need to actually solve the dynamical system (find the evolution or a form for the function Φ in Definition 1.6). Instead, all that is needed is to satisfy some basic properties of f to know that if you start sufficiently close to a root, the long-term (asymptotic) behavior of the iterates of any initial guess is a root. We will expand on this idea in Section 2.2.3.

Exercise 8 One can use the Intermediate Value Theorem of single-variable calculus to conclude that there is a root to the polynomial $f(x) = x^3 - 3x + 1$ in the unit interval $I = [0,1]$ (check this!). For starting values every tenth on I, iterate $g(x)$ to estimate this root to three decimal places (it converges quite quickly!). Now try to explain what is happening when you get to both $x_0 = 0.9$ and $x_0 = 1$.

Example 1.14 Autonomous ODEs: One can integrate the autonomous first-order ODE (in this IVP)

$$y' = f(y) = (y-2)(y+1), \quad y(0) = y_0,$$

since it is separable, and the integration will involve a bit of partial fraction decomposing. The solution is

(1.1.2)
$$y(t) = \frac{Ce^{3t} + 2}{1 - Ce^{3t}}.$$

Exercise 9 Calculate Equation 1.1.2 for the ODE in Example 1.14.

Exercise 10 Now find the evolution for the ODE in Example 1.14 (this means write the general solution in terms of y_0 instead of the constant of integration C.)

But, really, is the calculation for the explicit solution to the ODE in Example 1.14 even necessary? If only general behavior of solutions is important, then one can simply draw the *phase line* as in Figure 6. Recall from your ODE course that for any autonomous first-order ODE in one variable, $y' = f(y)$, its slope field in the ty-plane acts as a vector field whose integral curves are the solutions $y(t)$. For $f \in C^1$, these integral curves exist and are uniquely defined, meaning they cannot cross in the ty-plane. This is due to the Picard–Lindelöf Theorem (Theorem 2.44, commonly referred to just as the Existence and Uniqueness Theorem for a first-order ODE in one variable).

A phase line diagram can characterize much of the general behavior of solutions via a dipiction of the y-axis in the ty-plane together with diagrammatic indications of the slopes of the slope field lines along the axis. These indications include the following:

- Dots where equilibrium solutions cross the y-axis, indicating values where $f(y) = 0$;
- Leftward arrows on intervals outside of the dots where $f(y) < 0$, indicating where solutions decrease in y for all time;
- Rightward arrows on intervals between dots where $f(y) > 0$, indicating where solutions increase in y for all time.

From the schematic in Figure 6, the long-term tendencies of each solution can be readily gleaned. For instance, the equilibrium solutions occur at $y(t) \equiv -1$ and $y(t) \equiv 2$. And any solution whose initial value $y(0) = y_0 \in (-1, 2)$ will tend toward the equilibrium at -1. Given this diagram, we say that the equilibrium at -1 is asymptotically stable (the one at 2 is unstable). Thus, if long-term behavior is all that is necessary to understand the system, then we have

$$\lim_{t \to \infty} y(t) = \begin{cases} -1 & \text{if } y_0 < 2, \\ 2 & \text{if } y_0 = 2, \\ \infty & \text{if } y_0 > 2. \end{cases}$$

Figure 6 The phase line of $y' = (y - 2)(y + 1)$.

In both these last two examples, actually solving the dynamical system isn't necessary to gain important and possibly sufficient information about the solutions of the system.

1.2 The Viewpoint

In a certain sense, Dynamical Systems, as a field of study, is a type of mathematical analysis: the study of the formal properties of sets and the structures defined on them. Examples include functions and topological spaces and their properties. You encountered analysis in your calculus classes, although there one focuses on the more technical aspects of calculating and determining the properties of functions defined on a particular set, the real line $\mathbb{R}$. Indeed, the properties of functions and the spaces that serve as their domains and codomains are intimately intertwined in sometimes obvious and often subtle ways. For example, according to a celebrated 1912 theorem (Theorem 3.28 in this text) by Luitzen E. J. Brouwer [12], any continuous function from a compact, convex space to itself must contain at least one point where its image under the function is the same as the point itself (a *fixed point* of the function). This fact has been celebrated in various ways over the years in serious mathematical endeavors as well as simple fun facts; for example, visit a new city and open a map while there. According to Brouwer, there must always be a point on the map that precisely corresponds to your current position at the moment. Mathematically, this property has enormous implications for not simply the function we apply to the space, but for the space itself. The consequences of a theorem like this are evident even on the beginning stages of what can be called "higher math," like calculus and differential equations.

In general, studying how a map moves around the points of the space is to study the *dynamical content* of the map. Where the points go upon repeated iteration of a map on a space, or how solutions of a differential equation behave once their parameter domain is known, is to study the system *dynamically*. If most or all of the solutions tend to look alike, or if the diversity of the collection of iterates of a point under a map is small, then we say that the dynamics are *simple*. In essence, they are easy to describe, or it does not take a lot of information to describe the diversity of orbit behavior. In contrast, if different solutions to the ODE can do many different things, or if it takes a lot of information to describe how many different ways a map can move distinct points around in a space via iteration, then we say that the dynamics are *complex* or *complicated*. One may say that a dynamical system is more *interesting* if it is more complicated to describe, although that is certainly a subjective term.

The general goal of an analysis of a dynamical system is typically not to solve the system or to find an explicit expression for its evolution. Many nonlinear systems of ODEs are difficult, if not impossible, to solve. Rather, the goal of an analysis of a dynamical system is a general description of the movement of points under the map or the ODE.

In the following chapters, we will develop a language and methods of analysis to study the dynamical content of various kinds of dynamical systems. We will survey both discrete and continuous dynamical systems that exhibit a host of phenomena, and mine these situations for ways to classify and characterize the behavior of the iterates of a map or solutions of the ODE. We will show how the properties of the maps and the spaces they use as domains affect the dynamics of their interaction. We will start with situations that exhibit relatively simple dynamics, and progress through situations and applications of increasing complexity (complicated behavior). In all of these situations, we will keep the maps and spaces as easy to define and work with as possible, to keep the focus directly on the dynamics.

Perhaps the best way to end this chapter is on a more philosophical note, and allow a possible raison d'etre for why Dynamical Systems even exists as a field of study enmeshed in the world of analysis, topology, and geometry:

Definition 1.15 *Dynamical Systems is the study of the information contained in and the effects of groups of transformations of a space.*

For a discrete dynamical system defined by a map on a space, the properties of the map, as well as those of the space, will affect how points are moved around the space. As we will see, maps with certain properties can only do certain things, and if the space has particular properties, like the compactness and convexity properties of the space in Brouwer's Fixed-point Theorem, then certain things must be true (or may not be, like a fixed-point free transformation). Dynamics is the exploration of these ideas, and we will take this view throughout this text.

2 Simple Dynamics

2.1 Preliminaries

2.1.1 A Simple System

To motivate our first discussion and set the playing field for an exploration of some simple dynamical systems, recall some general theory of first-order autonomous ODEs in one dimension: Let

(2.1.1)
$$\dot{x} = f(x), \quad x(0) = x_0$$

be an IVP (again, an ODE with an initial value) where the function $f(x)$ is a differentiable function on all of $\mathbb{R}$. From any standard course in differential equations, we know that this means that solutions will exist and be uniquely defined for all values of $t \in \mathbb{R}$ near $t = 0$ and for all values of $x_0 \in \mathbb{R}$. Recall that the *general solution* of this ODE will be a 1-parameter family of functions $x(t)$ parameterized by x_0. In reality, one would first use some sort of integration technique (as best as one can) to find $x(t)$ parameterized by some constant of integration C. Indeed, recall that an autonomous ODE like Equation 2.1.1 is always separable, although $1/f(x)$ may not be easy to integrate. (A rather well-known example of a function not easily integrated is $f(x) = e^{x^2}$.) Then one would solve for the value of C given a value of x_0 to find a *particular solution* to the IVP, one that solves the ODE and is compatible with the initial value. However, one could also solve generally for C as a function of x_0 in the general solution to get

(2.1.2)
$$x : \mathbb{R} \times \mathbb{R} \to \mathbb{R}, \quad (t, x_0) \mapsto x(t, x_0)$$

as the evolution. Then, for each choice of x_0, we would get $x_{x_0}(t)$, a function from $\mathbb{R}$ to $\mathbb{R}$, as the particular solution to the IVP. This notation, using x_0 as a subscript, parameterizes the family of particular solutions to the ODE directly by the initial value. Specifying a value for x_0 means solving the IVP. Leaving x_0 unspecified means that if and when we ultimately have initial data, we will have a readily available particular solution. A visualization of this solution would be a graph of $x_{x_0}(t)$ in the tx-plane as a curve (the trajectory) passing through the point $(0, x_0)$. We sometimes say the solution "lives" in the tx-plane. Graphing a bunch of representative trajectories gives a good idea of what the evolution looks like. You did this in your differential equations course when you created *phase portraits*.

Example 2.1 Back to the Malthusian growth model of Example 1.7, let $\dot{x} = kx$, with $k \in \mathbb{R}$ a constant. Here, a general solution to the ODE is given by $x(t) = Ce^{kt}$. If, instead, we were given the IVP $\dot{x} = kx$, $x(0) = x_0$, the particular solution would be $x(t) = x_0 e^{kt}$. The trajectories would look like graphs of standard exponential functions (as long as $k \neq 0$) in the tx-plane. Figure 7 shows the three cases, which look substantially different from each other, when $k < 0$, $k = 0$, and $k > 0$.

Recall in higher dimensions the system introduced in Section 1.1.1, $\dot{\mathbf{x}} = \mathbf{f}(\mathbf{x})$. Visualizing solutions to this system, we typically do not graph them explicitly as functions of t. Rather, we use the t-parameterization of solutions $\mathbf{x}(t) = [x_1(t) \ x_2(t) \cdots x_n(t)]^T$ to trace out a curve directly in $\mathbf{x}$-space. This space, whose coordinates are the set of dependent variables $x_1, x_2, \ldots, x_n$, is called the *phase space* of the ODE, and stands in contrast to the domain of the evolution in Equation 2.1.2, the tx-plane, or tx-space, which we call the *trajectory space*. In fact, we will make this distinction explicit:

Definition 2.2 *For $\mathbf{x}(t)$, a solution to the ODE $\dot{\mathbf{x}} = \mathbf{f}(\mathbf{x})$, defined on an interval $(a, b) \subset \mathbb{R}$, a* trajectory *is the set*

$$\left\{ (t, \mathbf{x}) \in \mathbb{R} \times \mathbb{R}^n = \mathbb{R}^{n+1} \mid \mathbf{x} = \mathbf{x}(t), \ a \leq t \leq b \right\}.$$

An orbit *is the projection to $\mathbb{R}^n$, the last n coordinates, or the $\mathbf{x}$-subspace.*

We do note, though, that these two terms are often used interchangeably, and in context it is usually clear what the intentions are.

Example 2.3 The linear system IVP $\dot{x} = -y, \dot{y} = x, x(0) = 1, y(0) = 0$ has the particular solution $x(t) = \cos t, y(t) = \sin t$. Graphing the trajectory, according to the above

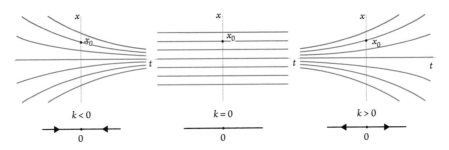

Figure 7 Sample solutions for $x_{x_0}(t) = x_0 e^{kt}$.

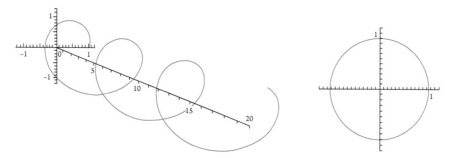

Figure 8 Solution curve $x(t) = \cos t$, $y(t) = \sin t$ in the txy-trajectory space and the xy-phase plane.

stipulations, means graphing the curve in txy-space, a copy of $\mathbb{R}^3$. While informative, it may be a little tricky to fully "see" what is going on. But the orbit, graphed in the xy-plane, which is the phase space, is the familiar unit circle a (circle of radius 1 centered at the origin). Here $t \in \mathbb{R}$ is the coordinate directly on the circle, and even the fact that it overwrites itself infinitely often is not a serious sacrifice to understanding. See Figure 8.

Of course, even for autonomous ODEs in one dependent variable, one could graph the orbits as subsets of $\mathbb{R}$. However, in this way, even when solutions may be uniquely defined, orbits can occupy the same points in $\mathbb{R}$ as long as the times are different, and the resulting graphs would be very confusing. Hence, for clarity, we usually graph trajectories for ODEs of one dependent variable, and orbits for ODEs of more than one dependent variable (or higher-order ODEs). However, even for ODEs of one dependent variable, the schematic phase-line diagram does give a good qualitative description of the behavior of orbits of $\dot{x} = f(x)$ in $\mathbb{R}$ without the clutter of orbits overwriting each other.

Going back to Figure 7 for a minute, note that the phase lines for $\dot{x} = kx$ for the three cases are pictured below the solution graphs. Also note that the phase line really is the phase space of the ODE, although we neglect to graph actual solutions and instead simply indicate directions of motion.

2.1.2 The Time-t Map

Again, for $\dot{x} = f(x)$, $x(0) = x_0$, the general solution $x : \mathbb{R} \times \mathbb{R} \to \mathbb{R}$, $x(t, x_0)$ is a 1-parameter family of solutions, written as $x_{x_0}(t)$, where the second argument is considered a parameter, and for each value x_0, the solution is simply a function of t. However, we can also think of this family of curves in a much more powerful way, namely, as a 1-parameter family of transformations of the phase space! To see this, rewrite the general solution as $\varphi : \mathbb{R} \times \mathbb{R} \to \mathbb{R}$, or $\varphi(t, x_0)$, instead of the possibly confusing notation $x(t, x_0)$. Now, instead of thinking of x_0 as the parameter, fixing the second argument and varying the first as the independent variable, do it the other way: fix a value of t, and allow the variable $x_0 = x$ (the starting point) to vary. Then we get, for $t = t_0$,

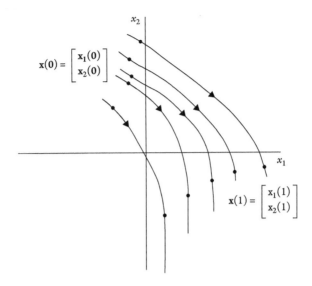

Figure 9 Some solutions to a planar ODE system at time $t = 0$ and time $t = 1$.

$$\varphi : \mathbb{R} \times \mathbb{R} \to \mathbb{R}, \quad \varphi(t_0, x) = \varphi^{t_0}(x),$$

and $\varphi^0(x) = x = x(0) \mapsto x(t) = \varphi^t(x)$. As t varies, every point $x \in \mathbb{R}$ (thought of as the initial point $x(0)$, gets "mapped" to its new position at $x(t)$. Since all solutions are uniquely defined for each value of t_0, this is a function on the "space of states" and will have some very nice properties. But this alternate way of looking at the solutions of an ODE, as a family of transformations of its phase space, is the true *dynamical view*, and one we will explore frequently.

Let X denote any particular topological space. For now, though, just think of X as some subset of the real space $\mathbb{R}^n$, something you are familiar with.

Definition 2.4 *For* $f : X \to X$ *a map, define the set*

$$\mathcal{O}_x = \left\{ y \in X \mid y = f^n(x), \ n \in \mathbb{N} \right\}$$

as the (forward) orbit of $x \in X$ *under* f.

Some notes:

- We define $\mathcal{O}_x$ as a set of points in X, but it is really more than just a set. It is a collection of points in X ordered, or parameterized, by the natural numbers: a sequence. Hence, we often write $\mathcal{O}_x = \left\{ x, f(x), f^2(x), \ldots \right\}$ to note the order, or, for $x_{n+1} = f(x_n)$, $\mathcal{O}_x = \{x_0, x_1, x_2, \ldots\}$.

- If f is invertible, we can also then define the backward orbit

$$\mathcal{O}_x^- = \left\{ y \in X \mid y = f^{-n}(x),\ n \in \mathbb{N} \right\}$$

and the full orbit

$$\mathcal{O}_x = \left\{ y \in X \mid y = f^n(x),\ n \in \mathbb{Z} \right\},$$

rewriting the forward orbit then as $\mathcal{O}_x^+$. However, at times, even for invertible maps, we are only concerned with the forward orbit and simply write $\mathcal{O}_x$, using context for clarity.

- If it is necessary to relate an orbit directly to the function of the dynamical system, we will use the notation $\mathcal{O}_x(f)$. And if we are interested in a partial orbit, say up to iterate n, we will use the notation

$$\mathcal{O}_x^n = \{x_0, x_1, \ldots, x_{n-1}\}.$$

Consider the discrete dynamical system $f : \mathbb{R} \to \mathbb{R}$, given by $f(x) = rx,\ r > 0$. What do the orbits look like? Basically, for $x \in \mathbb{R}$, we get

$$\mathcal{O}_x = \left\{ x, rx, r^2 x, r^3 x, \ldots, r^n x, \ldots \right\}.$$

In fact, we can "solve" this dynamical system by constructing the evolution

$$\Phi(n, x) = f^n(x) = r^n x.$$

Do the orbits change in nature as one varies the value of r? How about when r is allowed to be negative? How does this relate to the ordinary differential equation $\dot{x} = kx$?

Definition 2.5 *For $t \geq 0$, the time-t map of a continuous dynamical system is the transformation of state space that takes $x(0)$ to $x(t)$.*

Example 2.6 Again, back to Example 1.7, with $\dot{x} = kx$, $x(0) = x_0$ and $k < 0$. The state space is $\mathbb{R}$ (the phase space, as opposed to the trajectory space $\mathbb{R}^2$), and the general solution is $\Phi(t, x_0) = x_0 e^{kt}$ (the evolution of the dynamical system is $\Phi(t, x) = x e^{kt}$). Notice that

$$\Phi(0, x) = x, \quad \text{while} \quad \Phi(1, x) = e^k x.$$

Hence the time-1 map is simply multiplication by $r = e^k$. The time-1 map is the discrete dynamical system on $\mathbb{R}$ given by the function above, $f(x) = rx$. In this case, $r = e^k$, where $k < 0$, so that $0 < r = e^k < 1$. See Figure 10. Now how do the orbits behave?

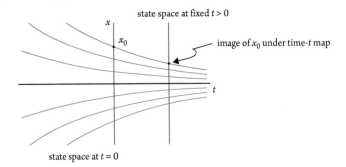

state space at fixed $t > 0$

x_0

image of x_0 under time-t map

t

state space at $t = 0$

Figure 10 The time-$t > 0$ map for $\dot{x} = kx$, $k < 0$.

Exercise 11 Given any dynamical system, describe the time-0 map.

Definition 2.7 *For a discrete dynamical system* $f : X \to X$, *a* fixed point *is a point* $x_* \in X$, *where* $f(x_*) = x_*$, *or where*

$$\mathcal{O}_{x_*} = \{x_*, x_*, x_*, \ldots\}.$$

The orbit of a fixed point is also called a trivial orbit. All other orbits are called non-trivial.

Definition 2.8 *For* $f : X \to X$ *a map, a point* $x \in X$ *is called* periodic (with period n) *if* $\exists n \in \mathbb{N}$ *such that* $f^n(x) = x$. *The smallest such natural number is called the* prime period *of* x.

Notes:

• If $n = 1$, then x is a fixed point.
• Define

$$\mathrm{Fix}(f) = \left\{ x \in X \mid f(x) = x \right\}$$
$$\mathrm{Per}_n(f) = \left\{ x \in X \mid f^n(x) = x \right\}$$
$$\mathrm{Per}(f) = \left\{ x \in X \mid \exists n \in \mathbb{N} \text{ such that } f^n(x) = x \right\}.$$

Keep in mind that these sets are definitely not mutually exclusive, and for $m, n \in \mathbb{N}$, $\mathrm{Per}_m(f) \subset \mathrm{Per}_n(f)$ precisely when $m \mid n$ (i.e., when n is an integer multiple of m.)

So, in our example above, $f : \mathbb{R} \to \mathbb{R}$, $f(x) = e^k x$, $k < 0$, we have $x = 0$ as the only fixed point. This corresponds nicely with the unique particular solution to the ODE $\dot{x} = kx$ corresponding to the equilibrium $x(t) \equiv 0$. Hence $\mathrm{Fix}(f) = \{0\}$. But it is also true that there are no *non-trivial* periodic points (points of prime period greater than 1: Can you show this?), so that $\mathrm{Fix}(f) = \mathrm{Per}_n(f)$ for each $n \in \mathbb{N}$, so that $\mathrm{Fix}(f) = \{0\} = \mathrm{Per}(f)$.

Exercise 12 Show that for $k \neq 0$, there are no non-trivial periodic points.

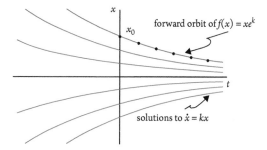

Figure 11 The orbit of a time-t map.

So what else can we say about the "structure" of the orbits? That is, what else can we say about the "dynamics" of this dynamical system? For starters, the forward orbit of a given x_0 will look like the graph of the discrete function $f_{x_0} : \mathbb{N} \to \mathbb{R}^2$, $f_{x_0}(n) = x_0 e^{kn}$. Notice how this orbit follows the trajectory of x_0 of the continuous dynamical system $\dot{x} = kx$. Here, f is the time-1 map of the ODE. Notice also that, as a transformation of phase space (the x-axis), f is not just a continuous function but a differentiable one, with $0 < f'(x) = e^k < 1, \forall x \in \mathbb{R}$. The orbit of the fixed point at $x = 0$, as a sequence, certainly converges to 0. But here <u>all</u> orbits have this property, and we can say

$$\forall x \in \mathbb{R}, \quad \lim_{n \to \infty} \mathcal{O}_x = 0, \text{ or } \mathcal{O}_x \longrightarrow 0.$$

This gives a sense of what we will mean by a dynamical system exhibiting *simple dynamics*: If with very little effort or additional structure, one can completely describe the nature of all of the orbits of the system. Here, there is one fixed point, and all orbits converge to this fixed point.

Definition 2.9 *For a discrete dynamical system, a smooth curve (or set of curves) ℓ in state space is called an* orbit line *if $\forall x \in \ell, \mathcal{O}_x \subset \ell$.*

Example 2.10 The orbit lines for time-t maps of ODEs are the trajectories of the ODE.

Exercise 13 Go back to Figure 7. Describe completely the orbit structure of the discrete dynamical system $f(x) = rx$ for the other two cases, when $r = 1$ and $r > 1$ (corresponding to $r = e^k$, for $k = 0$ and $k > 0$, respectively). That is, classify all possible different types of orbits, in terms of whether they are fixed or not, where they go as sequences, and such. You will find that, even here, the dynamics are simple, but at least for the $k > 0$ case, one has to be a little more careful about accurately describing where orbits go.

Exercise 14 As in the previous exercise, describe the dynamics of the discrete dynamical system $f(x) = rx$, when $r < 0$ (again, there are a number of cases here). In particular, what do the orbit lines look like in this case? You will find that this case does not, in general, correspond to a time-t map of the ODE $\dot{x} = kx$ for any value of k (why not?)

Exercise 15 Show that there does not exist a first-order, autonomous ODE where the map $f(x) = rx$ corresponds to the time-1 map, when $r < 0$.

Exercise 16 For $r < 0$ a constant, construct a second-order constant coefficient ODE whose time-1 map is $f(x) = rx$. (Hint: Work with the first-order system and then convert back at the end.)

Exercise 17 For the discrete dynamical system $f : \mathbb{R} \to \mathbb{R}$, $f(x) = rx + b$, calculate the evolution in closed form. Then completely describe the orbit structure when $b \neq 0$, noting in particular the different cases for different values of $r \in \mathbb{R}$.

Remark 2.11 *Is there really any difference in the dynamical content of the discrete dynamical systems given by the linear map $f : \mathbb{R} \to \mathbb{R}$, $f(x) = kx$, and the affine map $g(x) = kx + b$? By a linear change of variables (a translation here), one can be changed to the other in such a way that orbits go to orbits. And if this is so, is there any real need to study the more general affine map g once we know all of the characteristics of maps like f? We will explore this idea later in the concept of a topological conjugacy; the idea that two dynamical systems can be equivalent. For now, a simple exercise:*

Exercise 18 For $k \neq 1$, and f and g as in Remark 2.11, find a linear change of variables $h : \mathbb{R} \to \mathbb{R}$ that satisfies the condition that $g \circ h(x) = h \circ f(x)$ (equivalently, that $f(x) = h^{-1} \circ g \circ h(x)$.

Exercise 19 Given $\dot{x} = f(x)$, $f \in C^1(\mathbb{R})$, recall that an *equilibrium solution* is defined as a constant function $x(t) \equiv c$ that solves the ODE. It can be found by solving $f(x) = 0$. Instead, define an equilibrium solution $x(t)$ as follows: A solution $x(t)$ to $\dot{x} = f(x)$ is called an equilibrium solution if there exists $t_1 \neq t_2$ in the domain of $x(t)$ where $x(t_1) = x(t_2)$. Show that this new definition is equivalent to the old one.

Exercise 20 For the first-order autonomous ODE $dp/dt = p/2 - 450$, do the following:

- Solve the ODE by separating variables. Justify explicitly why the absolute value signs are not necessary when writing the general solution as a single expression.
- Calculate the time-1 map for this ODE flow.
- Discuss the simple dynamics of this discrete dynamical system given by the time-1 map.

And lastly, before we move on, one common theme that arises when discussing the orbit structure of a dynamical system, whether discrete or continuous, is to understand well the long-term (read: asymptotic) behavior of orbits. For orbits of discrete dynamical systems, sequences, this involves an analysis of the sequence's limit points (whether or not the sequence actually has a limit!) And for continuous dynamical systems, one can construct sequences within orbits via monotone sequences of parameter values that limit to the edge of the parameter space. We will expound on this later and throughout the text. But, for now, we offer a basic definition of a concept that will help us to encode the asymptotic nature of orbits:

Definition 2.12 *For $f : X \to X$ a continuous map of a metric space, the set*

$$\omega(x) = \bigcap_{n \in \mathbb{N}} \overline{\{f^i(x) \mid i \geq n\}}$$

is the set of all accumulation points of the orbit of x. It is called the ω-limit set of $x \in X$ with respect to f. For f an invertible map on X, the set

$$\alpha(x) = \bigcap_{n \in \mathbb{N}} \overline{\{f^{-i}(x) \mid i \leq n\}}$$

is called the α-limit set of x with respect to f.

Some notes:

- An alternative way to put this is that a point $y \in \omega(x)$ if and only if there is a strictly increasing sequence of natural numbers $\{n_k\}_{k \in \mathbb{N}}$ such that $f^{n_k}(x) \longrightarrow y$, as $k \longrightarrow \infty$.

- For orbits in flows, a point $y \in \omega(x)$ if and only if there is a strictly increasing sequence $\{t_n\}_{n \in \mathbb{N}}$ in $\mathbb{R}$ such that $\lim_{n \to \infty} t_n = \infty$ and $\lim_{n \to \infty} \varphi^{t_n}(x) = y$.

- It is possible for a point x to be in its own ω-limit set; all fixed and periodic points have this property. But there are more exotic examples, as we will see.

- One interesting property of these sets is that they always contain full orbits:

Proposition 2.13 *For $f : X \to X$ a continuous map, if $y \in \omega(x)$, then $\mathcal{O}_y \subset \omega(x)$.*

Proof Since $y \in \omega(x)$, $\lim_{k \to \infty} f^{n_k}(x) = y$. But then, for any $m \in \mathbb{N}$, we have

$$f^m(y) = f^m \left(\lim_{k \to \infty} f^{n_k}(x) \right) = \lim_{k \to \infty} f^{n_k + m}(x) = \lim_{k \to \infty} f^{m_k}(x)$$

for a new strictly increasing sequence of natural numbers $m_k = n_k + m$ with the same properties. Hence $\mathcal{O}_y = \{f^m(y)\} \in \omega(x)$. $\square$

Exercise 21 Reformulate Proposition 2.13 for flows and show that it remains true.

In our examples so far, the ω-limit sets of points have been quite simple. For example, $f : \mathbb{R} \to \mathbb{R}$, $f(x) = e^k x$, $k < 0$, we know that $\forall x \in X$, $\omega(x) = 0$. And for the ODE in Exercise 20, $\omega(x) = 900$ for all $x \in \mathbb{R}$.

Exercise 22 For general $a, b \in \mathbb{R}$, calculate all ω-limit sets of the ODE $\dot{x} = b - ax$, classified by initial values and values of the constants.

2.1.3 Metrics on Sets

The above questions are all good to explore. For now, the above example $f(x) = e^k x$, where $k < 0$, is an excellent example of a particular class of dynamical systems that we will discuss presently.

Definition 2.14 *A metric on a set X is a real-valued function $d : X \times X \to [0, \infty) \subset \mathbb{R}$ where*

1. $d(x,y) \geq 0, \forall x, y \in X$ and $d(x,y) = 0$ if and only if $x = y$;
2. $d(x,y) = d(y,x), \forall x, y \in X$;
3. $d(x,y) + d(y,z) \geq d(x,z), \forall x, y, z \in X$.

Remark 2.15 *Loosely speaking, a metric defines a notion of distance between points in a space. Obviously, this notion should be non-negative and symmetric, which are the first two properties in the definition above. Together, these comprise a property known as positive-definiteness. The third property is called the* triangle inequality. *Perhaps you can understand why.*

One such choice of metric is the "standard Euclidean distance" metric on a subset of $\mathbb{R}^n$,

$$d(\mathbf{x}, \mathbf{y}) = \sqrt{\sum_{i=1}^{n} (x_i - y_i)^2},$$

where $\mathbf{x} = (x_1, x_2, \ldots, x_n) \in \mathbb{R}^n$. Note that for $n = 1$, this metric reduces to $d(x,y) = \sqrt{(x-y)^2} = |x - y|$.

Exercise 23 Show explicitly that the standard Euclidean distance metric is indeed a metric by showing that it satisfies the three conditions.

Exercise 24 Show that $d(\mathbf{x}, \mathbf{y}) = \sum_{i=1}^{n} |x_i - y_i|$ also defines a metric on $\mathbb{R}^n$. (For $n = 2$, this is sometimes called the *taxicab* or the *Manhattan metric*. Can you see why?)

Exercise 25 Again, on $\mathbb{R}^n$, show that

$$d(\mathbf{x}, \mathbf{y}) = \max \left\{ |x_1 - y_1|, |x_2 - y_2|, \ldots, |x_n - y_n| \right\}$$

is a metric. (This metric is referred to as the *maximum metric, or max metric*.)

Exercise 26 Show that

$$d(x,y) = \begin{cases} 0, & x = y, \\ 1, & x \neq y \end{cases}$$

on an arbitrary set X defines a metric. (Note: this metric is called the *discrete metric*. Notice that none of the points in X are close to each other under this metric.)

Remark 2.16 *One can define notions of distance between points in vector spaces via vector norms. All three examples above are members of a family of distances defined by vector norms called the L^p-norms or simply p-norms. Euclidean distance corresponds to $p = 2$, the Manhattan distance is $p = 1$, and the maximum metric corresponds to the ∞-norm or maximum norm. All are defined via the definition of a p-norm*

$$||x||_p = \left(\sum |x_i|^p\right)^{1/p}.$$

The ∞-norm is so-named since it is the norm formed by letting $p \longrightarrow \infty$ and interpreting the limit appropriately.

Exercise 27 On $\mathbb{R}^2$, consider a notion of distance defined by

$$d(\mathbf{x},\mathbf{y}) = \begin{cases} |x_1 - y_1| & \text{if } x_1 \neq y_1, \\ |x_2 - y_2| & \text{if } x_1 = y_1. \end{cases}$$

This is similar to a lexicographical ordering of points in the plane. Show that this notion of distance is NOT a metric on $\mathbb{R}^2$.

Exercise 28 Show that $d(x,y) = |x^3 - y^3|$ is a metric on $\mathbb{R}$.

Exercise 29 The original definition of a circle as a planar figure comes directly from Euclid himself: A *circle* is the set of points in the plane equidistant from a particular point. Naturally, using the Euclidean metric, a circle is what you know well as a circle. Show that circles in the taxicab metric on $\mathbb{R}^2$ are squares whose diagonals are parallel to the coordinate axes. Then write down the formula for the unit circle centered at $\mathbf{x}_0 \in \mathbb{R}^2$.

Exercise 30 Following on from the previous exercise, construct a metric on $\mathbb{R}^2$ whose circles are squares whose sides are parallel to the coordinate axes. (Hint: Rotate the taxicab metric.)

Exercise 31 Let S be a circle of radius $r > 0$ centered at the origin of $\mathbb{R}^2$. Its circumference is $2\pi r$. Euclidean distance in the plane does restrict to a metric directly on the circle. Here instead, construct a metric on the circle using arclength, and verify that it is a metric. (Be careful about measuring distance correctly.)

Remark 2.17 *When discussing points in Euclidean space, it is conventional to denote scalars (elements of $\mathbb{R}$) as variables in italics, and vectors (elements of $\mathbb{R}^n$, $n > 1$) as variables in boldface. Thus $\mathbf{x} = [x_1, x_2, \ldots, x_n]^T$, where T denotes the transpose—or even $\mathbf{x} = (x_1, x_2, \ldots, x_n)$, although this is problematic for different reasons—(see the discussion in Section 4.5) In the above definition of a metric, we didn't specify whether X was a subset of $\mathbb{R}$ or something larger. In the absence of more information regarding a space X, we will always use simple italics for its points, so that $x \in X$, even if it is possible that $X = \mathbb{R}^5$, for example. We will only resort to the vector notation when it is assured that we are specifically talking about vectors of a certain size. This is common in areas of higher mathematics like topology.*

This notion of the shape of a circle in a metric space being dependent on the metric leads directly to the idea of *neighborhoods* of points in a metric space. To start, recall in any metric space X, we can define a small open set via a (considered small) positive number $\epsilon > 0$ and the strict inequality:

$$B_\epsilon(x) = \{y \in X \mid d(x,y) < \epsilon\}.$$

We typically call this an *open ball* of size ϵ about x, or an ϵ-ball about $x \in X$. Note immediately that, in the case that X is itself a subset of a larger space, there may be points within an ϵ distance of x but outside of X. Here, the ϵ-ball around x is only the part of the ball that is actually in X.

Definition 2.18 *A* neighborhood *of a point* $x \in X$ *is any subset* $U \subseteq X$ *such that there exists an* $\epsilon > 0$ *where* $B_\epsilon(x) \subset U$.

Essentially, if a set contains x and all of its nearest neighbors, it is considered a neighborhood of x.

Definition 2.19 *A subset* $U \subseteq X$ *is called* open *if* $\forall x \in U$, $\exists \epsilon > 0$ *such that* $B_\epsilon(x) \subset U$. *And* $U \subseteq X$ *is called* closed *if its complement*

$$U^c = \{x \in X \mid x \notin U\}$$

is open.

Hence, an open set in X is a neighborhood of any of its points. When a set U serves as a neighborhood of x, we call x an *interior point* of $U \subset X$. And for any subset $U \subset X$, the interior of U, denoted by $\mathring{U}$, is the set of all interior points of U. Here, $\mathring{U}$ is the largest open set inside U.

Definition 2.20 *A point* $x \in X$ *is called a* boundary point *of a subset* $U \subset X$ *if, for every* $\epsilon > 0$, *the set* $B_\epsilon(x)$ *contains at least one point in* U *and one point not in* U.

Here, it is important to see that boundary points of a set U may or may not actually be in U. For example, for $D = \{(x,y) \in \mathbb{R}^2 \mid x \geq 0, y > 0\}$, basically the first quadrant with an edge, all points on the positive x- and y-axes are boundary points. But those on the positive y-axis are in D, while those on the positive x-axis are not. And the origin? Here, D is neither open nor closed, but its interior is the first quadrant in the plane: $\mathring{D} = \{(x,y) \in \mathbb{R}^2 \mid x > 0, y > 0\}$.

Proposition 2.21 *A subset* $U \subset X$ *is closed in* X *if it contains all of its boundary points in* X.

Proof Suppose $x \in X$ is a boundary point of a closed set U, but $x \notin U$. By Definition 2.19, U^c is open and $x \in U^c$ is an interior point. But then there exists an $\epsilon > 0$ such that $B_\epsilon(x) \subset U^c$ and $B_\epsilon(x) \cap U = \emptyset$. Thus x cannot be a boundary point of U. This contradiction establishes the result. □

Here, we can go further, and define the *closure* of a set $U \subset X$, denoted by $\overline{U}$, as the union of U with the set of its boundary points. $\overline{U}$ is the smallest closed set containing U.

In a loose sense, one can think of a closed subset of real space as a set of solutions to either equations or inequalities of the form $\leq$ or $\geq$. In this fashion, curves in the plane and surfaces in $\mathbb{R}^3$ are closed sets, although ones without interior points. For example, let $I = [0,1]$ be the unit interval. For $I \subset \mathbb{R}$, I is closed, with interior $(0,1)$ and boundary points 0 and 1. But for $I \subset \mathbb{R}^3$, where I is the unit interval along the x-axis in xyz-space, I is still closed, but has no interior points. Often, however, in vector calculus, the closed sets constructed as domains for functions are open sets together with their closure. This makes the analysis of functions via limits, continuity, and derivatives easier to study from their definitions on interior points of domains.

Example 2.22 For $U(x) = B_\epsilon(x)$ the open ϵ-ball centered at $x \in \mathbb{R}^n$, its closure is

$$\overline{U}(x) = \overline{B}_\epsilon(x) = \left\{ y \in X \middle| d(x,y) \leq \epsilon \right\}.$$

The boundary of $\overline{U}(x)$, sometimes written $\partial \overline{U}$, is the set of all points $y \in \mathbb{R}^n$ where $d(x,y) = \epsilon$. From vector calculus, recall that this is the $(n-1)$-dimensional sphere in $\mathbb{R}^n$ of radius ϵ centered at x. We will construct and discuss such a space in more detail in Section 3.3.1. Now imagine what spheres look like in the taxicab metric on $\mathbb{R}^n$.

2.1.4 Lipschitz Continuity

Now with this rudimentary notion of a metric on a subset of Euclidean space, we can define a form of continuity of great import for our discussion of dynamics.

Definition 2.23 A map $f : X \subset \mathbb{R}^n \to \mathbb{R}^m$, is called Lipschitz continuous (with constant λ), or λ-Lipschitz, if

(2.1.3) $$d\big(f(x), f(y)\big) \leq \lambda d(x,y), \quad \forall x,y \in X.$$

Some notes:

- As a subset of $\mathbb{R}^n$, the set X can always "inherit" the metric on $\mathbb{R}^n$ simply by declaring that the distance between two points in X is defined by the their distance in $\mathbb{R}^n$ (See Exercise 31). So subsets of $\mathbb{R}^n$ are always metric spaces. This includes the subset of $\mathbb{R}^m$ that is the range of f, often called the *image* of f, denoted $f(X) \subset \mathbb{R}^m$. One can always define a different metric on X (or its image) if one wants. But the fact that X is a metric space comes for free, as we sometimes say.

- Note in the definition above that the metrics on the two sides of the inequality are different: the one on the left is a metric on X as a subset of $\mathbb{R}^n$, while the metric on the right is a metric on at least $f(X) \subset \mathbb{R}^m$. Usually, in dynamical systems, we will deal with maps whose domains and codomains are the same. In that case, the metric will usually be the same.

- λ is a bound on the stretching ability (comparing the distances between the images of points in relation to the distance between their original positions) of f on X. This is actually a stronger form of continuity (similar but not equal to uniform continuity): Lipschitz functions are always continuous (Exercise 32), but there are continuous functions that are not Lipschitz (Exercise 33). Basically, if the ratio of the distance between two images $f(x)$ and $f(y)$ to the distance between x and y is bounded by λ across all of X, then f is λ-Lipschitz.

- There is a local version of this property: a function f can be *locally Lipschitz* (see Definition 3.20 in Section 3.2.1), where for every x in the domain of f, there is a neighborhood $U(x)$ such that f, restricted to U, $f|_U$, is Lipschitz continuous. But only when the Lipschitz constant holds for the entire domain is the function Lipschitz continuous. We sometimes call this *uniformly Lipschitz continuous*.

- To get a better sense for Lipschitz continuity, consider the following: On a bounded interval in $\mathbb{R}$, polynomials are always Lipschitz continuous. But on $\mathbb{R}$ itself, only the constants and the linear polynomials are λ-Lipschitz. Rational functions, on the other hand, even though they are continuous and differentiable on their domains, are not Lipschitz continuous on any interval whose closure contains a vertical asymptote. The function $\sin x$ is 1-Lipschitz on $\mathbb{R}$, but $\tan x$ is not Lipschitz continuous on its domain. And a function like e^x?

- It should be obvious that $\lambda \geq 0$. Why?

- We can define

$$\text{Lip}(f) = \sup_{x \neq y} \frac{d\big(f(x),f(y)\big)}{d(x,y)},$$

which is the infimum of all λ's that satisfy Equation 2.1.3. When we speak of specific values of λ for a λ-Lipschitz function, we typically use $\lambda = \text{Lip}(f)$, if known.

Exercise 32 Show for $f : \mathbb{R} \to \mathbb{R}$ that Lipschitz continuity implies continuity.

Exercise 33 Let $f(x) = x^{-1}$. Show that f is Lipschitz continuous on any domain (a,b), $a > 0$, $a < b \leq \infty$, and, for any particular choice of a and b, produce the constant λ. Then show that f is not Lipschitz continuous on $(0,\infty)$.

Exercise 34 Show that $h(x) = |x|$ is Lipschitz continuous on R and produce $\text{Lip}(h)$.

Exercise 35 Show that $g(x) = x^2$ is not Lipschitz continuous on $\mathbb{R}$.

Exercise 36 For a given non-negative λ, construct a function whose domain is all of $\mathbb{R}$ and that is precisely λ-Lipschitz continuous on $I = (-\infty,2) \cup (2,\infty)$ but not Lipschitz continuous.

Exercise 37 Produce a function that is continuous on $I = [-1,1]$ but not Lipschitz continuous there.

It should be clear now that, for a real-valued function on $\mathbb{R}$, the Lipschitz condition $|f(x) - f(y)| \leq \lambda |x - y|$ implies (at least for all $y \neq x$)

$$\left| \frac{f(x) - f(y)}{x - y} \right| \leq \lambda.$$

If f is differentiable at x, then the left-hand side limits to $|f'(x)|$ as $y \to x$. Thus, for differentiable functions, the Lipschitz constant bounds the derivative, even through its definition says nothing about differentiability. So derivatives, when they exist, can say a lot about Lipschitz continuity. That is, they can bound the Lipschitz constant from below. One must be careful about the domain, however:

Proposition 2.24 *Let $f : I \subset \mathbb{R} \to \mathbb{R}$ be differentiable on an open interval I, where $\forall x \in I$, we have $|f'(x)| \leq \lambda$. Then f is λ-Lipschitz.*

Proof Really, this is simply an application of the Mean Value Theorem: For a function f differentiable on a bounded, open interval (a,b) and continuous on its closure, there is at least one point $c \in (a,b)$ where $f'(c) = [f(b)-f(a)]/(b-a)$, the average total change of the function over $[a,b]$. Here then, for any $x,y \in I$ (thus all of $[x,y] \in I$ even when I is neither closed nor bounded), there will be at least one $c \in I$ where

$$d\big(f(x),f(y)\big) = |f(x)-f(y)| = |f'(c)||x-y| \leq \lambda|x-y| = \lambda d(x,y). \qquad \square$$

Definition 2.25 *A λ-Lipschitz function $f : X \subset \mathbb{R}^n \to \mathbb{R}^m$ on a metric space X is called a* contraction *if $\lambda < 1$.*

Note that the definition here of a contraction, as well as the general definition above of Lipschitz continuity, both allow for the domain and range to be two different metric spaces. When using the function f as the fixed rule of a discrete dynamical system, however, we want the codomain and domain to be the same, and would define $f : X \to X$ to be a contraction if it is Lipschitz continuous with $\lambda < 1$.

Example 2.26 Back to the previous example $f : \mathbb{R} \to \mathbb{R}$, $f(x) = e^k x$, the time-1 map of the ODE $\dot{x} = kx$. Given that $f'(x) = e^k$ everywhere, in the case that $k < 0$, the map f is a contraction on ALL of $\mathbb{R}$. Indeed, using the Euclidean metric in $\mathbb{R}$, we have

$$d(f(x),f(y)) = |e^k x - e^k y| = |e^k(x-y)| = |e^k||x-y| = e^k|x-y| = \lambda|x-y|$$

for all $x,y \in \mathbb{R}$, where $\lambda = e^k < 1$.

Exercise 38 Without using derivative information, show that $f(x) = ax+b$ is a-Lipschitz on $\mathbb{R}$.

Exercise 39 Again without using derivative information, show that the monomial x^n, $n \in \mathbb{N}$, is na^{n-1}-Lipschitz on the interval $[0,a]$

Exercise 40 Show that $h(x) = \sqrt[3]{x}$ is not Lipschitz continuous on any interval whose closure contains 0. (As we will see in Exercise 95, though, h is locally Lipschitz continuous on the interval $(0,\infty)$.)

Exercise 41 Find a for the smallest interval $[0,a]$ where $f(x) = 3x^2 - 2$ is not a contraction.

Before we continue, we need to clarify some of the properties of the intervals we will be using in our dynamical systems. Here are a couple of definitions:

Definition 2.27 *A subset $U \in \mathbb{R}$ is called* bounded *if there exists a number $M > 0$ such that $\forall x \in U$, we have $|x| < M$.*

Based on Proposition 2.21, an interval $I \in \mathbb{R}$ is closed if it contains all of its limit points. If the interval is bounded, this means that I includes its endpoints. But closed intervals need not be bounded, and can take one of the forms $[a,b]$, $(-\infty, b]$, $[a, \infty)$ or $(-\infty, \infty)$, for $-\infty < a \leq b < \infty$.

Remark 2.28 *Recall that a function is continuous on a closed, bounded interval $I = [a,b]$ if it is continuous on the interior (a,b) and continuous from the appropriate side (interior to I) at each endpoint. And a function is differentiable on I if the derivative exists at every point in the interior and exists from the right at a and from the left at b. But one has to be quite careful here, as this definition of differentiability on a closed interval presents issues. For example, a function may be differentiable on an interval $[a,b]$, but not differentiable at b. And since the differentiability of a function at an endpoint of a closed interval vis-á-vis its interior is not usually important to the statement of a theorem or construction, often one sidesteps the issue with the more usual phrasing in many theorems in calculus—that a function is continuous on $[a,b]$ and differentiable on (a,b) (think Mean Value Theorem).*

Example 2.29 *While $f(x) = \sqrt[3]{x}$ is continuous on the closed interval $I = [0, \infty)$, and differentiable on its interior, it is not a differentiable function on I.*

Exercise 42 Produce a continuous function $g : \mathbb{R} \to \mathbb{R}$ that is differentiable on a closed $[a,b]$, but is not differentiable at b. Then produce a function that is differentiable on $[a,b]$ and $[b,c]$, but not differentiable on $[a,c]$.

Remark 2.30 *While a function may be deemed differentiable, this only means that the derivative exists at every point. It is not quite enough to ensure that the derivative is continuous. To be sure, the two functions*

$$
(2.1.4) \quad f(x) = \begin{cases} x^2 \sin\left(\dfrac{1}{x}\right), & x \neq 0, \\ 0, & x = 0 \end{cases} \quad and \quad g(x) = \begin{cases} x^3 \sin\left(\dfrac{1}{x}\right), & x \neq 0, \\ 0, & x = 0 \end{cases}
$$

are both differentiable on $\mathbb{R}$. But only one is continuously differentiable (i.e., has a derivative that is continuous). We leave it as Exercise 43 to determine which one. But to be clear, to call a function C^k means that it has derivatives at least up to order k that are all continuous functions on the domain of the original function.

Exercise 43 On $\mathbb{R}$, show that both f and g in Equation 2.1.4 are differentiable, but only one is continuously differentiable (of class C^1).

Proposition 2.31 *Let $f : I \to I$ be continuously differentiable on I, a closed, bounded interval, with $|f'(x)| < 1 \; \forall x \in I$. Then f is a contraction.*

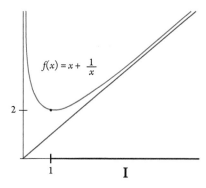

$f(x) = x + \dfrac{1}{x}$

Figure 12 Unbounded domain.

Proof $f'(x)$ is continuous on I, so it will achieve its maximum there by the Extreme Value Theorem, and

$$\max_{x \in I} |f'(x)| = \lambda < 1.$$

Now apply Proposition 2.24. ☐

The closed and bounded properties of the interval I in Proposition 2.31 are quite necessary. Without either, the proposition is not true.

Example 2.32 Let $f(x) = x + x^{-1}$ be defined on the closed, but not bounded, interval $I = [1, \infty)$. Here $f(x) > x \; \forall x \in I$ and $f'(x) = 1 - x^{-2} < 1$ on all of I. Yet what would be the value of $\lambda < 1$ for this Lipschitz function to be a contraction? See Figure 12.

And, for a good example for I bounded but not closed, see Exercise 44 below.

Exercise 44 Show that for $1 < b < 4$, the function $f(x) = 2\sqrt{x}$ is not a discrete dynamical system on $(1, b)$. Then show that $f(x)$ is not a contraction on $(1, b)$ for any $4 < b < \infty$.

2.2 The Contraction Principle

Lipschitz continuous functions are fairly well-behaved when it comes to iteration. Basically, the distance between the images of two points in a domain of a Lipschitz continuous function are bounded relative to the original distance between the two points. Thus, upon iteration in a dynamical system, when the domain and the codomain are the same, two orbits that start close together cannot diverge faster than exponentially (recall that near a vertical asymptote of a rational function, this is not so.) When the Lipschitz constant is less than 1, the images of points are actually closer together than their original distances and, upon iteration, converge exponentially. This has great consequences when trying to understand how orbits behave when viewing a map as a discrete dynamical system. Basically, in a Lipschitz contraction, all of the orbits can be described rather simply, as they all wind up going to the same place in the long term. We will say, in this case, that the map exhibits simple dynamics.

2.2.1 Contractions on Intervals

For discrete dynamical systems defined on (subsets of) the real line, we can generalize that contractions have some special properties. One very special feature is that the fixed-point set is always non-empty, but very small in size:

Theorem 2.33 (the Contraction Principle) *Let $I \subset \mathbb{R}$ be a closed interval, and $f : I \to I$ a λ-contraction. Then f has a unique fixed point x_0 and, $\forall x \in I$,*

$$\left| f^n(x) - x_0 \right| \leq \lambda^n \left| x - x_0 \right|.$$

Some notes before we prove this theorem:

- $\forall x \in I$, $\mathcal{O}_x \longrightarrow x_0$ as a sequence exponentially due to the factor λ^n, where $0 < \lambda < 1$.
- As stated, this is only a result valid on subsets of $\mathbb{R}$, for now.
- As a way to understand this theorem, first think about why a contraction cannot have more than one fixed point:

Exercise 45 Show, without using the Contraction Principle, that a contraction cannot have two fixed points.

- For the proof of the theorem, we will need a convenient fact from analysis: Euclidean space is *complete*. That is, the idea of a sequence converging in real space, with the Euclidean metric, is the same as the fact that the sequence is *Cauchy*:

Definition 2.34 *For $N \in \mathbb{N}$, a sequence $\{x_i\}_{i=1}^{\infty}$, $x_i \in \mathbb{R}^N$ is called Cauchy if $\forall \epsilon > 0$, $\exists A > 0$ such that $\forall m, n \geq A$, $d(x_m, x_n) < \epsilon$.*

Proposition 2.35 *A sequence in $\mathbb{R}^N$, $N \in \mathbb{N}$, converges iff it is Cauchy.*

In the present case, for $I \subset \mathbb{R}$ a closed interval, I, as a metric space, is also complete (this is not true for open intervals bounded on at least one side). Hence all Cauchy sequences in I converge.

Proof of Contraction Principle (in $\mathbb{R}$) The proof of this theorem will consists of three parts: (1) That the orbit of an arbitrary point converges to something in the interval; (2) that any two orbits also converge (point-wise); and (3) that the thing all of these orbits converge to is actually an orbit (the orbit of a fixed point).
To start, choose $x \in I$. It should be obvious since f is a map on I that $\mathcal{O}_x \subset I$. Now, for $m, n \in \mathbb{N}$, let's assume (without any loss of generality) that $m \geq n$. Then

$$\left| f^m(x) - f^n(x) \right| = \left| f^m(x) - f^{m-1}(x) + f^{m-1}(x) - f^{m-2}(x) + f^{m-2}(x) - \ldots \right.$$
$$\left. - f^{n+1}(x) + f^{n+1}(x) - f^n(x) \right|$$
$$= \left| \sum_{r=n}^{m-1} \left(f^{r+1}(x) - f^r(x) \right) \right|.$$

This is a common technique in analysis and is just a clever form of "addition by zero." Adding in (and then subtracting out) extra terms this way allows us to see additional structure not obvious in the calculation. Thus

$$\left|f^m(x) - f^n(x)\right| = \left|\sum_{r=n}^{m-1}\left(f^{r+1}(x) - f^r(x)\right)\right|$$

$$\leq \sum_{r=n}^{m-1}\left|f^{r+1}(x) - f^r(x)\right| \quad \text{(triangle inequality)}$$

$$\leq \sum_{r=n}^{m-1}\lambda^r\left|f(x) - x\right| \quad (r \text{ iterations of the Lipschitz condition})$$

$$= \left|f(x) - x\right|\sum_{r=n}^{m-1}\lambda^r$$

$$= \left|f(x) - x\right|\left(\frac{\lambda^n - \lambda^m}{1-\lambda}\right) \quad \text{(partial sum of a geometric series)}$$

$$\leq \frac{\lambda^n}{1-\lambda}\left|f(x) - x\right|.$$

The conclusion from all of this is that, as n gets large (under the assumption that $m \geq n$, this means that m gets large also), the right-hand side of the last inequality gets small. And n can always be chosen large enough that the distance between late terms in the sequence is less than some chosen ϵ. Hence, $\mathcal{O}_x$ is a Cauchy sequence. Hence it must converge to something. As I is closed, it must converge to something in I. Let's call this number x_0.

Now, applying the Lipschitz condition to the nth iterate of f leads directly to the condition that

$$\left|f^n(x) - f^n(y)\right| \leq \lambda^n\left|x - y\right|,$$

so that every two orbits also converge to each other. Hence every orbit converges to this number x_0.

And finally, we can show that x_0 is actually a fixed-point solution for f. To see this, again $\forall x \in I$, and $\forall n \in \mathbb{N}$,

$$\left|x_0 - f(x_0)\right| = \left|x_0 - f^n(x) + f^n(x) - f^{n+1}(x) + f^{n+1}(x) - f(x_0)\right|$$

$$\leq \left|x_0 - f^n(x)\right| + \left|f^n(x) - f^{n+1}(x)\right| + \left|f^{n+1}(x) - f(x_0)\right|$$

$$\leq \left|x_0 - f^n(x)\right| + \lambda^n\left|x - f(x)\right| + \lambda\left|f^n(x) - x_0\right|$$

$$= (1+\lambda)\left|x_0 - f^n(x)\right| + \lambda^n\left|x - f(x)\right|.$$

Again, the steps are straightforward. The first step is another clever addition of zero, and the second is the triangle inequality. The third involves using the Lipschitz condition on two of the terms, and the last is a clean up of the leftovers. However, this last inequality is the most important. It must be valid for every choice of $n \in \mathbb{N}$. Hence, choosing an increasing sequence of values for n, we see that as $n \to \infty$, both $|x_0 - f^n(x)| \longrightarrow 0$ and $\lambda^n \longrightarrow 0$. Hence, it must be the case that $|x_0 - f(x_0)| = 0$ or $x_0 = f(x_0)$. Thus, x_0 is a fixed-point solution for f on I and the theorem is established. □

Example 2.36 $f(x) = \sqrt{x}$ on $I = [1, \infty)$ is a $\frac{1}{2}$-contraction. Why? $f \in C^1$ on I and $0 < f'(x) = 1/(2\sqrt{x}) \leq \frac{1}{2}$ on I, and strictly less than $\frac{1}{2}$ on $(1, \infty)$ (the *interior* of I, sometimes denoted $\mathring{I} = I - \partial I$). Thus, by Proposition 2.24, f is $\frac{1}{2}$-Lipschitz. So, by the Contraction Principle, then, there is a unique fixed point for this discrete dynamical system. Can you find it?

Exercise 46 Without using any derivative information (i.e., without using Propositions 2.24 or 2.31), show that $f(x) = \sqrt{x}$ is a $\frac{1}{2}$-contraction on $[1, \infty)$.

Exercise 47 Find all periodic points (to an accuracy of $\frac{1}{1000}$) of the discrete dynamical system given by the map $f(x) = \ln(x-1) + 5$ on the interval $I = [2, 100]$.

Example 2.37 Recall $f(x) = x + x^{-1}$ on $I = [1, \infty)$ from Example 2.32 above. Here, $f'(x) = 1 - x^{-2} < 1$ on all of I, yet f is not a contraction. But this failure has an additional interesting side-effect: it also has no fixed point at all! While all orbits do satisfy $d(f(x), f(y)) < d(x, y)$, all orbits are also unbounded. The fact that the derivative is monotonically increasing, but bounded above by 1, coupled with the fact that $f(x) > x$ on all of the domain, creates an interesting situation.

Exercise 48 Show $g(x) = \ln(1 + e^x)$ satisfies $d(f(x), f(y)) < d(x, y)$ on $\mathbb{R}$, but is not a contraction.

2.2.2 Contractions in Several Variables

Generalizing the Contraction Principle to functions on Euclidean n-space is a very straightforward endeavor. The only difference, really, is that the absolute value signs are replaced by the more general metric in $\mathbb{R}^n$, in both the statement of the theorem as well as its proof.

Theorem 2.38 (The Contraction Principle) Let $X \subset \mathbb{R}^n$ be closed and $f : X \to X$ a λ-contraction. Then f has a unique fixed point $x_0 \in X$ and $\forall x \in X$,

$$d\left(f^n(x), x_0\right) \leq \lambda^n d\left(x, x_0\right).$$

Notes:

- Again, we say here that the "dynamics are simple." The orbit of every point in x does exactly the same thing: Converge exponentially to the fixed point solution x_0.
- This also means that every orbit converges to every other orbit also!
- What about periodic points? Can contractions have periodic points other than fixed points. The answer is no:

Exercise 49 Show that a contraction cannot have a non-trivial periodic point (prime period greater than 1).

- The Contraction Principle is also called the Contraction Mapping Theorem, and is a special case of the Banach Fixed-Point Theorem (Theorem 3.29), as we will see in Section 3.2.3.

Recall that in dimension 1, the derivative of f (if it exists) can help to define the Lipschitz constant (recall Propositions 2.24 and 2.31). In several dimensions, recall that for a C^1 function $f : X \subset \mathbb{R}^n \to \mathbb{R}^n$, the derivative at $x \in X$, $df_x : T_x\mathbb{R}^n \to T_{f(x)}\mathbb{R}^n$, is a linear map from the tangent space at x, $T_x\mathbb{R}^n$, to the tangent space of $f(x)$, $T_{f(x)}\mathbb{R}^n$. The points in the domain X where the $n \times n$ matrix df_x is of maximum rank are called *regular*. At a domain point, we can use the Euclidean norm for vectors to define a matrix norm for df_x as the maximal stretching ability of the unit vectors in the tangent space:

$$||df_x|| = \max_{||\mathbf{v}||=1} ||df_x(\mathbf{v})||.$$

This non-negative number is not difficult to calculate for an $n \times n$ matrix A. Indeed, recall from linear algebra:

Definition 2.39 *The* spectral radius *of a matrix A is the quantity $\rho(A)$ (alternatively ρ_A), where*

$$\rho(A) = \left\{ \max_i |\lambda_i| \mid \lambda_i \text{ is an eigenvalue of } A \right\}.$$

If A is symmetric, then $||A||$ is just the spectral radius $\rho(A)$; the largest magnitude of the eigenvalues of A. For general (finite-dimensional) A, $||A|| = \sqrt{\rho(A^T A)}$.

With this, there are two "derivative" versions of the Contraction Principle of note. The first is a sort of "global version" since the result holds over the entire domain. We will require that the space be strictly convex, however (see Figure 13).

Definition 2.40 *A subset $X \subseteq \mathbb{R}^n$ is* convex *if for any two points $x, y \in X$, the straight line segment joining x and y lies entirely in X. X is called* strictly convex *if for any two boundary points $x, y \in \overline{X}$, the line segment joining x to y intersects the boundary of X only at x and y.*

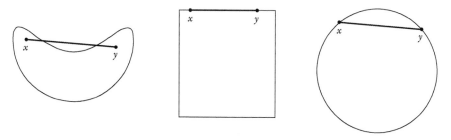

Figure 13 Convexity: non-convex, convex, but not strictly, and strictly convex sets, respectively.

Theorem 2.41 (Global version) *If X is the closure of a strictly convex set in $\mathbb{R}^n$, and $f : X \to \mathbb{R}^n$ a C^1-map with $\|df_x\| \leq \lambda < 1, \forall x \in X$, then f has a unique fixed point x_0, and $\forall x \in X$,*

$$d(f^n(x), x_0) \leq \lambda^n d(x, x_0).$$

Note here that, technically, we really want that f is differentiable on the interior of X and continuous on the boundary.

Exercise 50 For the affine map $f(x,y) = \left(\frac{1}{2}x + \frac{1}{3}y, -\frac{1}{2}x + \frac{1}{2}y + 1\right)$ on $\mathbb{R}^2$, find the fixed point and show that f is a contraction on $\mathbb{R}^2$.

This next version is a local version, and presents a very useful tool for the analysis of what happens near (in a neighborhood of) a fixed point:

Theorem 2.42 (Local version) *Let f be differentiable with a fixed point x_0 such that all of the eigenvalues of the derivative matrix df_{x_0} have absolute values less than 1. Then there exists $\epsilon > 0$ and an open neighborhood $U(x_0) = B_\epsilon(x_0)$ such that on the closure $\overline{U}, f(\overline{U}) \subset \overline{U}$ and f is a contraction on $\overline{U}$.*

Thus, on $\overline{U}$, which will be a strictly convex set, the global version above applies. To see these two versions in action, here are some applications:

2.2.3 Application: The Newton–Raphson Method

An iterative procedure for the location of a root of a C^2-function $f : \mathbb{R} \to \mathbb{R}$, called the Newton-Raphson method, is an application of the local version of the contraction principle. This method is part of a family of approximation methods that utilize the Taylor polynomials of a function to help identify important features of the function, and is typically found in most standard calculus texts as an application using the tangent line approximation to a function. Here, let f be C^1 near an unknown root x_*. Then, for a point x_0 near the root, the tangent line approximation to f at x_0 is

$$f(x) - f(x_0) = f'(x_0)(x - x_0).$$

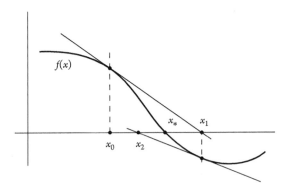

Figure 14 The first two iterates.

Under relatively mild conditions on f and for x_0 "close enough" to x_*, the tangent line function $f(x) = f(x_0) + f'(x_0)(x - x_0)$ will also have a root (its graph will have a horizontal intercept). Call this point x_1. Solving the tangent line function for x_1, we get

$$x_1 = x_0 - \frac{f(x_0)}{f'(x_0)}.$$

Again, under mild conditions on f, the number x_1 lies closer to x_* than x_0, and can serve as a new approximation to x_*.

Repeating this procedure yields a discrete dynamical system given by $x_{n+1} = g(x_n)$, where

$$g(x) = x - \frac{f(x)}{f'(x)}.$$

Some things to note:

- If f is C^2 near the root x_*, then, as long as $f'(x)$ does not vanish in a neighborhood of x_*, g is C^1 there. And

$$g'(x) = 1 - \frac{\left(f'(x)\right)^2 - f(x)f''(x)}{\left(f'(x)\right)^2} = \frac{f(x)f''(x)}{\left(f'(x)\right)^2}.$$

- If there exists an open interval containing x_*, in which there are positive constants δ, M, where $\left|f'(x)\right| > \delta$ (f' doesn't get too small), and $\left|f''(x)\right| < M$ (f'' doesn't get too large), then $g'(x)$ is bounded.
- Notice that if x_* is considered a root of f, then

$$g'(x_*) = \frac{f(x_*)f''(x_*)}{\left(f'(x_*)\right)^2} = 0$$

and

$$g(x_*) = x_* - \frac{f(x_*)}{f'(x_*)} = x_*.$$

Hence g fixes the root of f and, by continuity "near" x_*, $g'(x)$ will remain small in magnitude.

All this points to the contention that there will exist a small (closed) ϵ-neighborhood $\bar{B}_\epsilon(x_*)$ centered at x_*, where $\left|g'(x)\right| < 1$ for all $x \in \bar{B}_\epsilon(x_*)$. (Remember that the derivative of g is 0 at x_*, is continuous in a neighborhood of x_*, and cannot grow too quickly around x_* because of the two constraints on the derivatives of f. Once we have this, then, by Proposition 2.31, g will form a contraction on $\bar{B}$. Thus we have:

Proposition 2.43 *Let $f : \mathbb{R} \to \mathbb{R}$ be C^2 with a root x_*. If $\exists \delta, M > 0$, such that $|f'(x)| > \delta$ and $|f''(x)| < M$ on an open interval containing x_*, then*

$$g(x) = x - \frac{f(x)}{f'(x)}.$$

is a contraction near x_* with fixed point x_*.

2.2.4 Application: Existence and Uniqueness of ODE Solutions

A global version example of the contraction principle above involves the standard proof of the existence and uniqueness of solutions to the first-order IVP

(2.2.1) $$\dot{y} = f(t,y), \quad y(t_0) = y_0$$

in a neighborhood of (t_0, y_0) in the t, y-plane. The proof uses a dynamical approach, again with a first approximation and then successive iterations using a technique attributed to Charles Émile Picard and known as Picard iterations [39].

Theorem 2.44 (Picard–Lindelöf Theorem in $\mathbb{R}$) *Suppose $f(t,y)$ is continuous in some rectangle*

$$R = \left\{ (t,y) \in \mathbb{R}^2 \middle| \alpha < t < \beta, \gamma < y < \delta \right\},$$

containing the initial point (t_0, y_0), and f is Lipschitz continuous in y on R. Then, in some interval $t_0 - \epsilon < t < t_0 + \epsilon$ contained in $\alpha < t < \beta$, there is a unique solution $y = \phi(t)$ of Equation 2.2.1.

To prove this theorem, we will need to understand a bit about the structure of spaces of functions. To start, recall from linear algebra that a (real) *operator* is simply a function $f : U \to V$ whose domain and codomain are (real) vector spaces. A (real) operator is called *linear* if $\forall x, y \in U$ and $c_1, c_2 \in \mathbb{R}, f$ satisfies

$$f(c_1 x + c_2 y) = c_1 f(x) + c_2 f(y).$$

Linear operators where the dimensions of the domain $\dim(U) = n$ and codomain $\dim(V) = m$ are finite can be represented by matrices, so that for $f(x) = Ax$, A is an $m \times n$ matrix. Real-valued continuous functions on $\mathbb{R}$ also form a vector space; the sum of constant multiples of two continuous functions is always a continuous function, for example. But here the spaces are not finite-dimensional. One can form linear operators on spaces of functions like this one also, but the operator is not representable by a matrix. A good example is the derivative operator d/dx, which acts on the vector space of all C^∞, real-valued functions of one independent real variable, and takes them to other C^∞ functions. Think

$$\frac{d}{dx}(x^2 + \sin x) = 2x + \cos x.$$

This operator is linear owing to the Sum and Constant Multiple Rules for differentiation. There are numerous technical difficulties in discussing linear operators in general, but, for now, we shall adopt this general description for the present discussion.

Back to the case at hand, any possible solution $y = \phi(t)$ (if it exists) to Equation 2.2.1 must be a differentiable function that satisfies

(2.2.2)
$$\phi(t) = y_0 + \int_{t_0}^{t} f(s, \phi(s))\, ds$$

for all t in some interval containing t_0. Here, Equation 2.2.2 is called the integral equation associated with the ODE, and is, in general, an example of a *non-homogeneous, Volterra integral equation of the second kind*.

Exercise 51 Show that any function $\phi(t)$ that solves Equation 2.2.2 also solves Equation 2.2.1. (Hint: Think Fundamental Theorem of Calculus.)

Exercise 52 Solve the IVP $\dot{y} = ty, y(0) = 2$, and show that Equation 2.2.2 holds.

At this point, the existence of a solution to the ODE is *assured* in the case that $f(t,y)$ is continuous on R, as the integral will then exist at least on some smaller interval $t_0 - \epsilon < t < t_0 + \epsilon$ contained inside $\alpha < t < \beta$. Note the following:

- One reason a solution may not exist all the way out to the edge of R: What if the edge of R is an asymptote in the t variable?

- A function does not have to be continuous to be integrable (step functions are one example of integrable functions that are not continuous. However, the integral of a step function is continuous. And if $f(t,y)$ included a step-like function in Equation 2.2.1, solutions may still exist and be continuous.

As for uniqueness, suppose $f(t,y)$ is continuous as above, and consider the following operator T, whose domain is the space of all differentiable functions on R, which takes a function $\psi(t)$ to its image $T(\psi(t))$ (which we will denote by $T\psi$ to help remove some of the parentheses) defined by

$$(T\psi)(t) = y_0 + \int_{t_0}^{t} f(s, \psi(s))\, ds.$$

We can apply T to many functions $\psi(t)$, and the image will be a different function $T\psi$ (but still a function of t; see Example 2.45 below). However, looking back at Equation 2.2.2, if we apply T to an actual *solution* $\phi(t)$ to the IVP, the image $T\phi$ should be the same as ϕ. A solution will be a fixed point of the discrete dynamical system formed by T on the space of functions defined and continuous on R, since $T\phi = \phi$.

Exercise 53 Find all fixed points for the derivative operator d/dx whose domain is all infinitely differentiable functions on $\mathbb{R}$. Now find all period-2 and period-4 points (fixed points for the operator applied twice and four times, respectively.) Can you notice a pattern?

Hence, instead of looking for solutions to the IVP, we can look for fixed points of the operator T. How do we do this? Fortunately, this operator T has the nice property that it is a contraction.

Proof of Theorem By assumption, $f(t,y)$ is Lipschitz continuous in y on R. Hence there is a constant $M > 0$ where

$$\left|f(t,y) - f(t,y_1)\right| \le M\left|y - y_1\right|, \quad \forall y, y_1 \in R.$$

Choose a small number $\epsilon = C/M$, where $C < 1$. And define a distance within the set of continuous functions on the closed interval $I = [t_0 - \epsilon, t_0 + \epsilon]$ by

$$d(g,h) = \max_{t\in I}\left|g(t) - h(t)\right|.$$

You should check the triangle inequality to verify that this is, in fact, a metric on the space of continuous functions on I.

Then we have

$$(2.2.3) \qquad d(Tg, Th) = \max_{t\in I}\left|Tg(t) - Th(t)\right|$$

$$(2.2.4) \qquad = \max_{t\in I}\left|y_0 + \int_{t_0}^{t} f(s,g(s))\, ds - y_0 - \int_{t_0}^{t} f(s,h(s))\, ds\right|$$

$$(2.2.5) \qquad = \max_{t\in I}\left|\int_{t_0}^{t} f(s,g(s)) - f(s,h(s))\, ds\right|$$

$$(2.2.6) \qquad \le \max_{t\in I}\int_{t_0}^{t}\left|f(s,g(s)) - f(s,h(s))\right|\, ds$$

$$(2.2.7) \qquad \le \max_{t\in I}\int_{t_0}^{t} M\left|g(s) - h(s)\right|\, ds$$

$$(2.2.8) \qquad \le \max_{t\in I}\int_{t_0}^{t} M \cdot d(g,h)\, ds$$

$$(2.2.9) \qquad \le \max_{t\in I}\left\{M \cdot d(g,h) \cdot |t - t_0|\right\}$$

Exercise 54 Justify the logic in progressing from Step 2.2.5 to 2.2.6, Step 2.2.6 to 2.2.7, and Step 2.2.8 to 2.2.9. These are vector calculus concepts and theorems. Can you see now why the Lipschitz continuity of f is a necessary hypothesis to the theorem?

Exercise 55 Justify why the remaining steps are valid.

Now notice in the last inequality that since $I = [t_0 - \epsilon, t_0 + \epsilon]$, we have

$$|t - t_0| \le \epsilon = \frac{C}{M}.$$

Hence

$$d(Tg, Th) \leq \max_{t \in I} \{ M \cdot d(g,h) \cdot |t - t_0| \}$$

$$\leq M \cdot d(g,h) \cdot \frac{C}{M} = C \cdot d(g,h).$$

Hence T is a C-contraction and there is a unique fixed point ϕ (which is a solution to the original IVP) on the interval I. Here

$$\phi(t) = T\phi(t) = y_0 + \int_{t_0}^{t} f(s, \phi(s))\, ds.$$ □

We can actually use this construction to construct a solution to an ODE:

Example 2.45 Solve the IVP $y' = 2t(1 + y)$, $y(0) = 0$ using the above Picard iterations construction.

Here, $f(t,y) = 2t(1 + y)$ is a polynomial in both t and y, so that f is obviously continuous in both variables, as well as Lipschitz continuous in y, on the whole plane $\mathbb{R}^2$. Hence unique solutions exist everywhere. To actually find a solution, start with an initial guess. An obvious one is

$$\phi_0(t) = 0.$$

Notice that this choice of $\phi_0(t)$ does not solve the ODE. But since the operator T is a contraction, iterating will lead us to a solution. Here, $\phi_{n+1}(t) = T\phi_n(t)$. We get

$$\phi_1(t) = T\phi_0(t) = y_0 + \int_0^t 2s(1 + \phi_n(s))\, ds = \int_0^t 2s(1 + 0)\, ds = t^2$$

$$\phi_2(t) = T\phi_1(t) = y_0 + \int_0^t 2s(1 + \phi_1(s))\, ds = \int_0^t 2s(1 + s^2)\, ds = t^2 + \frac{1}{2}t^4,$$

$$\phi_3(t) = T\phi_2(t) = y_0 + \int_0^t 2s(1 + \phi_2(s))\, ds$$

$$= \int_0^t 2s\left(1 + s^2 + \frac{1}{2}s^4\right) ds = t^2 + \frac{1}{2}t^4 + \frac{1}{6}t^6,$$

$$\phi_4(t) = T\phi_3(t) = y_0 + \int_0^t 2s(1 + \phi_3(s))\, ds$$

$$= \int_0^t 2s\left(1 + s^2 + \frac{1}{2}s^4 + \frac{1}{6}s^6\right) ds = t^2 + \frac{1}{2}t^4 + \frac{1}{6}t^6 + \frac{1}{24}t^8.$$

Exercise 56 Find the pattern and write out a finite series expression for $\phi_n(t)$. Hint: Use induction.

Exercise 57 Find a closed-form expression for $\lim_{n\to\infty} \phi_n(t)$ and show that it is a solution of the IVP.

Exercise 58 Now rewrite the original ODE in a standard form as a first-order linear equation, and solve.

To understand why Lipschitz continuity in the dependent variable y is a necessary condition for uniqueness of solutions, consider the following example:

Example 2.46 Let $\dot{y} = y^{\frac{2}{3}}, y(0) = 0$, be a first-order, autonomous IVP. It should be clear that $y(t) \equiv 0$ is a solution. But so is

$$y_c(t) = \begin{cases} \frac{1}{27}(t+c)^3, & t < -c, \\ 0, & t \geq -c, \end{cases} \quad \forall c \geq 0.$$

There are lots of solutions passing through the origin in ty-trajectory space. Solutions exist but are definitely not unique here. What has failed in establishing uniqueness of solutions to this IVP in the Picard–Lindelöf Theorem? Here $f(y) = y^{\frac{2}{3}}$ is certainly continuous at $y = 0$, but it is not Lipschitz continuous there. In fact, $f'(y) = \frac{2}{3}y^{-\frac{1}{3}}$ is not defined at $y = 0$, and $\lim_{y\to 0^+} f'(y) = \infty$.

Exercise 59 Verify that the family of curves $y_c(t)$ above solve the IVP, and derive this family by solving the IVP as a separable ODE.

Exercise 60 Verify that $f(y) = y^{\frac{2}{3}}$ is not Lipschitz continuous at $y = 0$.

2.2.5 Application: Heron of Alexandria

Start with the beautiful idea that for two positive numbers $a, b > 0$, their *arithmetic mean* $\frac{1}{2}(a+b)$ is never less than their *geometric mean* $\sqrt{ab}$:

$$\frac{1}{2}(a+b) \geq \sqrt{ab},$$

with equality only when $a = b$. We will call this the AMGM Inequality.

Geometrically, one can visualize the AMGM Inequality via the following (see Figure 15): Let B_a and B_b be two disks of respective diameters $a, b > 0$ resting on the real line and touching (assume $a \geq b$ just for the sake of clarity of argument). Then the line connecting their center has length precisely $\frac{1}{2}(a+b)$, the arithmetic mean. Using this line as the hypotenuse of a right triangle by dropping a vertical from the center of the larger ball (in Figure 15, it is B_a), the vertical side has length $\frac{1}{2}(a-b)$ and the horizontal side has length $\sqrt{ab}$. Do you see the result now?

Exercise 61 Show that the lengths of the two sides adjacent to the right angle in the triangle of Figure 15 are the quantities $\frac{1}{2}(a-b)$ and $\sqrt{ab}$ specified.

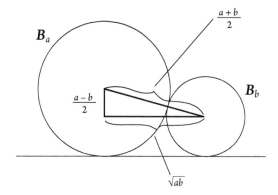

Figure 15 The geometric view of AMGM.

Exercise 62 Show algebraically that, for $a, b > 0$, $\frac{1}{2}(a+b) \geq \sqrt{ab}$, with equality iff $a = b$.

A Greek mathematician and engineer, Heron of Alexandria, is credited with being the first to write down an iterated method of approximating the numeric value of the principle square root of a positive number in the first century of the Common Era in his work "Metrica", an English translation of which appears in the Codex Constantinopolitanus [13]. The method itself evidently goes back to the Babylonians. The method of Heron (which we will call *Heron's method*) is simple enough: Let $N > 0$. We consider a way to approximate $\sqrt{N}$.

(1) Choose a whole number $a > 0$ near $\sqrt{N}$ as an approximation. A good choice would be the root of the closest perfect square. Then a^2 approximates N. But then N/a also approximates $\sqrt{N}$, and. in fact, lies "on the other side" of $\sqrt{N}$, meaning that $\sqrt{N}$ lies between the two numbers a and N/a, no matter which of these latter two is larger.

(2) The algebraic mean $\frac{1}{2}(a + N/a)$ is a new approximation to $\sqrt{N}$, which is closer to $\sqrt{N}$ than at least the farther of a and N/a.

(3) If a better approximation to $\sqrt{N}$ is desired, repeat with $\frac{1}{2}(a + N/a)$ as the new a.

In our modern terminology, for $N > 0$, let $f : \mathbb{R}_+ \to \mathbb{R}_+$ be the function $f(x) = \frac{1}{2}(x + N/x)$. For $x_0 > 0$ any guess, let $x_{i+1} = f(x_i)$. Then f has a unique fixed point at $x = \sqrt{N}$.

Now there are many ways to show that this method converges $\forall x_0 > 0$ and $\forall N > 0$. For example, one could calculate the length of the intervals $\ell_n = ||(x_n, N/x_n)||$ and show that $\lim_{n \to \infty} \ell_n = 0$, while $\sqrt{N} \in (x_n, N/x_n)$, $\forall n \in \mathbb{N}$.

Exercise 63 Show $\ell_{n+1} \leq \frac{1}{2}\ell_n$, $\forall n \in \mathbb{N}$. Thus the interval lengths decay exponentially.

Also, an interesting fact arises from the iterations of f: the new estimate at each stage is always an overestimate. This is due directly to the AMGM inequality, as the new estimate is the algebraic mean of a and N/a, while the geometric mean is precisely $\sqrt{N}$. Thus the sequence $\{x_i\}_{i=1}^{\infty}$ is a decreasing sequence, bounded below by $\sqrt{N}$. Thus, it must converge (by the Monotone Convergence Theorem.)

Exercise 64 Show that this is enough to establish that $\lim_{n\to\infty} x_n = \sqrt{N}$.

Exercise 65 Show that Heron's method is equivalent to the Newton–Raphson method when locating the positive root of $g(x) = x^2 - N$.

Exercise 66 Show that the AMGM inequality provides a constructive proof of the following optimization problems from calculus:

a. Among all positive numbers with a given product, the sum is minimal when the numbers coincide.

b. Among all positive numbers with a given sum, the product is maximal when the numbers coincide.

c. Among all quadrilaterals with the same perimeter, the square has the largest area.

Exercise 67 Show that, for $N > 0$, the family of functions

$$f_N : \mathbb{R}_+ \to \mathbb{R}_+, \quad f_N(x) = \frac{1}{2}\left(x + \frac{N}{x}\right)$$

are not contractions. Then find the largest interval $I_N \subset \mathbb{R}$, containing $\sqrt{N}$, where $f_N|_{I_N}$ is a $\frac{1}{2}$-contraction.

Exercise 68 Approximate $\sqrt{110}$ using Heron's method, to an accuracy of 0.001. Try this using a starting value of 10 and then again for a starting value of 1, noting the difference in convergence properties.

2.3 Interval Maps

The Contraction Principle above is a facet of some dynamical systems that display what is commonly referred to as "simple dynamics": with very little information about the system (map or ODE), one can say just about everything there is to say about the orbits of the system (the system's *orbit structure*. Another way to put this is to say that, in a contraction, all orbits do exactly the same thing. Which is that they all converge to the same fixed point (the equilibrium solution in the case of a continuous dynamical system.) We can build on this idea by now beginning a study of a relatively simple family of discrete dynamical systems that display slightly more complicated behavior.

Let $f : I \to I$ be a continuous map, where $I = [0, 1]$ (we will say f is a C^0-map on I, or $f \in C^0(I, I)$). The graph of f sits inside the unit square $[0, 1]^2 = [0, 1] \times [0, 1] \subset \mathbb{R}^2$, as in Figure 16, for example.

2.3.1 Cobwebbing

The map f has a graph that intersects the line $y = x$ at precisely the points where $y = f(x) = x$, or the fixed points of the dynamical system given by f on I. Recall that the dynamical system is formed by iterating f on I, and all of the forward iterates of x_0 under f comprise the orbit of x_0, $\mathcal{O}_{x_0}$, where

$$\mathcal{O}_{x_0} = \left\{ x_0,\ x_1 = f(x_0),\ x_2 = f(x_1) = f^2(x_0),\ \ldots \right\}.$$

One can track $\mathcal{O}_{x_0}$ in I (and in $[0,1]^2$) visually via the notion of *cobwebbing*. We use the example of $f(x) = x^2$ on I, in Figure 17, to illustrate:

Choose a starting value $x_0 \in I$. Under f, the next term in the orbit is $x_1 = f(x_0)$. Vertically, it is the height of the graph of f over x_0. Making it the new input value to f means finding its corresponding place on the horizontal axis. This is easy to see visually. The vertical line $x = x_0$ crosses the graph of f at the height $x_1 = f(x_0)$. The horizontal line $y = x_1 = f(x_0)$ crosses the diagonal $y = x$ precisely at the point (x_1, x_1). Taking the output x_1 and making it the new input to f (iterating the function) means finding where

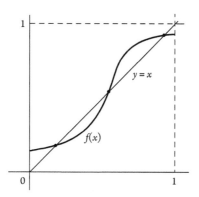

Figure 16 A C^0-map $f : [0, 1] \to [0, 1]$.

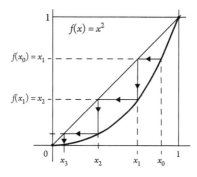

Figure 17 A cobweb of $f(x) = x^2$ on $[0, 1]$.

the vertical line through this point will again intersect the graph of f (at one point). That point will be at the height $x_2 = f(x_1)$ and constitutes the second value of the sequence $\mathcal{O}_{x_0}$. By only zig-zagging this way—moving vertically from the $y = x$ line to the graph of f and then horizontally back to the $y = x$ line—we can document the orbit of the point x_0 without actually calculating the explicit function values. While this technique is only as accurate as the drawing of the graph, it is an excellent way to "see" an orbit without calculating it.

Specific to this example $f(x) = x^2$ on I, we have two fixed points: $x = 0$ and $x = 1$ (the graph crosses the $y = x$ line at these points). And if x_0 is chosen to be strictly less than 1, then we can easily conclude via the cobweb that $\mathcal{O}_{x_0} \longrightarrow 0$. Visually, it makes sense. Analytically, it is also intuitive: squaring a number between 0 and 1 reduces its value as it remains positive. However, can you *prove* that every orbit goes to 0 except for the orbit $\mathcal{O}_1$? We will do something like this shortly.

Exercise 69 Cobweb the following functions and describe the dynamics as completely as possible:

a. $f(x) = x^3$ on $[-1, 1]$ b. $g(x) = \ln(x+1)$ on $[0, \infty)$

c. $h(x) = (x^2 - 3)/(x - 2)$ on $\mathbb{R}$ d. $k(x) = -\frac{5}{3}(x^2 - x)$ on $[0, 1]$.

e. $\ell(x) = -\frac{10}{3}(x^2 - x)$ on $[0, 1]$

2.3.2 Fixed-Point Stability

What happens to orbits near a fixed point of a discrete dynamical system (equilibrium solutions to a continuous one) is of profound importance in an analysis of a mathematical model. Often, the fixed points are the only easily discoverable orbits of a hard-to-solve system. They play the role of a "steady state" of the system that models the interplay of the variables representing measurable quantities. And, as for functions in general, knowledge of a function's derivatives at a point says important things about how a function behaves near a point. To begin this analysis, we will need some definitions that will allow us to talk about the nature of fixed points in terms of what happens around them. This language is a lot like the way we classified equilibrium solutions in any undergraduate differential equations course. For the moment, think of X as simply an interval in $\mathbb{R}$ with the metric just the absolute value of the difference between two points. These definitions hold for all metric spaces X, though. The only caveat here is that in higher dimensions, there are some more subtle things that can happen near a fixed point, as we will see.

Definition 2.47 Let x_0 be a fixed point of the C^0-map $f : X \to X$. Then x_0 is said to be

- Poisson stable if $\forall \epsilon > 0$, $\exists \delta > 0$ such that $\forall x \in X$, if $d(x, x_0) < \delta$, then $\forall n \in \mathbb{N}$ $d(f^n(x), x_0) < \epsilon$.

- asymptotically stable, an attractor, or a sink if the fixed point is Poisson stable and if $\exists \epsilon > 0$ such that $\forall x \in X$, if $d(x, x_0) < \epsilon$, then $\mathcal{O}_x \longrightarrow x_0$.

- a repeller or a source if $\exists \epsilon > 0$ such that $\forall x \in X$, if $0 < d(x,x_0) < \epsilon$, then $\exists N \in \mathbb{N}$ such that $\forall n > N$, $d(f^n(x),x_0) > \epsilon$.

Remark 2.48 *There are a couple of extensions to this classification that occur from time to time. For example, we call a fixed point semi-stable if on one side it is asymptotically stable, while on the other side it looks like a repeller. In context, this will be clear. But this also places a semi-stable fixed point, together with a source, into the category of unstable fixed points; those in which arbitrarily close to a fixed point there is at least point whose orbit moves away from the fixed point. Again, in context, this will be clear.*

The basic idea behind this classification is the following: Asymptotically stable means that there is a neighborhood of the fixed point where f restricted to that neighborhood is a contraction with x_0 as the sole fixed point. Poisson stable means that given any small neighborhood of the fixed point, we can choose a smaller neighborhood where if we start in the smaller neighborhood, the forward orbit never leaves the larger neighborhood. Asymptotically stable points are always Poisson stable, but not necessarily vice versa. And a fixed point is a repeller if, in a small neighborhood of the fixed point, all points that are not the fixed point itself have forward orbits that leave the neighborhood and never return.

Example 2.49 $f(x) = 1 - x$ on $[0,1]$ has a unique fixed point $x_* = \frac{1}{2}$ (see Figure 18). This fixed point is Poisson stable, but not asymptotically stable. To see this, simply let $\delta = \epsilon$ and write out the definition.

Remark 2.50 *Recall from your differential equations class (or wait until Chapter 4), that in the classification of 2×2, first-order, homogeneous, linear ODE systems, one can classify the type of the equilibrium solution at the origin via knowledge of the eigenvalues of the coefficient matrix. In this classification, the sink (negative eigenvalues) is the asymptotically stable equilibrium, the source is the repeller, and the center (recall where the two eigenvalues are purely imaginary complex conjugates) is the Poisson stable equilibrium.*

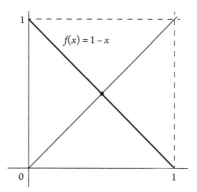

Figure 18 Poisson stable.

Example 2.51 Back to Figure 17, the graph of $f(x) = x^2$ on $[0, 1]$. This dynamical system has two fixed points. One can see visually via the cobweb that $x = 0$ is asymptotically stable. Also, $x = 1$ is unstable and a repeller.

Exercise 70 Show analytically that, for $f(x) = x^2$ on $[0, 1]$, $x = 0$ is an attractor while $x = 1$ is a repeller. That is, show that the fixed points satisfy the respective definitions.

Notice that we usually limit our discussion to maps $f : [0, 1] \to [0, 1]$ on the unit interval instead of a general closed, bounded $[a, b]$. There are basically two reasons for this: (1) They have applications beyond simple interval maps, and (2) maps of the unit interval are really all one need study when studying intervals. To see the second point, let $f : \mathbb{R} \to \mathbb{R}$, but suppose that there exists a closed interval with non-empty interior $[a, b]$, $b > a$ (see Figure 19), where

$$f|_{[a,b]} : [a, b] \to [a, b].$$

- Dynamically speaking, what happens to $f(x)$ under iteration on $[a, b]$ is no different from what happens to $g(y)$ on $[0, 1]$ under the linear transformation of coordinates $y = (x - a)/(b - a)$, where one must transform both the input and output variables appropriately. Note that this idea of changing a dynamical system to make it easier to study without changing its dynamics is that of a *conjugacy*, and will be a common theme in this text. Coordinate changes, like this one, are a form of this conjugacy.

Exercise 71 Find the map $g : [0, 1] \to [0, 1]$ on the unit interval in the plane that is equivalent dynamically to the map $f(x) = x^3$ on $[-1, 1]$.

- Let f be continuous on an unbounded interval I, either on one side or on both. Then, in the case that f has bounded image (possibly f has horizontal asymptotes, but this is not necessary), one can simply study the new dynamical system formed by f, where the domain is the interval $f(I)$, the set of first image points $f(x)$, for $x \in I$.

- Combining both of these items into one may also be useful: For example, a coordinate transformation can map an unbounded interval to a bounded one

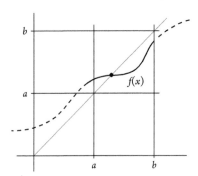

Figure 19 An interval map.

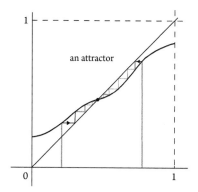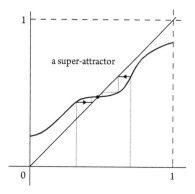

Figure 20 Follow the cobwebs to see how quickly nearby orbits converge respectively to a superattractor vis-á-vis an attractor.

(e.g., $g(x) = x^{-1}$, mapping $[1,\infty)$ onto $(0,1]$, or $h(x) = (2/\pi)\tan^{-1}(x)$, taking $\mathbb{R}$ to the interval $(-1,1)$). Under proper care with regard to the orbits, one can transform the dynamical system to one on the bounded interval. We shall elaborate on this later.

It is usually an $f(x)$ from above on some interval $[a,b]$ that appears in applications, and mathematically we usually only study maps like $g(y)$ on $[0,1]$. For an example, let's go back to the Newton–Raphson method for root location. First, a quick definition:

Definition 2.52 *A fixed point x_0 for $f : I \to I$ where $f \in C^1$ is called* superattracting *if $f'(x_0) = 0$. (Why? See Figure 20.)*

Proposition 2.53 *Let $f : \mathbb{R} \to \mathbb{R}$ be C^2 with a root r. If $\exists \delta > 0$, $M > 0$ such that $\left|f'(x)\right| > \delta$ and $\left|f''(x)\right| < M$ on a neighborhood of r, then r is a superattracting fixed point of*

$$F(x) = x - \frac{f(x)}{f'(x)}.$$

Proof As we have already calculated, $F'(r) = \dfrac{f(r)f''(r)}{[f'(r)]^2} = 0$. $\qquad\qquad\square$

We can go further: Since f is C^2 and $f'(r) \neq 0$, then we can find a small enough neighborhood where $f'(x) \neq 0$ where F is C^1. Calculating $F'(x)$ and knowing that it is both continuous and 0 at $x = r$, there will be a small, closed interval $[a,b]$, with $b > a$ and r in the interior, where $\left|F'(x)\right| < 1$. One can show that, restricted to this interval,

$$F\big|_{[a,b]} : [a,b] \to [a,b]$$

is a λ-contraction, with a superattracting fixed point at r. In this case, all orbits of F converge exponentially to r by λ^2, even though F is simply a λ-contraction.

Finally, before moving on to interval maps, a logical question to ask here is: Just how large is the interval $[a, b]$ on which F is a contraction? This is important for the Newton–Raphson method. Convergence is guaranteed when one chooses an approximation to a root that is "close enough." But what does "close enough" actually mean? Of course, this depends severely on the properties of the original function f. But to gauge the edges of an interval like $[a, b]$, we offer:

Definition 2.54 *A set* $Y \subset X$ *is invariant under a map* $f : X \rightarrow X$, *if*

$$f|_Y : Y \rightarrow Y.$$

Invariant subsets of the domain of a dynamical system will play a crucial role in our analysis. Essentially, f, restricted to an invariant $Y \subset X$, sets up a sub-dynamical system within the larger ambient one. We will soon see how important this is. For now, consider this: Every orbit in a dynamical system is an invariant set, as $f(\mathcal{O}_x) = \mathcal{O}_{f(x)} \subset \mathcal{O}_x$. Hence collections of orbits are also invariant. Thus, as a corollary to Proposition 2.13, we can immediately conclude the following:

Corollary 2.55 α-*limit sets and* ω-*limit sets are invariant.*

Here is another example of an invariant set:

Definition 2.56 *Let* $f : X \rightarrow X$ *be a discrete dynamical system with an attracting fixed point* $x \in X$. *Then the set*

$$\mathcal{B}_x = \{ y \in X \mid \mathcal{O}_y \longrightarrow x \}$$

is called the basin of attraction *of* x *for* f.

Essentially, $\mathcal{B}_x$ is the collection of all starting values of orbits that converge to x. Often, this set is quite easy to describe; For $f : X \rightarrow X$ a contraction with fixed point x_0, $\mathcal{B}_{x_0} = X$. Everything converges to the unique fixed point. However, as we will see, there are instances where this set is very complicated. As an indication of possible issues, consider the root search for a function $f : \mathbb{R} \rightarrow \mathbb{R}$ with at least two roots, both of which satisfy the conditions for convergence of, say, the Newton–Raphson method "near" each root. The individual basins of attraction for the two attractors should be open intervals, given the conditions for f. But must they come in one piece? The two basins cannot intersect, however. (Can you see why not?) But can they butt up against each other? And if a starting point is in neither basin, then where does its orbit go? In time, we will explore these issues and more. But for now, let's move on to a particular class of interval maps that are only slightly more complicated than contractions.

2.3.3 Monotonic Maps

Interval maps are quite general, and display tons of diverse and interesting behavior. To begin exploring this behavior, we will need to restrict our choice of maps to those exhibiting somewhat mild behavior. The first type designation is as follows:

Definition 2.57 *Let $f : [a, b] \to [a, b]$ be a C^0 map. We say f is*

- increasing *if for $x > y$, we have $f(x) > f(y)$;*
- non-decreasing *if for $x > y$, we have $f(x) \geq f(y)$;*
- non-increasing *if for $x > y$, we have $f(x) \leq f(y)$;*
- decreasing *if for $x > y$, we have $f(x) < f(y)$.*

In any of these cases, we say that f is a monotonic *map, and* strictly monotone *iff is either increasing or decreasing.*

It is intuitive to see how these definitions work. You should draw some examples to differentiate these types. You should also work to understand how these different types may affect the dynamics of a map (i.e., its orbit structure). For example, non-decreasing maps, which include all increasing maps, can have many fixed points (actually, the increasing map $f(x) = x$ has all of its points fixed!), while all non-increasing maps (and hence all decreasing maps also) can have only one fixed point each. Further, increasing maps cannot have points of period 2 (why not?), while there does exist a decreasing map with all of its points of period 2, although one point is not of prime period 2 (you have already seen this map in Figure 18). We will explore these in time. For now, we will start with a fact shared by all interval maps:

Proposition 2.58 *For the C^0 map $f : [a, b] \to [a, b]$, where $a, b \in \mathbb{R}$, f must have a fixed point.*

Visually, this should make sense. Imagine trying to draw a continuous curve from the left wall of the unit square to the right wall without crossing the diagonal from lower left to upper right. When you tire of trying, continue reading.

Proof Suppose for now that f has no fixed point on (a, b) (this seems plausible, since our example above $f(x) = x^2$ on $[0, 1]$ satisfies this criterion). Then, it must be the case that for all $x \in (a, b)$, either (1) $f(x) > x$ or (2) $f(x) < x$. This means that the entire graph of f lies respectively above the diagonal, or below it (see Figure 21).

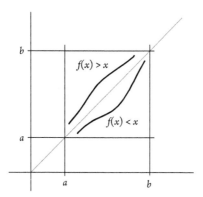

Figure 21 An interval map on $[a, b]$.

If we are in situation (1), then for any choice of $x \in (a,b)$, $\mathcal{O}_x$ will be an increasing sequence in $[a,b]$. It is also bounded above by b since the entire sequence lives in $[a,b]$. Recall from calculus the Monotone Convergence Theorem: A monotone sequence of real numbers has a finite limit iff it is bounded. For case (2), the orbit is strictly decreasing, and will be bounded below by a since the entire sequence lives in the closed interval $[a,b]$. Thus we can say in each instance that $\mathcal{O}_x \longrightarrow x_0$, for some $x_0 \in [a,b]$ (we can also say $\lim_{n\to\infty} f^n(x) = x_0$).

What can this point x_0 look like? Well, for starters, it must be a fixed point! Indeed,

$$f(x_0) = f\left(\lim_{n\to\infty} f^n(x)\right) = \lim_{n\to\infty} f^{n+1}(x) = \lim_{n\to\infty} f^n(x) = x_0.$$

Since there are no fixed points in (a,b) by assumption, it must be the case that either $x_0 = b$ (case (1)), or $x_0 = a$ (case (2)). □

Exercise 72 Reprove Proposition 2.58 using the Intermediate Value Theorem on the function $g(x) = f(x) - x$.

This last proof immediately tells us the following:

Proposition 2.59 Let the C^0-map $f : [a,b] \to [a,b]$ be non-decreasing, and suppose there are no fixed points on (a,b). Then either

- *exactly one endpoint is fixed and $\forall x \in [a,b]$, $\mathcal{O}_x$ converges to the fixed endpoint, or*
- *both endpoints are fixed, one is an attractor, and the other is a repeller.*

And if, in the second case above, f is also increasing, then $\forall x \in (a,b)$, $\mathcal{O}_x$ is forward asymptotic to one endpoint, and backward asymptotic to the other.

Example 2.60 Let $g(x) = \sqrt{x}$ on the closed interval $[1,b]$ for any $b > 1$. This is an example of the first situation in the proposition. Here, actually, we can let $g(x)$ have the closed interval $[1,\infty)$ as its domain, and we gain the same result. One does have to be careful here, though, as the function $k(x) = \ln(1 + e^x)$ on $[1,\infty)$ is monotonic on a closed interval, but has no fixed points at all.

Example 2.61 For the second case, think $f(x) = x^2$ on $[0,1]$, illustrated in Figure 17.

Exercise 73 For $h(x) = x^3$ on $I = [-1,1]$, both endpoints are fixed and repellers. But h does not satisfy the hypotheses of the proposition. (Why not?) Hence the proposition does not apply to this function. However, h may be seen as two separate dynamical systems existing side-by-side. Justify this and show that the proposition holds for each of these "sub"-dynamical systems.

Two important notes here:

- First, we can immediately see that the basin of attraction of the fixed point in the first case of Proposition 2.59 is the entire interval $[a,b]$, while in the second case it is everything except for the other fixed point (the repeller.) One property of a basin of attraction is that it is an open set (every point in the set contains a small open interval that is also completely in the set). This does not seem to be the case here. But it, in fact, is. The interval $[a,b]$, as a subset of $\mathbb{R}$, is closed (and bounded). But as a domain (not sitting inside $\mathbb{R}$ for the purpose of serving as the plug-in points for f), it is both open and closed as a topological space. In a sense, there is no outside to $[a,b]$ as a domain for f. It has open subsets, and these all look like one of $[a,c), (c,d)$, or $(d,b]$, for $a < c < d < b \in [a,b]$ and their various unions and finite intersections. But $[a,b]$ can be written as the union of two overlapping open sets. Hence it is open as a subset of $[a,b]$. And so is $(a,b]$, for example. Hence the basins of attraction in both of these cases are in fact open subsets of $[a,b]$.

- Recall from calculus (or even pre-calculus) that an increasing function $f : I \to \mathbb{R}$ is called *injective*, or *one-to-one*, if, for any two points $x, y \in I$, if $f(x) = f(y)$, then $x = y$. One would say its graph satisfies the Horizontal Line Test (remember this?) And one-to-one functions have inverses $f^{-1}(x)$, or we say that f is *invertible*. However, for $I = [a,b]$, a function $f : I \to I$, specifically from I back to itself, is only invertible if it satisfies certain precise criteria: it must be one-to-one and *onto*, or *surjective*, meaning for every $y \in I$, there is an $x \in I$ such that $f(x) = y$. A conclusion to draw from this is that if a function on an interval is both injective and surjective (we say it is then *bijective*), it must be the case that either $f(a) = a$ and $f(b) = b$, or $f(a) = b$ and $f(b) = a$. Using the above examples, $f(x) = x^2$ on $I = [0,1]$ and $h(x) = x^3$ on $I = [-1,1]$ are invertible, but $g(x)$ is not invertible on $[1,b]$ for any $b > 1$. And, the function given in Figure 16 is also not invertible, even though it is strictly increasing on $I = [0,1]$.

Exercise 74 Show that an invertible continuous map $f : I \to I$ on $I = [a,b]$, for $b > a$, must satisfy all of the following:

(a) f is injective (one-to-one);

(b) f is surjective (onto; the range must be all of I);

(c) f must satisfy either $f(a) = a$ and $f(b) = b$, or $f(a) = b$, and $f(b) = a$.

(Hint: All of these can be shown by assuming the property does not hold and then finding a contradiction.)

Hence we can revise the last statement of Proposition 2.59 as: *And if, in addition, f is invertible, then $\forall x \in (a,b)$, $\mathcal{O}_x$ is forward asymptotic to one endpoint, and backward asymptotic to the other.* Recall that if $\mathcal{O}_x$ is forward asymptotic to a point x_0, we write $\mathcal{O}_x^+ \to x_0$. If it is backward asymptotic, we write $\mathcal{O}_x^- \to x_0$. This notion of an orbit converging in both forward time and backward time is of sufficient importance in dynamics that we classify some differing ways in which this can happen.

2.3.4 Homoclinic/Heteroclinic Points

Using the notation for the forward and backward orbits introduced above, we can close in on a special property of non-decreasing interval maps. First, we can now say definitively that

- In the first case of Proposition 2.59, $\forall x \in [a,b]$, either $\mathcal{O}_x^+ \longrightarrow a$ or $\mathcal{O}_x^+ \longrightarrow b$. Here, f will certainly look like a contraction. Must it be? Keep in mind that the definition of a contraction is very precise, and maps that behave like contractions may not actually be contractions. Think $f(x) = x^2$ on the closed interval $[0,0.6]$. What is $\text{Lip}(f)$ here?

- For f in the second case of Proposition 2.59, and with f invertible (increasing), then $\forall x \in (a,b)$, either $\mathcal{O}_x^+ \longrightarrow a$ and $\mathcal{O}_x^- \longrightarrow b$, or $\mathcal{O}_x^+ \longrightarrow b$ and $\mathcal{O}_x^- \longrightarrow a$.

There is a term for points of an invertible map whose orbit is both forward and backward asymptotic to a point or points:

Definition 2.62 *Given $f : X \to X$, an invertible C^0-map, suppose $\exists x \in X$, where*

$$\mathcal{O}_x^- \longrightarrow a \text{ and } \mathcal{O}_x^+ \longrightarrow b.$$

Then x is said to be heteroclinic to a and b if $a \neq b$, and homoclinic to a if $a = b$. And the orbit $\mathcal{O}_x$ is called a heteroclinic (respectively homoclinic) orbit if x is a heteroclinic (respectively homoclinic) point.

Keep in mind here that the definitions of heteroclinic and homoclinic points require that f be invertible. Both the forward and backward orbits converge to something in each case. And both heteroclinic and homoclinic orbits are things you have seen before. Think of the phase portrait of the undamped pendulum on the left side of Figure 22. Here the *separatrices*—orbits of a dynamical system that separate phase space into regions where orbits in each region behave similarly—heteroclinic orbits from the unstable equilibrium solution at $(2n\pi,0)$, $n \in \mathbb{Z}$, and $(2(n \pm 1)\pi,0)$. However, in reality, there is a much more accurate picture of the phase space of the undamped pendulum. The vertical variable (representing the instantaneous velocity of the pendulum ball, actually) takes values in $\mathbb{R}$, while the horizontal variable (representing the angular position of the pendulum ball with respect to downward vertical position) is in reality 2π-periodic. In fact, it takes values in the circle S^1:

(2.3.1) $$S^1 = \text{unit circle in } \mathbb{R}^2 = \left\{ e^{2\pi i \theta} \in \mathbb{C} \middle| \theta \in [0,1) \right\}.$$

Thus, the phase space is really a cylinder, and in fact has only two equilibrium solutions; one at $(0,0)$ and the other at $(\pi,0)$ (see the right side of Figure 22). In this view, which we will elaborate upon in Section 6.2.4, there are only two separatrices (both shown in red at the right), and both are homoclinic to the unstable equilibrium at $(\pi,0)$. Also, it becomes clear once the picture is understood that all orbits of the undamped pendulum, except for the separatrices, are periodic. However, the period of these orbits is certainly not all the same. And there is no bound on how long a period may actually be. See if you can fully grasp this.

Exercise 75 Show that there cannot exist non-trivial homoclinic points for f a non-decreasing map on a closed, bounded interval.

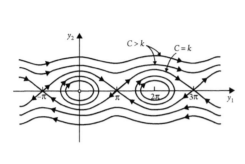
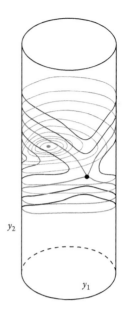

Figure 22 The phase space/cylinder of the pendulum.

Exercise 76 Construct an example (with an explicit expression) of a continuous C^0-map of S^1 that contains a homoclinic point. (Hint: We have already seen an example of an interval map that, when modified, will satisfy a C^0 (but not C^1) construction of a circle map. See Figure 17.)

Remark 2.63 *Any continuous map on the unit interval with both endpoints fixed can be viewed as a map on the circle by thinking of the interval and identifying 0 and 1 (you can use the map $x \mapsto e^{2\pi i x}$ explicitly to see this). But can you construct one that does not fix the endpoints? Can you construct one that is also differentiable on all of the circle? And how does one graph (visualize) this function?*

It turns out that forcing a map of an interval to be non-decreasing and forcing the interval to be closed really restricts the types of dynamics that can happen. We have:

Proposition 2.64 *Let $f : [a,b] \to [a,b]$ be C^0 and non-decreasing. Then $\forall x \in [a,b]$, either x is fixed, or asymptotic to a fixed point. And if f is also invertible (and thus increasing), then $\forall x \in [a,b]$, either x is fixed or heteroclinic to adjacent fixed points.*

And as an immediate corollary:

Corollary 2.65 *For $f : [a,b] \to [a,b]$ a C^0, non-decreasing map, there are no non-trivial periodic points. That is, Per$(f) = $ Fix(f).*

Clearly, the dynamics, although more complicated than for contraction maps, are nonetheless rather simple for non-decreasing interval maps (and even more so for

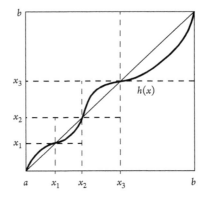

Figure 23 Trapped orbits.

invertible interval maps). Thus goes the second stop in our exploration of dynamical systems from simple to complex.

Before moving on, here are a few other things to think about: First, in the case of non-decreasing maps, the orbits of points in between fixed points are trapped in the interval bounded by the fixed points (see Figure 23). All of these closed intervals consisting of adjacent fixed points and the interval between them are invariant sets, and form sub-dynamical systems with the same f on the new restricted closed interval bounded by the fixed points. This has enormous implications, and severely restricts what orbits can do (we say it restricts the "orbit structure" of the map). It makes Proposition 2.59 above much more general and consequential, since any non-decreasing interval map is now simply a collection of disjoint interval maps, each of one of the types in the proposition. And second, it helps to establish not only the types of fixed points one can have for an interval map, but the way fixed points relate to each other within an interval map.

Exercise 77 Let f be a non-decreasing map on a closed interval. Show that if $x_0 \neq y_0$ are two fixed points of f, then $\forall x \in (x_0, y_0)$, $\mathcal{O}_x^+ \subset [x_0, y_0]$.

Exercise 78 Prove Corollary 2.65.

Exercise 79 Let f be a C^0 non-decreasing map on $[a, b]$, and $x_0 < y_0$ be two adjacent fixed points (i.e., there are no fixed points in (x_0, y_0)). Show that x_0 and y_0 cannot both be attractors or repellers.

Example 2.66 For $f(x) = \sqrt{x-1} + 3$ on $I = [1, \infty)$ in Figure 24, determine the set $\text{Per}(f)$.

We can address this question in a number of ways. First, notice that f is a strictly increasing function since $f'(x) = 1/(2\sqrt{x-1}) > 0$ on the interior of I (it is not defined at $x = 1$). Hence, every orbit is monotonic (not necessarily strictly, though: why is this true?) Hence, as a generalization of the conditions of Proposition 2.64, every point is either fixed or has an orbit asymptotic to a fixed point, or is unbounded (the consequence of

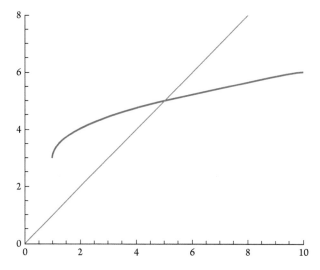

Figure 24 $f(x) = \sqrt{x-1} + 3$ on I.

an unbounded domain). At this point, you can conclude that there are no non-trivial periodic points, like in the above exercise. Now since the derivative is a strictly decreasing function, if there is a fixed point at all, it will be unique (think about this also). There is a fixed point x_0 of this map (the technique of Exercise 72 will work here). And, since $f(x) > x$ on $[1, x_0)$, and $f(x) < x$ on (x_0, ∞), the fixed point x_0 is an attractor. Hence $\text{Per}(f) = \{x_0\} = \{5\}$, where $x = x_0$ solves $f(x) = x$ algebraically.

Perhaps there is an easier way: $f(x)$ is certainly not a contraction, since for all pairs of points $x \neq y \in \left[1, \frac{5}{4}\right]$, the distance between images is actually greater than the original distances (check this!). In fact, on $[1, \infty)$, f is not even Lipschitz, although you should show this fact.

Exercise 80 Show that $f(x) = \sqrt{x-1} + 3$ on $I = [1, \infty)$ is not a contraction.

However, consider that the image of I, $f(I) = [3, \infty)$, and restricted to $f(I)$, f is actually a λ-contraction, with $\lambda = \frac{1}{2\sqrt{3}}$. Hence, one could simply start iterating after the first iterate, knowing that the long-term behavior of orbits, fixed and periodic points, convergent orbits and stability, will all be the same. Thus, the map $f : f(I) \to f(I)$ is a contraction, and hence will have a unique fixed point and no other periodic points. Thus the fixed point x_0 found above is precisely all of $\text{Per}(f)$.

Remark 2.67 *It is tempting to call the map f above eventually contracting, since it is a contraction on a forward iterate of the domain. However, this is not the case here. As we will see soon enough (Definition 3.30 in Section 3.2.3), there is a technical condition that makes a map an eventual contraction, and there is a pathology at $x = 1$ here (pay attention to the derivative as one approaches 1 from numbers larger than 1). Suffice it to say that we will have to settle for the following analysis: There is an invariant interval $I_0 \subset I$, containing the* *fixed point, where* $f\big|_{I_0}$ *is a contraction, and* $\forall x \in I$ *where* $x \notin I_0$, *there is a forward iterate of x which is in* I_0. *At that point, one can see that even though* f *is not a contraction, all orbits converge to a unique fixed point. We will elaborate more on this topic in Section 3.2.3 on fixed-point theorems.*

2.4 Bifurcations of Interval Maps

Now take an increasing C^1-interval map f on $I \subset \mathbb{R}$ and vary it slightly. Usually, the dynamical behavior of the "perturbed" map stays the same (the number and type of fixed points do not change, even though their positions may vary a bit). But sometimes, arbitrary small changes in a map produce new maps with different dynamical behavior. For example, in Figure 25, as the function changes so that the region below the $y = x$ line shrinks, a point is reached where it "appears" that the two adjacent fixed points defining this region coalesce, the newly created single fixed point forms a tangency with the $y = x$ line, and then the tangency and the fixed point vanish as the map continues to change, leaving only the endpoints of the interval fixed. We say the map in the middle is not *structurally stable*, in that a small change in the map may render either no interior fixed points or two. Of course, the notion of structural stability depends heavily on the types of perturbations that are allowed, and involves viewing the function as one such function in a space of functions of a particular type on the interval, and then studying whether all of the function's nearest neighbors in that space of functions behave the same way as that of the function. Among the more common spaces used to study a function's properties include the space of continuous functions on I, $C^0[I,I]$, and the space of continuously differentiable functions, $C^1[I,I]$. In a more restrictive sense, one can define a family of functions by introducing a parameter into the expression for f, and watch how the function behaves across an interval of values of that parameter. Indeed,

$$f_\alpha : I \to I, \, \alpha \in J \subset \mathbb{R}, \quad \text{or} \quad f : J \times I \subset \mathbb{R}^2 \to I, \, (\alpha, x) \mapsto f(\alpha, x).$$

The values of the parameter $\alpha_0 \in J$ that correspond to non-structurally stable maps (the ones where there exist maps in any neighborhood of that value in J that exhibit different dynamic behavior) are of particular interest, and are called *bifurcation values* of α.

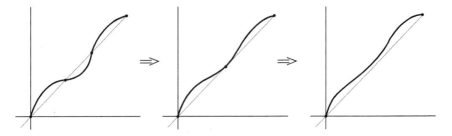

Figure 25 A bifurcation in an interval map.

(The term was first coined by Henri Poincaré [41] in 1885.) And the bifurcation value together with the fixed point where the bifurcation take place, $(\alpha_0, x_0) \in J \times I$, is called a *bifurcation point* of the map or system. In general, cataloging the types of bifurcations of maps on metric spaces is a rich and quite tricky endeavor. However, for interval maps with one parameter, there are only a few common types. Herein, we will detail the three most common bifurcation types for a one-parameter family of C^1-interval maps, using $\mathbb{R}$ as our interval, and study what are called their *normal forms* (prototypical examples.) All of these types that we will deal with currently rely on the same feature inherent to interval maps at a bifurcation point: a tangential intersection between the graph of f, as a function on an interval, and the line $y = x$. That is, for a map $f : J \times I \to I$ in this section, a necessary condition for the existence of a bifurcation point is a point $(\alpha_0, x_0) \in J \times I$ where $f(\alpha_0, x_0) = x_0$, and $(\partial f / \partial x)(\alpha_0, x_0) = 1$. Note, however, that in Section 7.5.1, we will detail the structure of a new type of bifurcation of an interval map that does not feature a tangential intersection between a function and the $y = x$ line. For now, though, all will have this property. The types we will highlight here are

- saddle-node, or fold, bifurcation;
- transcritical bifurcation; and
- pitchfork bifurcation.

The names are descriptive and refer to the shape of the fixed-point set, as seen in the αx-plane.

To begin, let $f_\alpha : \mathbb{R} \to \mathbb{R}$ be a family of C^1 maps, C^1-parameterized by $\alpha \in \mathbb{R}$, and rewritten so that $f : \mathbb{R}^2 \to \mathbb{R}$, $(\alpha, x) \mapsto f(\alpha, x)$ (we can restrict to subintervals in context later). The reformulation allows us to use some analytic geometry in our study. First, the fixed points we are interested in is the set

$$\text{Fix}(f_\alpha) = \left\{ (\alpha, x) \in \mathbb{R}^2 \,\middle|\, f(\alpha, x) = x \right\}.$$

In the αx-plane, we can graph these fixed points as the 0-level set of $F(\alpha, x) = f(\alpha, x) - x$.

Recall that, for a C^1-function $F : X \subset \mathbb{R}^2 \to \mathbb{R}$, and $c \in F(X)$, the *c-level set* of F is, generically, a C^1 curve in X with a well-defined tangent space defined by the gradient

$$\nabla F = \begin{bmatrix} \frac{\partial F}{\partial x}(x_0, y_0) \\ \frac{\partial F}{\partial y}(x_0, y_0) \end{bmatrix},$$

for $(x_0, y_0) \in f^{-1}(c)$, at least when $\nabla F \neq \mathbf{0}$.

Indeed, in the case when $\nabla F(x_0, y_0) \neq \mathbf{0}$ (that is, at a *regular* point of the level set,) the Implicit Function Theorem allows us to write the equation $F(x, y) = 0$, at least locally near (x_0, y_0), as the graph of an explicit function of at least one of the variables in terms of the other:

- As long as $(\partial F / \partial y)(x_0, y_0) \neq 0$, then, in a small neighborhood of $x = x_0$, we can find a $y(x)$ where $y(x_0) = y_0$ and $F(x, y(x)) = 0$, $\forall x$ near x_0.

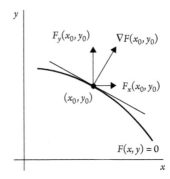

Figure 26 A regular point on a level curve.

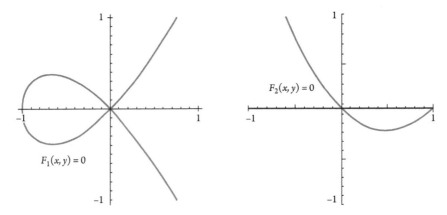

Figure 27 Irreducible (left) and reducible (right) curve double points.

- As long as $(\partial F/\partial x)(x_0, y_0) \neq 0$, then, in a small neighborhood of $y = y_0$, we can find an $x(y)$ where $x(y_0) = x_0$ and $F(y, x(y)) = 0$, $\forall y$ near y_0.

So, as long as $\nabla F(x_0, y_0) \neq \mathbf{0}$, there exists a unique curve of $F(x, y) = 0$ passing through (x_0, y_0).

In contrast, when $\nabla F(x_0, y_0) = \mathbf{0}$, the point (x_0, y_0) is called *singular*. Common singular points include local extrema and saddles of F. When $F(x, y) = 0$ contains a saddle (an ordinary double point, or a *crunode*), the level curve crosses itself in a non-tangential fashion, forming two distinct tangent directions at the crossing (x_0, y_0). (The term *ordinary* refers directly to the idea that the tangent directions are distinct. *Double* refers to the number of curves crossing at the point; see Figure 27). This is the type of singularity we will see when discussing bifurcations of interval maps, where one of the curves corresponds to a fixed point that is independent of the parameter α. In this case, $F(x, y)$ factors into $F(x, y) = G(x, y)H(x, y) = 0$ (locally when the ordinary double point resides in an irreducible curve, and globally when the level set consists of two separately defined curves (See Example 2.68). In either case, then, we can easily compute the derivatives of each of the curves. This will be very useful in our discussion.

Example 2.68 Let $F_1(x,y) = y^2 - x^2(x+1)$ and $F_2(x,y) = y^2 - y(x^2 - x)$. As you can see in Figure 27, both have ordinary double points (saddles) at the origin of the respective sets $F_1(x,y) = 0$ and $F_2(x,y) = 0$. Here, it is easy to see that the 0-curve of F_2 is reducible, and $F(x,y) = y(y - (x^2 - x))$, allowing us to graph each curve separately and calculate the tangent lines. But there is no global factoring of F_1. Locally, though, near the origin, we can write $F_1(x,y) = (y + x\sqrt{x+1})(y - x\sqrt{x+1})$. This allows us to (1) calculate tangent lines at the origin and (2) locally write each curve in the crossing as a local graph, as before, with regular points on each curve.

2.4.1 Saddle-Node Bifurcation

We start with an example: Let $f : \mathbb{R} \times \mathbb{R} \to \mathbb{R}, f(\alpha, x) = x^2 + x - \alpha$. As α plays the role of a simple vertical translation of the graph of f in $\mathbb{R}^2$, a parabola opening up with its vertex along the $x = -\frac{1}{2}$ vertical line, it should be clear geometrically that for some values of α, there will be no intersection with the $y = x$ line, while for others, there will be two. See Figure 28.

The fixed points of this family of maps satisfy $f(\alpha, x) = x$, or $x^2 - \alpha = 0$. We conclude that for each value of α, we have

- no solutions when $\alpha < 0$;
- one solution, at $x = 0$, when $\alpha = 0$;
- two solutions, $x = \pm\sqrt{\alpha}$, for $\alpha > 0$.

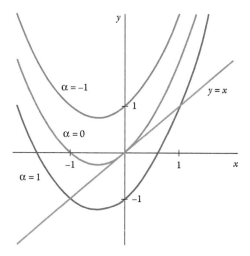

Figure 28 A fold bifurcation.

Graphing these solutions in the αx-plane yields two curves that cross uniquely st the origin. For any value of α, one can also use derivative information to establish the stability of these fixed points:

Exercise 81 For $\alpha = 36$, fully describe the dynamics of the map $f(36, x) = x^2 + x - 36$ on $\mathbb{R}$.

Notice that at the special $\alpha = 0$, the sole fixed point satisfies

$$f(0,0) = 0, \quad \frac{\partial f}{\partial x}(0,0) = 1.$$

This is the *bifurcation condition*, what establishes that the graph of f has a tangency with the $y = x$ line. Whether this tangency persists as α varies will depend on other properties of f. But this condition is necessary for a possible change in the dynamical nature of the system. In this case, we also see that $(\partial f/\partial \alpha)(0,0) = -1 \neq 0$ and $(\partial^2 f/\partial x^2)(0,0) = 2 \neq 0$. Thus the tangency does not persist, and $\alpha = 0$ is a bifurcation value for α, rendering $(\alpha, x) = (0,0)$ a bifurcation point in the αx-plane. By decorating this plane appropriately (by drawing phase lines as the vertical lines denoting the stability of all fixed points for each value of) α, we can draw the bifurcation diagram (Figure 29) for this system.

This type of bifurcation is characterized by the sudden appearance of a semi-stable fixed point in a region without fixed points, which immediately splits into two fixed points: one asymptotically stable and the other unstable. Some notes:

- Change this map to $f(\alpha, x) = x^2 + x + \alpha$, and the resulting bifurcation diagram looks like two distinct fixed points, which eventually coalesce into one and then vanish. In essence, the diagram and the bifurcation are backwards. We characterize the original as *supercritical* and the backwards as *subcritical*. Can you work out the relationship of this classification to the signs of the two non-zero derivatives $f_\alpha(0,0)$ and $f_{xx}(0,0)$?

- Besides the other name for this kind of bifurcation, saddle-node, a supercritical fold bifurcation is also sometimes called a *creation bifurcation* or a *blue-skies bifurcation*. A subcritical fold is also known as an *annihilation bifurcation*.

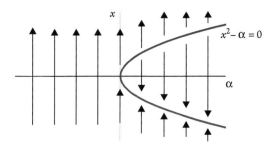

Figure 29 Bifurcation diagram for a fold bifurcation.

We can collect the information in this map to locate folds in other maps via a recipe:

Proposition 2.69 *Let $f : \mathbb{R} \times \mathbb{R} \to \mathbb{R}$, $(\alpha, x) \mapsto f(\alpha, x)$ satisfy at (α_0, x_0):*

(1) $f(\alpha_0, x_0) = x_0$, and $(\partial f / \partial x)(\alpha_0, x_0) = 1$;

(2) $(\partial f / \partial \alpha)(\alpha_0, x_0) = a \neq 0$; and

(3) $(\partial^2 f / \partial x^2)(\alpha_0, x_0) = b \neq 0$.

Then f has a saddle-node or fold bifurcation at $\alpha = \alpha_0$, which is supercritical when $ab > 0$ and subcritical when $ab < 0$.

Proof Create $F(\alpha, x) = f(\alpha, x) - x$, so that the fixed-point set of the family of functions corresponds to the 0-level set of F. Notice that $F(\alpha_0, x_0) = 0$, and, by the bifurcation condition of the first supposition,

$$\frac{\partial F}{\partial x}(\alpha_0, x_0) = \frac{\partial f}{\partial x}(\alpha_0, x_0) - 1 = 0.$$

Also, by the second supposition,

$$\frac{\partial f}{\partial \alpha}(\alpha_0, x_0) = \frac{\partial F}{\partial \alpha}(\alpha_0, x_0) = a \neq 0.$$

Hence, by the Implicit Function Theorem, we can (locally) write the 0-level set of F near (α_0, x_0) as the graph of $\alpha(x)$, where $\alpha(x_0) = \alpha_0$, and for all x near x_0, $F(\alpha(x), x) = 0$. Differentiate to get

$$\frac{d}{dx} F(\alpha(x), x) = 0 = \frac{\partial F}{\partial x}(\alpha(x), x) + \frac{\partial F}{\partial \alpha}(\alpha(x), x) \frac{d\alpha}{dx}.$$

Knowing that $(\partial F / \partial x)(\alpha_0, x_0) \neq 0$, we get

$$\left. \frac{d\alpha}{dx} \right|_{(\alpha_0, x_0)} = \alpha'(x_0) = \frac{-F_x(\alpha(x_0), x_0)}{F_\alpha(\alpha(x_0), x_0)} = 0.$$

Thus the 0-curve of F is tangent to the line $\alpha = \alpha_0$ at (α_0, x_0) in the αx-plane. To obtain a fold bifurcation, we will also need $\alpha''(x_0) \neq 0$, so that (1) this tangency does not persist and (2) (α_0, x_0) is not an inflection point (thus the entire 0-level curve, near x_0, remains on one side of the line $\alpha = \alpha_0$).

We have

$$0 = \frac{d^2}{dx^2} F(\alpha(x), x) = \frac{d}{dx} \left[\frac{\partial F}{\partial x}(\alpha(x), x) + \frac{\partial F}{\partial \alpha}(\alpha(x), x) \frac{d\alpha}{dx} \right]$$

$$= \frac{\partial^2 F}{\partial x^2}(\alpha(x), x) + \frac{\partial^2 F}{\partial \alpha \partial x}(\alpha(x), x) \frac{d\alpha}{dx}$$

$$+ \frac{\partial^2 F}{\partial x \partial \alpha}(\alpha(x), x) \frac{d\alpha}{dx} + \frac{\partial F}{\partial \alpha^2}(\alpha(x), x) \left(\frac{d\alpha}{dx} \right)^2$$

$$+ \frac{\partial F}{\partial \alpha}(\alpha(x), x) \frac{d^2\alpha}{dx^2}.$$

But, with $\alpha'(x_0) = 0$, this reduces, when evaluated at the bifurcation point, to

$$\alpha''(x_0) = \frac{-F_{xx}(\alpha_0, x_0)}{F_\alpha(\alpha_0, x_0)}.$$

With the third supposition, that $-F_{xx}(\alpha_0, x_0) \neq 0$, we achieve the desired form for the fixed-point set of $f(\alpha, x)$.

And finally, the concavity of the curve (in essence, which side of the $\alpha = \alpha_0$ line the curve resides on, is given by

$$\alpha''(x_0) = \frac{-F_{xx}(\alpha_0, x_0)}{F_\alpha(\alpha_0, x_0)} = -\frac{b}{a}.$$

Hence $\alpha(x)$ is concave-up, residing on the right of the vertical line (a supercritical or creation bifurcation,) when $ab < 0$, and a subcritical or annihilation when $ab > 0$. □

Exercise 82 For the following families of functions, find points that satisfy the bifurcation condition and determine whether a fold bifurcation occurs at these points. If the points do not exhibit a fold bifurcation, determine the conditions that fail.

(a) $f(\alpha, x) = x^3 - 2x - \alpha$, (b) $g(\alpha, x) = x^3 + x - \alpha$, (c) $h(\alpha, x) = x^2 + x - \alpha^2$,
(d) $i(\alpha, x) = \alpha - \cos(2x)$, (e) $j(\alpha, x) = x^3 - \alpha x + 2$.

2.4.2 Transcritical Bifurcation

As before, we start with a prototypical example: Let $f : \mathbb{R} \times \mathbb{R} \to \mathbb{R}$, $f(\alpha, x) = x^2 - \alpha x + x$. Here, the graph of f in $\mathbb{R}^2$ is again a parabola opening up. But in this case, the vertex moves as α varies. What does not move is one of the roots $x = 0$. Geometrically, except for the unique tangent intersection at $\alpha = x = 0$, there are always two intersections of the graph of f with the line $y = x$, and one of them is always at $x = 0$. See Figure 30. Hence it should be clear that $\alpha = 0$ is a bifurcation value for this family of functions, and the point $(\alpha_0, x_0) = (0,0)$ satisfies the bifurcation condition: $f(0,0) = 0$ and $(\partial f / \partial x)(0,0) = 2x + (1 - \alpha)|_{(0,0)} = 1$.

Fixed-point lines in the αx-plane correspond to the equation $f(\alpha, x) - x = 0 = x(x - \alpha)$. The two lines of fixed points are then $x = 0$ and $x = \alpha$. Stability criteria for the fixed points can determine the dynamics near the fixed points, and

$$\frac{\partial f}{\partial x}(\alpha, 0) = 1 - \alpha = \begin{cases} 1 & \text{for } \alpha = 0 : \text{ semi-stable,} \\ > 1 & \text{for } \alpha < 0 : \text{ unstable,} \\ < 1 & \text{for } \alpha > 0 : \text{ asymptotically stable,} \end{cases}$$

$$\frac{\partial f}{\partial x}(\alpha, \alpha) = \alpha + 1 = \begin{cases} 1 & \text{for } \alpha = 0 : \text{ semi-stable,} \\ < 1 & \text{for } \alpha < 0 : \text{ asymptotically stable,} \\ > 1 & \text{for } \alpha > 0 : \text{ unstable.} \end{cases}$$

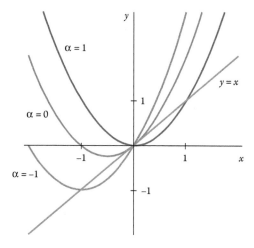

Figure 30 A transcritical bifurcation.

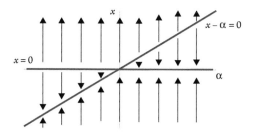

Figure 31 Bifurcation diagram for a transcritical bifurcation.

Hence the two fixed points seem to pass through each other at $\alpha_0 = 0$ and exchange their stability.

One can develop the bifurcation diagram in Figure 31 via this analysis. So, what are the properties of f in this case? We label them.

(1) The data at $(0,0)$:

- $f(\alpha,0) = 0$, for all $\alpha \in \mathbb{R}$;
- $\dfrac{\partial f}{\partial x}(0,0) = 1$;
- $\dfrac{\partial^2 f}{\partial x^2}(0,0) = 2 \neq 0$; and
- $\dfrac{\partial^2 f}{\partial \alpha \partial x}(0,0) = -1 \neq 0$.

(2) Since $x = 0$ is always a fixed point, there is always an intersection of the graph of f and the $y = x$ axis. And if the intersection is not a tangency, then there are always two distinct solutions. (Why?)

(3) In the case of two distinct solutions, whatever the stability of the fixed point $x = 0$, the other fixed point must have the opposite stability. (Why?)

(4) For any point (α_0, x_0), where the intersection is a tangency, so $(\partial f / \partial x)$ $(\alpha_0, x_0) = 1$, if the second partials do not vanish here, then the tangency will not persist and the fixed points will seems to pass through each other.

Again, we can collect up this information into a recipe for exposing a transcritical bifurcation in a more general family of maps:

Proposition 2.70 *Let $f : \mathbb{R} \times \mathbb{R} \to \mathbb{R}$, $(\alpha, x) \mapsto f(\alpha, x)$ satisfy at (α_0, x_0):*

(1) $f(\alpha, x_0) = x_0 \; \forall \alpha \in \mathbb{R}$, and $\dfrac{\partial f}{\partial x}(\alpha_0, x_0) = 1$;

(2) $\dfrac{\partial f}{\partial \alpha}(\alpha_0, x_0) = 0$;

(3) $\dfrac{\partial^2 f}{\partial \alpha \partial x}(\alpha_0, x_0) \neq 0$; and

(4) $\dfrac{\partial^2 f}{\partial x^2}(\alpha_0, x_0) \neq 0$.

Then f has a transcritical bifurcation at $\alpha = \alpha_0$.

Proof As previously, form $F(\alpha, x) = f(\alpha, x) - x$ as a function on the αx-plane, and whose 0-level set corresponds to the fixed point set of the family of functions defined by f. This fixed point will be composed of two curves intersecting (locally) only at (α_0, x_0):

- one curve the line $x = x_0$, and
- the other curve crossing $x = x_0$ that will be specified to satisfy the suppositions.

Remark 2.71 *Technically, the first supposition in the proposition, ensuring that one of the fixed-point curves be $x = x_0$, is not completely necessary. In a more general sense, one can use local generalized coordinates in a neighborhood of a potential bifurcation point to render one curve of the two intersecting curves as a (parallel of a) coordinate axis of a new coordinate system. This is similar to the standard substitution method in single-variable calculus, rendering the inside function of a composition as a new variable, allowing one to more easily integrate just the outside function. In the present case, allowing one fixed-point curve to be a constant function does facilitate the proof without sacrificing understanding.*

For these two curves to exist and intersect at the point (α_0, x_0), we must have

$$\frac{\partial F}{\partial \alpha}(\alpha_0, x_0) = 0 = \frac{\partial F}{\partial x}(\alpha_0, x_0)$$

or else only one level curve can exist at (α_0, x_0). In the present case, $F_x(\alpha_0, x_0) = f_x(\alpha_0, x_0) - 1$ and $F_\alpha(\alpha_0, x_0) = f_\alpha(\alpha_0, x_0)$. Thus, by the first two suppositions, $\nabla F(\alpha_0, x_0) = \mathbf{0}$. Now, given that one curve is $x - x_0 = 0$, we can write (locally) $F(\alpha, x) = (x - x_0) G(\alpha, x)$. Here

$$G(\alpha, x) = \frac{F(\alpha, x)}{x - x_0} = \frac{f(\alpha, x) - x}{x - x_0}$$

when $x \neq x_0$, and since

$$\frac{\partial F}{\partial x}(\alpha, x) = G(\alpha, x) + (x - x_0)\frac{\partial G}{\partial x}(\alpha, x),$$

we have that at $x = x_0$, $(\partial F / \partial x)(\alpha, x_0) = G(\alpha, x_0)$. Hence

$$G(\alpha, x) = \begin{cases} \dfrac{F(\alpha, x)}{x - x_0} = \dfrac{f(\alpha, x) - x}{x - x_0}, & x \neq x_0, \\[3mm] \dfrac{\partial F}{\partial x}(\alpha, x_0) = \dfrac{\partial f}{\partial x}(\alpha, x_0) - x_0, & x = x_0. \end{cases}$$

Here, $G(\alpha, x) = 0$ represents the other curve. In this case, $G_x(\alpha_0, x_0) = f_{xx}(\alpha_0, x_0) \neq 0$ by supposition (4), so we know the intersection is not tangent to the $x = x_0$ line. Then, by the Inverse Function Theorem, we can write this second curve as $\alpha(x)$ locally, so that $\alpha(x_0) = \alpha_0$ and $G(\alpha(x), x) = 0$ for all x near x_0. Differentiating this last implicit function of x, we get

$$\frac{d}{dx}G(\alpha(x), x) = 0 = \frac{\partial G}{\partial x}(\alpha(x), x) + \frac{\partial G}{\partial \alpha}(\alpha(x), x)\frac{d\alpha}{dx}(x).$$

Now $G_\alpha(\alpha_0, x_0)$ exists, and $(\partial G / \partial \alpha)(\alpha_0, x_0) = (\partial^2 f / \partial \alpha \partial x)(\alpha_0, x_0) \neq 0$ by supposition (3). And since $(\partial G / \partial x)(\alpha(x), x)$ doesn't vanish at x_0, it won't near x_0, and we can write

$$\alpha'(x) = \frac{d\alpha}{dx}(x) = \frac{-G_x(\alpha(x), x)}{G_\alpha(\alpha(x), x)},$$

concluding that $\alpha'(x_0) \neq 0$. This is enough to create the bifurcation diagram above, at least qualitatively with a possibly nonlinear curve representing the non-horizontal portion of the fixed-point set. □

Exercise 83 Draw a bifurcation diagram for $f(\alpha, x) = x^2 + (1 - \alpha^2)x$ and show that f does not have a transcritical bifurcation. Determine which of the stipulations fails.

Exercise 84 Determine conditions on the signs of both of the non-zero quantities $(\partial^2 f / \partial \alpha \partial x)(\alpha_0, x_0)$ and $(\partial^2 f / \partial x^2)(\alpha_0, x_0)$ so that the line $x = x_0$ is a line of unstable equilibria for $\alpha < \alpha_0$ and asymptotically stable fixed points for $\alpha > \alpha_0$.

2.4.3 Pitchfork Bifurcation

Lastly, let $f(\alpha, x) = (\alpha + 1)x - x^3$. In this case, we see that $f(\alpha, x) = x$ precisely when $\alpha x - x^3 = x(\alpha - x^2) = 0$. Hence we have

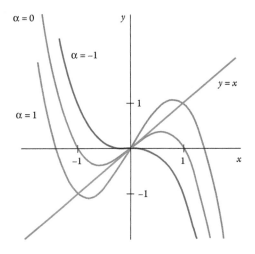

Figure 32 A pitchfork bifurcation.

- $x = 0$ when $\alpha \leq 0$ and
- $x = 0, \pm\sqrt{\alpha}$, for $\alpha > 0$.

Hence there is a change of behavior in α at $\alpha = 0$, with $x = 0$ as the only fixed point, and since $f_x(0,0) = 1$, the point $(0,0)$ is a bifurcation point for f. See Figure 32.

Remark 2.72 *This example is actually a bit more complicated than the previous two. To sidestep the extra complication (we will return to it in Section 7.5.1), we will limit ourselves to the parameter values $\alpha \in (-2,1)$. The nature of this complication will become clearer later, and we will return to this example then. For now, we continue our discussion.*

Here, we can easily see that there are two distinct curves of fixed points, crossing at the point $(0,0)$: one the line $x = 0$ and the other the curve $x^2 = \alpha$. Graphing these two lines in the αx-plane and evaluating derivative information, we see that $f_x(\alpha, x) = \alpha + 1 - 3x^2$, giving us the following:

- for $-2 < \alpha < 0$, $|f_x(\alpha, 0)| < 1$, and $x = 0$ is asymptotically stable;
- for $\alpha = 0$, $f_x(\alpha, 0) = 1$ and $x = 0$ is semi-stable; and
- for $0 < \alpha < 1$, $f_x(\alpha, 0) > 1$, rendering $x = 0$ as unstable and $0 < f_x(\alpha, \pm\sqrt{\alpha}) < 1$, so that each of $x = \sqrt{\alpha}$ and $x = -\sqrt{\alpha}$ are asymptotically stable.

This is enough information to create the bifurcation diagram in Figure 33.

So, what are the properties of this map? For one, the 0-level curve of $F(\alpha, x) = f(\alpha, x) - x$ has an ordinary double point at $(0,0)$, so, even though $\nabla F(0,0) = \mathbf{0}$, we can write $F(\alpha, x) = x(\alpha - x^2)$, with two distinct tangent lines at $(\alpha, x) = (0,0)$. But let's go

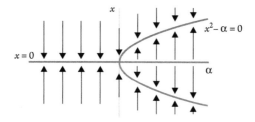

Figure 33 Bifurcation diagram for a pitchfork bifurcation.

directly to the generalized situation and establish just what properties characterize this type of bifurcation:

Proposition 2.73 *Let* $f : \mathbb{R} \times \mathbb{R} \to \mathbb{R}$, $(\alpha, x) \mapsto f(\alpha, x)$ *satisfy at* (α_0, x_0):

(1) $f(\alpha, x_0) = x_0 \; \forall \alpha \in \mathbb{R}$, *and* $\dfrac{\partial f}{\partial x}(\alpha_0, x_0) = 1$;

(2) $\dfrac{\partial f}{\partial \alpha}(\alpha_0, x_0) = 0$;

(3) $\dfrac{\partial^2 f}{\partial x^2}(\alpha_0, x_0) = 0$;

(4) $\dfrac{\partial^2 f}{\partial x \partial \alpha}(\alpha_0, x_0) = a \neq 0$; *and*

(5) $\dfrac{\partial^3 f}{\partial x^3}(\alpha_0, x_0) = b \neq 0$

Then f *has a pitchfork bifurcation at* $\alpha = \alpha_0$, *which is supercritical when* $ab < 0$ *and subcritical when* $ab > 0$.

Remark 2.74 *The determination of supercritical and subcritical is really just a convention, much like the idea of the positive x-axis in the xy-plane being on the right of the y-axis. The bifurcation diagram in Figure 33 represents a supercritical pitchfork bifurcation, where the lone fixed point for* $\alpha < 0$ *bifurcates into three fixed points, as an increasing* α *passes through* $\alpha = 0$. *Flipping this diagram about the vertical axis creates a diagram of a subcritical bifurcation.*

Proof Again, let $F(\alpha, x) = f(\alpha, x) - x$ be the fixed-point set of f, written as the 0-level set of a function on the αx-plane. And again, in this situation, this 0-level set will consist (locally, near (α_0, x_0)) of two curves intersecting locally only at (α_0, x_0). One of the curves is $x = x_0$ (note Remark 2.71 above) and the other remains unspecified for the moment except for the isolated intersection. This intersection point is only possible when $\nabla F(\alpha_0, x_0) = \mathbf{0}$, so both $F_x(\alpha_0, x_0) = 0$ and $F_\alpha(\alpha_0, x_0) = 0$. This is in agreement with the first two suppositions in the proposition, as $F_x(\alpha_0, x_0) = 0 = f_x(\alpha_0, x_0) - 1$ by supposition (1) and $F_\alpha(\alpha_0, x_0) = f_\alpha(\alpha_0, x_0) = 0$ by supposition (2).

Write $F(\alpha, x) = (x - x_0)G(\alpha, x)$, at least locally, so that, as in the proof of Proposition 2.70, we can write

$$
G(\alpha, x) = \begin{cases} \dfrac{F(\alpha, x)}{x - x_0} = \dfrac{f(\alpha, x) - x}{x - x_0}, & x \neq x_0, \\[3mm] \dfrac{\partial F}{\partial x}(\alpha, x_0) = \dfrac{\partial f}{\partial x}(\alpha, x_0) - 1, & x = x_0. \end{cases}
$$

Thus $G_\alpha(\alpha_0, x_0) = f_{\alpha x}(\alpha_0, x_0) \neq 0$, by supposition (4), and the intersection of the two curves is not tangential. Hence we can find $\alpha(x)$ such that $\alpha(x_0) = \alpha_0$ and along this curve $G(\alpha(x), x) = 0$ for all x near x_0. Differentiate this implicit equation of x to get

$$
\frac{d}{dx}G(\alpha(x), x) = 0 = G_x(\alpha(x), x) + G_\alpha(\alpha(x), x)\frac{d\alpha}{dx}(x),
$$

so that

$$
\alpha'(x) = \frac{-G_x(\alpha(x), x)}{G_\alpha(\alpha(x), x)},
$$

knowing that this is well-defined near x_0 since $G_\alpha(\alpha_0, x_0) \neq 0$. But for the curve $\alpha(x)$ to be twice differentiable and to persist on only one side of the $\alpha = \alpha_0$ vertical line, we need $\alpha'(x_0) = 0$ and $\alpha''(x_0) \neq 0$. But we know

$$
\alpha'(x_0) = \frac{-G_x(\alpha(x_0), x_0)}{G_\alpha(\alpha(x_0), x_0)} = 0 = -\frac{f_{xx}(\alpha(x_0), x_0)}{f_{x\alpha}(\alpha(x_0), x_0)},
$$

since $f_{xx}(\alpha_0, x_0) = 0$ by supposition 3.

And we also want $\alpha(x)$ to persist on only one side of $\alpha = \alpha_0$ line, so we require that $\alpha''(x_0) \neq 0$. Here,

$$
\begin{aligned}
\frac{d^2}{dx^2}G(\alpha(x), x) = {}& G_{xx}(\alpha(x), x) + G_{\alpha x}(\alpha(x), x)\alpha'(x) \\
& + G_{\alpha x}(\alpha(x), x)\alpha'(x) + G_{\alpha\alpha}(\alpha(x), x)\left(\alpha'(x)\right)^2 \\
& + G_\alpha(\alpha(x), x)\alpha''(x).
\end{aligned}
$$

Since $\alpha'(x_0) = 0$, we get

$$
\alpha''(x_0) = \frac{-G_{xx}(\alpha_0, x_0)}{G_\alpha(\alpha_0, x_0)}.
$$

Since, from supposition (5), $f_{xxx}(\alpha_0, x_0) \neq 0$, and $f_{xxx}(\alpha_0, x_0) = G_{xx}(\alpha_0, x_0)$, we know then that $\alpha''(x_0) \neq 0$. □

2.5 First Return Maps

Recall From Section 2.1.2 that the time-1 map of an ordinary differential equation defines a discrete dynamical system on the phase space (Figure 9). Indeed, for $\mathbf{x} \in \mathbb{R}^n$, the system

$\dot{\mathbf{x}} = \mathbf{f}(\mathbf{x})$ defines the map $\phi^1 : \mathbb{R}^n \to \mathbb{R}^n$, where $\phi^1 : \mathbf{x}(0) \mapsto \mathbf{x}(1)$ is a transformation of $\mathbb{R}^n$. Really, $t = 1$ is only one such example, and any t will work, so long as the system solutions are defined and unique.

There is another kind of discrete dynamical system that comes from a continuous one: the *first return map* or *Poincaré map* (sometimes called a *Poincaré recurrence map*). One can, in a sense, view the first return map as a local version (only defined near interesting orbits) of the more globally defined time-*t* map defined over all of phase space, although we caution that this is not an entirely accurate comparison. For now, since we are dealing only with interval maps, we will highlight only a 2-dimensional version. Consider the planar flow given by the first-order system of ODEs in polar coordinates:

$$(2.5.1) \qquad \begin{aligned} \dot{r} &= r(1-r), \\ \dot{\theta} &= 1. \end{aligned}$$

Without solving this system (although this is not difficult as the equations are uncoupled; see Exercise 85), we can say a lot about how solutions behave:

- The system is autonomous, so when you start does not matter, and the vector field is static, or constant over time.
- The only equilibrium solution is at the origin. The second equation in the system really states that, off the origin in the plane, no point is fixed. But, owing to the special nature of the origin in terms of the coordinate system, even with $\dot{\theta} \neq 0$, the solution at the origin is an equilibrium as long as $\dot{r} = 0$ there.
- Considering only the first equation in the system, $\dot{r} = r(1-r) = f(r)$, for a minute, we find equilibrium solutions at $r(t) \equiv 0$ and $r(t) \equiv 1$, where $f(r) = 0$ (see the left side of Figure 34). However, any solution that starts with initial value $r(0) = 1$ is only fixed in r. It is a periodic solution called a *cycle*. What is the period?
- $r(t) \equiv 1$ is asymptotically stable as a cycle, and is an example of a *limit cycle*: a periodic solution in which at least one other solution is asymptotic to it in either the forward or backward direction. Can you see why?

Indeed, define a closed, bounded interval I along the positive vertical axis in the $r\theta$-plane by

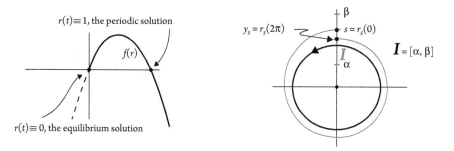

Figure 34 Graph of the vector field $f(r)$ (left) and a Poincaré section (right).

$$I = \left\{ \left(s, \frac{\pi}{2}\right) \subset \mathbb{R}^2 \mid \alpha \le s \le \beta, 0 < \alpha < 1 < \beta \right\}.$$

Of course, then, I is parameterized by s. For each $s \in I$, call $r_s(t)$ the solution of Equation 2.5.1 passing through s at $t = 0$, so that $s = r_s(0)$. Let y_s be the point in I that corresponds to the earliest positive time at which the resulting $r_s(t)$ again crosses I. (1) It must cross again (why?), and (2), really $y_s = r_s(2\pi)$, due to the constant angular velocity throughout the vector field. Then the map $\phi : s \mapsto y_s$ defines a discrete dynamical system on I, as in the right side of Figure 34.

Some properties of this discrete dynamical system should be clear:

- The dynamics are simple on I: There is a unique fixed point at $s = 1$ corresponding to the limit cycle crossing. This fixed point is asymptotically stable so that $\forall s \in I$, $\mathcal{O}_s \longrightarrow 1$. Thus this discrete dynamical system is a contraction on I. Can you determine the Lipschitz constant for this contraction?

- The same can be said for the system

(2.5.2)
$$\dot{r} = g(r) = r\left(\tfrac{1}{2} - r\right)(r - 1)\left(\tfrac{3}{2} - r\right),$$
$$\dot{\theta} = -1,$$

but only if I is chosen more carefully (see Figure 35): Here choose

$$I = \left\{ \left(s, \frac{\pi}{2}\right) \subset \mathbb{R}^2 \mid \alpha \le s \le \beta, \tfrac{1}{2} < \alpha < 1 < \beta < \tfrac{3}{2} \right\}.$$

- You should draw pictures to verify this. In this last system, what happens near the cycles $r(t) \equiv \tfrac{1}{2}$ and $r(t) \equiv \tfrac{3}{2}$? Is there some kind of discrete dynamical system in the form of a first return map near there also?

Exercise 85 Solve the system in Equation 2.5.1. (Hint: It is uncoupled, so you can solve each equation separately.)

A Poincaré map (first return map) is a powerful tool of analysis when studying the stability properties of flows near a periodic solution. Indeed:

- It associates with an n-dimensional flow an $n - 1$-dimensional discrete dynamical system. This is a form of reduction of order that can locally capture a lot of qualitative information about the flow. The resulting discrete map has as its domain a local cross-section to the periodic solution, called a *Poincaré section*. This is an $n - 1$-dimensional (or codimension-1) subspace in phase space. One takes care to ensure only that this subspace is never tangent to the flow where it is defined. We would say the Poincaré section is *transverse* to the flow everywhere if it is not tangent to the flow anywhere.

- In the above examples, the orbits that start on I all re-intersect I at $t = 2\pi$. Restricted to I, this coincides with the time-2π map of the flow. In general, this need not be the case. In fact, there need not be a bound on when a trajectory may again

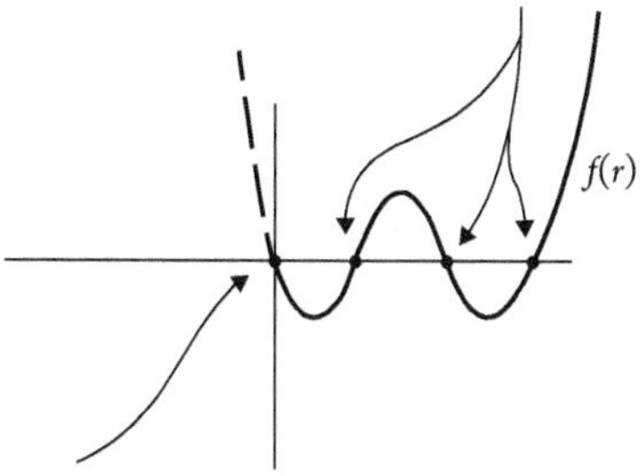

$r(t) \equiv 1$, $r(t) \equiv .5$, $r(t) \equiv 1.5$ are
periodic solutions

$f(r)$

$r(t) \equiv 0$, is the equilibrium solution

Figure 35 A more complicated system.

intersect a Poincaré section. For example, draw a Poincaré section along the vertical axis through the origin of the "eye" in the pendulum phase plane of Figure 22. This section can reach all the way from the top separatrix to the bottom, but cannot include either. And then at either "end" of this codimension-1 subspace, the trajectories have arbitrarily long periods. But still are guaranteed to cross. We will explore this situation in detail later when we highlight the pendulum.

For now, let's leave this discussion and move on to another interval map that will expose in time many interesting concepts in this text. We will return to this idea of first return maps once we study flows in $\mathbb{R}^n$.

2.6 A Quadratic Interval Map: The Logistic Map

Like linear functions defined on the unit interval, discrete dynamical systems constructed via maps whose expression is a quadratic polynomial have many interesting properties. The ideal model for a study of quadratic maps of the interval is the logistic map, first popularized by Robert May in 1976 [34] as a discrete version of a similar ODE that is now often called the Verhulst equation, a model for population growth first published in 1838 by Pierre Verhulst [56]. Before defining it, however, let's motivate its importance.

Consider the standard linear map on the real line, $f : \mathbb{R} \to \mathbb{R}, f(x) = rx$. As a model for population growth (or decay), we restrict the domain of f to be non-negative (to model a population size) and the values for the parameter $r \in \mathbb{R}$ to be positive, so that $f_r : [0, \infty) \to [0, \infty)$, where $r > 0$. Hence the recursive model is $x_{n+1} = f(x_n) = rx_n$, and again $\mathcal{O}_x = \{y \in [0, \infty) \mid y = f^n(x) = r^n x, n \in \mathbb{N}\}$. It is a good, simple model for population growth and is the discrete version of the Malthusian growth model in Examples 1.7 and 2.1 where $r = e^k$ for k the growth rate of the population and $p(t) = p_0 e^{kt}$ the evolution. When the population size is not affected by any environmental conditions or resource access, or when resources are to be considered infinite in size and scope, one

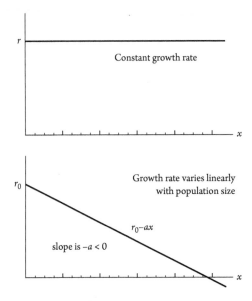

Figure 36 Constant and linear growth factors.

can speak of "ideal" growth, and this model reflects that contention. One way to view this is to say that, in this case, the growth factor r is constant and independent of the size of the population (see Figure 36).

However, realistically speaking, unlimited population growth is unsustainable in any limited environment, and hence the actual growth factor winds up being dependent on the actual size of the population. Things like crowding and the finite allocation of resources typically mean that larger population sizes usually experience a dampened growth factor over time vis-á-vis small populations (think of a small number of fish in a large pond as opposed to a very large number of fish in the same pond). Hence a better model to simulate populations over time is to allow the growth factor to vary with the population size. The easiest way to do this is to replace the constant growth factor r with one that varies linearly with population size. Thus, here, r is replaced with the expression $r_0 - ax$, where r_0 is an ideal growth factor (for very small populations near 0) and a is a positive constant (see the bottom graph in Figure 36). The model becomes

$$f : \mathbb{R} \to \mathbb{R}, \quad f(x) = (r_0 - ax)x,$$

or, with a change in variables,

$$f : \mathbb{R} \to \mathbb{R}, \quad f(y) = \lambda y(1 - y).$$

Keep in mind, however, the limitations of the model as a guide to studying populations. For λ a positive constant, f is positive only on the interval $[0, 1]$. And really only some values of λ make this a good model for populations. To understand the last statement, you will need to actually see how λ relates to the constants r_0 and a, and to study the graph of $r_0 - ax$ in Figure 36 as it relates to a population x.

Exercise 86 Do the change of variables that takes $f(x) = (r_0 - ax)x$ to $f(y) = \lambda y(1 - y)$, writing λ as a function of a and r_0.

Hence we will begin to study the dynamics of the map $f : [0,1] \to [0,1]$, $f(x) = \lambda x(1 - x)$, which we call the (discrete) logistic map. We will eventually see just how rich and complex is the dynamical system given by f on $[0,1]$. For now, however, we will only spend time on the values of λ where the dynamics are simple to describe. First some general properties:

- f is only a map on the unit interval when $\lambda \in [0,4]$. Why does it fail for other values of λ?

Exercise 87 Show that the logistic map $f(x) = \lambda x(1 - x)$ does not produce a dynamical system on the interval $[0,1]$, for $\lambda \notin [0,4]$.

- λ is sometimes called the *fertility constant* in population dynamics.
- We will use the notation f_λ to emphasize the dependence of f on the parameter.

Proposition 2.75 *For $\lambda \in [0,1]$, $\forall x \in [0,1]$, we have $\mathcal{O}_x(f_\lambda) \longrightarrow 0$.*

Visually, the graph of f_λ is a parabola opening down with horizontal intercepts at $x = 0, 1$. The vertex is at $\left(\frac{1}{2}, \frac{\lambda}{4}\right)$. And for $\lambda \in [0,1]$, the entire graph of f_λ, restricted to $[0,1]$ lies below the diagonal $y = x$ (see Figure 37). Cobweb to see where the orbits go.

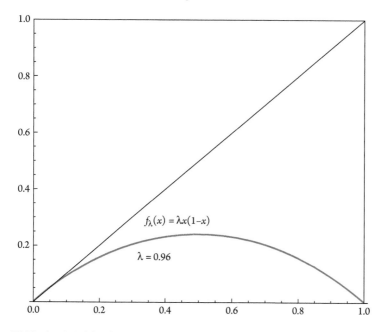

Figure 37 The Logistic Map for $0 < \lambda < 1$.

Proof The fixed points of f_λ satisfy $f_\lambda(x) = x$, or $\lambda x(1 - x) = x$. This is solved by either $x = 0$ or $x = (\lambda - 1)/\lambda = 1 - \lambda^{-1}$. Hence for $0 \leq \lambda \leq 1$, the only fixed point on the interval $[0, 1]$ is $x = 0$. Also, $\forall x \in [0, 1], f_\lambda(x) < x$. This implies that $\mathcal{O}_x$ is a decreasing sequence. As it is obviously bounded below, it must converge.

Now choose a particular $x \in [0, 1]$ and notice that $f_\lambda(x) < \frac{1}{2}$. Thus, after one iteration of the map, every orbit of $f_\lambda|_{[0,1]}$ lies inside the subinterval $[0, \frac{1}{2}]$. So, for studying orbit behavior, one need only study $f_\lambda|_{[0,\frac{1}{2}]}$, where the map is non-decreasing and has no fixed points on the interior $(0, \frac{1}{2})$. Then, by Proposition 2.59, the only fixed point is $x = 0$ and all orbits converge to it. $\qquad \square$

Some notes:

- Both conditions—that the interval be closed and that the map be non-decreasing—are necessary for the application of Proposition 2.59. Since the original map and none of its iterates are non-decreasing, we needed to modify the situation a bit to fit the lemma. The structure of the graph of f_λ allowed for this by the observation that a future iteration of the function places the image of the entire domain within a non-decreasing region. This is a common idea, and while it is not the same as a map being *eventually non-decreasing*, where a power of a map is non-decreasing, it is a strong analytical tool to study orbit structure. And here it is easy to also see that every orbit, for every $\lambda \in [0, 1]$, is decreasing as a sequence. We will explore these ideas a bit more in Section 3.2.3. But look for this sort of behavior in other maps.

- The orbit $\mathcal{O}_1$ is special:

$$\mathcal{O}_1 = \{1, 0, 0, 0, \ldots\}.$$

The point $x = 1$ is called a *pre-image* of the fixed point $x = 0$. This is often seen in maps that are not one-to-one. The point x is called *eventually fixed*. One similarly defines *eventually periodic* points also. Neither of these can exist in invertible maps (why?), but it is easy to see that f_λ is not invertible on $[0, 1]$.

- If one were to use this logistic map as a model for populations, then with this range of the parameter λ, one can conclude immediately the following:

All starting populations are doomed!

Think about that.

Now, let's change our parameter range a bit, and consider some higher parameter values:

Proposition 2.76 *For $\lambda \in [1, 3)$, $\forall x \in (0, 1)$, $\mathcal{O}_x \longrightarrow 1 - \lambda^{-1}$.*

Remark 2.77 *If this is true, then $\lambda = 1$ is a bifurcation value for the family of maps f_λ, since*

- *for $\lambda \in [0, 1]$, the sole fixed point $x = 0$ is an attractor for f_λ on $[0, 1]$; and*

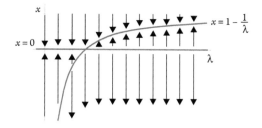

Figure 38 A transcritical bifurcation in the logistic map.

- for $\lambda \in [1,3)$, f_λ on $[0,1]$ has two fixed points: $x = 0$ is a repeller and $x = 1 - \lambda^{-1}$ is an attractor.

This is not a new situation, but just an example of a transcritical bifurcation (See Section 2.4.2.) Indeed, expand the map $f_\lambda : \mathbb{R} \to \mathbb{R}$. Then, for $\lambda \neq 0$, the equation $\lambda x(1 - x) = x$, always has two solutions $x = 0$ and $x = 1 - \lambda^{-1}$. So let $\lambda \in (0,\infty)$. Then these two solutions always lie within $[0,1]$, and coincide when $\lambda = 1$. We can graph these two solutions and get Figure 38. At the point $(\lambda_0, x_0) = (1,0)$, for $f(\lambda, x)$, we have

(1) $f(\lambda, 0) = 0, \forall x \in \mathbb{R}$, and $f_x(1,0) = (\lambda - 2\lambda x)\big|_{(1,0)} = 1$;

(2) $(\partial f / \partial \lambda)(1,0) = 0$;

(3) $f_{xx}(1,0) = -2 \neq 0$, and $f_{\lambda x}(1,0) = 1 \neq 0$.

By Proposition 2.70, the point $(1,0)$ is a transcritical bifurcation. But this is somewhat hidden if viewed only as a function on $[0,1]$.

Exercise 88 Verify that the diagram in Figure 38 is correct by assessing the stability of the fixed points on either side of the bifurcation value.

The idea of the proof of Proposition 2.76 is that on this range of values for λ, the graph of f_λ intersects the line $y = x$ at two places, and these places are the two solutions of $x = \lambda x(1 - x)$ (see the proof of Proposition 2.75 above.) Figure 39 illustrates the graphs for three typical logistic maps, for $\lambda = 1.5$ (bottom curve), $\lambda = 2$ (middle curve), and $\lambda = 2.5$ (top curve). It turns out that showing the fixed point $x_\lambda = 1 - \lambda^{-1}$ is attractive is straightforward and left as an exercise.

Exercise 89 Show that for $\lambda \in (1,3)$, there is an attracting fixed point of the logistic map $f_\lambda(x)$ at $x = 1 - \lambda^{-1}$.

However, showing that almost every orbit, in fact every orbit off the endpoints of $[0,1]$, converges to x_λ is somewhat more involved.

To prove Proposition 2.76, let's start with a lemma:

Lemma 2.78 For $\lambda \in [2,3)$, the map $f_\lambda^2(x)\big|_{I}$, on the interval $I = \left[1 - x_\lambda, f_\lambda\left(\frac{1}{2}\right)\right]$, is a contraction.

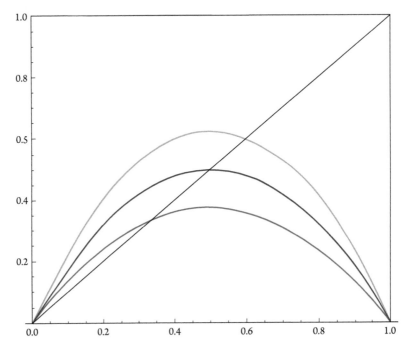

Figure 39 The logistic map for $1 < \lambda < 3$.

Proof To start, create $I = I_1 \cup I_2$, where $I_1 = [1 - x_\lambda, x_\lambda]$ and $I_2 = [x_\lambda, f\left(\frac{1}{2}\right)]$ as in the left graph of Figure 40. Note that $I = [1/\lambda, \lambda/4]$. We claim that $f^2(I) \subset I$. To see this, it should be clear that $f_\lambda(I_1) = I_2$, so it remains to show that $f_\lambda(I_2) \subset I_1$. Note that $f_\lambda|_{I_2}$ is strictly decreasing and so is injective with a minimum at $x = \lambda/4$. So we show that, for $\lambda \in (2,3), f_\lambda(\lambda/4) > 1 - x_\lambda = 1/\lambda$.

Here $f_\lambda(\lambda/4) = \lambda^2/4 - \lambda^3/16$, so the inequality to check is

$$\frac{\lambda^2}{4} - \frac{\lambda^3}{16} > \frac{1}{\lambda}, \text{ or } \frac{\lambda^3}{4} - \frac{\lambda^4}{16} > 1.$$

For $h(\lambda) = \lambda^3/4 - \lambda^4/16$ a function of λ, we have $h(2) = 2 - 1 = 1$, and $h'(\lambda) = \frac{3}{4}\lambda^2 - \frac{1}{4}\lambda^3 = \lambda^2\left(\frac{3}{4} - \frac{1}{4}\lambda\right)$. $h'(\lambda) > 0$ on $[2,3)$ so that $h(\lambda)$ is an increasing function there. Hence $h(\lambda) > 1$ on $(2,3)$.

Next we claim that $g_\lambda(x) = f_\lambda^2(x)$ is a contraction on I, for $\lambda \in (2,3)$. Indeed, we will show that $\left|g_\lambda'(x)\right| < 1$ on I for this interval of λ (see Figure 40 at right). Here

$$g_\lambda(x) = \lambda\left(\lambda x(1 - x)\right)\left(1 - \lambda x(1 - x)\right)$$
$$= -\lambda^3 x^4 + 2\lambda^3 x^3 - (\lambda^2 + \lambda^3)x^2 + \lambda^2 x,$$

so $g_\lambda'(x) = -4\lambda^3 x^3 + 6\lambda^3 x^2 - 2(\lambda^2 + \lambda^3)x + \lambda^2$ and $g_\lambda'(x) = 0$ when $x = \frac{1}{2}$ and

$$x_0^\pm = \frac{1}{2} \pm \frac{\sqrt{\lambda^2 - 2\lambda}}{2\lambda}.$$

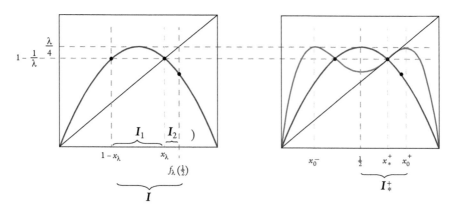

Figure 40 $f_\lambda(x)$ and $I = I_1 \cup I_2$ (left) and $g_\lambda(x) = f_\lambda^2(x)$ on I_*^+ (right).

Thus $g_\lambda''(x) = -12\lambda^3 x^2 + 12\lambda^3 x - 2(\lambda^2 + \lambda^3) = 0$ at

$$x_*^{\pm} = \frac{1}{2} \pm \frac{\sqrt{3}\sqrt{\lambda^2 - 2\lambda}}{6\lambda}.$$

Some properties of $g_\lambda(x)\big|_I$ are that

- symmetric with respect to the $x = \frac{1}{2}$ line;
- $I \subset [x_0^-, x_0^+]$.

Indeed, we will leave this last assertion as Exercise 90.

Hence we are left only with establishing that $|g_\lambda'(x)| < 1$ on $[x_0^-, x_0^+]$ for $\lambda \in (2,3)$. Due to the symmetry of $g_\lambda(x)$, we only need to show this for $x \in I_0^+ = [\frac{1}{2}, x_0^+]$. Indeed, we know the following:

- For $\lambda \in (2,3)$, $x = \frac{1}{2}$ is a local minimum (check that $g_\lambda''\left(\frac{1}{2}\right) > 0$ here). Hence $g_\lambda'(x) > 0$ on $\left(\frac{1}{2}, x_0^+\right)$ since the endpoints are consecutive zeros of g_λ'.
- Hence $\max\limits_{I_0^+} g_\lambda'(x)$ is at the inflection point of g_λ on I_0^+, which is at

$$x_*^+ = \frac{1}{2} + \frac{\sqrt{3}\sqrt{\lambda^2 - 2\lambda}}{6\lambda},$$

and here, as a function of λ, we have

$$g_\lambda'(x_*^+) = \left(\frac{\lambda^2 - 2\lambda}{3}\right)^{\frac{3}{2}}, \text{ on } \lambda \in (2,3).$$

But here

$$\frac{d}{d\lambda}\bigg|_{\lambda=2} g_\lambda'(x_*^+) = 0, \text{ and } \frac{d}{d\lambda} g_\lambda'(x_*^+) > 0 \text{ on } \lambda \in (2,3).$$

Hence

$$\max_{I_0^+} g_\lambda'(x) = \left(\frac{\lambda^2 - 2\lambda}{3}\right)^{\frac{3}{2}}\bigg|_{\lambda=3} = 1.$$

Hence we are done. □

Exercise 90 Show that $I \subset [x_0^-, x_0^+]$ for $\lambda \in (2,3)$ by showing that $x_0^- < 1 - x_\lambda$ and $f_\lambda\left(\frac{1}{2}\right) = \frac{\lambda}{4} < x_0^+$.

Proof of Proposition 2.76 To begin, note that the case $\lambda = 1$ is covered in Proposition 2.75 and we already know by Exercise 89 that $x_\lambda = 1 - \lambda^{-1}$ is an attractor on all of $\lambda \in [1,3)$. So to prove this we will consider three cases: First, case 1, where $\lambda \in [1,2]$. Here the proof is straightforward. $f_\lambda\left([0,1]\right) \subseteq \left[0,\frac{1}{2}\right]$ and $f|_{\left[0,\frac{1}{2}\right]}$ is an increasing map. Hence, by Proposition 2.64, every point $\left[0,\frac{1}{2}\right]$ is either fixed or tends to a fixed point. Since x_λ is the only attractive fixed point, its basin of attraction is everything that is not 0 or its pre-image 1. Hence the proposition is proved for $\lambda \in [1,2]$. Next, we consider the second case for $\lambda \in (2,3)$. Here, $x_\lambda > \frac{1}{2}$, so the previous method will not work. Instead, notice that

$$0 < 1 - x_\lambda < \frac{1}{2} < x_\lambda < f_\lambda\left(\frac{1}{2}\right) < 1.$$

We claim here that the interval $I = \left[1 - x_\lambda, f_\lambda\left(\frac{1}{2}\right)\right]$ is invariant under f_λ. Indeed, Lemma 2.78 establishes this, along with the idea that every orbit that enters I is not only trapped, but is asymptotic to x_λ. So it only remains to establish that every orbit in $(0,1)$ eventually enters I. To this end, notice that for $x \in \left[f_\lambda\left(\frac{1}{2}\right),1\right)$, $f_\lambda(x) \in (0,x_\lambda]$. And $f_\lambda|_{(0,1-x_\lambda]}$ is an increasing function without fixed points such that $f_\lambda(x) > x$. Hence the orbit of every x is increasing until it enters I. Hence $\forall x \in (0,1)$, $\mathcal{O}_x \longrightarrow x_\lambda$. □

As a final note, once we reach and surpass the value $\lambda = 3$ for $\lambda \in [0,4]$, things get trickier. We will suspend our discussion of interval maps and the logistic map here for a bit and develop some more machinery first.

3 The Objects of Dynamics

Before moving on to more general and more complicated dynamical systems, now is a good time to develop a better understanding of the pieces that make up a dynamical system, either continuous or discrete. The spaces that serve as the state space are certainly not always Euclidean, although in this text they will be easily recognizable as spaces that locally look so. Sometimes two spaces that look different are actually the same for the purpose of the dynamical system. And even on spaces that are the same topologically, changing the notion of distance between points, a metric, can have surprising results. In this chapter, we will stop exploring new dynamical systems directly and instead create a better sense of what makes a set a topological space, and how the notions of maps, continuity, and metrics look like outside of the standard Euclidean spaces. After this chapter, we will have more tools to work with. We start with the idea of just what is a topology, and follow with some more subtle notions of continuity and metrics. Then we finish with a description of some of the more common and non-Euclidean spaces we will work with in this text.

3.1 Topology on Sets

Roughly speaking, a topology on a set X is a well-defined notion of what constitutes an open subset of X. So, for now, let X be any set of objects or elements, and we will call the elements of X its points.

Definition 3.1 *A topology on X is a collection $\mathcal{T}_X$ of subsets of X that satisfy the following:*

- *$\emptyset$ and X are in $\mathcal{T}_X$;*
- *the union of the elements of any subcollection of $\mathcal{T}_X$ is in $\mathcal{T}_X$; and*
- *the intersection of the elements in any finite subcollection of $\mathcal{T}_X$ is in $\mathcal{T}_X$.*

For a topology $\mathcal{T}_X$ on X, the elements of $\mathcal{T}_X$ are called *open sets*. Also, any set X that is given a topology is then said to have a *topological structure* on it, and a set with a topological structure is then called a *topological space*.

Example 3.2 The set of all open intervals $(a, b) \subset \mathbb{R}$ constitutes a topology on $\mathbb{R}$, called the *standard* topology $\mathcal{T}_\mathbb{R}$, for $-\infty \leq a \leq b \leq \infty$. It should be obvious that this allows all of $\mathbb{R}$ to be in $\mathcal{T}_\mathbb{R}$, since one can write $\mathbb{R}$ as $(-\infty, a) \cup (a - 1, \infty)$ for any $a \in \mathbb{R}$. And if we let $a = b$, then the element $(b, b) = \emptyset$ is also in $\mathcal{T}_\mathbb{R}$. The union of any collection of open intervals is certainly open also. However, without the third condition, we would have a problem: Suppose we allowed that the intersection of any subcollection of $\mathcal{T}_\mathbb{R}$ to be in $\mathcal{T}_\mathbb{R}$. Then the set

$$\bigcap_{n=1}^{\infty} \left(-\frac{1}{n}, \frac{1}{n} \right) = \{0\}$$

would have to be open. But then all individual points would also be open, and thus, by the second condition, any subset of X would be open! You can see why the third condition is necessary.

Remark 3.3 *Incidentally, there is a topology on $\mathbb{R}$ (or any other set), where each of the points is considered open (and hence any subset is considered an open subset). It is called the* discrete *topology on the set. Since every subset of X is open in the discrete topology, every subset is also closed (since its complement will be open). And although it works via the definitions, it is said to be the* largest topology *or the* finest topology *a set can have, as any other topology will be a subcollection of this one. This renders the discrete topology as a rather extreme example of a topology of a space. In a certain sense, the discrete topology on a set has an opposite, sometimes called the* indiscrete *topology on a space, or more often the* trivial *topology. The indiscrete topology contains only the empty set and the entire set as open subsets of X. Thus we can say that the indiscrete topology is the smallest topology (or the* coarsest*) a set can possess, and here one cannot use open sets to distinguish points at all. You should check that these two extreme examples fully satisfy the conditions in Definition 3.1.*

Some other examples of topologies on some familiar sets:

- **The metric topology:** For any set X with a metric, define the open sets (neighborhoods) to be open balls of a size centered at the points of X. This is one of the easy topologies to visualize, and corresponds to the standard topology on Euclidean space $\mathbb{R}^n$.

- **The compact complement topology:** On $\mathbb{R}$, declare a set to be open if it is either empty, or if its complement is a closed, bounded interval in the Euclidean topology. Notice here that almost all open sets are unbounded, and any two non-trivial open sets have non-empty intersection.

- **The cofinite topology:** Here a set is declared open iff it is either empty or if its complement is a finite set of points.

- **The lower limit topology:** A topology on $\mathbb{R}$ whereby the open sets are the half-open intervals $[a, b)$. One fairly interesting feature of this topology is that the union of two disjoint open sets can be connected (form one piece: $[a, c) \cup [c, b) = [a, b)$. This is not the case in the standard Euclidean topology. We will see this topology again when we discuss rotations of the circle in Section 5.1.

- **The subspace topology:** Here, for a subset $X \subset Y$ of a topological space, declare a set in X to be open in X if it is the intersection of X with an open set in Y. For instance, using the Euclidean metric topology in the plane, any ϵ-ball around a point in the plane that intersects the unit circle in the plane results in a open interval in the circle. One can use this to define a topology on $S^1 \in \mathbb{R}^2$ where open sets are simply open intervals.

- **The product topology:** When a set X is defined as the Cartesian product of two or more topological spaces (for example, the plane $\mathbb{R}^2 = \mathbb{R} \times \mathbb{R}$), then one can define a topology on the product by defining open sets to be products of open sets for some or all of the factors. The number of factors in the product can be finite (like $\mathbb{R}^n = \mathbb{R} \times \mathbb{R} \times \cdots \times \mathbb{R}$), or infinite (like the space $\Omega_{\mathcal{M}}$ in Section 1.1.3, which we will study more closely later in Sections 7.1.4 and 7.1.5). Defining and understanding this topology is a little tricky, though, since what constitutes an open set may depend on whether there are a finite number of factors or not. This topology is sometimes called the *Tychaonoff* topology and is related to what is called the *box* topology, at least when the number of factors is finite.

Some facts:

Definition 3.4 *A function $f : X \to Y$ between two topological spaces is continuous if whenever $V \in \mathcal{T}_Y$, then $f^{-1}(V) \in \mathcal{T}_X$. A continuous function between two topological spaces is called a* map.

In more familiar terms, a function $f : X \to Y$ is continuous if the inverse image of every open set in Y is open in X. This definition allows us to talk about maps being continuous between arbitrary topological spaces without the use of metrics, in a way that is entirely compatible with what you already learned as the definition of continuity between spaces like subsets of $\mathbb{R}$ in single variable calculus or subsets of $\mathbb{R}^n$ in multivariable calculus. There, you simply assumed the standard topologies on Euclidean space, and the notion of "nearness" that is at the center of continuity came out of the little ϵ-balls used to define continuity via the metrics. These ϵ-balls are all elements of the standard topology on $\mathbb{R}^n$, but since they are defined via the metric, they allow for the description you well remember from calculus classes.

We can alter this definition of continuity to better fit the notion with which you are already familiar (see Figure 41):

Definition 3.5 *A function $f : X \to Y$ is continuous at some point $x \in X$ if and only if for any neighborhood V of $f(x)$, there is a neighborhood U of x such that $f(U) \subseteq V$.*

In topology, any open set in X (a member of $\mathcal{T}_X$) is a neighborhood of any of its points (recall Definition 2.18). Hence again, this definition depends on the topologies of

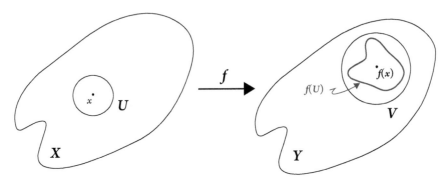

Figure 41 A continuous map.

X and Y. It basically says that no matter how "small" we choose an open set V containing $f(x)$, we can always find an open set U, containing x, where $f(U)$ sits entirely inside V.

Definition 3.6 *A continuous bijection (a continuous map that is injective, or one-to-one, and surjective, or onto) $f : X \to Y$ with a continuous inverse is called a* homeomorphism. *A continuously differentiable homeomorphism $f : X \to Y$ with a continuously differentiable inverse is called a* diffeomorphism.

Remark 3.7 *We will, on occasion, need or use the fact that a map is differentiable or is a diffeomorphism. Mostly, however, continuity will be sufficient. But do keep in mind that in dynamical systems, calculus often plays a critical role in understanding structure, and homeomorphisms where a map and its inverse are also continuously differentiable can be a crucial property.*

Example 3.8 For any metric space (or any topological space in general!) X, the *identity map* on X ($f : X \to X$, $f(x) = x$) is a homeomorphism. It is obviously continuous (using the metric definition of continuity, for any $\epsilon > 0$, choose $\delta = \epsilon$), one-to-one, and onto, and it is its own inverse.

Example 3.9 The map $h : [0, 1) \to S^1$ given by $h(x) = e^{2\pi i x}$ is continuous, one-to-one, and onto. It also has an inverse, but the inverse is not continuous. See Figure 42.

Exercise 91 Show that h in Example 3.9 is continuous, one-to-one, and onto. Construct h^{-1} and show that it is not continuous.

Exercise 92 Let $h : \mathbb{R} \to \mathbb{R}$ be defined by

$$h(x) = \begin{cases} x^2, & x \geq 0, \\ -x^2, & x < 0. \end{cases}$$

Show that h is a homeomorphism on $\mathbb{R}$ but not a diffeomorphism.

Example 3.10 Recall that affine maps $f : \mathbb{R} \to \mathbb{R}$, $f(x) = ax + b$ are, of course, invertible, as long as $a \neq 0$. They are also one-to-one and onto all of $\mathbb{R}$. However, be careful

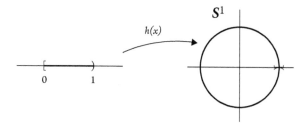

Figure 42 The half-open unit interval in $\mathbb{R}$ is not homeomorphic to the unit circle in $\mathbb{C}$.

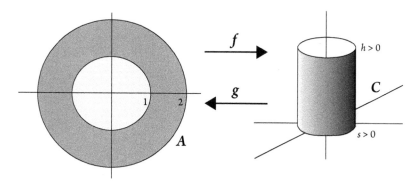

Figure 43 The annulus $A \in \mathbb{R}^2$ and the cylinder $C \in \mathbb{R}^3$ are homeomorphic spaces.

when restricting the domain: Let $f(x) = \frac{1}{2}x + \frac{1}{2}$ on $I = [0,1]$. Here, f is certainly injective (it is non-decreasing, and a contraction!). But it cannot have an inverse on I, since it is not onto I. Recall Exercise 74 and the discussion preceding it. However, at least for affine maps of I, if the inverse exists, it will be continuous.

Remark 3.11 *When a homeomorphism exists between two topological spaces, the two spaces are called* homeomorphic *and topologically they are considered equivalent, or the same space. Anything defined on a space or with it can be defined or used on any other space that is homeomorphic to it. It is one of the chief ways for mathematicians to classify topological spaces according to their properties.*

Exercise 93 Given the annulus A and the cylinder C in Figure 43 (both include their boundary circles, although the dimensions of C are given only as radius $s > 0$ and height $h > 0$), show that they are homeomorphic by explicitly constructing the maps $f : \mathbb{R}^2 \to \mathbb{R}^3$ that takes A to C and $g : \mathbb{R}^3 \to \mathbb{R}^2$ that takes C to A. (Hint: Use the obvious coordinate systems for the Euclidean spaces.)

Definition 3.12 *Given $f : X \to X$ and $g : Y \to Y$ two discrete dynamical systems, suppose there is an onto map $h : X \to Y$ that satisfies $h \circ f(x) = g \circ h(x) \; \forall x \in X$. Then h is called* a topological semiconjugacy *and we say f and g are topologically semiconjugate. If h is a homeomorphism, then h is a* topological conjugacy *and we say f and g are topologically conjugate.*

This last definition will be very important in our future discussions. In essence, when two dynamical systems are (topologically) conjugate, they are in effect equivalent (much like when two spaces are homeomorphic, they are basically the same). By the very definition, we can see that, at least in a conjugacy, we have $f(x) = h^{-1} \circ g \circ h(x)$, and thus $f^n(x) = h^{-1} \circ g^n \circ h(x)$, so that orbits are mapped to orbits via h. Hence the orbit structure remains the same. We will see that the implications of this equivalency are strong, as it will allow us sometimes to change a dynamical system without actually changing the orbit structure. For example, in Remark 2.11 and Exercise 18 for affine maps, as well as Exercise 86 for the logistic map, we have already seen that we can change coordinates without changing the orbit structure to make a map easier to study. Changing coordinate systems on a space is a homeomorphism from a space to itself, creating a conjugacy between a space with one coordinate system and the same space with another. We will see many examples of this and in more general settings throughout this text, and chiefly in Section 8.1.

3.2 More on Metrics

There are easy-to-describe-and-visualize dynamical systems that occur on subsets of Euclidean space that are not Euclidean. As long as we have a metric on the space, it remains easy to discuss how points move around by their relative distances from each other. So let's generalize a bit and talk about metric spaces without regard to how they sit in a Euclidean space. To this end, let X be a metric space.

Definition 3.13 *A map $f : X \to X$ on a metric space X is called an* isometry *if*

$$\forall x, y \in X, \quad d\big(f(x), f(y)\big) = d(x, y).$$

We can generalize this last definition (and indeed the definitions of Lipschitz continuity and just continuity) to maps where the domain and the range are two different spaces: $f : X \to Y$, where both X and Y are metric spaces with respective metrics, d_X and d_Y:

Definition 3.14 *A map $f : X \to Y$ between two metric spaces is called an* isometry *if*

$$\forall x_1, x_2 \in X, \quad d_Y\big(f(x_1), f(x_2)\big) = d_X(x_1, x_2).$$

Definition 3.15 *A map $f : X \to Y$ between two metric spaces is* Lipschitz continuous *with constant λ, or simply λ-Lipschitz if*

$$\forall x_1, x_2 \in X, \quad d_Y\big(f(x_1), f(x_2)\big) \leq \lambda d_X(x_1, x_2).$$

And once again, in this context, we define continuity:

Definition 3.16 *A map* $f : X \to Y$ *between two metric spaces is called* continuous *at* $x \in X$ *if* $\forall \epsilon > 0, \exists \delta > 0$, *such that if for every* $y \in X$ *where* $d_X(x,y) < \delta$, *then* $d_Y\big(f(x), f(y)\big) < \epsilon$.

As an illustration of this last definition, go back to Figure 41 and let $y \in U(x) = B_\delta(x)$ and $f(y) \in V\big(f(x)\big) = B_\epsilon\big(f(x)\big)$. This gives us an easy way to define what makes a function continuous at a point when the spaces are not Euclidean but are metric spaces. We will need this as we talk about common spaces in dynamical systems that are not like $\mathbb{R}^n$ but still allow metrics on them. That a map on a space is continuous is a vital property to possess if we are to be able to study how orbits behave under iteration of the map. Incidentally, this last definition of continuity is precisely the one developed and used in calculus. In single variable calculus, this is the standard ϵ-δ definition of a limit, altered for continuity by switching the L for $f(x)$ and where $X = Y = \mathbb{R}$ with the standard Euclidean metric.

We will end this discussion with one more definition, which will be useful later:

Definition 3.17 *A map* $f : X \to Y$ *between two metric spaces is called* uniformly continuous *if* $\forall \epsilon > 0, \exists \delta > 0$ *such that for every* $x,y \in X$, *if* $d_X(x,y) < \delta$, *then* $d_Y\big(f(x), f(y)\big) < \epsilon$.

In essence, continuity of a map is a local condition, while uniform continuity is a global condition of a map; the required δ-closeness between points in the domain will result in the ϵ-closeness of the image points across the entire range of the function. The δ is a function of the ϵ but not of the points in the domain. A common place to see uniform continuity is when the domain is compact, as one can then locate a sufficiently small but positive δ that will work across the entire domain. In fact, this is a famous theorem:

Theorem 3.18 (Heine–Cantor Theorem) *If* $f : X \to Y$ *is a continuous map on metric spaces and* X *is compact, then* f *is uniformly continuous.*

Example 3.19 $f : \mathbb{R} \to \mathbb{R}$, $f(x) = e^x$ is certainly continuous on $\mathbb{R}$, but not uniformly continuous. However, on any closed, bounded interval $I \in \mathbb{R}$, f is uniformly continuous on I.

3.2.1 More on Lipschitz Continuity

The definition of a Lipschitz continuous function back in Section 2.1.4 worked for the task at hand. However, we will need to generalize for a more thoughtful study of more complicated dynamical systems. We start with the idea that even functions that are not Lipschitz continuous on a domain may look Lipschitz continuous near each point:

Definition 3.20 *A function* $f : X \to X$ *on a metric space is called* locally Lipschitz continuous *if for every* $x \in X$, $\exists \epsilon > 0$ *such that* f, *restricted to* $B_\epsilon(x)$, *is Lipschitz continuous.*

In a sense, the original definition was the definition of a function being *globally Lipschitz continuous*; on all of X, the function was λ-Lipschitz. The bound held everywhere on X. For locally Lipschitz functions, the bound can depend on the choice of x (and ϵ).

Example 3.21 Let $f(x) = e^x$ on $\mathbb{R}$ with the usual metric. Here, f is not a Lipschitz continuous function. Indeed, suppose there is a bound $M > 0$ on all of $\mathbb{R}$ where

$$\left|e^x - e^y\right| = d(f(x), f(y)) \le M d(x, y) = M\left|x - y\right|.$$

This would have to work for any choice of $x, y \in \mathbb{R}$. So choose $y = 0$. Then, the inequality stipulates that $\forall x \in \mathbb{R}$, $|e^x - 1| \le M|x|$. But then, for $x > 1$, $|e^x - 1| = e^x - 1$, so that the inequality renders

$$e^x \le Mx + 1 < (M+1)x, \quad \text{or} \quad \frac{e^x}{x} < (M+1).$$

But, as any calculus student learning L'Hospital's Rule will know, the function e^x/x is unbounded as x goes to infinity. Thus no bound M exists. It is true, however, that f is locally Lipschitz.

Exercise 94 Show that any degree-$d > 1$ polynomial $p(x)$ is locally Lipschitz on $\mathbb{R}$ but not Lipschitz on $\mathbb{R}$. Then show that p, restricted to any closed, bounded interval I, is a Lipschitz function.

Exercise 95 Show $f(x) = \sqrt[3]{x}$ is locally Lipschitz on $I = (0, \infty)$. You will recall, though, from Exercise 40 that f is not Lipschitz on I.

It should be clear that a Lipschitz continuous function is also locally Lipschitz. What may be less clear are the following, whose proofs are left as exercises:

Proposition 3.22 *Every locally Lipschitz function is continuous.*

Proposition 3.23 *Every continuously differentiable function is locally Lipschitz.*

Exercise 96 Prove Proposition 3.22.

Exercise 97 Prove Proposition 3.23

As before, be careful on this last exercise. We refer you back to Remark 2.30 on the difference between differentiable and continuously differentiable.

Exercise 98 For the functions on $\mathbb{R}$ given in Equation 2.1.4 in Section 2.1.4, show that f is locally Lipschitz, but g is not.

3.2.2 Metric Equivalence

Typically, on a metric space X, there are many metrics that one can define. However, like the above notion of homeomorphism, many of them are basically the same, and can be treated as equivalent—others, maybe not. To understand this better:

Definition 3.24 *Let d_1 and d_2 be two metrics on a metric space X. Then d_1 and d_2 are called isometric if $\forall x, y \in X$, $d_1(x, y) = d_2(x, y)$.*

Definition 3.25 *Two metrics d_1 and d_2 on a metric space X are called (uniformly) equivalent if the identity map, taking X with d_1 to X with d_2 and its inverse are both Lipschitz continuous.*

To elaborate on this last definition, we consider $f : X \to X, f(x) = x$, to be the map that takes points in X using the metric d_1 to points in x using the other metric d_2. This is like considering X as two different metric spaces, one with d_1 and the other with d_2. In essence, then the definition says that $\exists C, K \geq 0$ such that $\forall x, y \in X$, both (1) $d_2\left(f(x), f(y)\right) \leq Cd_1(x,y)$ and (2) $d_1\left(f^{-1}(x), f^{-1}(y)\right) \leq Kd_2(x,y)$ hold. This simplifies using the identity map to (1) $d_2(x,y) \leq Cd_1(x,y)$ and (2) $d_1(x,y) \leq Kd_2(x,y)$ everywhere. Of course, what this really means is simply that there are global bounds (over the space X, that is) on how the two metrics differ.

Some notes:

- The notion of uniform equivalence is sometimes also called *strong equivalence*.

- We can rewrite the conditions for uniform equivalence as one statement: d_1 and d_2 are uniformly equivalent if there exist constants m, $M > 0$ such that

$$md_1(x,y) \leq d_2(x,y) \leq Md_1(x,y), \quad \forall x, y \in X.$$

This one pair of constants works for each pair of points on the entire space X.

- Continuity (or differentiability) of a function is preserved if either the metric on the domain or range is changed to an equivalent one.

- If a space X is complete under a metric, it is complete under all equivalent metrics.

- A function, however, may be a contraction under one metric and not under an equivalent one.

Example 3.26 On $\mathbb{R}^2$, the Euclidean metric and the Manhattan metric are uniformly equivalent. Recall from Remark 2.16 that these are the metrics that come, respectively, from the $(\mathbf{p} = 2)$-norm and the corresponding 1-norm. Call d_2 the Euclidean metric and d_1 the Manhattan metric. Then we can show uniform equivalence via the following:

$$\left(d_1(\mathbf{x}, \mathbf{y})\right)^2 = \left(|x_1 - y_1| + |x_2 - y_2|\right)^2 \geq |x_1 - x_2|^2 + |y_1 - y_2|^2 = \left(d_2(\mathbf{x}, \mathbf{y})\right)^2.$$

Hence $\left(d_2(\mathbf{x}, \mathbf{y})\right)^2 \leq \left(d_1(\mathbf{x}, \mathbf{y})\right)^2$, so that $d_2(\mathbf{x}, \mathbf{y}) \leq d_1(\mathbf{x}, \mathbf{y})$.

Going the other way is a little trickier. Given the Cauchy–Schwartz inequality from linear algebra, $(x \cdot y)^2 \leq ||x|| \cdot ||y||$, we can say

$$d_1(\mathbf{x}, \mathbf{y}) = |x_1 - y_1| + |x_2 - y_2|$$
$$= |x_1 - y_1| \cdot 1 + |x_2 - y_2| \cdot 1$$
$$\leq \sqrt{(x_1 - y_1)^2 + (x_2 - y_2)^2}\sqrt{1^2 + 1^2} = \sqrt{2}\, d_2(\mathbf{x}, \mathbf{y}).$$

Hence

$$d_1(\mathbf{x}, \mathbf{y}) \leq \sqrt{2}\, d_2(\mathbf{x}, \mathbf{y}) \text{ and } d_2(\mathbf{x}, \mathbf{y}) \leq d_1(\mathbf{x}, \mathbf{y}).$$

Exercise 99 Going back to Remark 2.16, show that the Euclidean and maximum metrics are also uniformly equivalent on $\mathbb{R}^2$. (In fact, all **p**-norm metrics are equivalent.)

Example 3.27 Let $d_3(x,y) = \left| x^3 - y^3 \right|$ be a metric on $\mathbb{R}$ (Exercise 28 establishes that this is a metric.) Then d_3 is not uniformly equivalent to $d(x,y) = |x - y|$. Indeed, recall that $(x^3 - y^3) = (x - y)(x^2 + xy + y^2)$. Then, we see that $d_3(x,y) = |x - y| \cdot \left| x^2 + xy + y^2 \right|$. It is obvious, then, that if there existed an $M > 0$ such that $d_3(x,y) \le Md(x,y)$, for all $x, y \in \mathbb{R}$, then $\left| x^2 + xy + y^2 \right| \le M$ for all $x \ne y \in \mathbb{R}$. Choose $x = 1$ and $y = M$ to show such a global M cannot exist.

3.2.3 Fixed-Point Theorems

Fixed points (and periodic points) of maps will play a central role in any study of a discrete dynamical system since even without any hope of finding the evolution that solves the dynamical system, we can still study the more direct and simple features of a map, namely the possibly few orbits that do not move much. The Contraction Principle of Chapter 2 is an example of what is called a *fixed-point theorem*: an existence theorem that helps us to identify particularly simple properties of some general and complicated dynamical systems. Sometimes, even in dynamical systems that are not simple at all, we can still show that certain simple orbits do exist. Perhaps the most widely used and credited fixed-point theorem is attributed to Luitzen E. J. Brouwer [12]:

Theorem 3.28 (Brouwer Fixed-Point Theorem) *A continuous map $f : X \to X$ on a compact, convex subset of Euclidean space has a fixed point.*

This theorem allows us to quickly conclude that any continuous map from, say, a ball in Euclidean space to itself will possess at least one fixed point. You have already seen the 1-dimensional version of this in Proposition 2.58, for maps on a closed, bounded interval. Strictly speaking, of course, convexity is sufficient but not entirely necessary; one can take a nice convex ball in $\mathbb{R}^n$ and, with a homeomorphism, turn it into a non-convex mess. But the original map composed with the homeomorphism is a map on the new space that has a fixed point. One can use this to prove that this theorem holds for any compact, connected (exists in one piece) and simply connected (has no holes) region in $\mathbb{R}^n$.

The Contraction Principle, from Chapter 2, is really just a restricted version of a more general theorem, first stated by Stefan Banach [5] in 1922:

Theorem 3.29 (Banach Fixed-Point Theorem) *Let $f : X \to X$ be a contraction on a complete metric space. Then f has a unique fixed point x_*. Furthermore, $\forall x \in X$, $\mathcal{O}_x \longrightarrow x_*$ and convergence is exponential.*

But, as we have already seen, there are many maps that are not contractions, but do have unique and attractive fixed points. Zvi Artstein [4] calls these maps *contraction-like*: mappings of metric spaces in which the iterates all converge to a stable fixed point. He shows that if a contraction-like mapping $f : X \to X$ on a metric space X is *exponential*— that there exist constants $C > 0$ and $0 < \lambda < 1$ where $d(f^n(x), f^n(y)) \le C\lambda^n d(x,y)$, for all x and y—then there is a metric, uniformly equivalent to the original, in which f is a contraction. This new inequality is close to the original definition of a contraction, but includes an allowance for a distortion, at least during the early iterates. This is the idea of a map being *eventually contracting*. The equivalent metric Artstein constructs is called a Lyapunov metric because of its similarity with Lyapunov functions.

Definition 3.30 *A map $f : X \to Y$ is called* eventually contracting *if $\exists C > 0$, such that $\forall x, y \in X$ and $\forall n \in \mathbb{N}$,*

$$d\left(f^n(x), f^n(y)\right) \leq C\lambda^n d(x, y)$$

for some $0 < \lambda < 1$.

Exercise 100 Go back to Example 2.66, $f(x) = \sqrt{x-1} + 3$ on the interval $I = [1, \infty)$. Show that f is not an eventual contraction. (Hint: Try to find a value for C in Definition 3.30 that works in a neighborhood of $x = 1$.) Now show that f is an eventual contraction on any closed interval $[b, \infty)$, for $1 < b \leq 5$.

It turns out that eventually contracting maps behave a lot like contracting maps. In fact, one can generalize the Contraction Principle to include maps that are eventually contracting. The following proposition is a simplification of that of Artstein. The new metric constructed in the proof, though, is different and is a construction of Hasselblatt and Katok [26]:

Proposition 3.31 *If $f : X \to X$ is eventually contracting on a metric space X, with $C, \lambda > 0$ satisfying*

$$d(f^n(x), f^n(y)) \leq C\lambda^n d(x, y), \quad \forall x, y \in X \text{ and } n \in N,$$

then there exists a metric d_* on X, uniformly equivalent to d, such that

$$d_*(f(x), f(y)) \leq \rho d_*(x, y), \quad \forall x, y \in X$$

for any $\lambda < \rho < 1$.

Proof Choose a natural number $n \in \mathbb{N}$ such that $C(\lambda/\rho)^n < 1$. Then define

(3.2.1)
$$d_*(x, y) := \sum_{i=0}^{n-1} \frac{d(f^i(x), f^i(y))}{\rho^i}.$$

That this is a metric is left as Exercise 101. The map f is a contraction under this new metric, since

$$d_*(f(x), f(y)) = \sum_{i=1}^{n} \frac{d(f^i(x), f^i(y))}{\rho^{i-1}} = \rho \sum_{i=1}^{n} \frac{d(f^i(x), f^i(y))}{\rho^i}$$

$$= \rho \left(d_*(x, y) + \frac{d(f^n(x), f^n(y))}{\rho^n} - d(x, y) \right)$$

$$\leq \rho \left(d_*(x, y) + C\left(\frac{\lambda}{\rho}\right)^n d(x, y) - d(x, y) \right)$$

$$= \rho d_*(x, y) + \left(C\left(\frac{\lambda}{\rho}\right)^n - 1 \right) d(x, y) \leq \rho d_*(x, y),$$

since $(C(\lambda/\rho)^n) < 1$. And d_* is uniformly equivalent to d since, by the definition, we have $d(x, y) \leq d_*(x, y)$ and

$$d_*(x,y) \le \sum_{i=0}^{n-1} \frac{1}{\rho^i} d(f^i(x), f^i(y))$$

$$= \sum_{i=0}^{n-1} \frac{1}{\rho^i} C\lambda^i d(x,y) = Cd(x,y) \sum_{i=0}^{n-1} \left(\frac{\lambda}{\rho}\right)^i$$

$$\le Cd(x,y) \sum_{i=0}^{\infty} \left(\frac{\lambda}{\rho}\right)^i = C\left(\frac{1}{1-\lambda/\rho}\right) d(x,y). \qquad \square$$

Thus we can rework the Contraction Principle to a sort of Eventual Contraction Principle:

Theorem 3.32 *Let X be a complete metric space and $f : X \to X$ eventually contracting. Then f has a unique fixed point $x_* \in X$ and $\forall x \in X, \mathcal{O}_x \longrightarrow x_*$ and convergence is exponential.*

It also seems rather obvious that not all contraction-like mappings can be made contractions by a suitably constructed uniformly equivalent metric. At least, even to Artstein, this seemed quite possibly not true. But a good question is: to what extent is there a converse to the Contraction Principle? One rather obvious facet of the (Eventual) Contraction Principle is that if f is an (eventual) contraction, then for $n \in \mathbb{N}, f^n$ also has a unique fixed point. That contractions cannot have non-trivial periodic points is the result of Exercise 49. The Polish mathematician Czeslaw Bessaga [8] took this as the right context and created the following in 1958:

Proposition 3.33 *Suppose $f : X \to X$ is a map on an abstract set X such that every iterate of f has a unique fixed point. Then, for any $0 < k < 1$, there exists a complete metric d_k on X such that f is a contraction with Lipschitz constant k.*

We will not prove this here, but it is an interesting fact. Indeed, it is quite surprising:

Example 3.34 The map $f : \mathbb{R} \to \mathbb{R}, f(x) = 3x$ has $x = 0$ as its unique fixed point. This also applies to each of $f^n(x) = 3^n x$, for all $n \in \mathbb{N}$. Hence there exists a metric on $\mathbb{R}$ such that f is a contraction! Can you construct such a metric?

Exercise 101 Show that, for d a metric on X in the proof of Proposition 3.31, the d_* in Equation 3.2.1 is indeed a metric.

Example 3.35 The map $g : \mathbb{R} \to \mathbb{R}, g(x) = 1 - x$ has a unique fixed point at $x = \frac{1}{2}$. But it is not a contraction, and cannot be made so by a change of metric. Here f^2 is the identity map on $\mathbb{R}$, and hence does not have a unique fixed point.

But even without changing a metric, maps can behave like contractions over time (iteration). When this happens, the conclusion of the Contraction Principle may hold even if the map itself is not a contraction:

Proposition 3.36 *Let $f : X \to X$ be a continuous map on a complete metric space, where $\exists n \in \mathbb{N}$ such that f^n is a contraction. Then f has a unique fixed point $x_0 \in X$, and $\forall x \in X$, $\mathcal{O}_x(f) \longrightarrow x_0$.*

Proof Since f^n is a contraction, $x_0 = \text{Fix}(f^n)$. And since

$$f(x_0) = f(f^n(x_0)) = f^n(f(x_0)),$$

we have that $f(x_0) \in \text{Fix}(f^n)$. But, for f^n a contraction, its fixed point is unique. Hence $f(x_0) = x_0$ and $x_0 \in \text{Fix}(f)$.

Now let $y \in X$. Since for any choice of $r \in \{0, 1, \ldots, n-1\}$ and for $k \in \mathbb{N}$, we have

$$f^{kn}\left(f^r(y)\right) = \left(f^n\right)^k\left(f^r(y)\right),$$

we know that under f^n, $\mathcal{O}_{f^r(y)}(f^n) \longrightarrow x_0$. But this is true for all choices of r. Hence the entire orbit of y under f converges to x_0, or $\mathcal{O}_y(f) \longrightarrow x_0$. $\qquad\square$

Example 3.37 Let $f_2(x) = 2x(1-x)$ be the $\lambda = 2$-logistic map, and restrict it to the subinterval $I = \left[\frac{1}{10}, \frac{9}{10}\right] \subset [0,1]$. This cuts out the repelling fixed point at 0 and its pre-image 1. But it remains a dynamical system, since $f\left(\frac{1}{10}\right) = f\left(\frac{9}{10}\right) = \frac{18}{100} > \frac{1}{10}$.

Recall that this is the value of the parameter in which the logistic map has a super-attracting fixed point at $x = \frac{1}{2}$. Note that f_2 is definitely not a contraction. You can see this visually by inspecting the graph in Figure 44. But even analytically, the derivative has a maximum at the left endpoint, so that f_2 is $\frac{8}{5}$-Lipschitz on I. This property that the derivative is a maximum at the left endpoint remains true in its iterates also (can you show this?), so that by the fourth iterate the entire graph has flattened out and f^4 is a contraction on I, as seen in Figure 44. In fact, one can show that for any $a \in (0, 1)$, f restricted to $I_a = [a, 1-a]$ will have an iterate f^n for some $n \in \mathbb{N}$ that is a contraction. Indeed, f on

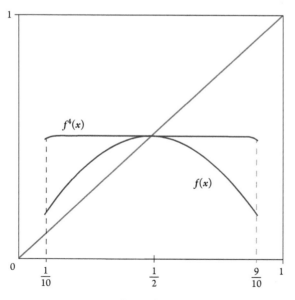

Figure 44 The eventual contraction f_2 on $\left[\frac{1}{10}, \frac{9}{10}\right]$.

$[0,1]$ is smooth and 2-Lipschitz. Hence its iterates are, at most, 2^n-Lipschitz. But, by the chain rule on the iterates at a, one can show that eventually the derivative decays to 0, so that at some point, an iterate of the function is a contraction on I_a. We will leave the analysis to the reader here.

Exercise 102 Show that $f(x) = 1/(1+x)$ is not a contraction on $[0,1]$ with the usual metric. It does have a unique fixed point though. Find it. Then show that an early iterate of f is a λ-contraction, and find λ.

Exercise 103 Show that no iterate of $g(x) = x/(1+x)$ is a contraction on $[0,1]$, even though each iterate has a unique fixed point.

Exercise 104 For $X \subset \mathbb{R}^n$ a closed and bounded metric space, and $f : X \to X$ satisfying $d(f(x), f(y)) < d(x,y)$, $\forall x, y \in X$, show f has a unique fixed point $x_0 \in X$, such that $\forall x, \mathcal{O}_x \longrightarrow x_0$.

We leave this section with a corollary that will be useful later:

Corollary 3.38 *Let $Y \subset X$ be a closed, invariant subset of a complete metric space, with a contraction f. Then the unique fixed point is in Y.*

3.3 Some Non-Euclidean Metric Spaces

Here we construct some of the more common non-Euclidean metric spaces encountered in dynamical systems. First, recall from vector calculus the definition of a *parameterization* of a space in $\mathbb{R}^n$:

Definition 3.39 *Let $U \subset \mathbb{R}^m$ be a connected, open region, possibly together with some or all of its boundary points, and $V \subset \mathbb{R}^n$. A parameterization of V is a continuous surjection $\varphi : U \to V$ which is injective at least on the interior $\mathring{U}$.) Here, we say V is parameterized by φ, and the coordinates of U become the parameters to distinguish points in V.*

Note that, as in Figure 45, $\varphi(u,v) = (x(u,v), y(u,v), z(u,v))$ is a point in $V \in \mathbb{R}^3$, and u and v become parameters used to distinguish points on V in terms of the parameters instead of the coordinates in $\mathbb{R}^n$. This is often used to place coordinates directly on a subspace and give the subspace a sense of size and shape as it sits inside its ambient space. And, where defined, the inverse of the parameterization $\varphi^{-1} : V \to U$ is called a *coordinate system*.

Example 3.40 Given the IVP $\dot{\mathbf{x}} = \mathbf{F}(\mathbf{x})$, $\mathbf{x}(t_0) = \mathbf{x}^0$, where $\mathbf{F} : \mathbb{R}^n \to \mathbb{R}^n$ is C^1, a particular solution is a function $\mathbf{x}(t) : I \to \mathbb{R}^n$ defined on some open interval $I \in \mathbb{R}$, where $t_0 \in I$. The image of $\mathbf{x}(t)$ is a curve in $\mathbb{R}^n$ parameterized by a single parameter t.

Example 3.41 Given a real-valued function $f : U \in \mathbb{R}^n \to \mathbb{R}$, its graph is a function $\mathbf{u} \mapsto (\mathbf{u}, f(\mathbf{u})) \in \mathbb{R}^{n+1}$. This graph is a parameterization of the image of U, as the domain points distinguish points on the graph due to that fact that f is a function.

As in the last example, graphs of functions are common places to see parameterizations. However, parameterizations are much more general than simply graphs of functions.

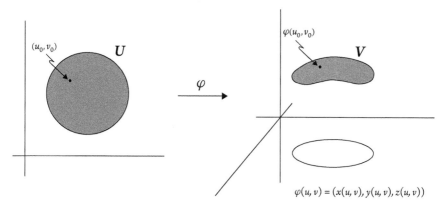

Figure 45 The map φ is a parameterization on $V \in \mathbb{R}^3$.

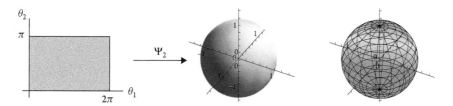

Figure 46 A common parameterization of the 2-sphere S^2.

3.3.1 The n-Sphere

The *n*-dimensional sphere

$$S^n = \left\{ \mathbf{x} \in \mathbb{R}^{n+1} \mid ||\mathbf{x}|| = 1 \right\}.$$

Here, the $n \in \mathbb{N}$ denotes the "size" of the space and not the space in which it exists in this definition. In fact, the space called the *n*-sphere doesn't actually exist in any space unless we define it that way. We typically place the *n*-sphere in $\mathbb{R}^{n+1}$ so that we can place coordinates on it easily. For example, we can parameterize the 2-sphere via the function $\Psi_2 : D \rightarrow \mathbb{R}^3$, where $D = [0, 2\pi] \times [0, \pi]$, by

(3.3.1) $$\Psi_2(\theta_1, \theta_2) = (\cos\theta_1 \sin\theta_2, \sin\theta_1 \sin\theta_2, \cos\theta_2),$$

as in Figure 46. Notice immediately that Ψ_2 is onto its image, and one-to-one on the interior of D. In fact, the entire top edge of D goes to the point $(0,0,1)$ at the top of the image in $\mathbb{R}^3$. The bottom edge entirely maps to $(0,0,-1)$. And the left and right edges both comprise the same half great circle joining the north pole and the south. (Can you draw it into the figure?) In general, for $n \in \mathbb{N}$, the spherical coordinate system in $\mathbb{R}^n$ provides a ready parameterization for S^n: In spherical coordinates (ρ, θ), the unit sphere is simply the set of all points in $\mathbb{R}^{n+1}$ with first coordinate 1. Hence all of the other (angular) coordinates $\theta = (\theta_1,\ldots,\theta_n)$ parameterize the n-sphere.

Exercise 105 Find the pattern in parameterizing S^1 by $\Psi_1 : [0,2\pi] \to \mathbb{R}^2$, where $\Psi_1(\theta) = (\cos\theta,\sin\theta)$ and S^2 by Equation 3.3.1, to construct an explicit parameterization Ψ_n of S^n using the n-angles of the spherical coordinate system in $\mathbb{R}^{n+1}$. What are the ranges of each of your angular coordinates θ_i, $i = 1,\ldots,n$?

Exercise 106 For the parameterization in Equation 3.3.1, one can view this as a wraparound of the box at left of the figure onto the 2-sphere at right. Identify where the four edges of the box go under the map Ψ_2, and draw the images of these four edges (the "seam" of the parameterization) onto the 2-sphere.

3.3.2 The Unit Circle

Really this is the 1-dimensional sphere

$$S^1 = \{\mathbf{x} \in \mathbb{R}^2 \mid ||\mathbf{x}|| = 1\}.$$

However, we also can interpret the circle as the unit-modulus complex numbers

$$S^1 = \{z \in \mathbb{C} \mid |z| = 1\},$$
$$= \{e^{i\theta} \in \mathbb{C} \mid \theta \in [0,2\pi)\},$$

(compare this with Equation 2.3.1) and also in a more abstract sense as

$$S^1 = \{x \in \mathbb{R} \mid x \in [0,1] \text{ where } 0 = 1\}.$$

This last definition requires a bit of explanation. From set theory, we have the following:

Definition 3.42 *Given a set X, a set partition $\mathcal{P}_X$ is a set of disjoint, exhaustive subsets of X.*

Remark 3.43 *You are already familiar with the term "partition" from calculus. For example, in the development of the notion of a definite integral of a function defined on a closed, bounded interval $I = [a,b] \in \mathbb{R}$, $a < b$, one defines a partition of I into smaller intervals via a finite set of points in I, $a = x_0 < x_1 < \ldots < x_{n-1} < x_n = b$, so that*

$$[a,b] = [a = x_0,x_1] \cup [x_1,x_2] \cup \ldots \cup [x_{n-1},x_n = b].$$

This is different from our current definition in that the intervals overlap on their edges. For our current purpose, we require that the partition elements be mutually exclusive and exhaustive and hence will distinguish a set partition from what you used in calculus.

Definition 3.44 *An equivalence relation R on a set X is a collection of elements of $X \times X$, denoted (x,y) or $x \sim_R y$ (or simply $x \sim y$, when R is understood,) such that R is*

(1) reflexive: $x \sim_R x$, $\forall x \in X$;

(2) *symmetric: $x \sim_R y$ iff $y \sim_R x$, $\forall x, y \in X$; and*

(3) *transitive: if $x \sim_R y$ and $y \sim_R z$, then $x \sim_R z$, $\forall x, y, z \in X$.*

Each $x \in X$ is an element of a unique equivalence class $[x]$, a subset of X where

$$[x] = \{ y \in X \mid y \sim_R x \}.$$

And the set of all equivalence classes of X under R form a set partition of X.

Furthermore, the set of all set partition elements form a new set, called the *quotient set* of the equivalence relation. Given X a topological space, one can always make the quotient set into a space using the topology of X (it is called the *quotient topology*). But it is a much deeper question mathematically exactly when the quotient set, made into a space using the topology of X, has similar properties to that of X. But for now, we say that for X a set with an equivalence relation R, the quotient set is denoted $Y = X/R$, or sometimes $Y = X/\sim$.

Example 3.45 Parity defines an equivalence relation on the integers, with two equivalence classes familiarly known as *odd* and *even*. Indeed, for $x \in \mathbb{Z}$, we have

$$[x] = \left\{ y \in \mathbb{Z} \;\middle|\; \frac{x-y}{2} \in \mathbb{Z} \right\}.$$

One can also denote the two equivalence classes as $[2k]$ and $[2k+1]$, for $k \in \mathbb{Z}$. Another way is to define the the *modular relation*

$$x \equiv_2 y \quad \text{iff} \quad 2 \mid (x-y).$$

Then the evens are all integers equivalent to 0, denoted by $[0]_2$, and the odds are all numbers equivalent to 1, members of the class $[1]_2$. This generalizes immediately to any natural number $m \in \mathbb{N}$, so that for each m, we have the equivalence relation with equivalence classes defined by

$$[x]_m = \{ y \in \mathbb{Z} \mid x \equiv_m y \} = \{ y \in \mathbb{Z} \mid m \mid (x-y) \},$$

and there are m of them, $[i]_m$, $i = 1, \ldots, m-1$.

Exercise 107 Show that for $m \in \mathbb{N}$, the modular relation "$\equiv_m$" is an equivalence relation.

Example 3.46 The partial ordering "$\leq$" on $\mathbb{R}$ is not an equivalence relation. It is certainly reflexive and transitive. But it is not symmetric.

Example 3.47 Declare two natural numbers to be in the same equivalence class if they have a common factor greater than 1. This relation is symmetric and reflexive. However, this fails to be an equivalence relation because it is not transitive; As an example, 3 and 12 have a common factor greater than 1, and so does 12 and 4, but 3 and 4 do not.

Exercise 108 For $S = \{p/q \mid p,q \in \mathbb{Z}\}$ the set of all ratios of integers, show that the relation R, given by

$$\frac{a}{b} \sim_R \frac{c}{d} \quad \text{if} \quad ad = bc$$

is an equivalence relation. Then describe the quotient set. (Hint: it is a set with which you are most likely already familiar.)

Exercise 109 Prove the *Fundamental Theorem of Equivalence Relations*: That with every set partition of a set X, one can associate an equivalence relation compatible with the set partition, and for every equivalence relation on X, one can construct a set partition. (We say that there is a natural bijection between the set of all set partitions of a set X and the set of all equivalence relations on X.)

And before moving on, one more example that will play a vital role in our constructions of non-Euclidean spaces from Euclidean ones.

Example 3.48 Any function $f : X \to Y$ between sets defines an equivalence relation on X. Each element of the set partition is simply the collection of all points that map to the same point in the range of X:

$$[x] = \{y \in X \mid f(y) = f(x)\}.$$

Recall that in calculus we defined the inverse image of a point in the range of a function as the set

$$f^{-1}(y) = \{x \in X \mid f(x) = y\} \subset X.$$

Hence we can say here that the equivalence class of a point $x \in X$ given by the function $f : X \to \mathbb{R}$ is simply the inverse image of the image of x, or $[x] = f^{-1}(f(x))$. This is well-defined regardless of whether f even has an inverse, and regardless of whether f is onto or not. Think about this.

Using this last example, we have one more definition of S^1: Namely, let $r : \mathbb{R} \to S^1$ be a function $r(x) = e^{2\pi i x}$. Then $r(x) = r(y)$ iff $x - y \in \mathbb{Z}$.

Exercise 110 Show that for r defined here, $r(x) = r(y)$ iff $x - y \in \mathbb{Z}$.

In this case, each point on the circle has as its inverse image under r all of the points in the real line that are the same distance from the next highest integer (see Figure 47). Thus the map r looks like the real line $\mathbb{R}$ infinitely coiled around the circle. In this way, we commonly write

$$S^1 = \mathbb{R}/\mathbb{Z},$$

and the right side is the real numbers modulo the integers and called "$\mathbb{R}$ mod $\mathbb{Z}$".

Remark 3.49 *Now the more abstract definition*

$$S^1 = \{x \in \mathbb{R} \mid x \in [0,1] \text{ where } 0 = 1\}$$

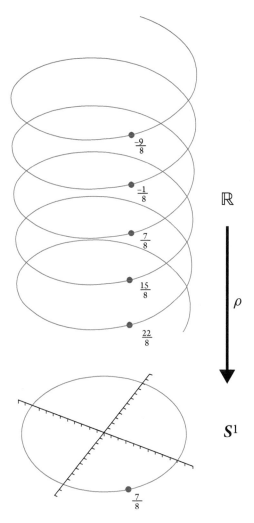

$-\frac{9}{8}$

$-\frac{1}{8}$

$\frac{7}{8}$

$\frac{15}{8}$

$\frac{22}{8}$

$\mathbb{R}$

ρ

S^1

$\frac{7}{8}$

Figure 47 The map $\rho : \mathbb{R} \to S^1$.

should make more sense. In a way, one can take the unit interval in $\mathbb{R}$, pluck it out of $\mathbb{R}$, and curve it around to the point where one can "join" its two endpoints together to make the circle. We say the two endpoints are now identified, and the space S^1, while still having a well-defined parameter on it, isn't sitting in an ambient space anymore. Note also that the new space is still closed and bounded, but it has no boundary! Intervals in $\mathbb{R}$ cannot have these properties simultaneously. But the circle, still one-dimensional, is not a subset of $\mathbb{R}$.

Another interesting consequence of this idea involves how to view functions on S^1 vis-á-vis those on $\mathbb{R}$. Let $f : \mathbb{R} \to \mathbb{R}$ be any function that is T-periodic (and thus satisfies $f(x+T) = f(x), \forall x \in \mathbb{R}$). Then f induces a function $g : S^1 \to \mathbb{R}$ on S^1, given by $g(t) =$

$f(tT)$. Conversely, any function defined on S^1 may be viewed as a periodic function on $\mathbb{R}$, a tool that will prove very useful later on.

Exercise 111 Show that a T-periodic, continuous map $f : \mathbb{R} \to \mathbb{R}$ will induce a continuous map $g : S^1 \to \mathbb{R}$. Then show that a T-periodic, differentiable map $f : \mathbb{R} \to \mathbb{R}$ will induce a differentiable map $g : S^1 \to \mathbb{R}$. (Hint: Parameterize the circle carefully and correctly.)

Exercise 112 Show that you cannot have a continuous surjective contraction on S^1. However, construct a continuous, non-trivial contraction on the circle S^1. (Hint: A continuous map cannot break the circle in its image or it would not be continuous at the break. But it can fold the circle.)

3.3.3 The Cylinder

Relatively simple-to-describe *product spaces* show up often as the state spaces of dynamical systems. Define the cylinder as $C = S^1 \times I$, where $I \subset \mathbb{R}$ is some interval. Here I can be closed, open, or half-closed, and can be bounded or all of $\mathbb{R}$. In fact, by the above discussion, any function $f : \mathbb{R}^2 \to \mathbb{R}^2$ that is T-periodic in one of its variables may be viewed as a function on a cylinder (think of the phase space of the undamped pendulum). Sometimes we call a cylinder whose linear variable is all of $\mathbb{R}$ the *infinite cylinder*.

3.3.4 The 2-Torus

The 2-dimensional torus $\mathbb{T}^2$ (or just the torus $\mathbb{T}$ when there is no confusion) $\mathbb{T} = S^1 \times S^1$ is another surface. Like before, any function $f : \mathbb{R}^2 \to \mathbb{R}^2$ that is periodic in each of its variables may be viewed as a function on a torus. Conversely, a function on the torus may be studied instead as a doubly periodic function on $\mathbb{R}^2$. We will have occasion to use this fact later. For now, consider the parameterization of this surface as a subset of $\mathbb{R}^3$ by $\Phi_2 : [0, 2\pi] \times [0, 2\pi] \to \mathbb{R}^3$, where

$$(3.3.2) \qquad \Phi_2(\theta_1, \theta_2) = ((2 + \cos\theta_1)\cos\theta_2, (2 + \cos\theta_1)\sin\theta_2, \sin\theta_1).$$

Figure 48 shows the parameterization, along with the images of the bottom and left edges of the parameter space.

Exercise 113 Show that the function $g(x, y) = (2\cos 2x, 4y^2 - y^4)$ from the plane to itself can be made into a function on the standard infinite cylinder $C = S^1 \times \mathbb{R}$. Show also that by limiting the domain appropriately, one can use g to construct a continuous function on the torus $T = S^1 \times S^1$.

One thing to note: Using Figure 48, locate the images of the top and right edges of the rectangle at left in the torus at right. Since both parameters are 2π-periodic, they have the same images, respectively, as the bottom and left edges. As a mental exercise, do the following: Envision taking the square piece of paper at left, bending it into a cylinder to

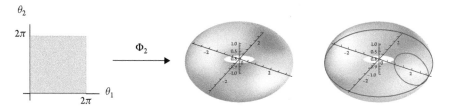

Figure 48 One parameterization of the 2-torus $\mathbb{T}^2$.

identify the left and right edges and then bending the cylinder to identify the top and bottom circles. That is a torus. And if you were an ant walking on the torus and you approached the thick line and crossed it, how would your shadow path look back in the parameter square? Understanding this will be very important at some point soon. Think about it.

We end this chapter with one last example of a particular set that has an interesting property constructed via a contraction map. For now, we will only define the set and identify the property. In time, we shall return to this set, since it is quite ubiquitous in dynamical systems theory.

3.4 A Cantor Set

Simple dynamics also allows us to define some of the physical properties of the spaces that a map can "act" on. For example, consider the following subset of the unit interval $[0, 1]$, defined by what is called a *finite subdivision rule*—a method of recursively dividing a polygon or other geometric shape into smaller and smaller pieces.

Define $C_0 = [0, 1]$. On C_0, remove the open middle third interval (this is the finite subdivision rule) $\left(\frac{1}{3}, \frac{2}{3}\right)$ and call the remainder C_1. Hence $C_1 = \left[0, \frac{1}{3}\right] \cup \left[\frac{2}{3}, 1\right]$. Continue removing the open middle third from each remaining closed interval from the previous C_n to construct C_{n+1} (see Figure 49). Then define

$$C = \bigcap_{n=0}^{\infty} C_n.$$

C is called the ternary Cantor set, after Georg Cantor [14], who studied its properties. It has the following rather remarkable properties:

- There are no positive-length intervals in C. Indeed, at each step, one removes open intervals whose total length is exactly $\frac{1}{3}\left(\frac{2}{3}\right)^{n-1}$. Sum these lengths over the natural numbers (this forms a geometric series) and you get

Figure 49 The first few steps in the construction of the ternary Cantor set.

$$\sum_{i=1}^{\infty} \frac{1}{3}\left(\frac{2}{3}\right)^{i-1} = \frac{1}{3}\left(\frac{1}{1-\frac{2}{3}}\right) = 1.$$

- It is easy to see that, although all intervals of points of $[0,1]$ are not in C, there are many individual points that remain. Any point in $[0,1]$ that is a multiple of a power of $\frac{1}{3}$ (the endpoints of the intervals at each stage) is in C. What is more surprising is that there are many others. One way to see this is to use an alternate description of C as the set of points in the unit interval whose ternary expansion (the ternary or base-3 number system is like our decimal number system but uses only 0's 1's, and 2's) has no 1's. Indeed, recall our standard decimal expansion of points in the unit interval: For $s \in [0,1]$,

(3.4.1) $\quad s = 0.x_1 x_2 \ldots x_i \ldots = \displaystyle\sum_{i=1}^{\infty} \frac{x_i}{10^i}, \quad$ where $\quad x_i \in \{0,1,2,\ldots,9\}, \quad \forall i \in \mathbb{N}.$

Exercise 114 Show the series in Equation 3.4.1 always converges.

In a ternary expansion, we have the use of only the digits 0, 1, and 2. However, we can still write any $s \in [0,1]$ as

(3.4.2) $\quad s = (0.x_1 x_2 \ldots x_i \ldots)_3 = \displaystyle\sum_{i=1}^{\infty} \frac{x_i}{3^i}, \quad$ where $\quad x_i \in \{0,1,2\}, \quad \forall i \in \mathbb{N}.$

Note that, as in the decimal expansion, one can represent some numbers on the unit interval in more than one way: If the ternary expansion of $s \in [0,1]$ is such that there exists a largest $i_* \in \mathbb{N}$ where x_{i_*} is non-zero, then one can write s as

(3.4.3) $\quad s = \left(0.x_1 \ldots x_{i_*} 0000 \ldots\right)_3 = \left(0.x_1 \ldots (x_{i_*} - 1)2222 \ldots\right)_3,$

or, using the standard practice of including a bar in a truncated expansion over a repeating pattern,

$$s = \left(0.x_1 \ldots x_{i_*}\overline{0}\right)_3 = \left(0.x_1 \ldots (x_{i_*} - 1)\overline{2}\right)_3.$$

Of course, there are only a countable number of these points that are representable in more than one (two) ways, and they are only the ones whose expansion terminates, or has a finite number of terms.

Exercise 115 Show that $\frac{1}{3} = (0.1\overline{0})_3 = (0.0\overline{2})_3$ in the ternary expansion, and that $1.0 = 0.\overline{9}$ in decimal expansion.

So why do elements of C have a ternary expansion with no 1's? Because we take them out. Think about this: In a ternary expansion of points in $[0,1]$, the middle third (every point in the open interval $\left(\frac{1}{3}, \frac{2}{3}\right)$) will have an expansion $(0.1 * * * * *...)_3$ where not all of the *'s are 0. These are the points that are one-third plus some other quantity strictly less than one-third, no? But these points are not in C. The same applies to all points of the form $(0.*1 * * * * *...)_3$, since these numbers represent elements in the intervals $\left(\frac{1}{9}, \frac{2}{9}\right)$, namely, those where $x_1 = 0$, and $\left(\frac{7}{9}, \frac{8}{9}\right)$ those with $x_1 = 2$. Make sense?

Exercise 116 Show that the number $x = \frac{1}{4} \in C$ via its ternary expansion (it is not a boundary point of any open subinterval removed at any intermediate stage C_n.)

And note that the problem with some numbers having two ternary representations from Equation 3.4.3 is not an issue in C, since for all of these points, at least one of the representations includes a 1. Removing all numbers with an expansion that includes a 1 also removes this discrepancy.

- It is also easy to see that there are infinitely many points in C (all of the multiples of powers of $\frac{1}{3}$ that are the boundary points of removed intervals.) What is more interesting is that the number of points in C is uncountable (so not like the natural number kind of infinity. This is more like the number of points in $[0,1]$ kind of infinity.) In fact, there are as many points left in C as there were in the original $[0,1]$! (Wrap your head around that mathematical fact!)

One can give C a topology so that it is a space, like $[0,1]$ is a space (the subspace topology it inherits from $[0,1]$ will do.) Then one can say that anything homeomorphic to a Cantor set is a Cantor set. Hence this one example will share its properties with all other Cantor sets defined similarly. In fact, we can define a Cantor set in a much more general fashion:

Definition 3.50 *A non-empty subset of an interval I is called a* Cantor set *if it is a closed, totally disconnected, perfect subset of I.*

Definition 3.51 *A non-empty subset $C \subset I$ is* perfect *if, for every point $x \in C$, there exists a sequence of points $x_i \in C$, $x_i \neq x$, $i \in \mathbb{N}$, where $\{x_i\}_{i \in \mathbb{N}} \longrightarrow x$.*

Definition 3.52 *A non-empty subset $C \subset I$ is* totally disconnected *if, for every $x, y \in C$, $x \neq y$, the closed interval $[x, y] \not\subset C$.*

Roughly, there are no isolated points in a perfect set. And there are no closed, positive-length intervals in a totally disconnected subset of an interval.

Exercise 117 Use the ternary expansion description of the ternary Cantor set to show that is it perfect and totally disconnected. (Hint: For totally disconnected, show that between any two points in C, there is a point in I whose unique ternary expansion has a 1 in it.)

Continuing on the ternary expansion theme, with

$$C = \left\{ x \in \mathbb{R} \;\middle|\; x = (0.x_1 x_2 x_3 \ldots)_3 = \sum_{n \in \mathbb{N}} x_n \cdot 3^{-n}, \quad x_n = 0 \text{ or } 2 \right\},$$

there is a wonderful, real-valued function defined on $[0,1]$ which illuminates some of the properties of C. This function, the *Cantor–Lebesgue function*, relies on this series description of Cantor numbers: Define $F : [0,1] \to [0,1]$, by

$$F(y) = \sum_{n \in \mathbb{N}} x_n 2^{-(n+1)}, \quad \text{where} \quad x = (0.x_1 x_2 x_3 \ldots)_3 = \min_{x \in C} x \geq y.$$

In essence, all points in $[0,1]$ situated in a positive-length gap between points in C are mapped to a constant, this constant being a multiple of a power of $\frac{1}{2}$ corresponding uniquely to that gap and the function value of the Cantor point at the upper end of the gap. The resulting graph of F is an example of what is called a *Devil's Staircase* (see Figure 50). It is an example of a continuous, surjective function whose derivative is almost everywhere defined and equal to 0! To see the surjectivity, let $z \in [0,1]$. Then its binary expansion is $z = (0.a_1 a_2 a_3 a_4 \ldots)_2$, where $z = \sum_{n \in \mathbb{N}} a_n 2^{-n}$. But then $y = \sum_{n \in \mathbb{N}} (2a_n) 3^{-n} \in C$ and $F(y) = z$. Conclusion? There are at least as many points in C as there are in $[0,1]$. There cannot be more, so the cardinality of C and $[0,1]$ are the same! Cantor sets are quite popular in analysis owing to their ability to provide counterexamples to seemingly intuitive, but untrue, beliefs.

The map $f : C \to C, f(x) = \frac{1}{3}x$ is a contraction whose sole fixed point is at 0.

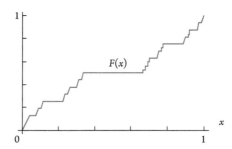

Figure 50 A Devil's Staircase: The graph of a Cantor function.

Exercise 118 Show that for $f(x) = \frac{1}{3}x, f(C) \subset C$.

This last point is special:

Definition 3.53 *A topological space that has a contraction map that is a homeomorphism onto its image has the property of* self-similarity.

The set is also called *rescalable*. This is easy to see for a contraction on $\mathbb{R}$.

Exercise 119 Find an expression for a contraction on C whose fixed point is $x = 1$.

Exercise 120 Show that the map $f(x) = 1 - \frac{1}{3}x$ is a contraction on C, and verify that its fixed point is in C.

In fact, all of these contractions are maps that take the interval C_0 onto one of the subintervals of C_1 (the last exercise reversing the order of points), indicating that the subset of C inside each of the subintervals in C_1 looks exactly like the parent set C in C_0. In this way, one can build a (linear) map taking C_0 onto any subinterval in any C_n. And, in this way, the part of C in any subinterval of C_n looks exactly the same as C. That such a complicated set can be defined by such a simple single finite subdivision rule is quite remarkable.

Example 3.54 Let $f : \mathbb{R} \to \mathbb{R}$ be defined by the function $f(x) = (2/\pi) \arctan x$. Here, f is a homeomorphism onto its image the open interval $(0, 1)$ (see Figure 51). Viewed as a dynamical system, it is easy to see from the picture that iterating the map quickly leads to the conclusion that f is a contraction. What is the Lipschitz constant in this case? (Hint: use the derivative!)

Exercise 121 In fact, every open interval $I = (a, b) \in \mathbb{R}$, for $a < b$, is homeomorphic to every other open interval in $\mathbb{R}$: Show that the map

$$f : I \to \mathbb{R}, \quad f(x) = \frac{x - \frac{1}{2}(a + b)}{(x - a)(b - x)}$$

is a homeomorphism.

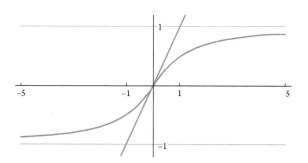

Figure 51 $f(x) = \frac{2}{\pi} \arctan x$ on $\mathbb{R}$.

Hence, $\mathbb{R}$ is a self-similar set. More interesting, though, are self-similar sets defined via some finite subdivision rules, like the Cantor set. Repeating the rule on smaller and smaller regions inside some original set creates the self-similarity. Then the contraction used to verify self-similarity is simply the map taking one stage in the recursive construction to a future stage. Iterating the map uncovers the recursive structure on finer and finer scales, and ultimately all orbits converge to a single point in the set. Self-similarity, in this case constructed via a recursive rule, is one of the features of mathematical objects called *fractals*, of which the Cantor set is one.

Remark 3.55 *One can say that the Cantor set C has a fractal structure. The term* fractal *is actually quite difficult to define formally (read: mathematically), and usually the term refers to various objects that have certain key features. One of these features is self-similarity. Another feature, at least for a continuous curve, is that a fractal is nowhere differentiable. Odd sets, like the Cantor set, the Peano curve, the Julia sets and the famous Mandelbrot set that we will see in Section 7.6.1, are all considered fractals, even though the term is not yet completely defined. We will not spend a lot of time refining any particular definition of fractal for this text. However, we will use the term on occasion to refer to objects like those mentioned. Note that the first use of the term is commonly attributed to Benoit Mandelbrot in 1975 [32], when, basing the word on the Latin word for fractured,* fractum *he used it to describe mathematical objects as descriptions of or approximations to natural phenomena. In fact, one quote of his is [31]: "A fractal is a mathematical object that is irregular or fragmented at all scales." It is now mostly understood that fractals are mathematical objects that tend to scale fractionally; doubling the dimension of the lengths of a fractal object tends to scale its spatial content by a fractional amount. This is related to notions of the dimension of an object, where a fractal may be defined as a geometric object that has a fractional dimension. We will spend more time on this in Section 8.2.3. For now, consider another quote from Mandelbrot. In a popular forum on scientific discovery, Mandelbrot was challenged to come up with a clear summary of a scientific concept that anyone can understand in seven words. His response was, "beautiful, damn hard, increasingly useful. That's fractals" [33].*

Solely for a bit more exposure to the playful nature of some of these constructions, we will detail a few more common and easily describable fractal objects presently. Later, in Chapter 8, we will discuss more formally the properties of some fractal constructions that arise from dynamical systems.

3.4.1 The Koch Curve

In 1904, the Swedish mathematician Helge von Koch [58] constructed a special planar curve that was considered an oddity at the time. The curve, continuous everywhere but differentiable nowhere, was described via a finite subdivision rule, and it seems to be the first fractal description in mathematics. To construct it, let $K_0 = [0, 1]$ be the unit interval, and again think of K_0 being divided into three equal parts. Then construct K_1 by removing the middle third of K_0, and adding in an equilateral triangle minus the base. Then K_1 looks like four connected equal line segments with a peaked point in the middle, as in the top left part of Figure 52. Then, on each of these four line segments of K_1, remove the middle

Figure 52 The first few iterates of the Koch curve (left) and the curve itself (right).

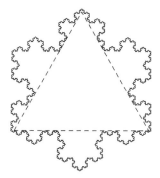

Figure 53 The Koch Snowflake.

third and again add in the upper two sides of an equilateral triangle, to create K_2. Repeat this procedure for all $n \in \mathbb{N}$ to create $K = \lim_{n \to \infty} K_n$. The result is on the right in Figure 52. Can you see the contraction that makes the curve self-similar?

Exercise 122 Show that the Koch curve has infinite length. (Hint: Calculate how much length is added as each stage of the construction.)

Now take an equilateral triangle and create a Koch curve on each side. This construction is called the *Koch snowflake* and is a compact region in the plane bounded by a continuous curve that is differentiable nowhere and is of infinite perimeter. See Figure 53 for this rather interesting example!

3.4.2 Sierpinski Carpet

Another interesting 2-dimensional Cantor-like set is given by the *Sierpinski Carpet S,* described by Waclaw Sierpiński in 1916 [51] (but actually attributed by Sierpiński to a construction of one of his students Stefan Mazurkiewicz, in 1913). This planar figure is described by the finite division rule of removing the middle ninth of the unit square in the plane. The resulting figure looks like a square version of an annulus. Considering the remaining solid region as eight remaining solid squares, remove the middle ninth from each. Continue this rule indefinitely, and take the intersection of what remains. Figure 54 illustrates the first few iterates, and Figure 55 is the result.

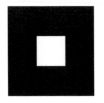  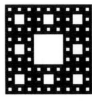 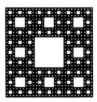

Figure 54 The first few steps in the construction of the Sierpinski Carpet.

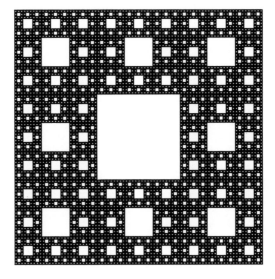

Figure 55 The Sierpinski Carpet.

Here, the coordinates of a point in S are $(x,y) \in [0,1]^2 \subset \mathbb{R}^2$ where both $x = (x_1 x_2 x_3 \ldots)_3$ and $y = (y_1 y_2 y_3 \ldots)_3$ do not simultaneously have a 1 in the same position; i.e., $\forall i \in \mathbb{N}$, $(x_i, y_i) \neq (1,1)$. These are precisely the open sets removed at each stage. Sierpiński used the carpet as a means to catalog all compact 1-dimensional topological objects in the plane, in the sense that for any curve in the plane, there is a set homeomorphic to the Sierpinski Carpet that contains the curve. Thus, the Sierpinski Carpet is a *universal plane curve*. One way to describe this is to say that any curve in the plane that is *locally connected* (any neighborhood of any point on the curve has an open subset that is connected) and has no "local cut points" (points at which if one removes the point, the resulting curve is not connected) is homeomorphic to a Sierpinski Carpet. This is not important for this text, but it does render this space quite useful.

An obvious feature of this space is the following:

Exercise 123 Show that the total area of a Sierpinski Carpet is zero. (Hint: Show that the total area removed is 1.)

So there are no open intervals of each coordinate, hence no open bands across or down the carpet. But there is a Canter set of lines spanning the carpet in each direction.

Figure 56 The Menger sponge.

Exercise 124 Construct a contraction from S to itself that takes the carpet to the smaller copy of itself that is the middle square along the top row in the first stage of the construction (the box of size $\frac{1}{9}$ of the original).

3.4.3 The Sponges

The construction of S, the Sierpinski Carpet, has an obvious n-dimensional version. For $n = 3$, we call the fractal the *Sierpinski Sponge*, formed by considering the unit cube in $\mathbb{R}^3$ as a set of 27 smaller, equal sized cubes, and removing the middle one. This procedure is repeated on each of the other 26 cubes, and the process is continued. From the outside, the cube still looks solid. Cutting through the construction, however, would reveal its intricate structure as a form of sponge. Related to this, and more interesting to view, is the *Menger Sponge M*, named for Karl Menger, who introduced the space in 1926 [35]. To construct M, again take the unit cube in 3-space and again break it up into 27 smaller cubes each of volume $\frac{1}{27}$. Now remove the middle cube, along with the middle cubes along each face (ultimately removing 7 cubes). Repeat this on each of the remaining 20 cubes of the original 27. See Figure 56.

Exercise 125 Show that the Menger sponge has zero volume.

Exercise 126 Show that the Menger sponge has infinite surface area. (Hint: Find an expression for the increase in surface area at each stage.)

4 Flows and Maps of Euclidean Space

4.1 Linear, First-order ODE Systems in the Plane

Almost every standard undergraduate course in differential equations includes the concepts and theory of first-order, linear, homogeneous, autonomous differential equations in more than one variable (usually two). The generalization of the theory of a single linear first-order ODE to a possibly coupled system of two such ODEs already provides a rich environment to study. Any higher-order ODE can be rewritten as a system of first-order ODEs, as Example 4.1 illustrates below, and if the ODE is linear, then so is the resulting system. And most likely the special case of a first-order ODE system where the coefficient functions are constants is well known to the reader. However, for the sake of completeness, and to tie first-order systems of linear ODEs into the theory of linear maps of the plane, we will review some of the theory of linear, constant-coefficient, systems in the plane (mostly) and in $\mathbb{R}^n$ (on occasion) to place them in the proper context of dynamical systems.

4.1.1 General Homogeneous, Linear Systems in Euclidean Space

From your ODE course, you learned that a first-order linear, homogeneous, autonomous system in $\mathbb{R}^n$ has the form

(4.1.1) $$\dot{\mathbf{x}}(t) = A(t)\mathbf{x}(t),$$

where $\mathbf{x} = \begin{bmatrix} x_1 \\ \vdots \\ x_n \end{bmatrix} \in \mathbb{R}^n$, and $A(t)$ is a family of $n \times n$ matrices whose entries $a_{ij}(t)$ are real-valued functions of t continuous on a common interval $I \subset \mathbb{R}$. A solution is a vector-valued function $\mathbf{x}(t)$ that satisfies the ODE. One example of such a system is an nth-order, linear, homogeneous ODE, which can be converted to such a system in Equation 4.1.1.

Example 4.1 For the general nth-order, linear, homogeneous ODE

$$y^{(n)} + a_{n-1}(t)y^{(n-1)} + \ldots + a_1(t)y' + a_0(t)y = 0,$$

the introduction of new dependent variables

$$x_1 = y, \quad x_2 = y', \quad \ldots, \quad x_n = y^{(n-1)}$$

yields the system

$$\dot{\mathbf{x}} = \begin{bmatrix} \dot{x}_1 \\ \vdots \\ \dot{x}_n \end{bmatrix} = \begin{bmatrix} x_2 \\ x_3 \\ \vdots \\ x_n \\ -a_0(t)x_1 - a_1(t)x_2 - \ldots - a_{n-1}(t)x_n \end{bmatrix}$$

$$= \begin{bmatrix} 0 & 1 & 0 & 0 & \cdots & 0 \\ 0 & 0 & 1 & 0 & \cdots & 0 \\ & & \vdots & & \ddots & \vdots \\ 0 & 0 & 0 & 0 & \cdots & 1 \\ -a_0(t) & -a_1(t) & -a_2(t) & -a_3(t) & \cdots & -a_{n-1}(t) \end{bmatrix} \begin{bmatrix} x_1 \\ \vdots \\ x_n \end{bmatrix}.$$

In general, linear ODEs of high order and ODE systems like that in Equation 4.1.1 are difficult, if not impossible, to solve. However, when the coefficient functions are relatively mild in form, or when some solutions are already known, ODE systems of this type do have a lot of structure that can be exploited in their study. One fundamental fact about the solutions to such ODEs or systems is that the solutions behave linearly: They satisfy the *Principle of Superposition*.

Proposition 4.2 (The Superposition Principle) *Let $\mathbf{x}(t), \mathbf{y}(t) : I \to \mathbb{R}^n$ both solve the ODE system in Equation 4.1.1 on I. Then $\forall c_1, c_2 \in \mathbb{R}$, the function*

$$\mathbf{z}(t) = c_1 \mathbf{x}(t) + c_2 \mathbf{y}(t)$$

is also a solution defined on I.

Proof This is just a calculation:

$$\dot{\mathbf{z}}(t) = \frac{d}{dt}[a\mathbf{x}(t) + b\mathbf{y}(t)] = a\dot{\mathbf{x}}(t) + b\dot{\mathbf{y}}(t)$$

$$= aA(t)\mathbf{x}(t) + bA(t)\mathbf{y}(t) = A(t)\left(a\mathbf{x}(t) + b\mathbf{y}(t)\right) = A(t)\mathbf{z}(t). \qquad \square$$

Note that homogeneity is absolutely critical here, since the proposition is generally not true when there are forcing functions present in the system. Another qualitative fact that, by simple inspection, is very powerful is that solutions exist and are unique whenever $A(t)$ is defined and continuous. Indeed, existence and uniqueness is assured for rather mild conditions on $A(t)$ by the n-dimensional form of the Picard–Lindelöf Theorem. Let $\dot{\mathbf{x}} = \mathbf{F}(t, \mathbf{x})$, $\mathbf{x}(t_0) = \mathbf{x}^0$.

Theorem 4.3 (Picard–Lindelöf Theorem in $\mathbb{R}^n$) *Suppose $\mathbf{F} : R_{a,b} \to \mathbb{R}^n$ is continuous on*

$$R_{a,b} = \left\{ (t, \mathbf{x}) \in \mathbb{R} \times \mathbb{R}^n \mid |t - t_0| \le a, \ d(\mathbf{x}, \mathbf{x}^0) \le b \right\},$$

bounded by M and λ-Lipschitz in $\mathbf{x}$, with the same λ for all $t \in [t_0 - a, t_0 + a]$. Let $\alpha = \min(a, b/M)$. Then the system $\dot{\mathbf{x}} = \mathbf{F}(t, \mathbf{x})$, $\mathbf{x}(t_0) = \mathbf{x}^0$ has a unique solution $\mathbf{x}(t)$ defined on $[t_0 - \alpha, t_0 + \alpha]$.

The proof of this is quite similar to that of the $\mathbb{R}$-version and can be found, for example, in Philip Hartman's 1964 book *Ordinary Differential Equations* [25]. But for linear, homogeneous systems like Equation 4.1.1, where $\mathbf{F} = A(t)\mathbf{x}$, then at every point in the domain of continuity in t, one can certainly find an appropriate $R_{a,b}$ that satisfies the suppositions: $\mathbf{F}$ is linear in $\mathbf{x}$ and hence Lipschitz, and $\mathbf{F}$, by continuity in both t and $\mathbf{x}$, will be locally bounded where continuous. Hence, for these kind of systems, we have a somewhat stronger result:

Proposition 4.4 *Let $A(t)$ be a family of $n \times n$ matrices whose entries $a_{ij}(t)$ vary continuously in t on some common interval $I \subset \mathbb{R}$. Then $\forall t \in I$ and $\forall \mathbf{x} \in \mathbb{R}^n$, the IVP in Equation 4.1.1 has a unique solution defined on I.*

It can also be shown that if $\mathbf{F}(t, \mathbf{x})$ is C^r, $r \ge 1$ on the interior of $R_{a,b}$, then so is $\mathbf{x}(t)$, both in t, as well as in the initial data. And in this case, we know more: Recall that n real-valued functions $f_1(t), \dots, f_n(t)$ are *linearly independent* on an interval $I \subset \mathbb{R}$ if the equation

$$\sum_{i=1}^{n} c_i f_i(t) = 0$$

has only the (trivial) solution $c_1 = c_2 = \dots = c_n = 0$, for at least one point $t \in I$. For a system of first-order ODEs in n variables, with n solution functions $\mathbf{x}_i(t)$, $i = 1, \dots, n$ (each with n coordinate functions), the dependence equation is the vector equation

$$\sum_{i=1}^{n} c_i \mathbf{x}_i(t) = \mathbf{0},$$

which can be written as a matrix equation

(4.1.2) $$X(t)\mathbf{c} = \mathbf{0},$$

for $X(t) = [\mathbf{x}_1(t) \ \mathbf{x}_2(t) \ \dots \ \mathbf{x}_n(t)]$. As a homogeneous linear system for each t, $\mathbf{c} = \mathbf{0}$ is the only solution iff $\det X(t) \neq 0$. Here, $\det X(t)$ is called the *Wronskian (determinant)* of the set of solutions that form the columns of $X(t)$, and $X(t)$ is called a *fundamental (solution) matrix* of the system. Then

$$\mathbf{x}(t) = X(t)\mathbf{c}$$

is a general solution of the system: for an IVP with initial data $\mathbf{x}(t_0) = \mathbf{x}^0$, one can solve for the constants $\mathbf{c}$ via

$$\mathbf{x}(t_0) = \mathbf{x}^0 = X(t_0)\mathbf{c}, \quad \text{or} \quad \mathbf{c} = X^{-1}(t_0)\mathbf{x}^0$$

as long as $\det X(t) \neq 0$. Fortunately, if $\det X(t) \neq 0$ at some point where solutions are uniquely defined (where $A(t)$ is continuous), then $\det X(t) \neq 0$ everywhere solutions are uniquely defined:

Proposition 4.5 *Let $\dot{\mathbf{x}} = A(t)\mathbf{x}$, where $A(t)$ is continuous on some interval $I \subset \mathbb{R}$, and let $X(t)$ be a fundamental matrix solution on I. Then either $\det X(t) = 0$ for every $t \in I$ or $\det X(t) \neq 0$ for every $t \in I$.*

Proof Suppose that there exists a $t_0 \in I$ where $\det X(t_0) = 0$. Then there exists a nontrivial vector solution to the homogeneous matrix equation $X(t_0)\mathbf{c} = \mathbf{0}$. Call this nontrivial solution $\mathbf{c}_0$. Then $\mathbf{x}(t) = X(t)\mathbf{c}_0$ is a solution to the ODE system. But solutions are unique at $t_0 \in I$, and since $\mathbf{x}(t) \equiv \mathbf{0}$ is also a solution passing through $\mathbf{0}$ at $t = t_0$, $X(t)\mathbf{c}_0 = \mathbf{0}, \forall t \in I$, and hence $\det X(t) = 0, \forall t \in I$. $\qquad\square$

In fact, we can go one step further and incorporate the calculation for the constants $\mathbf{c}$ directly into the general solution: Since $\mathbf{c} = X^{-1}(t_0)\mathbf{x}^0$, we can write

$$\mathbf{x}(t) = X(t)X^{-1}(t_0)\mathbf{x}^0 = \Psi(t)\mathbf{x}^0,$$

where $\Psi(t) = X(t)X^{-1}(t_0)$ is a special fundamental matrix that satisfies $\Psi(t_0) = I_n$. Some notes:

- For now, we will call $\Psi(t)$ a *normalized fundamental matrix*. This choice is special since the general solution $\mathbf{x}(t) = \Psi(t)\mathbf{x}^0$ is the actual evolution of the continuous dynamical system given by the system. For all $t \in I$ where solutions are defined, the flow is a time-t map.

- Any matrix $X(t)$, where $\det X(t) \neq 0, \forall t \in I$, that satisfies the matrix version of the ODE system $\dot{X}(t) = A(t)X(t)$ is a fundamental matrix, since, by the very definition of matrix multiplication, restricting the system to each column of $X(t)$ results in a standard system where the column is a solution.

- For a choice of fundamental matrix $X(t)$, the equation

$$\mathbf{x}^0 = X(t)\mathbf{c}$$

is just a calculation of the vector $\mathbf{x}^0$ in the coordinates $\mathbf{c}$ of the basis of $\mathbb{R}^n$ given by the columns of $X(t)$ (the columns do form a basis, since $\det X(t) \neq 0$.) Hence "choosing" $\Psi(t)$ as the fundamental matrix means keeping the coordinate system compatible with the standard basis in $\mathbb{R}^n$. Think about this.

Exercise 127 For the following systems with solutions given, first convert the given ODE into a 2 × 2 linear system and resolve, forming a fundamental matrix solution $X(t)$. (Hint: The new vector solutions are related to $x_1(t)$ and $x_2(t)$.) Then, find the normalized fundamental matrix that satisfies the given vector initial data.

(a) $t^2 y'' - 2y = 0, t > 0, y_1(t) = t^2, y_2(t) = d\frac{1}{t}$, and $\mathbf{x}(1) = \begin{bmatrix} 1 \\ 5 \end{bmatrix}$.

(b) $(1 - t)y'' + ty' - y = 0, 0 < t < 1, y_1(t) = t, y_2(t) = e^t$, and $\mathbf{x}\left(\frac{1}{2}\right) = \begin{bmatrix} 1 \\ 0 \end{bmatrix}$.

(c) $ty'' - (1 + t)y' + y = 0, t > 0, y_1(t) = 1 + t, y_2(t) = e^t$, and $\mathbf{x}(1) = \begin{bmatrix} 1 \\ 1 \end{bmatrix}$.

Exercise 128 Consider the two vector functions $\mathbf{x}(t) = \begin{bmatrix} t \\ 1 \end{bmatrix}$ and $\mathbf{x}(t) = \begin{bmatrix} t^2 \\ 2t \end{bmatrix}$ as solutions to a linear, first order, homogeneous, 2-dimensional ODE system. Compute the Wronskian and determine the intervals of linear independence. Then construct an IVP where these functions form a general solution and write out the evolution for the initial valus $\mathbf{x}(1) = \begin{bmatrix} 1 \\ 1 \end{bmatrix}$.

4.1.2 Autonomous Linear Systems

Suppose now that our system is also autonomous, so that $A(t) = A_{n \times n}$ is a constant matrix. Then the system is

(4.1.3) $$\dot{\mathbf{x}} = A\mathbf{x}.$$

Here, since the system is autonomous (time-invariant), we will confine our initial data to $t_0 = 0$, so $\mathbf{x}(0) = \mathbf{x}^0$. Now, we are in a position to look for solutions $\mathbf{x}(t) : \mathbb{R} \to \mathbb{R}^n$, whose coordinate function derivatives are linear combinations of the coordinate functions. (Since A is constant, it is a continuous function of t on all of $\mathbb{R}$. Hence solutions exist and are unique on $\mathbb{R}$.) This suggests, as in the 1-dimensional case $\dot{x} = ax$, that exponential functions will play a significant role in solution construction. Indeed, a good guess at a solution would be

$$\mathbf{x}(t) = \mathbf{v}e^{rt},$$

for $\mathbf{v} \in \mathbb{R}^n, r \in \mathbb{R}$ unknown but to-be-determined variables.
To be a solution, we would need

$$\dot{\mathbf{x}}(t) = r\mathbf{v}e^{rt} = A\mathbf{x}(t) = A\left(\mathbf{v}e^{rt}\right),$$

or, after dividing out by the non-vanishing exponential, $r\mathbf{v} = A\mathbf{v}$. This is simply the eigenvalue/eigenvector equation for A. Hence it follows immediately that solutions are related to the properties of A, and when available, reflect the eigenvalue/eigenvector pairs of A. Hence, in the case where A has a full set of n linearly independent eigenvectors $\mathbf{v}_1, \ldots, \mathbf{v}_n$,

$$\mathbf{x}(t) = \sum_{i=1}^{n} \mathbf{v}_i e^{r_i t}$$

is a general solution to Equation 4.1.3 when each r_i is the eigenvalue corresponding to the eigenvector $\mathbf{v}_i$.

However, when A does not have a full set of linearly independent eigenvectors, we do need to adjust our analysis:

- If the corresponding eigenvalue equation $\det(A - rI_n) = 0$ contains solutions that are simple complex conjugate pairs $r = \lambda \pm i\mu$, one can also construct associated complex solutions to the eigenvector equation so that $\mathbf{v} = \mathbf{a} \pm i\mathbf{b}$. From these complex conjugate solutions, one can construct pairs of linearly independent real solutions

$$\mathbf{x}_1(t) = e^{\lambda t}(\mathbf{a}\cos\mu t - \mathbf{b}\sin\mu t),$$
$$\mathbf{x}_2(t) = e^{\lambda t}(\mathbf{a}\sin\mu t + \mathbf{b}\cos\mu t).$$

- For *defective matrices*, i.e., those with at least one eigenvalue whose algebraic multiplicity (its multiplicity as a solution to the characteristic equation) is greater than its geometric multiplicity (the dimension of the null space of $A - rI_n$ for a choice of r), one must revert to the construction of *generalized eigenvectors* to "fill" the defective eigenspaces. Indeed, if for $A_{2\times 2}$ we have the two eigenvalues $r_1 = r_2 = r$ and $\mathbf{v}$ is the only independent eigenvector, then $\mathbf{x}_1(t) = \mathbf{v}e^{rt}$ is the only independent solution constructed in the usual fashion. A second solution can be constructed of the form

(4.1.4) $$\mathbf{x}_2(t) = \mathbf{v}te^{rt} + \mathbf{w}e^{rt}$$

via a reduction-of-order technique, where $\mathbf{w}$ is the generalized eigenvector that satisfies

(4.1.5) $$(A - rI_2)\mathbf{w} = \mathbf{v}.$$

Exercise 129 Show that for r and $\mathbf{v}$ an eigenvalue/eigenvector pair for a 2×2 matrix A, if $\mathbf{x}_2(t)$ in Equation 4.1.4 solves the ODE $\dot{\mathbf{x}} = A\mathbf{x}$, then $\mathbf{w}$ satisfies Equation 4.1.5.

Exercise 130 Show that $(A - rI_2)\mathbf{v} = \mathbf{0}$ has only one linearly independent solution iff $(A - rI_2)$ is 2-nilpotent (i.e., $(A - rI_2)^2 = 0_{2\times 2}$).

Now, from Exercise 130, any vector $\mathbf{w}$ that is independent of $\mathbf{v}$ can serve in the construction of $\mathbf{x}_2(t)$, so that

$$\mathbf{x}(t) = c_1 \mathbf{v} e^{rt} + c_2 (\mathbf{v} t e^{rt} + \mathbf{w} e^{rt})$$

is the general solution in the case of a linear, homogeneous, autonomous 2-dimensional system with a defective matrix A.

In any case, one can always construct a set of n-linearly independent solutions to Equation 4.1.3. Now place each of these solutions $\mathbf{x}_i(t)$, $i = 1, \ldots, n$ as the columns of $X(t)$ and calculate $X(t)X^{-1}(0)$. Notice that for two different fundamental matrices $X_1(t)$ and $X_2(t)$, they both solve the same system where solutions are known to be unique and defined on all of $\mathbb{R}$. Hence

$$X_1(t)X_1^{-1}(0) = \Psi(t) = X_2(t)X_2^{-1}(0)$$

is well-defined.

Recall that in 1 dimension, with IVP $\dot{x} = ax$, $x(0) = x_0$, the evolution was $x(t) = x_0 e^{at}$. In n dimensions, it is tempting to do the same thing and write the evolution of $\dot{\mathbf{x}} = A\mathbf{x}$, $\mathbf{x}(0) = \mathbf{x}^0$ as $\mathbf{x}(t) = \mathbf{x}^0 e^{At}$. This would make sense if (1) the exponential of a matrix made sense, and (2) its derivative satisfied the usual rule for the derivative of an exponential, namely $(d/dx)e^{At} = Ae^{At}$. It turns out that there is a well-defined notion of the exponential of a matrix.

4.1.3 The Matrix Exponential

Define

(4.1.6) $$e^{At} = I_n + \frac{A}{1!}t + \frac{A^2}{2!}t^2 + \cdots + \frac{A^k}{k!}t^k + \cdots = \sum_{k=0}^{\infty} \frac{A^k}{k!}t^k.$$

This expression, the matrix exponential, figures heavily in the linear systems theory of ODEs. But to be able to use it, we must first ensure that it indeed makes sense as an expression. To this end, denote the ijth entry of A^k, the product of A with itself k times, by $a_{ij}(k)$ and extend this to the entries of e^{At} as $a_{ij}(\infty)$. Then, by inspection of the definition alone, we can see the following:

(1) e^{At}, when it exists, is an $n \times n$ matrix, being an (infinite) sum of scalar multiples of powers of A.

(2) The series in Equation 4.1.6 converges absolutely iff the series defining each entry of e^{At} does.

(3) Each entry

$$a_{ij}(\infty) = \sum_{k=0}^{\infty} \frac{a_{ij}(k)}{k!}t^k.$$

Note that when $A_{n \times n}$ is diagonal, this is an easy calculation, since $a_{ij}(\infty) = 0 = a_{ij}(k)$, $\forall k \in \mathbb{N}$ when $i \neq j$. And $a_{ii}(k) = (a_{ii}(1))^k$. Hence, for all $i, j = 1, \ldots, n$, we have $a_{ij}(\infty) = e^{a_{ij}(1)t}$ in this case. A form for the entries of e^{At} is less clear in the case where A is not diagonal.

Then we have:

Proposition 4.6 *For $A_{n \times n}$, the series in Equation 4.1.6 converges absolutely $\forall t \in \mathbb{R}$.*

Proof Let $a = \max_{ij} |a_{ij}(1)|$. Then

$$\left|a_{ij}(2)\right| = \left|\sum_{\ell=1}^{n} a_{i\ell}(1) \cdot a_{\ell j}(1)\right| \leq \sum_{\ell=1}^{n} \left|a_{i\ell}(1) \cdot a_{\ell j}(1)\right| \leq n(a \cdot a) < n^2 a^2.$$

Note that the extra n in the bound above is superfluous, but will make a later calculation much easier to see. Also see that

$$\left|a_{ij}(3)\right| = \left|\sum_{\ell=1}^{n} a_{i\ell}(2) \cdot a_{\ell j}(1)\right| \leq \sum_{\ell=1}^{n} \left|a_{i\ell}(2) \cdot a_{\ell j}(1)\right| \leq n(n^2 a^2 \cdot a) < n^3 a^3.$$

So inductively, suppose $\left|a_{ij}(k-1)\right| < n^{k-1} a^{k-1}$. Then

$$\left|a_{ij}(k)\right| = \left|\sum_{\ell=1}^{n} a_{i\ell}(k) \cdot a_{\ell j}(1)\right| \leq \sum_{\ell=1}^{n} \left|a_{i\ell}(k) \cdot a_{\ell j}(1)\right| \leq n(n^{k-1} a^{k-1} \cdot a) < n^k a^k.$$

Then we can write

$$\left|\sum_{k=0}^{\infty} \frac{a_{ij}(k)}{k!} t^k\right| \leq \sum_{k=0}^{\infty} \left|\frac{a_{ij}(k)}{k!} t^k\right| \leq \sum_{k=0}^{\infty} \frac{|a_{ij}|(k)}{k!} t^k$$

$$\leq \sum_{k=0}^{\infty} \frac{(na)^k}{k!} t^k = e^{na}. \qquad \square$$

Now that we know that the expression is well-defined, we can immediately calculate:

Proposition 4.7 $\dfrac{d}{dt}\left(e^{At}\mathbf{x}^0\right) = Ae^{At}\mathbf{x}^0.$

Proof Really, this is just the definition of a derivative:

$$\frac{d}{dt} e^{At} = \lim_{h \to 0} \frac{e^{A(t+h)} - e^{At}}{h} = \lim_{h \to 0} \frac{e^{At}e^{Ah} - e^{At}}{h} = e^{At} \lim_{h \to 0} \frac{e^{Ah} - I_n}{h}$$

$$= e^{At} \lim_{h \to 0} \frac{1}{h}\left(\sum_{n=0}^{\infty} \frac{(Ah)^n}{n!} - I_n\right) = e^{At} \lim_{h \to 0} \frac{1}{h} \sum_{n=1}^{\infty} \frac{(Ah)^n}{n!}$$

$$= e^{At} \lim_{h \to 0} \sum_{n=1}^{\infty} \frac{A^n h^{n-1}}{n!} = e^{At} \lim_{h \to 0} A \sum_{n=1}^{\infty} \frac{A^{n-1} h^{n-1}}{n!}$$

$$= A e^{At} \lim_{h \to 0} \left(\frac{I_n}{1!} + \frac{Ah}{2!} + \frac{A^2 h^2}{3!} + \frac{A^3 h^3}{4!} + \cdots \right).$$

At this point, every term in the remaining series has an h in it except for the $n = 1$ term, which is I_n. So

$$\frac{d}{dt} e^{At} = A e^{At} \lim_{h \to 0} \sum_{n=1}^{\infty} \frac{A^{n-1} h^{n-1}}{n!} = A e^{At} \cdot I_n = A e^{At}. \qquad \square$$

Proposition 4.8 *The general solution to the IVP* $\dot{\mathbf{x}} = A\mathbf{x}$, $\mathbf{x}(0) = \mathbf{x}^0$ *is* $\mathbf{x}(t) = e^{At}\mathbf{x}^0$.

Proof $\dot{\mathbf{x}}(t) = \frac{d}{dt}\left(e^{At}\mathbf{x}^0\right) = A e^{At}\mathbf{x}^0 = A\mathbf{x}(t).$ $\qquad \square$

Corollary 4.9 *For any fundamental matrix solution to* $\dot{\mathbf{x}} = A\mathbf{x}$, *we have* $e^{A(t-t_0)} = X(t)$ $X^{-1}(t_0)$.

Exercise 131 Prove Corollary 4.9.

Hence we can immediately write out the flow for the system $\dot{\mathbf{x}} = A\mathbf{x}$, $\mathbf{x}(0) = \mathbf{x}^0$ as

$$\varphi^t(\mathbf{x}) = e^{At}\mathbf{x}^0.$$

Some properties of this flow are apparent immediately. For example, the equilibria of the system correspond to the solutions to the homogeneous linear system $A\mathbf{x} = \mathbf{0}$. When $\det A \neq 0$, $\mathbf{x} = \mathbf{0}$ is the only equilibrium. From linear algebra, the determinant of A is the product of the eigenvalues of A (with multiplicity). Hence $\det A \neq 0$ iff all eigenvalues are non-zero. Thus we get the obvious results:

Proposition 4.10 $\dot{\mathbf{x}} = A\mathbf{x}$ *has*

(1) *a unique equilibrium at the origin of* $\mathbb{R}^n$ *if* $r = 0$ *is not an eigenvalue of* A, *and*

(2) *a linear subspace of equilibria if* $r = 0$ *is an eigenvalue of* A.

The following is another facet of the flow that is rather immediate: A solution $\mathbf{x}(t) = \mathbf{v}e^{rt}$ to $\dot{\mathbf{x}} = A_{n \times n}\mathbf{x}$, where r and $\mathbf{v}$ are a real eigenvalue/eigenvector pair is confined to the subspace of $\mathbb{R}^n$ spanned by $\mathbf{v}$ for all $t \in \mathbb{R}$. Indeed, e^{rt} is scalar-valued, so for any $t \in \mathbb{R}$, $\mathbf{v}e^{rt}$ is a scalar multiple of $\mathbf{v}$. Hence a solution that starts in span $\{\mathbf{v}\}$ remains in span $\{\mathbf{v}\}$. And if $r = \lambda + i\mu$ is complex, then so is the eigenvector $\mathbf{v} = \mathbf{a} + i\mathbf{b}$. Then the 2-dimensional subspace of $\mathbb{R}^n$ spanned by $\mathbf{a}$ and $\mathbf{b}$ will be invariant under the flow. Indeed, all of the eigenspaces of A are invariant under the flow. And we can characterize them according to the properties of the eigenvalues r:

Definition 4.11 *For* $\dot{\mathbf{x}} = A_{n \times n} \mathbf{x}$, *the subspaces of the phase space* $\mathbb{R}^n$

    i. $\mathbb{E}^s = \text{span} \left\{ \mathbf{v}^s \in \mathbb{R}^n \mid \mathbf{v}^s \text{ is a generalized eigenvector for} \right.$
        $\left. \text{an eigenvalue } r_s \text{ where } \text{Re}(r_s) < 0 \right\}$,

    ii. $\mathbb{E}^u = \text{span} \left\{ \mathbf{v}^u \in \mathbb{R}^n \mid \mathbf{v}^u \text{ is a generalized eigenvector for} \right.$
        $\left. \text{an eigenvalue } r_u \text{ where } \text{Re}(r_u) > 0 \right\}$,

    iii. $\mathbb{E}^c = \text{span} \left\{ \mathbf{v}^c \in \mathbb{R}^n \mid \mathbf{v}^c \text{ is a generalized eigenvector for} \right.$
        $\left. \text{an eigenvalue } r_c \text{ where } \text{Re}(r_c) = 0 \right\}$

are the asymptotically stable, unstable, *and* center *subspaces, respectively, of the system.*

The names are related to the behavior of solutions within the subspaces, noting that solutions in the stable subspace $\mathbb{E}^s$ all tend in the limit to the origin as $t \to \infty$, and are unbounded as $t \to -\infty$, either monotonically or in an oscillatory fashion, while orbits in $\mathbb{E}^u$ do exactly the opposite. Trajectories in $\mathbb{E}^c$, at least when the eigenvalues are simple, are closed and periodic, or centered at the origin (when r is purely imaginary), or equilibria (when $r = 0$). We note that when complex eigenvalues are not simple, orbits in $\mathbb{E}^c$ may experience growth or decay, but this is polynomial and not exponential. We will not elaborate on that here. But this does lead to a general qualitative result on the stability of the origin as an equilibrium of the system:

Proposition 4.12 *For the system* $\dot{\mathbf{x}} = A\mathbf{x}$,

    (1) if all eigenvalues of A have negative real part, then $\mathbf{x}(t) \equiv 0$ *is an asymptotically stable equilibrium;*

    (2) if at least one eigenvalue of A has positive real part, then the equilibrium $\mathbf{x}(t) \equiv \mathbf{0}$ *is unstable.*

Exercise 132 Prove Proposition 4.12.

4.1.4 Two-Dimensional Classification

Now the general behavior of orbits for the system $\dot{\mathbf{x}} = A_{n \times n} \mathbf{x}$ when $n > 2$ can be quite complicated, even though the qualitative behavior can be mostly gleaned from the data of A. To get a better sense of this, let's restrict now to the $n = 2$ case, and study linear homogeneous and autonomous systems in the plane. Here the classification of types and stability of the equilibrium a the origin results in only a few cases.

For $A_{2 \times 2}$, solutions to the eigenvalue/eigenvector equation $A\mathbf{v} = r\mathbf{v}$ comprise the kernel of the matrix $A - rI_n$ for any particular choice of r. Of course, the homogeneous system $(A - rI_n)\mathbf{v} = \mathbf{0}$ has no non-trivial solutions when $\det(A - rI_n) \neq 0$. So our first step is to solve the equation $\det(A - rI_n) = 0$ for those values of r that render a positive dimension kernel of $A - rI_n$. This equation is called the *characteristic equation* of A. For $A_{n \times n}$, this will be a nth-degree polynomial in r with coefficients in the entries of A. For $A_{2 \times 2}$, we have

$$r^2 - (\text{tr}\,A)r + (\det A) = 0,$$

where $\text{tr}\,A$ is the *trace* of the matrix A, or the sum of the elements on the main diagonal. Counting multiplicities and possible complex conjugates, there are always two solutions, and the possibilities are determined mainly by the discriminant of the quadratic formula for solutions:

(4.1.7) $$r = \frac{\text{tr}\,A \pm \sqrt{(\text{tr}\,A)^2 - 4\det A}}{2}.$$

The three possibilities are whether the discriminant $(\text{tr}\,A)^2 - 4\det A$ is positive, zero, or negative. For now, let's assume that $\det A \neq 0$. We can return to the zero-determinant case later.

Suppose, as a start, that both $\det A > 0$ and $(\text{tr}\,A)^2 - 4\det A > 0$. Then we know that both eigenvalues r_1, r_2 are (1) of the same sign, (2) real, (3) distinct, and (4) non-zero. Solutions can be parameterized as a linear combination of $x_1 = v_1 e^{r_1 t}$ and $x_2 = v_2 e^{r_2 t}$, where $r_1 \neq r_2$ and v_i is a choice of eigenvector (a non-trivial solution to $(A - rI_2)v = 0$) for each r_i. And since the eigenvalues are distinct, one can show that x_1 and x_2 are independent as functions from each other, so that all solutions are linear combinations of these two. Hence the general solution to the ODE system in this case is

$$x(t) = c_1 v_1 e^{r_1 t} + c_2 v_2 e^{r_2 t}.$$

Suppose further that $r_1 > r_2 > 0$. The only equilibrium here is at the origin (Proposition 4.10, since $\det A \neq 0$). A choice of vector $c = \begin{bmatrix} c_1 \\ c_2 \end{bmatrix}$ determines the curve, parameterized by t, that serves as the particular solution passing through that point in the plane at $t = 0$. Here, all solutions, except for the equilibrium, tend to behave the same way:

- For every $c = \begin{bmatrix} c_1 \\ c_2 \end{bmatrix} \neq \begin{bmatrix} 0 \\ 0 \end{bmatrix}$, $\lim_{t \to -\infty} x(t) = 0$, and $\lim_{t \to \infty} x(t)$ is unbounded, and
- motion is straight along the eigendirections (when either c_1 or c_2 is 0). Otherwise, solutions paths are curved (why?)

In this case, we call the origin an unstable node, or a *source*. If $r_1 < r_1 < 0$, then the origin would be an asymptotically stable node, or a *sink*. In either case, these are examples of what is called a *node*, when the eigenvalues are real and distinct and of the same sign. Note that Figure 57 denotes a source. For a sink, the drawing would look basically the same, but with the arrows reversed.

Example 4.13 (A source node.) Let $\dot{x} = Ax$, where $A = \frac{1}{4}\begin{bmatrix} 5 & 1 \\ 3 & 7 \end{bmatrix}$. Here, the characteristic equation is $r^2 - 3r + 2 = 0$, so that the eigenvalues are $r_1 = 1$ and $r_2 = 2$. Eigenvectors can be easily computed, and effective choices are $v_1 = \begin{bmatrix} 1 \\ -1 \end{bmatrix}$ and

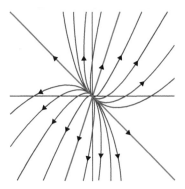

Figure 57 A source node.

$\mathbf{v}_2 = \begin{bmatrix} 1 \\ 3 \end{bmatrix}$. Since both eigenvalues are greater than 0 and distinct, we have a source at the origin, as in Figure 57. Note that the solution curves off of the invariant eigenspaces are curved, owing to the different rates of growth in the two coordinates.

Now suppose that the discriminant of Equation 4.1.7 is 0, so that $(\mathrm{tr}\,A)^2 = 4\det A$. Then the eigenvalues $r = r_1 = r_2 = \frac{1}{2}\mathrm{tr}\,A$ are equal, and here is where a bit of linear algebra comes in. For starters, $\mathbf{x}_1 = c_1\mathbf{v}_1 e^{r_1 t}$ and $\mathbf{x}_2 = c_2\mathbf{v}_2 e^{r_2 t}$ are not necessarily linear independent as functions, unless $\mathbf{v}_1$ and $\mathbf{v}_2$ are as vectors, so if the r-eigenspace is not defective (again, if there are enough linearly independent eigenvectors to serve as a basis), then, even with the same exponential terms, the two solutions will be linearly independent as functions. In this case, we can use $\mathbf{v}_1$ and $\mathbf{v}_2$ and write out a complete set of solutions as

$$\mathbf{x}(t) = c_1\mathbf{v}_1 e^{rt} + c_2\mathbf{v}_2 e^{rt} = (c_1\mathbf{v}_1 + c_2\mathbf{v}_2)\,e^{rt} = \begin{bmatrix} k_1 \\ k_2 \end{bmatrix} e^{rt}.$$

Geometrically, motion is along all straight lines through the origin. (Do keep in mind, though, that even though motion is straight, velocity is not constant!) The origin, in this case, is also a node, and called a *proper node*, or also a *star node*, with any associated stability included as an adjective. See the left-hand side of Figure 58, for an example of a star sink.

And if the matrix A is defective? Then, again, one must "fill out" the eigenspace by constructing a second, generalized eigenvector, independent of the first, but still independent from any other eigenspaces. This is the vector $\mathbf{w}$ in Equation 4.1.4, which solves the generalized eigenvector equation given in Equation 4.1.5. However, in a 2-dimensional system with a repeated eigenvalue and an insufficient basis, really any independent vector in the eigenspace will do (see Exercise 133). In this case, straight-line motion in the plane occurs only along the line through the origin corresponding to the eigenvector $\mathbf{v}$ of the defective A. See the right side drawing of Figure 58 for an example of what we call an *improper node*, or sometimes a *degenerate node*.

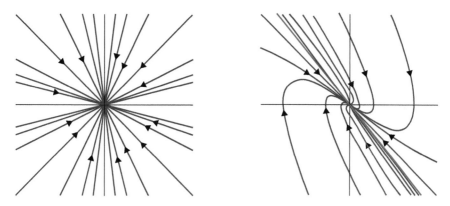

Figure 58 A star node (left) and degenerate node (right), both sinks.

Exercise 133 Show that in a 2×2 system $\dot{\mathbf{x}} = A_{2\times 2}\mathbf{x}$ with A defective, the generalized eigenvector $\mathbf{w}$ can be chosen arbitrarily, as long as it is independent of the eigenvector $\mathbf{v}$. (Hint: Use Exercise 130.)

Example 4.14 (A degenerate sink.) Let $A = \begin{bmatrix} 0 & 1 \\ -1 & -2 \end{bmatrix}$ be the matrix in a linear system The characteristic equation is then $r^2 + 2r + 1 = 0$, with $r_1 = r_2 = r = -1$. But the only linearly independent solution to the eigenvector equation

$$A\mathbf{v} = r\mathbf{v} = -\mathbf{v} \quad\Longrightarrow\quad \begin{bmatrix} 0 & 1 \\ -1 & -2 \end{bmatrix}\begin{bmatrix} v_1 \\ v_2 \end{bmatrix} = -\begin{bmatrix} v_1 \\ v_2 \end{bmatrix}$$

is $v_2 = -v_1$. Choosing $v_1 = 1$, we get the general solution

$$\mathbf{x}(t) = c_1 \begin{bmatrix} 1 \\ -1 \end{bmatrix}e^{-t} + c_2 \left(\begin{bmatrix} 1 \\ -1 \end{bmatrix}te^{-t} + \begin{bmatrix} 0 \\ 1 \end{bmatrix} \right).$$

Note here that $(A - rI_2)\mathbf{w} = \mathbf{v}$, used to find the second vector in the general solution, comes from

$$\left(\begin{bmatrix} 0 & 1 \\ -1 & -2 \end{bmatrix} + \begin{bmatrix} 1 & 0 \\ 0 & 1 \end{bmatrix} \right)\begin{bmatrix} w_1 \\ w_2 \end{bmatrix} = \begin{bmatrix} 1 \\ -1 \end{bmatrix},$$

with $w_1 + w_2 = 1$ the only equation to satisfy. We chose $w_1 = 0$ and $w_2 = 1$. The phase portrait is on the right side of Figure 58.

Example 4.15 (A star sink.) Related to Example 4.14 is the system with coefficient matrix $A = \begin{bmatrix} -2 & 0 \\ 0 & -2 \end{bmatrix}$. Here, it is obvious that the eigenvalues are $r_1 = r_2 = r = -2$, and the eigenvector equation

$$(A - rI_2)\mathbf{v} = \left(\begin{bmatrix} -2 & 0 \\ 0 & -2 \end{bmatrix} - (-2) \begin{bmatrix} 1 & 0 \\ 0 & 1 \end{bmatrix} \right) \begin{bmatrix} v_1 \\ v_2 \end{bmatrix} = \begin{bmatrix} 0 \\ 0 \end{bmatrix}$$

is fully degenerate, so that any two independent choices of vectors for $\mathbf{v}_1$ and $\mathbf{v}_2$ will work. For example, one set of choices leads to a general solutions

$$\mathbf{x}(t) = c_1 \begin{bmatrix} 1 \\ 2 \end{bmatrix} e^{-2t} + c_2 \begin{bmatrix} 3 \\ -1 \end{bmatrix} e^{-2t} = \begin{bmatrix} w_1 \\ w_2 \end{bmatrix} e^{-2t} = \mathbf{w} e^{-2t},$$

where $\mathbf{w}$ is any vector in $\mathbb{R}^2$, since the two vectors chosen form a basis for $\mathbb{R}^2$. See the left side of Figure 58.

Another tack for solving systems whose matrix has deficient eigenspaces, however, goes back to our guess-and-check scheme. Exponentials typically solve $\dot{\mathbf{x}} = A\mathbf{x}$. But in the case where this idea fails, another function independent of an exponential that has derivatives closely related to the original function is te^{kt}, since $(d/dt)(te^{kt}) = e^{kt} + kte^{kt}$. One can produce a new independent solution through a technique involving a *reduction of order* of a higher-order ODE where a solution is known: Given $y'' + p(t)y' + q(t)y = 0$, with $y_1(t)$ a known solution, one can form a guess for a second solution $y_2(t) = u(t)y_1(t)$. As long as $u(t)$ is not a constant, the new solution will be independent from the first. Substituting this new solution into the original differential equation yields a new differential equation in $u(t)$:

$$u'' + (p + y_1 q)u' = 0,$$

a first-order ODE in $u'(t)$, which can be solved for and then integrated to obtain $u(t)$. In this current linear system case (with vector solutions), this technique always yields $u'(t)$ as a constant, so that $u(t) = t$, up to another constant. Converting the second-order ODE into a system, one can readily see how the function te^{rt} appears in the "second" part of the general solution, given in Equation 4.1.4.

Next, suppose that the discriminant of Equation 4.1.7 is negative, so $(\operatorname{tr} A)^2 - 4\det A < 0$. Then the solutions to the characteristic equation are complex conjugates $r_1 = \lambda + i\mu$ and $r_2 = \lambda - i\mu$ (since both their product and their sum must be real). Using these values of r, the vector solutions to the eigenvector equation will also have complex entries and also be complex conjugates:

$$\mathbf{v}_1 = \begin{bmatrix} a_1 + ib_1 \\ a_2 + ib_2 \end{bmatrix} = \mathbf{a} + i\mathbf{b},$$

(4.1.8)
$$\mathbf{v}_2 = \begin{bmatrix} a_1 - ib_1 \\ a_2 - ib_2 \end{bmatrix} = \mathbf{a} - i\mathbf{b}.$$

By superposition again, any linear combination of these two solutions is also a solution. Hence, by creatively taking multiples of these two complex solutions, one can draw out real solutions. In short, then, the general solution for $\dot{\mathbf{x}} = A\mathbf{x}$ when A has complex eigenvalues is

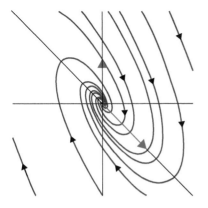 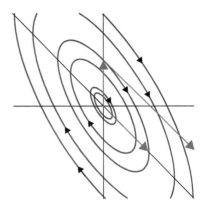

Figure 59 A spiral sink (left) and center (right).

(4.1.9) $$x(t) = e^{\lambda t} [c_1 (\mathbf{a} \cos \mu t - \mathbf{b} \sin \mu t) + c_2 (\mathbf{a} \sin \mu t + \mathbf{b} \cos \mu t)].$$

Exercise 134 Show that by taking creative linear combinations of the complex solutions $\mathbf{x}_1(t) = \mathbf{v}_1 e^{r_1 t}$ and $\mathbf{x}_2(t) = \mathbf{v}_2 e^{r_2 t}$, where $r_1 = \lambda + i\mu$, $r_2 = \lambda - i\mu$ and $\mathbf{v}_i$ is as in Equation 4.1.8, one can arrive at the real general solution in Equation 4.1.9.

In this case of complex conjugate solutions to the characteristic equation of A, the equilibrium at the origin is classified as a *spiral point* or a *focus* (or simply a *spiral*) when $\lambda \neq 0$, and its stability is naturally determined by the sign of λ. Figure 59 shows a spiral sink, the phase portrait of the system given by the matrix in Example 4.16 below. When the eigenvalues are purely imaginary, the general solution in Equation 4.1.9 reduces to just the trigonometric parts. Here, the orbits are closed, and form concentric ellipses centered around the origin, called a *center*, shown in the phase portrait on the right side of Figure 59. Notice how (1) the direction of travel is reflected in the structure of the solutions, and (2) how the vectors in this example, $\mathbf{a} = \begin{bmatrix} 1 \\ -1 \end{bmatrix}$ and $\mathbf{b} = \begin{bmatrix} 0 \\ 1 \end{bmatrix}$ affect the shapes of the orbits of the spiral and the ellipses of the center.

Example 4.16 (A spiral sink.) Let $A = \begin{bmatrix} 1 & 2 \\ -4 & -3 \end{bmatrix}$. Then the system that has A as its coefficient matrix, with characteristic equation $r^2 + 2r + 5 = 0$, will have general solution

$$x(t) = e^{-t} \left[c_1 \left(\begin{bmatrix} 1 \\ -1 \end{bmatrix} \cos 2t - \begin{bmatrix} 0 \\ 1 \end{bmatrix} \sin 2t \right) + c_2 \left(\begin{bmatrix} 1 \\ -1 \end{bmatrix} \sin 2t + \begin{bmatrix} 0 \\ 1 \end{bmatrix} \cos 2t \right) \right].$$

The phase portrait is on the left side of Figure 59.

Exercise 135 Verify that the solution given in Example 4.16 is correct by calculating the eigenvalues and eigenvectors of A.

Example 4.17 (A center.) For $A = \begin{bmatrix} 2 & 2 \\ -4 & -2 \end{bmatrix}$, the general solution is

$$\mathbf{x}(t) = c_1 \left(\begin{bmatrix} 1 \\ -1 \end{bmatrix} \cos 2t - \begin{bmatrix} 0 \\ 1 \end{bmatrix} \sin 2t \right) + c_2 \left(\begin{bmatrix} 1 \\ -1 \end{bmatrix} \sin 2t + \begin{bmatrix} 0 \\ 1 \end{bmatrix} \cos 2t \right).$$

Solutions form the closed orbits on the right side of Figure 59. Notice two things here:

(1) The two vectors $\mathbf{a} = \begin{bmatrix} 1 \\ -1 \end{bmatrix}$ and $\mathbf{b} = \begin{bmatrix} 0 \\ 1 \end{bmatrix}$, superimposed on the phase portrait, affect the shape and orientation of the ellipses. All solutions wind up being linear combinations of these two vectors. It is not generally the case that either of the two vectors is on the axes of the ellipse.

(2) The direction of movement along the orbits is not hard to determine. For example, here, note that the period of all of these solutions is $T = \pi$. Then one can choose a particular solution like $c_1 = 0$ and $c_2 = 1$ and watch. Then at $t = 0$, we are at the point $(0, 1)$ in the plane. Then, at $t = \frac{1}{8}\pi$, when the two trig functions are the same value, we are at the point $\frac{\sqrt{2}}{2}(1, 0)$. Thus the direction is clockwise. Of course, one can also do nothing more than evaluate the slope field at $\begin{bmatrix} 0 \\ 1 \end{bmatrix}$, which is just $\begin{bmatrix} 2 & 2 \\ -4 & -2 \end{bmatrix} \begin{bmatrix} 0 \\ 1 \end{bmatrix} = \begin{bmatrix} 2 \\ -2 \end{bmatrix}$. On the right side of Figure 59, this element of the vector field in marked in green.

This last case exhausts the cases with a positive determinant of the coefficient matrix A. Of course, there are linear systems with a zero determinant, and these are treated in exactly the same fashion. If an eigenvalue is zero, but the other eigenvalue is non-zero, then the eigenvalues are distinct and again the general solution is a linear combination of the two solutions $\mathbf{x}_1(t) = \mathbf{v}_1 e^{r_1 t}$ and, in this case,

$$\mathbf{x}_2(t) = \mathbf{v}_2 e^{r_2 t} = \mathbf{v}_2 e^{0t} = \mathbf{v}_2.$$

This produces a phase portrait with an entire linear subspace of equilibria along the line given by $\text{span}(\mathbf{v}_2)$, and renders solutions that start off of this line restricted to lines in the plane parallel to $\text{span}(\mathbf{v}_1)$.

Example 4.18 (A zero eigenvalue.) Let $A = \begin{bmatrix} -1 & 2 \\ -2 & 4 \end{bmatrix}$. Then the ODE system has general solution given by $\mathbf{x}(t) = c_1 \begin{bmatrix} 1 \\ 2 \end{bmatrix} e^{3t} + c_2 \begin{bmatrix} 2 \\ 1 \end{bmatrix}$. The center space here is 1-dimensional and $\mathbb{E}^c = \text{span} \begin{bmatrix} 2 \\ 1 \end{bmatrix}$, marked by the dotted line in the phase portrait in Figure 60. All solutions off of this line are unbounded. The origin is unstable, even though it is not an isolated equilibrium. Note that if the arrows were reversed, the origin would be classified as stable, but not asymptotically so, since in any neighborhood of the origin there are solutions that neither scale nor converge to the origin. Think about this.

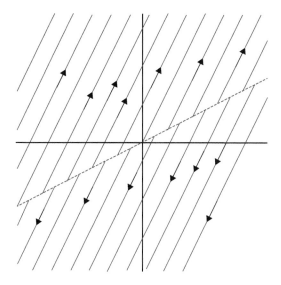

Figure 60 *A* has a 0-eigenvalue.

And lastly, the case where the determinant of a matrix A is less than 0 is all encompass-ing. In this case, both eigenvalues are always non-zero, real, and distinct. Each will then correspond to a unique 1-dimensional eigenspace, and the general solution will always be a simple linear combination of the two solutions $x_1(t) = v_1 e^{r_1 t}$ and $x_2(t) = v_2 e^{r_2 t}$. How-ever, since one eigenvalue is negative, the exponential term decays to 0 as t grows without bound, while the other grows without bound. We are left with positive-dimensional stable and unstable invariant spaces $\mathbb{E}^s$ and $\mathbb{E}^u$. The result is that, for solutions off of these invariant subspaces, motion is asymptotic to both, one in forward time and the other in reverse time. We call the origin in such a system a *saddle point*, or simply a *saddle*, noting that a saddle is always unstable.

Example 4.19 (A saddle.) Let $A = \frac{1}{3}\begin{bmatrix} 1 & -2 \\ -4 & -1 \end{bmatrix}$. Then the characteristic equation is $r^2 - 1 = 0$, so that the two eigenvalues are $r_1 = 1$ and $r_2 = -1$. Calculating a choice of eigenvectors as $v_1 = \begin{bmatrix} 1 \\ -1 \end{bmatrix}$ and $v_2 = \begin{bmatrix} 1 \\ 2 \end{bmatrix}$, we have the phase portrait in Figure 61, where the invariant eigenspaces are in red.

The relationship between the type and stability of the origin as an equilibrium of a lin-ear 2-dimensional system and the properties of the coefficient matrix A are evident: The trace and determinant of A completely determine the type and hence the stability of the equilibrium at the origin. And since slightly changing the entries of a matrix (perturbing the matrix) only slightly changes these two quantities (they are polynomials in the entries of A) certain types of equilibria should appear stable under these perturbations. In fact, we

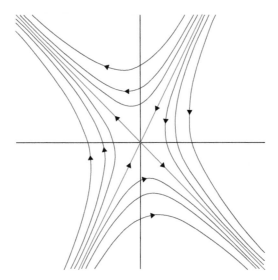

Figure 61 An unstable saddle at the origin.

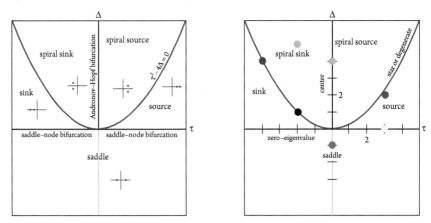

Figure 62 The trace–determinant plane.

can collect up all of the values of these two quantities into a space of possibilities. Denote $\tau = \text{tr}(A)$ and $\Delta = \det(A)$ and consider these two quantities as the coordinates of $\mathbb{R}^2$, the *trace–determinant plane*, shown in Figure 62. Then all of the above analysis can be conveniently displayed in this plane. Note here that we have marked some of the previous examples of phase portraits in the diagram:

- Saddle: Example 4.19 has a coefficient matrix A with $\text{tr}\,A = 0$ and $\det A = -1$. This is marked in red.
- Source: Example 4.13 is marked in blue.
- Spiral: Example 4.16 is marked in green.

- Degenerate sink: Example 4.14, with a defective matrix A is marked in black.
- Star Sink: Example 4.15, with a diagonal matrix A is marked in purple.
- Center: Example 4.17, with the eigenvalues of A purely imaginary, is marked in orange.
- Zero Eigenvalue: Example 4.18, with a defective matrix A, is marked in yellow.

4.2 Bifurcations in Linear Planar Systems

Take another look at the trace–determinant plane in Figure 63. Imagine a family of matrices A parameterized by a single parameter. This family, in many cases, will trace out a curve in this plane. Should the curve meet or cross any of the boundaries in this plane, it will then include systems whose phase portraits include more than one type of equilibrium at the origin, and the stability of the equilibria at the origin may be of more than one type across the family. In other words, the parameter values may include bifurcation values. Some of these may be obvious enough to have names and others not. And, depending on the parameterized curve, there may be many bifurcations or one or none. Herein, we highlight just a couple of examples of the types of bifurcations one may see, given a particular family of systems.

4.2.1 Linearized Poincaré–Andronov–Hopf Bifurcation

Let

(4.2.1) $$\dot{\mathbf{x}} = A_\alpha \mathbf{x}, \quad \text{where} \quad A = \begin{bmatrix} 0 & 1 \\ -1 & \alpha \end{bmatrix},$$

be a family of ODE systems paramaterized by $\alpha \in [-3,3]$. Here, $\operatorname{tr} A_\alpha = \alpha$ and $\det A_\alpha = 1$. This family of matrices traces out the red horizontal line in the trace–determinant plane in Figure 63. One can easily see the region crossings in this diagram,

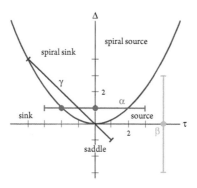

Figure 63 Two families of systems with bifurcations.

and these crossings correspond to the values $\alpha = -2, 0, 2$. As the characteristic equation for the family is $r^2 - \alpha r + 1 = 0$, we have solutions

(4.2.2)
$$r = \frac{\alpha}{2} \pm \frac{\sqrt{\alpha^2 - 4}}{2},$$

with eigenvectors satisfying

$$\begin{bmatrix} 0 & 1 \\ -1 & \alpha \end{bmatrix} \begin{bmatrix} x \\ y \end{bmatrix} = r \begin{bmatrix} x \\ y \end{bmatrix}, \quad \text{or} \quad x = (\alpha - r)y.$$

Thus, for r_i, $i = 1, 2$, we have the eigenlines

$$\ell_i = \text{span} \left\{ \begin{bmatrix} \alpha - r_i \\ 1 \end{bmatrix} \right\}.$$

With this structure, it is easy to see what the phase portrait looks like and what is the type and stability of the equilibrium at the origin for certain ranges of α:

- $\alpha \in [-3, -2)$: Here $r_2 < r_1 < 0$ and the origin is a sink.
- $\alpha \in (-2, 0)$: Here $r_2 = \overline{r_1} \in \mathbb{C}$, where the bar signifies the complex conjugate, and $\text{Re}(r_1) < 0$. The origin here is a spiral sink.
- $\alpha \in (0, 2)$: Here $r_2 = \overline{r_1} \in \mathbb{C}$, and $\text{Re}(r_1) > 0$. The origin here is a spiral source.
- $\alpha \in (2, 3]$: Here $0 < r_2 < r_1$ and the origin is a source.

But it is in the transition values of α, that $\alpha = -2, 0, 2$, that are interesting. In Figure 63, the red line corresponds to this α-parameterized family of ODE systems.

For $\alpha = -2$, $r_1 = \frac{1}{2}\alpha + \frac{1}{2}\sqrt{\alpha^2 - 4} = -1 = r_2$. The eigenspaces, again, are given by

$$\ell_1 = \text{span} \left\{ \begin{bmatrix} \alpha - r_1 \\ 1 \end{bmatrix} \right\} = \text{span} \left\{ \begin{bmatrix} r_2 \\ 1 \end{bmatrix} \right\},$$

$$\ell_2 = \text{span} \left\{ \begin{bmatrix} \alpha - r_2 \\ 1 \end{bmatrix} \right\} = \text{span} \left\{ \begin{bmatrix} r_1 \\ 1 \end{bmatrix} \right\},$$

since it is always true that $\text{tr} A = r_1 + r_2 = \alpha$. These eigenspaces were distinct for $\alpha \in [-3, 2)$, but are the same at $\alpha = -2$. Hence these eigenlines have coalesced into the single line given by the equation $y = -x$ in $\mathbb{R}^2$. So, in the case of one eigenvalue and only one linearly independent eigenvector, the matrix A_{-2} is defective, and we have a degenerate sink at $\alpha = -2$.

Indeed, for $A_{-2} = \begin{bmatrix} 0 & 1 \\ -1 & -2 \end{bmatrix}$, the eigenvector equation is

$$\begin{bmatrix} 0 & 1 \\ -1 & -2 \end{bmatrix} \begin{bmatrix} x \\ y \end{bmatrix} = - \begin{bmatrix} x \\ y \end{bmatrix},$$

with $y = -x$ the only linearly independent solution. To see this phase portrait, simply refer back to Example 4.14, since this was the same matrix. The phase portrait is in Figure 58 on the right.

Exercise 136 Solve the system in Equation 4.2.1 at the parameter values $\alpha = -3$, $-2, -1$ and sketch phase portraits, verifying that your phase portrait for the system at $\alpha = -2$ is the one in Figure 58.

Exercise 137 Do the same analysis at $\alpha = 2$ and sketch the phase portrait of the resulting degenerate node there.

At $\alpha \in (-2,2)$, the eigenvalues are complex conjugates, since the discriminant in Equation 4.2.2 is strictly less than 0. We have

- for $\alpha < 0, \mathrm{Re}(r_1) < 0$,
- for $\alpha = 0, \mathrm{Re}(r_1) = 0$, and
- for $\alpha > 0, \mathrm{Re}(r_1) > 0$.

Hence the phase portrait transitions from a spiral sink through a center to a spiral source as α passes through 0. This is the linearized version of a *linear Poincaré–Andronov–Hopf bifurcation*, in that this would be the local linear analysis of the equilibrium as the parameter α passes through its bifurcation value. We will leave this until we treat this type of bifurcation in Section 4.6.5.

Exercise 138 Add to your series of phase portraits from Exercises 136 and 137 for the system in Equation 4.2.1 by sketching the phase portraits for $\alpha = 0$ and 1.

Exercise 139 Show that the 1-parameter family of linear systems given by $\dot{\mathbf{x}} = A_a \mathbf{x}$, where $A_a = \begin{bmatrix} a & -1 \\ 1 & a \end{bmatrix}$, has a linear Poincaré–Andronov–Hopf bifurcation at $a = 0$. Document the type and stability of the equilibrium at the origin as a function of a, and draw representative phase portraits of the system both at $a = 0$ and on either side of $a = 0$. Then, in the trace–determinant plane, draw the parameterized line corresponding to this family for $a \in [-2,2]$. Note that we will return to this system in Section 4.6.5.

Exercise 140 For the 1-parameter family of linear systems given by $\dot{\mathbf{x}} = B_a \mathbf{x}$, where $B_a = \begin{bmatrix} a-1 & 1 \\ -a & -1 \end{bmatrix}$, do the same analysis as in Exercise 139. (Note that this is a parameterized form of a special case of the Jacobian of the *brusselator*, a model for a chemical oscillator given by the 2-parameter ODE system in Equation 4.6.7 in Section 4.6.5.)

4.2.2 Saddle-Node Bifurcation

Now consider the green line in Figure 63. This curve in the trace–determinant plane corresponds to the family of systems $\dot{\mathbf{x}} = A_\beta \mathbf{x}$, with matrices $A_\beta = \begin{bmatrix} 4 & -1 \\ \beta & 0 \end{bmatrix}$ where, parameterized by $\beta \in [-1,1]$, we have $\operatorname{tr} A_\beta = 4$ and $\det A_\beta = \beta$. The characteristic equation is $r^2 - 4r + \beta = 0$, and the solutions are

$$r_1 = 2 + \sqrt{4 - \beta}, \quad \text{and} \quad r_2 = 2 - \sqrt{4 - \beta}.$$

Based on this data alone, one can classify the origin as:

- a saddle for $-1 \leq \beta < 0$, and
- a source for $0 < \beta \leq 1$.

Hence we have a bifurcation at $\beta = 0$, with the equilibrium at the origin transitioning from a saddle point to a node. This is an example of the aptly named *saddle–node bifurcation*. What, in detail, is happening here?

- For either $r_1, i = 1, 2$, one can calculate the eigenspaces ℓ_{r_1} and ℓ_{r_2} (the eigenvalues are distinct for every $\beta \in [-1,1]$, so there will always be two linearly independent eigenvectors) to get

$$\ell_{r_i} = \operatorname{span} \left\{ \begin{bmatrix} 1 \\ r_i - 4 \end{bmatrix} \right\}.$$

  However, also note that $r_1 - 4 = -r_2$ and $r_2 - 4 = -r_1$, which simplifies the calculations somewhat.
- Here $r_1 = 2 + \sqrt{4 - \beta} > 0, \forall \beta \in [-1,1]$. Hence $\ell_{r_1} \subset \mathbb{E}^u$.
- However, $r_2 < 0$ only when $\beta \in [-1, 0)$ and $r_2 > 0$ when $\beta \in (0, 1]$.
- At $\beta = 0$, $r_1 = 4$ and $\ell_{r_1} = \operatorname{span} \left\{ \begin{bmatrix} 1 \\ 0 \end{bmatrix} \right\}$ and $\ell_{r_2} = \operatorname{span} \left\{ \begin{bmatrix} 1 \\ -4 \end{bmatrix} \right\}$, and we get a special system where there is a line of equilibria passing through the origin along ℓ_{r_2}.
- For $-1 \leq \beta < 0$, $\mathbb{E}^s = \ell_{r_2}$ and $\mathbb{E}^u = \ell_{r_1}$. For $0 < \beta \leq 1$, $\mathbb{E}^u = \mathbb{R}^2$. And for $\beta = 0$, $\mathbb{E}^u = \ell_{r_1}$, and $\mathbb{E}^c = \ell_{r_2}$.
- The eigenspaces, spans of vectors whose entries are dependent on the values of the eigenvalues, vary in direction as β varies, but throughout the interval $[-1,1]$, the origin remains unstable. This is true even at $\beta = 0$, where, arbitrarily close to the origin, there are other equilibria.
- In Figure 64, where only the eigenspaces are graphed, one can see the saddle configuration when $\beta = -1$, the zero-eigenvalue situation when $\beta = 0$, and the source when $\beta = 1$. The eigenline directions vary linearly in direction, and one can see how the movement toward the origin along the stable eigenline slows down as β approaches 0, stops at $\beta = 0$, and reverses when $\beta > 0$.

With regard to this last bullet point, we will see this type of bifurcation later in Section 4.7.2 in a nonlinear first-order system of ODEs commonly called the Competing Species Model. Figure 84 details geometrically the phase portraits as the parameter passes through the bifurcation value of $\delta = \frac{1}{2}$.

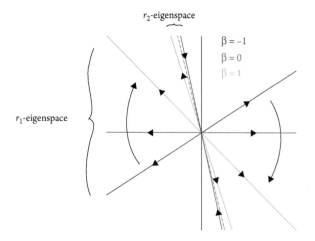

Figure 64 The eigenspaces through a saddle–node bifurcation.

In terms of β, one can write the general solution as

$$\mathbf{x}(t) = c_1 \begin{bmatrix} 1 \\ 2 - \sqrt{4-\beta} \end{bmatrix} e^{(2+\sqrt{4-\beta})t} + c_2 \begin{bmatrix} 1 \\ 2 + \sqrt{4-\beta} \end{bmatrix} e^{(2-\sqrt{4-\beta})t}.$$

Exercise 141 The purple line in Figure 63 runs from $(-4, 4)$ to $(1, -1)$. Find a family of matrices A_γ parameterized by γ along this line. Recognize that both the trace and the determinant are changing along this line. Identify any and all bifurcation points and detail what is happening in the phase portraits corresponding to the systems $\dot{\mathbf{x}} = A_\gamma \mathbf{x}$ as γ passes through these points.

4.3 Linear Planar Maps

Recall for the moment the linear map of $\mathbb{R}$ defined by $f(x) = \lambda x$ (this can also be written $x \xrightarrow{f} \lambda x$). One can classify the dynamical behavior of this map by the magnitude of λ, neglecting the reflection of the real line given by $f(x) = -x$. This is because it is the magnitude that determines the dynamical structure of the linear map: We classify by whether $|\lambda| < 1$, $|\lambda| = 1$, or $|\lambda| > 1$, for which, respectively, the origin is a sink and asymptotically stable, all points are either fixed ($\lambda = 1$) or at most 2-periodic ($\lambda = -1$) and all are stable but not asymptotically stable, or the origin is a source. However, the actual orbit structure is also affected by the sign of λ. Indeed, when $\lambda < 0$, the sequence becomes an alternating sequence, and successive terms of the orbit will flip in sign (see Figure 65).

To see that, for this map, orbits will converge or diverge regardless of the sign of λ, create a new dynamical system using the square of the map, $f^2(x) = \lambda^2 x$. Then the new factor $\lambda^2 > 0$ always. The orbits of f^2 will stay on the side of the origin they started on.

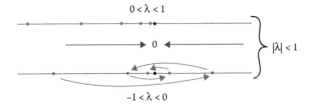

Figure 65 1-dimensional linear map $f(x) = \lambda x$, for $0 < |\lambda| < 1$.

For f, however, the orbits will flip back and forth from one side of the origin to the other (see Figure 65).

Moving toward linear maps of the plane, we can classify dynamical behavior in a similar fashion into a few distinct types. Indeed, let $f : \mathbb{R}^2 \to \mathbb{R}^2$ be linear. Then, for $\mathbf{v} = \begin{bmatrix} x \\ y \end{bmatrix}$, we have $\mathbf{f}(\mathbf{v}) = A\mathbf{v}$, or

$$\begin{bmatrix} x \\ y \end{bmatrix} \overset{f}{\longmapsto} A \begin{bmatrix} x \\ y \end{bmatrix}, \quad \text{where } A = \begin{bmatrix} a & b \\ c & d \end{bmatrix}.$$

- Here, $\mathbf{v}$ is an eigenvector of A, with eigenvalue λ if $\mathbf{v}$ satisfies the vector equation $A\mathbf{v} = \lambda \mathbf{v}$, equivalently $(A - \lambda I_2)\mathbf{v} = \mathbf{0}$, where I_2 is the 2-dimensional identity matrix.

- Recall the characteristic equation of A, $\det(A - \lambda I) = 0$, written as

$$\lambda^2 - (\text{tr}\,A)\lambda + \det A = 0.$$

Solutions to the characteristic equation are the eigenvalues of A.

A good question to ask is: What information is conveyed by $\mathbf{v}$ and λ about the discrete dynamical system formed by iterating $\mathbf{f}$ on $\mathbb{R}^2$?

There is an easy classification of matrix types for A, determined by the eigenvalue data of A:

I. Two real distinct eigenvalues $\lambda \neq \mu$.

II. One real eigenvalue λ. In this case, there are two possibilities:

- $A = \lambda I$. Here A is called a *homothety* or a *scaling*.

- A is conjugate to $\begin{bmatrix} \lambda & 1 \\ 0 & \lambda \end{bmatrix}$. When $\lambda = 1$, this map is often called a *shear*.

III. Two complex conjugate eigenvalues $\lambda = a + ib = \rho e^{i\theta}$ and $\mu = a - ib = \rho e^{-i\theta}$, where $\rho^2 = a^2 + b^2$, and $\tan\theta = b/a$. (What happens when λ is purely imaginary?) Here, the effect of A is by rotation by θ and a scaling by ρ. When $\rho = 1$, the effect is a pure rotation (see below).

Note here that in the case of a shear above, the fact that the upper right-hand entry is non-zero is vital, while the actual numerical value of this entry is less important:

Exercise 142 Show if a 2×2 matrix A is conjugate to $\begin{bmatrix} \lambda & 1 \\ 0 & \lambda \end{bmatrix}$, it is conjugate to $\begin{bmatrix} \lambda & s \\ 0 & \lambda \end{bmatrix}$ for any non-zero $s \in \mathbb{R}$.

Exercise 143 Show that any matrix that is conjugate to $\begin{bmatrix} 1 & s \\ 0 & 1 \end{bmatrix}$ is not conjugate to $\begin{bmatrix} -1 & s \\ 0 & -1 \end{bmatrix}$.

Remark 4.20 *The set of all 2×2 matrices that have non-zero determinant form a group via matrix multiplication (composition when the matrices are viewed as linear transformations of $\mathbb{R}^2$). The general linear group is*

$$GL(2,\mathbb{R}) = \left\{ \begin{bmatrix} a & b \\ c & d \end{bmatrix} \,\middle|\, a,b,c,d \in \mathbb{R},\ ad - bc \neq 0 \right\},$$

it has identity element I_2, and for every element A, the inverse of A, denoted by A^{-1}, satisfies $AA^{-1} = A^{-1}A = I_2$. Of particular interest to dynamicists is a subgroup of this group: the set of all 2×2 matrices of determinant 1, or the special linear group

$$SL(2,\mathbb{R}) = \left\{ \begin{bmatrix} a & b \\ c & d \end{bmatrix} \,\middle|\, a,b,c,d \in \mathbb{R},\ ad - bc = 1 \right\}.$$

These matrices correspond to linear transformations that preserve both area and orientation in the plane. A useful classification of special linear matrices centers around the trace *of the matrix (the sum of its main diagonal elements):*

I. **Hyperbolic:** $|\text{tr}(A)| > 2$. *A has two real distinct eigenvalues λ and μ where $\lambda = 1/\mu$ (necessarily $|\lambda| > 1 > |\mu|$). Thus A is diagonalizable over the real numbers and*
$$A \cong \begin{bmatrix} \lambda & 0 \\ 0 & \mu \end{bmatrix}.$$

II. **Parabolic:** $|\text{tr}(A)| = 2$. *A has one real eigenvalue $|\lambda| = 1$ (of multiplicity 2) and may or may not be diagonalizable. In the case where A is not diagonalizeable, A is conjugate to a shear, one of the four* $\begin{bmatrix} \pm 1 & \pm s \\ 0 & \pm 1 \end{bmatrix}$ *for some non-zero $s \in \mathbb{R}$.*

III. **Elliptic:** $|\text{tr}(A)| \leq 2$. *A has two, unit-modulus complex conjugate eigenvalues $\lambda = e^{i\theta}$, and $\mu = e^{-i\theta}$ for $\theta \in [0, 2\pi)$. With the exception of the special values $\theta = 0$ or π, A is not diagonalizable (over the real numbers) and A is only conjugate to* $\begin{bmatrix} a & b \\ -b & a \end{bmatrix}$, *for $a, b \in \mathbb{R}$, $a^2 + b^2 = 1$. The special cases of $\theta = 0$ or π correspond to $a = \pm 1$ and $b = 0$, respectively $A = \pm I_2$, and are generally considered parabolic matrices.*

Exercise 144 Exercise 142 showed that within $GL(2, \mathbb{R})$, any two shears are conjugate. Now show that, if we limit ourselves to $SL(2, \mathbb{R})$, then for A conjugate to $\begin{bmatrix} 1 & 1 \\ 0 & 1 \end{bmatrix}$, it is also conjugate to $\begin{bmatrix} 1 & s \\ 0 & 1 \end{bmatrix}$ for any non-zero, positive $s \in \mathbb{R}$, but not conjugate to such a matrix with $s \leq 0$.

Now realize the following:

- Every non-degenerate 2×2 matrix A is a scaled version of a matrix of determinant (± 1), since $(\sqrt{\det A})^{-1} A = C$, where $|\det C| = 1$. Here, then, we can view a transformation of the plane corresponding to A as a composition of C with a pure scaling matrix $S = \begin{bmatrix} s & 0 \\ 0 & s \end{bmatrix}$ so that $A = \begin{bmatrix} \sqrt{|\det A|} & 0 \\ 0 & \sqrt{|\det A|} \end{bmatrix} C$.

- Every determinant-(-1) matrix C is a composition of a determinant-1 matrix B with the reflection $R = \begin{bmatrix} 1 & 0 \\ 0 & -1 \end{bmatrix}$, so that $B = RC$ and $\det B = 1$.

Hence many of the features of linear maps on $\mathbb{R}^2$ are inherent already in $SL(2, \mathbb{R})$. We will return to these types of linear transformations later.

Some additional notes:

- Both $I_2 = \begin{bmatrix} 1 & 0 \\ 0 & 1 \end{bmatrix}$ and its negative $\begin{bmatrix} -1 & 0 \\ 0 & -1 \end{bmatrix}$ show up in this classification as parabolic matrices. In a sense, they are elliptic also, the two eigenvalues being $1 = e^{i \cdot 0} = e^{-i \cdot 0}$, or $-1 = e^{i\pi} = e^{-i\pi}$. In either case, $\pm I_2$ are considered quite special and are usually treated separately.

- Compose a hyperbolic matrix with the reflection R from above and you still get something that looks hyperbolic, but the two eigenvalues only satisfy $\lambda = -1/\mu$. Now compose a parabolic matrix with the reflection R and you again get one that may be parabolic, but may also be hyperbolic, but again with eigenvalues reciprocals of different sign. But when you compose an elliptic matrix with R, do you get another elliptic matrix? Think about this.

Exercise 145 Calculate the angle of rotation for an elliptic matrix. Then construct an elliptic matrix with the inverse rotation and show that a rotation matrix and its inverse are not conjugate within $SL(2, \mathbb{R})$.

Exercise 146 Show that the trace of a 2×2 matrix $A \in SL(2, \mathbb{R})$ completely determines the classification into hyperbolic, parabolic, and elliptic sets.

Exercise 147 Find the elliptic matrix that rotates the plane through an angle of $\pi/6$ radians.

Geometrically, choose a determinant-1 representative from each case above. Then it will be easier to see the effect on points in the plane. Figure 66 is useful here, noting the following: in the left-hand case, $A = \begin{bmatrix} \lambda & 0 \\ 0 & \mu \end{bmatrix}$, for $\lambda > 1 > \mu > 0$; in the middle case,

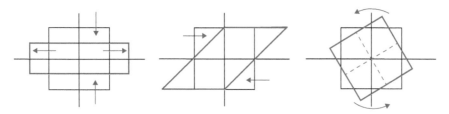

Figure 66 $f(x) = Ax$ for A hyperbolic, parabolic, and elliptic, respectively.

$$A = \begin{bmatrix} 1 & 1 \\ 0 & 1 \end{bmatrix}; \text{ and in the right-hand case, } A = \begin{bmatrix} a & b \\ -b & a \end{bmatrix}, \text{ for eigenvalues } \lambda = a \pm ib,$$
and $a^2 + b^2 = 1$.

We will presently embark on a detailed study of linear maps of the plane corresponding to these basic types. However, let's begin by returning to a previous notion to dispel an incorrect but seemingly intuitive belief about linear planar maps.

Recall from linear algebra (or Definition 2.39) that the spectral radius of a matrix $\rho(A)$ is related to the matrix norm $||A|| = \max_{||\mathbf{v}||=1} ||A\mathbf{v}||$, but they are not equal in general (they are equal when the matrix A is symmetric, though). The fact that the matrix norm is not necessarily equal to the modulus of its largest eigenvector is a clue to the following:

Proposition 4.21 *A linear transformation A of $\mathbb{R}^n$ is eventually contracting if $\rho_A < 1$.*

Exercise 148 Construct an explicit example of a linear map on $\mathbb{R}^2$ that is eventually contracting but NOT a contraction. (Hint: Pay attention to the types in the classification of matrices above.)

Exercise 149 Show that no determinant-1 linear map on $\mathbb{R}^2$ can be eventually contracting. (Note: This has enormous implications in the mathematical models of physics and engineering, since it restricts the stability classification of both fixed points of maps and a equilibrium and periodic solutions of ODE systems.)

Some things to consider:

- Any positive-determinant linear map on $\mathbb{R}^2$ can be written as a linear combination of the three types in Figure 66 (sum of scalar multiples of the three, that is). And any negative-determinant linear map will have a square that is positive and, as with linear maps of $\mathbb{R}$, the dynamics (up to flipping across some line) will be similar in nature. Hence a detailed study of these three determinant-1 types and their scaled versions is necessary to explore the dynamical structure of linear maps of the plane.

- Diagonalizing a matrix (conjugating it, as best as one can, to one where the eigenvalues are prominent) is really just a linear coordinate change. Hence this process can be viewed simply as a change of the metric on $\mathbb{R}^2$. However, the new metric is always uniformly equivalent to the old one, as in Exercise 150 below. The process of diagonalization **does not** change the dynamical structure of the system! It is a topological conjugacy.

Exercise 150 Show that, for a non-degenerate 2×2 matrix A, the metric $d_A(\mathbf{x}, \mathbf{y}) = d(A\mathbf{x}, A\mathbf{y})$ is uniformly equivalent to the Euclidean metric d on $\mathbb{R}^2$.

Hence like maps on closed intervals, where we only needed to study maps on the unit interval, we can always limit our analysis to certain types of linear maps to capture all of the possible dynamical behavior of the entire family of linear planar maps. With this in mind, we begin our survey.

Let $\mathbf{v} \in \mathbb{R}^2$ and consider $\mathcal{O}_\mathbf{v}$ under the linear map $f(\mathbf{v}) = A\mathbf{v}$. On iteration,

$$\mathbf{v} \longmapsto A\mathbf{v} \longmapsto A(A\mathbf{v}) = A^2\mathbf{v} \longmapsto \cdots \longmapsto A^n\mathbf{v} \longmapsto \cdots .$$

Hence, the orbit of $\mathbf{v}$ will depend critically on the data associated to A.

4.3.1 Nodes: Sinks and Sources

Suppose that the two eigenvalues of A are real and distinct, so that $\lambda \neq \mu$. Then there exists a matrix B where $A \overset{\text{conj}}{\cong} B = \begin{bmatrix} \lambda & 0 \\ 0 & \mu \end{bmatrix}$. Suppose further that $0 < |\lambda| < |\mu| < 1$ (A is nondegenerate—we will treat the 0-eigenvalue case separately). Then, by the previous proposition, the origin is a sink and all orbits tend to $\mathbf{0}$. That is, $\forall \mathbf{v} \in \mathbb{R}^2$, $\mathcal{O}_\mathbf{v} \longrightarrow \mathbf{0} = \begin{bmatrix} 0 \\ 0 \end{bmatrix}$. A deeper question is, however, how the orbit evolves as it tends to the origin.

For this, let's restrict further to the case where both eigenvalues are positive, so $0 < \lambda < \mu < 1$. Then, for $\mathbf{v} = \begin{bmatrix} v_1 \\ v_2 \end{bmatrix}$, the nth term in the orbit sequence is $B^n\mathbf{v} = \begin{bmatrix} \lambda^n & 0 \\ 0 & \mu^n \end{bmatrix} \mathbf{v} = \begin{bmatrix} \lambda^n v_1 \\ \mu^n v_2 \end{bmatrix}$. A typical orbit would live entirely within one quadrant of the plane, like the black dots in Figure 67. Changing the sign of λ and/or μ without changing the magnitude would create orbits like the red dots in Figure 67. Can you work out the signs of the two eigenvalues to create the orbit in the figure? Some observations:

- Since we have diagonalized the matrix A to get B, the nth term is easy to calculate as we have uncoupled the coordinates.
- The smaller eigenvalue λ means that the first coordinate sequence in the orbit will have a faster decay toward 0 than that of the second coordinate. In the plane, this will imply a curved path toward the origin, with the orbit line (recall the definition of the orbit line?) bending toward the μ-eigendirection. (Why is this the case? Think about which coordinate is decaying faster.)
- Given the orbit lines in the figure, do you recognize the phase portrait? From the classification of 2-dimensional first-order, homogeneous linear systems (with constant coefficients), systems with this phase portrait have a node at the origin that is asymptotically stable. This is a sink.
- In Figure 67, the first few elements of $\mathcal{O}_\mathbf{v}$ are plotted in black.

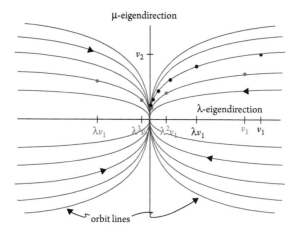

μ-eigendirection

v_2

λ-eigendirection

λv_1 $\lambda^3 v_1$ $\lambda^2 v_1$ λv_1 v_1 v_1

orbit lines

Figure 67 Phase Portrait for $f(v) = Bv$, where $0 < \lambda < \mu < 1$.

- How does the phase portrait change if one or both of the eigenvalues of B are negative? As a hint, the orbit lines do not change at all. But the orbits, themselves? In Figure 67, the first few elements of $\mathcal{O}_v$ in the case that $0 < -\lambda < \mu < 1$ are plotted in black. Do you see the effect?

We can actually calculate the equations for the orbit lines: Let $x = v_1$ and $y = v_2$. Then the orbit lines satisfy the equation

$$(4.3.1) \qquad\qquad |y| = C|x|^\alpha, \quad \text{where} \quad \alpha = \frac{\log|\mu|}{\log|\lambda|}.$$

Exercise 151 Derive the formula in Equation 4.3.1 for the orbit lines.

Exercise 152 The map above, $f(v) = Bv$ (the original one above, where $0 < \lambda < \mu < 1$, that is), is the time-1 map of a first-order, linear homogeneous 2×2 system of ODEs. Find such a system and compare the matrix in the ODE system with B.

Exercise 153 Sketch the phase portrait in the case that $|\lambda| > |\mu| > 1$.

4.3.2 Star or Proper Nodes

Suppose now that the linear map has a matrix with only one eigenvalue but two independent eigenvectors. That is, $A \overset{\text{conj}}{\cong} B = \lambda \begin{bmatrix} 1 & 0 \\ 0 & 1 \end{bmatrix} = \lambda I_2$, a homothety. Here, it should be clear that for any starting vector $v \in \mathbb{R}^2$, the nth orbit element is $B^n v = \lambda^n v$. In the case that $0 < |\lambda| < 1$, we have $\mathcal{O}_v \longrightarrow 0$, for every $v \in \mathbb{R}^2$. What does the motion look like in this case?

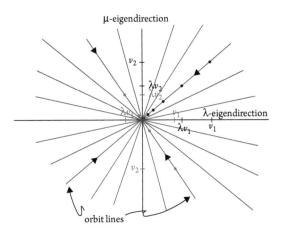

Figure 68 Star node phase portrait for $f(v) = Bv$, where $0 < \lambda < 1$.

Since we are simply re-scaling the initial vector **v**, motion will remain on the line through the origin determined by **v**. All orbits will decay exponentially along these orbit lines. The phase portrait is that of a *star node* in this case, which is a sink, or implosion, when $|\lambda| < 1$, and a source, or explosion, when $|\lambda| > 1$. Again, think about what the orbits look like when $\lambda < 0$. Would anything change if the eigenvectors were $\mathbf{v}_1 = \begin{bmatrix} 1 \\ 0 \end{bmatrix}$, and $\mathbf{v}_2 = \begin{bmatrix} 0 \\ -1 \end{bmatrix}$?

4.3.3 Degenerate or Improper Nodes

Suppose now that the linear map has a matrix with only one linearly independent eigenvector. In this case, $A \overset{\text{conj}}{\cong} B = \begin{bmatrix} \lambda & \lambda \\ 0 & \lambda \end{bmatrix}$. Recall by Exercise 142 that A is conjugate to a matrix with λ's on the main diagonal and any non-zero real number in the upper right-hand corner. We choose λ again here so that the calculations simplify a bit without losing detail. Then $B^n = \begin{bmatrix} \lambda^n & n\lambda^n \\ 0 & \lambda^n \end{bmatrix} = \lambda^n \begin{bmatrix} 1 & n \\ 0 & 1 \end{bmatrix}$, and

$$\mathbf{v} = \begin{bmatrix} v_1 \\ v_2 \end{bmatrix} \overset{B^n}{\longmapsto} \lambda^n \begin{bmatrix} 1 & n \\ 0 & 1 \end{bmatrix} \begin{bmatrix} v_1 \\ v_2 \end{bmatrix} = \lambda^n \begin{bmatrix} v_1 + nv_2 \\ v_2 \end{bmatrix}.$$

Here, the presence of the summand nv_2 has a twisting effect on $\mathcal{O}_\mathbf{v}$ even though the exponential factor λ^{n-1} still dominates the long-term orbit behavior.

One can see this twisting effect in the orbit lines. Indeed, for ease of argument, assume $\lambda > 0$ and combine the two equations defining points in the plane along the curve containing $x = v_1$ and $y = v_2$ at $n = 0$ through n: $x = \lambda^n (v_1 + nv_2)$, and $y = \lambda^n v_2$.

Solving the second for n, we get $n = (\ln y - \ln v_2)/\ln \lambda$. With substitution into the first, we have

$$x = \lambda^n (v_1 + n v_2)$$

$$= \frac{y}{v_2} \left(v_1 + \left(\frac{\ln y - \ln v_2}{\ln \lambda} \right) v_2 \right)$$

$$= y \frac{v_1}{v_2} + \frac{y \ln y}{\ln \lambda} - \frac{y \ln v_2}{\ln \lambda}$$

$$= y \left(\frac{v_1}{v_2} - \frac{\ln v_2}{\ln \lambda} + \frac{\ln y}{\ln \lambda} \right) = y \left(C + \frac{\ln y}{\ln \lambda} \right).$$

Thus, if $0 < \lambda < 1$, then $\forall v$, $\mathcal{O}_v \longrightarrow 0$, but the orbit lines twist toward, but do not rotate around, the origin!

This phase portrait exhibits a *degenerate node* or *improper node*, pictured in Figure 69. Notice a few things here: First, the equations for the invariant lines cannot express the solutions for initial points that include $v_2 = 0$. But the original equations would allow such a solution. Here $x = \lambda^n v_1$ and $y = 0$, for all $n \in \mathbb{N}$. These straight-line solutions along the λ-eigendirection are extraneous, in that they are hidden by an assumption in the method of solution (namely dividing by v_2 in solving for n). Be careful not to neglect such solutions.

Second, the one-dimensional eigenspace in this case is the only subspace comprising "linear motion" (compare the nodes above). In solving for the general solution to the corresponding linear 2×2 system of first-order ODEs, one would employ a calculation involving a *generalized eigenvector*, detailed in Section 4.1.2 to construct the solution. And lastly, while it is easy to also understand the case where $\lambda > 1$, what changes in the calculation of the orbit curves for $\lambda < 0$?

Exercise 154 Recalculate the equations for the invariant curves of a degenerate node in the case that $\lambda < 0$.

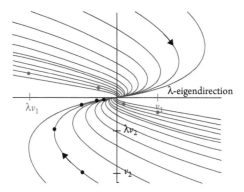

Figure 69 Degenerate node phase portrait for $f(v) = Bv$, where $0 < \lambda < 1$.

4.3.4 Spirals and Centers

Suppose that the linear map has two complex conjugate eigenvalues $\lambda = \rho e^{i\theta} = a + ib$, and $\mu = \rho e^{-i\theta} = a - ib$. Here, in general, $\rho^2 = a^2 + b^2$. Then

$$
A \cong B \cong \rho \begin{bmatrix} \frac{a}{\rho} & \frac{b}{\rho} \\ -\frac{b}{\rho} & \frac{a}{\rho} \end{bmatrix},
$$

where B is a constant ρ times a pure rotation. This scaling affects the rotational effect of the map. The orbit lines are

- spirals toward the origin if $0 < \rho < 1$;
- spirals away from the origin if $\rho > 1$; and
- concentric circles if $\rho = 1$ (the eigenvalues then are purely imaginary).

The equilibrium at the origin is then either a *spiral point* of a *focus* if $\rho \neq 1$ or a *center* if $\rho = 1$.

Exercise 155 Write down an explicit expression for the orbit lines in this case.

4.3.5 Saddle Points

Now suppose A is a 2×2 matrix with eigenvalues $0 < |\mu| < 1 < |\lambda|$. Then $A \overset{\text{conj}}{\cong} B = \begin{bmatrix} \lambda & 0 \\ 0 & \mu \end{bmatrix}$ like the other examples in Case I, but the orbit lines are different. In fact, writing out the nth term in $\mathcal{O}_{\mathbf{v}}$ for a choice of $\mathbf{v} \in \mathbb{R}^2$, we see that there are four types:

(1) $\mathcal{O}_{\mathbf{v}}^+ \longrightarrow \begin{bmatrix} 0 \\ 0 \end{bmatrix}$ and $\mathcal{O}_{\mathbf{v}}^- \longrightarrow \infty$;

(2) $\mathcal{O}_{\mathbf{v}}^+ \longrightarrow \infty$ and $\mathcal{O}_{\mathbf{v}}^- \longrightarrow \infty$;

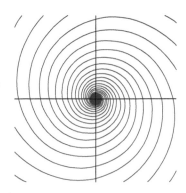

Figure 70 Spiral sink phase portrait for $f(v) = Bv$, where $\lambda = a + ib$, and $|\lambda| < 1$.

(3) $\mathcal{O}_v^+ \longrightarrow \infty$ and $\mathcal{O}_v^- \longrightarrow \begin{bmatrix} 0 \\ 0 \end{bmatrix}$; and

(4) $\mathcal{O}_v^+ \longrightarrow \begin{bmatrix} 0 \\ 0 \end{bmatrix}$ and $\mathcal{O}_v^- \longrightarrow \begin{bmatrix} 0 \\ 0 \end{bmatrix}$.

With B as our matrix, the eigenvectors $\mathbf{v}_\lambda = \begin{bmatrix} 1 \\ 0 \end{bmatrix}$ and $\mathbf{v}_\mu = \begin{bmatrix} 0 \\ 1 \end{bmatrix}$ lie on the coordinate axes, and, for a choice of $\mathbf{v} \in \mathbb{R}^2$, the nth term is again $B^n \mathbf{v} = \begin{bmatrix} \lambda^n v_1 \\ \mu^n v_2 \end{bmatrix}$. Can you envision the orbit lines and motion along them?

Do you recognize the phase portrait in Figure 71? Can you classify the type and stability of the origin? This is, again, a *saddle point* and the orbit lines in this case are simple hyperbolas, at least when $\mu = 1/\lambda$.

Exercise 156 Show the orbit lines for $f(\mathbf{v}) = B\mathbf{v}$, for $B = \begin{bmatrix} \lambda & 0 \\ 0 & \mu \end{bmatrix}$, and $0 < |\mu| < 1 < |\lambda|$ are $y = Cx^\gamma$, where $\gamma = \ln|\mu|/\ln|\lambda|$. Note that when $|\lambda| = 1/|\mu|$, $\gamma = -1$.

For a more concrete example, consider now the hyperbolic matrix $A = \begin{bmatrix} 0 & 1 \\ 1 & 1 \end{bmatrix}$. Here the characteristic equation is $r^2 - r - 1 = 0$, which is solved by $r = \frac{1 \pm \sqrt{5}}{2}$, giving us the eigenvalues

$$\lambda = \frac{1 + \sqrt{5}}{2} > 1, \quad \text{and} \quad \mu = \frac{1 - \sqrt{5}}{2} \in (-1, 0).$$

The eigenspace of λ is the line $y = \frac{1+\sqrt{5}}{2}x = \lambda x$, and, for an eigenvector, we choose $\mathbf{v}_\lambda = \begin{bmatrix} 1 \\ \lambda \end{bmatrix}$.

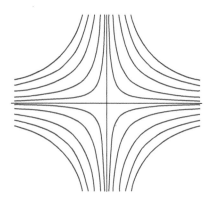

Figure 71 A saddle point phase portrait for $f(v) = Bv$, where $0 < |\mu| < 1 < |\lambda|$.

Now let $f : \mathbb{R}^2 \to \mathbb{R}^2$ be the linear map $f(\mathbf{v}) = A\mathbf{v} = \begin{bmatrix} 0 & 1 \\ 1 & 1 \end{bmatrix} \mathbf{v}$. Then, for $\mathbf{v} = \begin{bmatrix} 1 \\ 1 \end{bmatrix}$, we get

$$\mathcal{O} = \left\{ \begin{bmatrix} 1 \\ 1 \end{bmatrix}, \begin{bmatrix} 1 \\ 2 \end{bmatrix}, \begin{bmatrix} 2 \\ 3 \end{bmatrix}, \begin{bmatrix} 3 \\ 5 \end{bmatrix}, \begin{bmatrix} 5 \\ 8 \end{bmatrix}, \begin{bmatrix} 8 \\ 13 \end{bmatrix}, \ldots \right\}.$$

Do you see the patterns? Call $\mathbf{v}_n = \begin{bmatrix} x_n \\ y_n \end{bmatrix} = \begin{bmatrix} x_n \\ x_{n+1} \end{bmatrix}$ and the sequences $\{x_n\}_{n\in\mathbb{N}}$ and $\{y_n\}_{n\in\mathbb{N}}$ are Fibonacci with $y_n = x_{n+1}$. Notice that the sequence of ratios

$$\left\{ \frac{y_n}{x_n} \right\}_{n\in\mathbb{N}} = \left\{ \frac{x_{n+1}}{x_n} \right\}_{n\in\mathbb{N}}$$

has a limit, and $\lim_{n\to\infty}(y_n/x_n) = \frac{1+\sqrt{5}}{2} = \lambda$.

Recall how to find this limit: Use the second-order recursion inherent in the Fibonacci sequence, namely $a_{n+1} = a_n + a_{n-1}$, and the ratio to calculate a first-order recursion. This first-order recursion will correspond to a map, which one can study dynamically. Indeed, let $r_{n+1} = x_{n+1}/x_n$, Then

$$r_{n+1} = \frac{x_{n+1}}{x_n} = \frac{x_n + x_{n-1}}{x_n} = 1 + \frac{1}{x_n/x_{n-1}} = 1 + \frac{1}{r_n}.$$

So $r_{n+1} = f(r_n)$, where $f(x) = 1 + x^{-1}$. The only non-negative fixed point of this map is the sole positive solution to $x = f(x) = 1 + x^{-1}$, or $x^2 - x - 1 = 0$ (do you recognize the characteristic equation of the original A here?), which is $x = \frac{1+\sqrt{5}}{2}$. Note that really there are two solutions, and the other one is indeed μ. However, since we are talking about populations, the negative root doesn't really apply to the problem.

Exercise 157 Recall the Opossums Problem in Exercise 3. There, the second-order recursion you were encouraged to construct is $a_{n+1} = 2a_n + 2a_{n-1}$. Show that the sequence of ratios of successive terms $\{a_{n+1}/a_n\}_{n\in\mathbb{N}}$ has the limit $1 + \sqrt{3}$.

Here are two rhetorical questions:

(1) What is the meaning of these limits?

(2) How does the hyperbolic matrix in the above Fibonacci sequence example help in determining the limit?

To answer these, let's start with the Fibonacci sequence

$$\{b_n\} = \{1, 1, 2, 3, 5, 8, 13, 21, \ldots\}.$$

As before, we see

$$\mathbf{v}_{n+1} = \begin{bmatrix} b_{n+1} \\ b_{n+2} \end{bmatrix} = \begin{bmatrix} b_{n+1} \\ b_{n+1} + b_n \end{bmatrix} = \begin{bmatrix} 0 & 1 \\ 1 & 1 \end{bmatrix} \begin{bmatrix} b_n \\ b_{n+1} \end{bmatrix} = A \begin{bmatrix} b_n \\ b_{n+1} \end{bmatrix} = A\mathbf{v}_n,$$

where $A = \begin{bmatrix} 0 & 1 \\ 1 & 1 \end{bmatrix}$. This is precisely the matrix that (1) moves the second entry into the first entry slot, and (2) creates a new entry two by summing the two entries.

Here, we have associated with the second-order recursion $b_{n+2} = b_{n+1} + b_n$ the matrix $A = \begin{bmatrix} 0 & 1 \\ 1 & 1 \end{bmatrix}$ and the first-order *vector* recursion $\mathbf{v}_{n+1} = A\mathbf{v}_n$.

Remark 4.22 *This is basically a reduction-of-order technique, much like the manner with which one would reduce a second-order ODE to a system of two first-order ODES, written as a single vector ODE.*

Remark 4.23 *Note here that the second-order recursion is not a dynamical system, since one needs not only the previous state to determine the next state, but the previous two states. However, transformed into a first-order vector recursion, the new linear system is now a dynamical system.*

We can use the first-order vector recursion form of a higher-order recursion to construct a function that gives the nth term of a Fibonacci sequence in terms of n (rather than only in terms of the $(n-1)$th term); In essence, we can produce the evolution of the dynamical system given by the first-order system that is the functional form of the second-order recursion.

Proposition 4.24 *Given the second-order recursion $b_{n+2} = b_{n+1} + b_n$ with the initial data $b_0 = b_1 = 1$, we have*

$$b_n = \frac{\lambda^{n+1} - \mu^{n+1}}{\lambda - \mu},$$

where $\lambda = \frac{1+\sqrt{5}}{2}$ and $\mu = \frac{1-\sqrt{5}}{2}$.

We leave the proof of this as Exercise 158 for now. However, it is also simply a special case of the upcoming Proposition 4.26.

Exercise 158 Prove Proposition 4.24. (Hint: One way is to first show that the quantities themselves λ and μ satisfy the second-order recursion.)

We showed that λ and μ were the eigenvalues of a matrix $A = \begin{bmatrix} 0 & 1 \\ 1 & 1 \end{bmatrix}$, and that the linear map on $\mathbb{R}^2$ given by A, $\mathbf{v}_{n+1} = A\mathbf{v}_n$, is in fact the first-order vector recursion for the second-order recursion in the proposition under the assignment $\mathbf{v}_n = \begin{bmatrix} b_n \\ b_{n+1} \end{bmatrix}$. This reduction-of-order technique for the study of recursions is quite similar to (and is the discrete version of) the technique for studying the solutions of a single, second-order, homogeneous, ODE with constant coefficients by instead studying the system of two first-order, linear, constant-coefficient, homogeneous ODEs. In fact, this analogy is much more robust, which we will see in a minute.

First, a couple of notes:

- For very large n,

$$b_n = \frac{\lambda^{n+1} - \mu^{n+1}}{\lambda - \mu} \sim K\lambda^{n+1},$$

since $0 < |\mu| < 1 < |\lambda|$. Thus the growth rate of terms in the Fibonacci sequence is not exponential. It does, however, tend to look more and more exponential as n gets large. In fact, we can say the Fibonacci sequence displays *asymptotic exponential growth*, or that the sequence grows *asymptotically exponentially*.

- Start with the initial data $\mathbf{v}_0 = \begin{bmatrix} 1 \\ 1 \end{bmatrix}$, and plot $\mathcal{O}_{\mathbf{v}_0}$ in the plane. What you will find is that the iterates of $\mathcal{O}_{\mathbf{v}}^+$ will live on two curves of motion, as there will be flipping here across the $y = \lambda x$ eigenline due to the sign of μ. See Figure 72. And $\mathcal{O}_{\mathbf{v}}^+$ will tend toward the λ-eigenline as they grow off of the page. Getting closer to the λ-eigenline means that the growth rate is getting closer to the growth rate on the λ-eigenline. But, on this line, growth is purely exponential, with growth factor $\lambda > 1$.

Exercise 159 If we neglect the application to a rabbit population, the discrete dynamical system we constructed above is invertible. Calculate the first few pre-images of the vector $\mathbf{v} = \begin{bmatrix} 1 \\ 1 \end{bmatrix}$, and plot them on Figure 72. Then calculate the orbit line equations for the orbit line on which the sequence lives. Hint: You may need to solve the original second-order ODE to do this.

Now, start with any $\mathbf{v}_0 = \begin{bmatrix} x_0 \\ y_0 \end{bmatrix} \in \mathbb{R}^2$ as initial data for the second-order recursion (or the first-order vector version), and begin taking iterates. Plot them and you will see that these orbits will also live on either a single curve or will flip between two curves of motion and the phase diagram. Follow the orbit lines in Figure 72, and the ultimate fate of the orbit is apparent.

Exercise 160 For the Fibonacci vector recursion, find non-zero, explicit starting data for which the limit is not infinity (i.e., that lead to a sequence that does not run off of the page as n goes to infinity.)

Exercise 161 Show that for any non-zero, integer-valued initial values $\mathbf{v} = \begin{bmatrix} v_1 \\ v_2 \end{bmatrix}$, $\mathcal{O}_{\mathbf{v}} \longrightarrow \infty$ for the Fibonacci vector recursion.

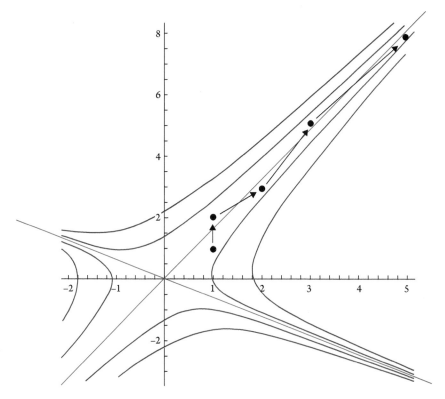

Figure 72 A saddle orbit.

With Exercise 161, we have the following:

Proposition 4.25 *All populations governed by the second-order Fibonacci recursion experience asymptotic exponential growth limiting to a growth factor of $\frac{1+\sqrt{5}}{2}$.*

More generally, let $a_{n+2} = pa_n + qa_{n+1}$ (careful of the order of the terms in this expression). Then we can construct a first-order vector recursion

$$\mathbf{v}_{n+1} = \begin{bmatrix} a_{n+1} \\ a_{n+2} \end{bmatrix} = \begin{bmatrix} 0 & 1 \\ p & q \end{bmatrix} \begin{bmatrix} a_n \\ a_{n+1} \end{bmatrix} = A\mathbf{v}_n, \text{ for } A = \begin{bmatrix} 0 & 1 \\ p & q \end{bmatrix}.$$

The characteristic equation of A is $r^2 - qr - p = 0$, with solutions $r = \frac{1}{2}(q \pm \sqrt{q^2 + 4p})$.

Proposition 4.26 *If $\begin{bmatrix} 0 & 1 \\ p & q \end{bmatrix}$ has two distinct eigenvalues $\lambda \neq \mu$, then every solution to the second-order recursion $a_{n+2} = pa_n + qa_{n+1}$ is of the form*

$$a_n = x\lambda^n + y\mu^n,$$

where $x = \alpha v_1$ *and* $y = \beta w_1$, $\mathbf{v} = \begin{bmatrix} v_1 \\ v_2 \end{bmatrix}$ *and* $\mathbf{w} = \begin{bmatrix} w_1 \\ w_2 \end{bmatrix}$ *are respective eigenvectors of* λ *and* μ, *and* α *and* β *satisfy the vector equation*

$$\begin{bmatrix} a_0 \\ a_1 \end{bmatrix} = \alpha \mathbf{v} + \beta \mathbf{w}.$$

Remark 4.27 *Hence the general second-order recursion and the first-order vector recursion carry the same information, and the latter provides all of the information necessary to completely understand the former. The method of solution is quickly discernable: Given a second-order recursion, calculate the data from the matrix A in the corresponding first-order vector recursion, including the eigenvalues and a pair of respective eigenvectors. Use this matrix data along with the initial data given with the original recursion to calculate the parameters in the functional expression for* a_n.

Here is an example going back to our Fibonacci Rabbits Problem. In essence, we use Proposition 4.26 to essentially prove Proposition 4.24.

Example 4.28 Go back to the original Fibonacci recursion $a_{n+2} = a_{n+1} + a_n$, with initial data $a_0 = a_1 = 1$. The matrix $A = \begin{bmatrix} 0 & 1 \\ 1 & 1 \end{bmatrix}$ has $\lambda = \frac{1+\sqrt{5}}{2}$ and $\mu = \frac{1-\sqrt{5}}{2}$ (as before) and, using the notation of Proposition 3.1.13, one can calculate representative eigenvectors as $\mathbf{v} = \begin{bmatrix} 1 \\ \lambda \end{bmatrix}$ and $\mathbf{w} = \begin{bmatrix} 1 \\ \mu \end{bmatrix}$. Thus $v_1 = w_1 = 1$. To calculate α and β, we have to solve the vector equation

(4.3.2) $$\begin{bmatrix} a_0 \\ a_1 \end{bmatrix} = \alpha \mathbf{v} + \beta \mathbf{w}, \quad \text{or} \quad \begin{bmatrix} 1 \\ 1 \end{bmatrix} = \alpha \begin{bmatrix} 1 \\ \lambda \end{bmatrix} + \beta \begin{bmatrix} 1 \\ \mu \end{bmatrix}.$$

This is solved by

$$\alpha = \frac{1-\mu}{\lambda - \mu}, \qquad \beta = \frac{\lambda - 1}{\lambda - \mu}.$$

Exercise 162 Verify that

$$\alpha = \frac{1-\mu}{\lambda - \mu} \qquad \beta = \frac{\lambda - 1}{\lambda - \mu}$$

solves the system in Equation 4.3.2.

Hence we have

$$x = \alpha v_1 = \frac{1-\mu}{\lambda - \mu}$$

and

$$y = \beta w_1 = \frac{\lambda - 1}{\lambda - \mu},$$

and our formula for the nth term of the sequence is

$$a_n = \frac{(1-\mu)\lambda^n + (\lambda - 1)\mu^n}{\lambda - \mu}.$$

This does not look like the form in Proposition 4.24, however. But considering that the term

$$1 - \mu = \frac{2}{2} - \frac{1-\sqrt{5}}{2} = \frac{1+\sqrt{5}}{2} = \lambda,$$

and similarly $\lambda - 1 = -\mu$, we wind up with

$$a_n = \frac{(1-\mu)\lambda^n + (\lambda - 1)\mu^n}{\lambda - \mu} = \frac{\lambda \cdot \lambda^n + (-\mu) \cdot \mu^n}{\lambda - \mu} = \frac{\lambda^{n+1} - \mu^{n+1}}{\lambda - \mu},$$

and we recover Proposition 4.24 precisely.

Exercise 163 Perform this calculation for the second-order recursion in the Opossums Problem (Exercises 3 and 157), and use it to calculate the population of opossums today, given that the initial population was a single pair in 1990.

And lastly, at least for this section, we have a way to ensure when a second-order recursion with non-zero initial data exhibits asymptotically exponential growth—namely, when the phase portrait of the associated first-order vector recursion has a saddle point at the origin.

Proposition 4.29 *Let* $a_{n+2} = pa_n + qa_{n+1}$ *be a second-order recursion, where* $p, q \in \mathbb{N}$ *and* $0 < p \leq q$. *Then all populations with non-zero initial conditions exhibit asymptotically exponential growth with asymptotic growth factor given by the spectral radius of*

$$A = \begin{bmatrix} 0 & 1 \\ p & q \end{bmatrix}.$$

Proof By the discussion above, the only thing to ensure is that the two eigenvalues of A are $\lambda = \frac{1}{2}(q + \sqrt{q^2 + 4p}) > 1$ and $-1 < \mu = \frac{1}{2}(q - \sqrt{q^2 + 4p}) < 0$ and $\lambda \notin \mathbb{Q}$. It is obvious that $\lambda > 1$, since $q \geq 1$. And since $\det A = -p < 0$, $\mu < 0$. So we claim $\mu > -1$. Since $q \geq p > 0$ by supposition, we have $q + 1 > p$ implies $4q + 4 > 4p$, which implies $q^2 + 4q + 4 = (q+2)^2 > q^2 + 4p$, which implies $q + 2 > \sqrt{q^2 + 4p}$, which implies $-2 < q - \sqrt{q^2 + 4p}$, which implies $-1 < \mu$. $\qquad\square$

4.4 Linear Flows versus Linear Maps

These calculations lead to a very important discussion on the relationship between the matrices found in first-order, 2-dimensional homogeneous linear systems (with constant coefficients) of ODEs and the corresponding matrices of the discrete, time-1 maps of those systems. The central question may be: Why is it that for a ODE system with coefficient matrix A, the **signs** of the eigenvalues determine the stability of the equilibrium solution at the origin, while for a linear map of $\mathbb{R}^n$, it is the **magnitudes** of the eigenvalues that determine the stability of the fixed point at the origin? The matrix of the time-1, ODE system is not the same matrix as the coefficient matrix of the system. The two matrices are certainly related, but they are not identical. Furthermore, any ODE system has a time-1 map. But only certain types of linear maps correspond to the time-1 maps of ODE systems. To understand better why, let's start with an example:

Example 4.30 Calculate the time-1 map of the ODE system

$$\dot{x} = \begin{bmatrix} 2 & 0 \\ 0 & -1 \end{bmatrix} x, \quad x(0) = x^0 = \begin{bmatrix} x_1^0 \\ x_2^0 \end{bmatrix}.$$

This system is uncoupled and straightforward to solve. Using linear system theory, the eigenvalues of the matrix $A = \begin{bmatrix} 2 & 0 \\ 0 & -1 \end{bmatrix}$ are $\lambda = 2$ and $\mu = -1$, and, since A is diagonal, we can choose the vectors $v_\lambda = \begin{bmatrix} 1 \\ 0 \end{bmatrix}$ and $v_\mu = \begin{bmatrix} 0 \\ 1 \end{bmatrix}$. Hence the general solution is

$$x(t) = c_1 \begin{bmatrix} 1 \\ 0 \end{bmatrix} e^{2t} + c_2 \begin{bmatrix} 0 \\ 1 \end{bmatrix} e^{-t},$$

or $x_1(t) = c_1 e^{2t}$ and $x_2(t) = c_2 e^{-t}$. For the choice of any initial data, the particular solution is $x_1(t) = x_1^0 e^{2t}$ and $x_2(t) = x_2^0 e^{-t}$, and the evolution of this continuous dynamical system is

$$\varphi(t,x) = x_1 \begin{bmatrix} 1 \\ 0 \end{bmatrix} e^{2t} + x_2 \begin{bmatrix} 0 \\ 1 \end{bmatrix} e^{-t} = \begin{bmatrix} e^{2t} & 0 \\ 0 & e^{-t} \end{bmatrix} \begin{bmatrix} x_1 \\ x_2 \end{bmatrix} = \begin{bmatrix} e^{2t} & 0 \\ 0 & e^{-t} \end{bmatrix} x.$$

The time-1 map is then $\varphi(1,x) = \varphi^1(x) : x(0) \longmapsto x(1)$, or the linear map

$$\varphi^1 : \mathbb{R}^2 \to \mathbb{R}^2, \quad \varphi^1(x) = Bx,$$

where $B = \begin{bmatrix} e^2 & 0 \\ 0 & e^{-1} \end{bmatrix}$ is the matrix associated with the linear map.

Do you see the relationship between the ODE matrix A and the time-1 linear map matrix B? The type and stability of the equilibrium solution at the origin of this linear system given by A is that of a saddle, and unstable. The time-1 map must also be a saddle, since the orbit lines of the time-1 map coincide precisely with the solution curves of the ODE system. It is the sign of the eigenvalues (non-zero entries of A in this case) that determine the type and stability of the origin of the ODE system. However, it is the "sizes" (moduli) of the eigenvalues of B that determine the type and stability of the fixed point at the origin in the linear map given by B.

Indeed, notice that the *exponential map*, $\exp : x \mapsto e^x$, takes $\mathbb{R}$ to $\mathbb{R}^+$ and maps all non-negative numbers $\mathbb{R}_0^+ = \{0\} \cup \mathbb{R}^+$ to the interval $[1, \infty)$ and all negative numbers to $(0, 1)$, as in Figure 73. This is no accident, and exposes a much deeper meaning of the

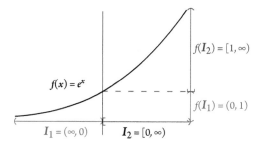

Figure 73 The domain and range of $f(x) = e^x$.

exponential map. With this, one might conclude that there could not be a time-1 map of a linear, constant coefficient, homogeneous ODE system with negative eigenvalues. And you would be correct in this hyperbolic case. But the general case is a little more complicated, as one can see in Exercise 164. Further, one may also want to conclude that for any 2×2 matrix A, the associated time-1 map B would simply be the exponentials of each of the entries of A. Here, you must definitely be much more careful, as we shall see presently.

Exercise 164 Let $f : \mathbb{R}^2 \to \mathbb{R}^2$ be the linear map $f(\mathbf{x}) = B\mathbf{x}$, where $B = \begin{bmatrix} a & 0 \\ 0 & b \end{bmatrix}$ and both $a > 0$ and $b > 0$. Determine a linear, 2-dimensional ODE system that has f as its time-1 map. For $a > 0$, show that B cannot correspond to a time-1 map of a linear ODE system if $b \leq 0$. Can B correspond to a time-1 map of a linear ODE system if both $a < 0$ and $b < 0$? Hint: The answer is yes, if $a = b \leq 0$. Show this.

The expression e^{At} does behave a lot like the exponential of a scalar. In fact, it solves the vector ODE $\dot{\mathbf{x}} = A\mathbf{x}$, with initial condition $\mathbf{x}(0) = \mathbf{x}^0$. However, contrary to Example 4.30, it is not in general true that the exponential of a matrix is simply the matrix of exponentials of the entries.

Example 4.31 Find the evolution for $\dot{\mathbf{x}} = \begin{bmatrix} 4 & -2 \\ 3 & -3 \end{bmatrix} \mathbf{x}$.

Here, the characteristic equation is $r^2 - r - 6 = 0$, with solutions giving eigenvalues of $\lambda = 3$ and $\mu = -2$. Calculating eigenvectors, we choose $\mathbf{v}_\lambda = \begin{bmatrix} 2 \\ 1 \end{bmatrix}$ and $\mathbf{v}_\mu = \begin{bmatrix} 1 \\ 3 \end{bmatrix}$. Thus the general solution is

(4.4.1) $$\mathbf{x}(t) = c_1 \begin{bmatrix} 2 \\ 1 \end{bmatrix} e^{3t} + c_2 \begin{bmatrix} 1 \\ 3 \end{bmatrix} e^{-2t} = \begin{bmatrix} 2e^{3t} & e^{-2t} \\ e^{3t} & 3e^{-2t} \end{bmatrix} \begin{bmatrix} c_1 \\ c_2 \end{bmatrix}.$$

Writing this in terms of $\mathbf{x}^0$ (in essence, finding the evolution), we get the linear system

$$\begin{bmatrix} x_1^0 \\ x_2^0 \end{bmatrix} = \begin{bmatrix} 2c_1 + c_2 \\ c_1 + 3c_2 \end{bmatrix} = \begin{bmatrix} 2 & 1 \\ 1 & 3 \end{bmatrix} \begin{bmatrix} c_1 \\ c_2 \end{bmatrix}.$$

Solving for c_1 and c_2 in terms of the initial conditions involves inverting the matrix, and
$$\begin{bmatrix} 2 & 1 \\ 1 & 3 \end{bmatrix}^{-1} = \tfrac{1}{5} \begin{bmatrix} 3 & -1 \\ -1 & 2 \end{bmatrix}. \text{ Hence the evolution is}$$

$$\mathbf{x}(t) = \begin{bmatrix} 2e^{3t} & e^{-2t} \\ e^{3t} & 3e^{-2t} \end{bmatrix} \begin{bmatrix} c_1 \\ c_2 \end{bmatrix}$$

$$= \begin{bmatrix} 2e^{3t} & e^{-2t} \\ e^{3t} & 3e^{-2t} \end{bmatrix} \begin{bmatrix} \tfrac{3}{5} & -\tfrac{1}{5} \\ -\tfrac{1}{5} & \tfrac{2}{5} \end{bmatrix} \begin{bmatrix} x_1^0 \\ x_2^0 \end{bmatrix}$$

$$= \begin{bmatrix} \tfrac{6}{5}e^{3t} - \tfrac{1}{5}e^{-2t} & -\tfrac{2}{5}e^{3t} + \tfrac{2}{5}e^{-2t} \\ \tfrac{3}{5}e^{3t} - \tfrac{3}{5}e^{-2t} & -\tfrac{1}{5}e^{3t} + \tfrac{6}{5}e^{-2t} \end{bmatrix} \begin{bmatrix} x_1^0 \\ x_2^0 \end{bmatrix}.$$

Hence we can also say now that

$$e^{At} = \begin{bmatrix} \tfrac{6}{5}e^{3t} - \tfrac{1}{5}e^{-2t} & -\tfrac{2}{5}e^{3t} + \tfrac{2}{5}e^{-2t} \\ \tfrac{3}{5}e^{3t} - \tfrac{3}{5}e^{-2t} & -\tfrac{1}{5}e^{3t} + \tfrac{6}{5}e^{-2t} \end{bmatrix}, \quad \text{for} \quad A = \begin{bmatrix} 4 & -2 \\ 3 & -3 \end{bmatrix}$$

and that the time-1 map of this ODE is the linear map given by

$$e^A = \begin{bmatrix} \tfrac{6}{5}e^3 - \tfrac{1}{5}e^{-2} & -\tfrac{2}{5}e^3 + \tfrac{2}{5}e^{-2} \\ \tfrac{3}{5}e^3 - \tfrac{3}{5}e^{-2} & -\tfrac{1}{5}e^3 + \tfrac{6}{5}e^{-2} \end{bmatrix}.$$

So how does one square these calculations into a general understanding of e^A? via the properties of of a matrix exponential and a bit of standard linear algebra:

Proposition 4.32 Let $A_{n \times n}$ be diagonalizeable. Then $A = SBS^{-1}$, where

- $B_{n \times n}$ is diagonal, and
- the columns of $S_{n \times n}$ form an eigenbasis of A.

Proposition 4.33 If $A_{n \times n}$ is diagonalizeable, then $e^A = Se^B S^{-1}$, where both B and e^B are diagonal.

Proof Note that since

$$e^A = \sum_{n=1}^{\infty} \frac{A^n}{n!} \quad \text{and} \quad (SAS^{-1})^n = SA^n S^{-1},$$

we have

$$Se^B S^{-1} = S \left(\sum_{n=1}^{\infty} \frac{B^n}{n!} \right) S^{-1} = \sum_{n=1}^{\infty} \frac{SB^n S^{-1}}{n!} = \sum_{n=1}^{\infty} \frac{(SBS^{-1})^n}{n!} = \sum_{n=1}^{\infty} \frac{A^n}{n!} = e^A. \quad \square$$

And finally, we have

Proposition 4.34 *If* $\mathbf{v}$ *is an eigenvector of A corresponding to the eigenvalue r, then* $\mathbf{v}$ *is an eigenvector of* e^A *corresponding to the eigenvalue* e^r.

Proof For r and $\mathbf{v}$ satisfying $A\mathbf{v} = r\mathbf{v}$, we have

$$e^A \mathbf{v} = \lim_{n \to \infty} \left(\sum_{k=0}^n \frac{A^k}{k!} \mathbf{v} \right) = \lim_{n \to \infty} \left(\sum_{k=0}^n \frac{r^k}{k!} \mathbf{v} \right) = \lim_{n \to \infty} \left(\sum_{k=0}^n \frac{r^k}{k!} \right) \mathbf{v} = e^r \mathbf{v}. \qquad \square$$

Example 4.35 Back to the previous system, with $\dot{\mathbf{x}} = A\mathbf{x}$, and $A = \begin{bmatrix} 4 & -2 \\ 3 & -3 \end{bmatrix}$. The general solution, written in Equation 4.4.1, was

$$
\begin{aligned}
\mathbf{x}(t) &= \begin{bmatrix} 2e^{3t} & e^{-2t} \\ e^{3t} & 3e^{-2t} \end{bmatrix} \begin{bmatrix} c_1 \\ c_2 \end{bmatrix} \\
&= \begin{bmatrix} 2e^{3t} & e^{-2t} \\ e^{3t} & 3e^{-2t} \end{bmatrix} \begin{bmatrix} \frac{3}{5} & -\frac{1}{5} \\ -\frac{1}{5} & \frac{2}{5} \end{bmatrix} \begin{bmatrix} x_1^0 \\ x_2^0 \end{bmatrix} \\
&= \begin{bmatrix} \frac{6}{5}e^{3t} - \frac{1}{5}e^{-2t} & -\frac{2}{5}e^{3t} + \frac{2}{5}e^{-2t} \\ \frac{3}{5}e^{3t} - \frac{3}{5}e^{-2t} & -\frac{1}{5}e^{3t} + \frac{6}{5}e^{-2t} \end{bmatrix} \begin{bmatrix} x_1^0 \\ x_2^0 \end{bmatrix} = e^{At} \mathbf{x}^0.
\end{aligned}
$$

But the middle equality in the last grouping can easily be written

$$
\begin{aligned}
\mathbf{x}(t) &= \begin{bmatrix} 2e^{3t} & e^{-2t} \\ e^{3t} & 3e^{-2t} \end{bmatrix} \begin{bmatrix} \frac{3}{5} & -\frac{1}{5} \\ -\frac{1}{5} & \frac{2}{5} \end{bmatrix} \begin{bmatrix} x_1^0 \\ x_2^0 \end{bmatrix} \\
&= \begin{bmatrix} 2 & 1 \\ 1 & 3 \end{bmatrix} \begin{bmatrix} e^{3t} & 0 \\ 0 & e^{-2t} \end{bmatrix} \begin{bmatrix} \frac{3}{5} & -\frac{1}{5} \\ -\frac{1}{5} & \frac{2}{5} \end{bmatrix} \begin{bmatrix} x_1^0 \\ x_2^0 \end{bmatrix} = Se^{Bt}S^{-1}\mathbf{x}^0,
\end{aligned}
$$

where S is the matrix whose columns form an eigenbasis of A, and e^{Bt} is the exponential of the diagonal matrix B. Hence, as in the proposition, $e^{At} = Se^{Bt}S^{-1}$.

Exercise 165 Show that the time-1 map of the ODE system $\dot{\mathbf{x}} = \begin{bmatrix} \lambda & 1 \\ 0 & \lambda \end{bmatrix} \mathbf{x}$ is given by the linear map $f(\mathbf{x}) = B_1 \mathbf{x}$, where $B_1 = \begin{bmatrix} e^\lambda & e^\lambda \\ 0 & e^\lambda \end{bmatrix}$, but the time-$t$ map in general is not given by $B_t = \begin{bmatrix} e^{\lambda t} & e^{\lambda t} \\ 0 & e^{\lambda t} \end{bmatrix}$.

Exercise 166 Find the time-1 map of the IVP $\dot{\mathbf{x}} = \begin{bmatrix} 0 & \alpha \\ -\alpha & 0 \end{bmatrix} \mathbf{x}$, and use it to construct a form for the exponential of a matrix with purely imaginary eigenvalues.

4.5 Local Linearization and Stability of Equilibria

Nonlinear systems of first-order autonomous ODEs are inherently difficult to study or explore, given their complexity and the lack of general techniques for discovering solutions. Still, as for functions in general, the fact that differentiability allows for a simplifying "local" structure (a dominant linear component to a Taylor expansion) means that one can study a system by linearizing it locally and then using the linear theory above as a model to study orbit behavior. Herein, we explore this situation, leading to the very powerful Poincaré–Lyapunov and Hartman–Grobman Theorems, and giving us a method of classification of continuous flow near a critical point of a system into a relatively small number of types and possibilities.

Following this, we will continue our nonlinear exploration by examining the stability of isolated periodic solutions in a multidimensional system, leading up to another celebrated existence theorem for periodic orbits, the Poincaré–Bendixson Theorem. We will also highlight a new kind of bifurcation in a simple system, the Poincaré–Andronov–Hopf (or simply the Hopf) bifurcation, with its creation of a non-trivial periodic solution (a periodic solution that is not an equilibrium). One of the more beautiful effects about systems in the plane is that non-trivial periodic solutions are *Jordan curves*, closed curves that separate the plane into two disjoint regions. In dynamical systems, the existence of such a curve has the consequence that inside such a curve we have an invariant dynamical system that leaves the boundary invariant also. Such a system restricted to an invariant compact region is governed by the Brouwer Fixed-Point Theorem, so that inside any non-trivial periodic solution to a flow there lies an equilibrium.

Remark 4.36 *Local linearization, in its full glory, relies on the notion of equivalency (topological conjugacy) between two different dynamical systems, although in this case the conjugacy is only local to an equilibrium. In essence, one can linearize the original flow near an equilibrium, although there is no need to explicitly do so in many cases, as we will see.*

To start, consider the autonomous, first-order system $\dot{\mathbf{x}} = \mathbf{F}(\mathbf{x})$, for $\mathbf{x} \in \mathbb{R}^n$. Some facts regarding a system like this:

(1) Recall that autonomous means that neither the vector field nor its flow vary in time.

(2) Solutions, in general, are difficult or impossible to calculate.

(3) Critical points of the system are points in $\mathbb{R}^n$ where $\mathbf{F}(\mathbf{x}) = \mathbf{0}$ (called zeros of the vector field, or elements of the *zero-set* of $\mathbf{F} : \mathbb{R}^n \to \mathbb{R}^n$).

(4) Technically speaking, $\mathbf{F}$, as a vector field, is a function that takes points to vectors, thereby associating a vector with each point. For a point $(x_1^0, \ldots, x_n^0) \in \mathbb{R}^n$, we can always place its coordinate elements in vector form and write $\mathbf{x}^0 = \begin{bmatrix} x_1^0 \\ \vdots \\ x_n^0 \end{bmatrix}$. But these are two different notions, even if there is an isomorphism from $\mathbb{R}^n$ to $\mathbb{R}^n$ taking points to vectors in this way and preserving the operation of addition on

each copy of $\mathbb{R}^n$. The presence of the isomorphism does make the spaces (and the group structure in each) the same. But they play different roles. Thus $(x_1^0, \ldots, x_n^0)$ is a critical point of the system precisely when $\mathbf{x}(t) \equiv \mathbf{x}^0$ is an equilibrium solution (we use the sign $\equiv$ to denote that the solution is a constant one for all time). However, we will often "abuse notation" and confuse the association of $(x_1^0, \ldots, x_n^0)$ with $\mathbf{x}^0$ when it is not important to the discussion. (When is it not critical?)

The idea of locally linearizing a system near a critical point revolves around the notion of a vector field being *almost linear* near the point; that is, the function describing the vector field is well-approximated by a linear function. You have seen this in your calculus courses, where a function is well-approximated by a linear function whenever the derivative exists and varies continuously. One way to describe a system as being almost linear at a critical point is the following:

Definition 4.37 *A system* $\dot{\mathbf{x}} = \mathbf{F}(\mathbf{x})$ *in* $\mathbb{R}^n$ *with an isolated critical point at* $\mathbf{x} = \mathbf{0}$ *is almost linear in a neighborhood of* $\mathbf{0}$ *if there exists a non-degenerate* $n \times n$ *matrix* A *and a continuous function* $\mathbf{g} : \mathbb{R}^n \to \mathbb{R}^n$ *where* $\mathbf{F}(\mathbf{x}) = A\mathbf{x} + \mathbf{g}(\mathbf{x})$ *and*

$$\lim_{\mathbf{x} \to 0} \frac{||\mathbf{g}(\mathbf{x})||}{||\mathbf{x}||} = 0.$$

Some notes:

- Of course, the constraint that the isolated critical point is at the origin is not limiting, since one can always employ a translation transformation $\mathbf{u} = \mathbf{x} - \mathbf{a}$, creating a new system where a critical point at $\mathbf{a}$ in the $\mathbf{x}$-coordinate system is now at the origin in the new $\mathbf{u}$-coordinate system.

- Here, we would say that the function $\mathbf{g}(\mathbf{x})$ must be small near $\mathbf{a}$ in comparison with $\mathbf{x}$.

- Often, it is useful to check the limit for $\mathbf{g}(\mathbf{x})$ component-wise by an easy switch to spherical coordinates, at least when using the Euclidean norm in $\mathbb{R}^n$: For

$$\mathbf{g}(\mathbf{x}) = \begin{bmatrix} g_1(\mathbf{x}) \\ \vdots \\ g_n(\mathbf{x}) \end{bmatrix}, \text{ we have}$$

$$\lim_{\mathbf{x} \to 0} \frac{g_i(\mathbf{x})}{||\mathbf{x}||} = \lim_{\mathbf{x} \to 0} \frac{g_i(\mathbf{x})}{\sqrt{x_1^2 + \cdots + x_n^2}} = \lim_{r \to 0} \frac{g_i(r, \theta)}{r} = 0,$$

where $i = 1, \ldots, n$, and $r^2 = \sum_{i=1}^n x_i^2$.

Example 4.38 For $\dot{\mathbf{x}} = \mathbf{F}(\mathbf{x})$ in the plane, with a critical point at the origin and where $\mathbf{F} = \begin{bmatrix} F(x,y) \\ G(x,y) \end{bmatrix}$ and both F and G are polynomials, A is readily isolated as the linear part of the polynomials of $\mathbf{F}$: Consider $\dot{x} = x - x^2 - xy$ and $\dot{y} = -y + y^2 + 2xy$. Then

$$\dot{\mathbf{x}} = \begin{bmatrix} 1 & 0 \\ 0 & -1 \end{bmatrix} \mathbf{x} + \begin{bmatrix} -x^2 - xy \\ y^2 + 2xy \end{bmatrix} = A\mathbf{x} + \mathbf{g}(\mathbf{x}),$$

and with

$$\lim_{r \to 0} \frac{g_1(r,\theta)}{r} = \lim_{r \to 0} \frac{-r^2 \cos^2\theta - r^2 \cos\theta \sin\theta}{r} = 0$$

and

$$\lim_{r \to 0} \frac{g_2(r,\theta)}{r} = \lim_{r \to 0} \frac{r^2 \sin^2\theta + r^2 \cos\theta \sin\theta}{r} = 0,$$

the system is almost linear near the origin. Note that this will always be the case when the vector field has polynomial coordinate functions, since $\mathbf{g}(\mathbf{x})$ will then always contain only quadratic and/or higher-order monomial terms.

For more general $\mathbf{F}(\mathbf{x})$, things are a bit more subtle:

Example 4.39 Let $\dot{x} = (1+x)\sin y$ and $\dot{y} = 1 - x - \cos y$. Here, there is a critical point at $\mathbf{x} = \begin{bmatrix} 2 \\ \pi \end{bmatrix}$, and a translational change of variable yields the system

(4.5.1)
$$\begin{aligned} \dot{u} &= -(u+3)\sin v, \\ \dot{v} &= -u - 1 + \cos v. \end{aligned}$$

Exercise 167 Show that the original system above is topologically conjugate to Equation 4.5.1 through the proper coordinate change.

While there does not seem to be a linear part to the first equation in Equation 4.5.1, there is under the Taylor expansions of the trigonometric functions. The system is then

$$\begin{aligned} \dot{u} &= -(u+3)\left(v - \frac{v^3}{3!} + \frac{v^5}{5!} - \cdots \right), \\ \dot{v} &= -u - 1 + \left(1 - \frac{v^2}{2!} + \frac{v^4}{4!} - \cdots \right) \end{aligned}$$

so that

$$\dot{\mathbf{u}} = \begin{bmatrix} 0 & -3 \\ -1 & 0 \end{bmatrix} \begin{bmatrix} u \\ v \end{bmatrix} + \begin{bmatrix} -uv + (u+3)\dfrac{v^3}{3!} - \cdots \\ -\dfrac{v^2}{2!} + \dfrac{v^4}{4!} - \cdots \end{bmatrix},$$

so $A = \begin{bmatrix} 0 & -3 \\ -1 & 0 \end{bmatrix}$.

Of course, for vector fields that are C^1, one can establish almost linearity directly through the derivative. Recall that for a real-valued function of one variable $f(x)$ to be differentiable at x_0, the limit

$$f'(x_0) = \frac{df}{dx}\Big|_{x=x_0} = \lim_{x \to x_0} \frac{f(x) - f(x_0)}{x - x_0}$$

must exist. Defining the difference between the difference quotient and the derivative at x_0,

$$\eta(x) = \frac{f(x) - f(x_0)}{x - x_0} - f'(x_0),$$

we get a sort of measure of the higher-order nonlinearity of $f(x)$, at least near x_0. It is obvious that $\eta(x) \to 0$ as $x \to x_0$. But this formulation also allows us to rewrite $f(x)$ as

$$f(x) = f(x_0) + f'(x_0)(x - x_0) + \eta(x)(x - x_0),$$

which, when we stay close to x_0, gives the familiar tangent linear approximation to $f(x)$ near x_0,

$$f(x) \approx f(x_0) + f'(x_0)(x - x_0), \text{ for } x \approx x_0.$$

Note that this works whether $f'(x_0)$ is zero or not.

In higher dimensions, we can play the same game, as is usually the case in any good vector calculus course: For functions of two variables, $f(x, y)$ is differentiable at (x_0, y_0) if the partial derivatives f_x and f_y are continuous in some open neighborhood of $(x_0, y_0) \in \mathbb{R}^2$. Then, there is a function $\eta(x, y)$ such that

$$f(x, y) = f(x_0, y_0) + f_x(x_0, y_0)(x - x_0)$$
$$+ f_y(x_0, y_0)(y - y_0) + \eta(x, y)\|(x - x_0, y - y_0)\|,$$

where $\eta(x, y) \to 0$ as $(x, y) \to (x_0, y_0)$. Thus a function is well-approximated by its tangent linear function near (x_0, y_0).

Back to our system, let's again stay with a 2-dimensional system $\dot{x} = F(x, y)$, $\dot{y} = G(x, y)$, where $F, G \in C^1$ at (x_0, y_0), or $\dot{\mathbf{x}} = \mathbf{F}(x, y)$, where $\mathbf{F}(x, y) = \begin{bmatrix} F(x, y) \\ G(x, y) \end{bmatrix}$. Then

$$\dot{\mathbf{x}} = \begin{bmatrix} \dot{x} \\ \dot{y} \end{bmatrix} = \begin{bmatrix} F(x, y) \\ G(x, y) \end{bmatrix}$$

$$
= \begin{bmatrix} F(x_0,y_0) + F_x(x_0,y_0)(x - x_0) + F_y(x_0,y_0)(y - y_0) + \eta_F(x,y) \left\| \begin{bmatrix} x - x_0 \\ y - y_0 \end{bmatrix} \right\| \\ G(x_0,y_0) + G_x(x_0,y_0)(x - x_0) + G_y(x_0,y_0)(y - y_0) + \eta_G(x,y) \left\| \begin{bmatrix} x - x_0 \\ y - y_0 \end{bmatrix} \right\| \end{bmatrix}
$$

$$
= \begin{bmatrix} F(x_0,y_0) \\ G(x_0,y_0) \end{bmatrix} + A \begin{bmatrix} x - x_0 \\ y - y_0 \end{bmatrix} + \mathbf{H}(x,y) \left\| \mathbf{x} - \mathbf{x}^0 \right\|
$$

where $A = DF(x_0,y_0)$, $\mathbf{x}^0 = \begin{bmatrix} x_0 \\ y_0 \end{bmatrix}$ is just the vector form of the initial data, and

$$
\mathbf{H}(x,y) = \begin{bmatrix} \eta_F(x,y) \\ \eta_G(x,y) \end{bmatrix}.
$$

Thus we have:

Proposition 4.40 *A system* $\dot{\mathbf{x}} = \mathbf{F}(\mathbf{x}) = \begin{bmatrix} F(x,y) \\ G(x,y) \end{bmatrix}$ *is almost linear at an isolated critical point* $\mathbf{x} = \mathbf{x}^0$ *if both F and G are* C^1, *and the Jacobian*

$$
DF(\mathbf{a}) = \begin{bmatrix} \frac{\partial F}{\partial x}\big|_{\mathbf{x}=\mathbf{x}^0} & \frac{\partial F}{\partial y}\big|_{\mathbf{x}=\mathbf{x}^0} \\ \frac{\partial G}{\partial x}\big|_{\mathbf{x}=\mathbf{x}^0} & \frac{\partial G}{\partial y}\big|_{\mathbf{x}=\mathbf{x}^0} \end{bmatrix}
$$

is non-singular.

Exercise 168 Explicitly prove Proposition 4.40 by packaging the previous argument above.

Now with a non-degenerate A, and $\lim_{\mathbf{x} \to \mathbf{x}_0} \mathbf{H}(x,y) = 0$, a good approximation to the system is the linear system

$$
\dot{\mathbf{x}} = \begin{bmatrix} F(x_0,y_0) \\ G(x_0,y_0) \end{bmatrix} + A \begin{bmatrix} x - x_0 \\ y - y_0 \end{bmatrix}
$$

$$
= \mathbf{F}(\mathbf{x}^0) + A(\mathbf{x} - \mathbf{x}^0), \text{ when } \mathbf{x} \approx \mathbf{x}^0.
$$

And when $\mathbf{x}^0$ is a critical point, so $\mathbf{F}(\mathbf{x}) = \mathbf{0}$, we have

$$
\dot{\mathbf{x}} = A(\mathbf{x} - \mathbf{x}^0).
$$

Via the translation of coordinates (placing the critical point at the origin) $\mathbf{u} = \begin{bmatrix} u \\ v \end{bmatrix} = \begin{bmatrix} x - x_0 \\ u - y_0 \end{bmatrix} = \mathbf{x} - \mathbf{x}^0$, where $\dot{\mathbf{u}} = \dot{\mathbf{x}}$, we have

(4.5.2) $$\dot{\mathbf{u}} = A\mathbf{u},$$

which we call the *linear system* associated with the original system at $\mathbf{x}^0$.

Example 4.41 Back to Example 4.39 and the system $\dot{x} = (1 + x)\sin y$ and $\dot{y} = 1 - x - \cos y$. At the critical point $\mathbf{a} = \begin{bmatrix} 2 \\ \pi \end{bmatrix}$, we have

$$A = D\mathbf{F}(\mathbf{a}) = \begin{bmatrix} \sin \pi & (1+2)\cos \pi \\ -1 & \sin \pi \end{bmatrix} = \begin{bmatrix} 0 & -3 \\ -1 & 0 \end{bmatrix}.$$

The new linear system in Equation 4.5.2 is called the *local linearization* of $\dot{\mathbf{x}} = \mathbf{F}(\mathbf{x})$ near the critical point $\mathbf{x}^0$, and the original system $\dot{\mathbf{x}} = \mathbf{F}(\mathbf{x})$ is said to be *locally linear* at $\mathbf{x}^0$.

Now, given a linear system that approximates a locally linear system at an equilibrium, we can use linear system theory to aid us in our analysis of the behavior of solutions near an equilibrium in our nonlinear system. We start with a simple statement of stability:

Theorem 4.42 (The Poincaré–Lyapunov Theorem (for autonomous systems)) *Suppose that, for $r \geq 1$, $\mathbf{x}^0$ is a critical point of a C^r autonomous system $\dot{\mathbf{x}} = \mathbf{F}(\mathbf{x})$ and suppose that all of the eigenvalues of the local linearization matrix have negative real part. Then $\mathbf{x}^0$ is locally asymptotically stable.*

In the flavor of Proposition 4.12, we can immediately add the following:

Theorem 4.43 *Suppose, for $r \geq 1$, that $\mathbf{x}^0$ is a critical point of a C^r autonomous system $\dot{\mathbf{x}} = \mathbf{F}(\mathbf{x})$ and suppose that at least one of the eigenvalues of the local linearization matrix has positive real part. Then $\mathbf{x}^0$ is unstable.*

There are many statements to make about these two theorems:

- They only describe local behavior of solutions near $\mathbf{x}^0$. They say very little about the actual phase space portraits. For example, they determine when $\mathbf{x}^0$ is a sink or not, but not whether the node of the nonlinear system is proper or not. That cannot be determined via local linearization.

- The actual theorem attributed to Poincaré and Lyapunov is a bit more robust. Indeed, consider the more general non-autonomous system

(4.5.3) $$\dot{\mathbf{x}} = A\mathbf{x} + B(t)\mathbf{x} + \mathbf{f}(t, \mathbf{x}), \quad \mathbf{x}(t_0) = \mathbf{x}^0, \quad t \in \mathbb{R},$$

where $B(t)$ and $f(t, \mathbf{x})$ are relatively well-behaved functions in a neighborhood of $\mathbf{x} = \mathbf{0}$, an isolated equilibrium. In detail, the theorem then is:

Theorem 4.44 (The Poincaré–Lyapunov Theorem) *For the IVP in Equation 4.5.3, suppose A is a constant $n \times n$ matrix with all eigenvalues having negative real part, $B(t)$ is a continuous family of $n \times n$ matrices that satisfy*

$$\lim_{t \to \infty} ||B(t)|| = 0,$$

and $f(t, \mathbf{x})$ is continuous in both t and $\mathbf{x}$ and Lipschitz continuous in $\mathbf{x}$, satisfying

$$\lim_{||\mathbf{x}|| \to 0} \frac{||\mathbf{f}(t, \mathbf{x})||}{||\mathbf{x}||} = 0 \text{ uniformly in } t.$$

Then there exist constants C, δ, and $r > 0$, such that when $||\mathbf{x}^0|| < \delta$, the solution $\mathbf{x}(t)$ satisfies

$$||\mathbf{x}(t)|| \leq C||\mathbf{x}^0||e^{-r(t-t_0)}, \quad t > t_0.$$

Thus for $\mathbf{x}^0 \in N_\delta(\mathbf{0})$, $\mathbf{x}(t) \to \mathbf{0}$ and the equilibrium at $\mathbf{0}$ is asymptotically stable.

It is often the case in textbooks (see, e.g., Ricardo[47]), however, that a much more detailed classification of a nonlinear equilibrium is attached to this theorem, at least in two dimensions. Indeed, it is discussed how the nonlinear sink is a node or spiral or proper node iff the linearized system is also. But if the linearized system had an improper node, for example, it would not be clear if the original nonlinear equilibrium was proper or not. Furthermore, if a linearized system displayed a center, then not even the stability of the original critical point could be determined from the linear system. Looking back at the trace–determinant classification diagram in Figure 62, it is easy to see that classification by stability and type is possible when the linearized system has data that places it in the interior of the stability regions, but is not possible when it is on the boundaries. This diagram is worth a detailed study. But the general discussion revolves on a much more detailed conclusion that one can draw about the behavior near an equilibrium in a nonlinear system via the linearized version. This more detailed conclusion better exposes the relationship between the actual phase portraits. First, a definition:

Definition 4.45 *A critical point $\mathbf{x}^0$ of a C^r, $r \geq 1$ autonomous system $\dot{\mathbf{x}} = \mathbf{F}(\mathbf{x})$ is called hyperbolic if the local linearization matrix associated with $\mathbf{x}^0$ has no eigenvalues with zero real part.*

Remark 4.46 *Recognize that this safely eliminates the nuanced classification types in the linear system classification of the last section. In effect, hyperbolic critical points are either sinks (all eigenvalues have negative real part), sources (all eigenvalues have positive real part), or saddles (with all eigenvalues having non-zero real part but at least one with positive real part and at least one with negative real part).*

Theorem 4.47 (The Hartman–Grobman Theorem) *Suppose $\mathbf{x}^0$ is a hyperbolic critical point. Then the phase portrait of the nonlinear system at $\mathbf{x}^0$ and the phase portrait of the associated linear system are locally homeomorphic in a neighborhood of $\mathbf{x}^0$.*

So what does it mean for two phase portraits to be locally homeomorphic? Basically, it means that there is a homeomorphism that takes trajectories to trajectories, which is close to the identity map near the equilibria.

Consider, in this case, that the theorem establishes the link directly between the phase portraits of the original nonlinear equilibrium and the linearized one at the origin. Once done, stability follows immediately. These flows are topologically conjugate.

Some other notes:

- For non-hyperbolic equilibria, the phase portraits may not match.
- This local homeomorphism actually gives us more: Solution curves of the nonlinear phase portrait are asymptotically tangent to their respective counterparts of the linear phase portrait at the critical points. To see this, though, we need a bit more structure:

Example 4.48 The ODE system

(4.5.4)
$$\dot{x} = 2y,$$
$$\dot{y} = 3x^2 + 2x$$

is an autonomous, nonlinear (but certainly almost linear) system. One way to study this system is to write it out as a single ODE

$$\frac{dy}{dx} = \frac{dy/dt}{dx/dt} = \frac{3x^2 + 2x}{2y},$$

or

$$3x^2 + 2x - 2y\frac{dy}{dx} = 0.$$

This last ODE is exact, and can be integrated to obtain

(4.5.5)
$$\varphi(x,y) = y^2 - x^2(x+1) = c.$$

as an implicitly defined general solution.

Exercise 169 Solve the system in Equation 4.5.4 to obtain the implicit general solution function in Equation 4.5.5.

Here, the 0-level set $\varphi(x,y) = 0$ is the irreducible curve in Figure 27 (on the left) in Section 2.4, part of the discussion of Example 2.68. The curve $\varphi(x,y) = 0$ does not factor globally, but, local to $(0,0)$, the level set factors to

$$y = \pm x\sqrt{x+1}.$$

Thus the tangent lines to the two local branches have slopes

$$\frac{dy}{dx}(0) = \pm\left(\sqrt{x+1} + \frac{x}{2\sqrt{x+1}}\right)\Bigg|_{x=0} = \pm 1.$$

So the equations of the tangent lines are $y = x$ and $y = -x$.

As for the original system in Equation 4.5.4, we can also linearize at the origin to obtain

$$\dot{x} = \begin{bmatrix} 0 & 2 \\ 2 & 0 \end{bmatrix}\mathbf{x}.$$

The coefficient matrix $A = \begin{bmatrix} 0 & 2 \\ 2 & 0 \end{bmatrix}$ has characteristic equation $r^2 - (\operatorname{tr} A)r + \det A = r^2 - 4 = 0$ with solutions $r_1 = -2$ and $r_2 = 2$, with corresponding eigenvectors

$$\mathbf{v}_1 = \begin{bmatrix} 1 \\ -1 \end{bmatrix} \quad \text{and} \quad \mathbf{v}_2 = \begin{bmatrix} 1 \\ 1 \end{bmatrix}.$$

Hence the eigenlines, $\ell_1 : y = -x$ for r_1 and $\ell_2 : y = x$ for r_2, agree with the tangent lines to $\varphi(x,y) = 0$.

And lastly, implicitly defined solutions to the ODE system do not automatically indicate the direction of motion along the level set curve. This direction is lost as one "factors out" t. But the original slope field recovers this information: At the point $(-1,0)$ of the 0-level set of φ, we have $\dot{x} = 0$ and $\dot{y} = 1$. This is compatible with the eigenline data of the local linearization, as in Figure 74.

Definition 4.49 *Let $\mathbf{p}$ be critical for $\dot{\mathbf{x}} = \mathbf{F}(\mathbf{x})$ and $U(\mathbf{p})$ a small open neighborhood of $\mathbf{p}$ in phase space. The* local stable manifold *of $\mathbf{p}$ in U is*

$$\mathcal{W}^s_{loc}(\mathbf{p}) = \left\{ \mathbf{q} \in U \mid \varphi^t(\mathbf{q}) \in U, t > 0, \text{ and } d(\varphi^t(\mathbf{q}), \mathbf{p}) \to 0 \text{ as } t \to \infty \right\}.$$

The local unstable manifold *of $\mathbf{p}$ in U is*

$$\mathcal{W}^u_{loc}(\mathbf{p}) = \left\{ \mathbf{q} \in U \mid \varphi^{-t}(\mathbf{q}) \in U, t > 0, \text{ and } d(\varphi^{-t}(\mathbf{q}), \mathbf{p}) \to 0 \text{ as } t \to \infty \right\}.$$

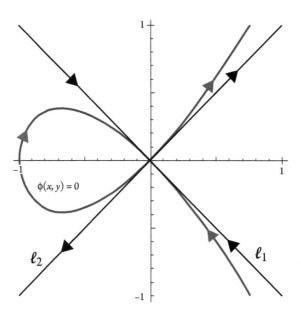

Figure 74 The set $\varphi(x,y) = 0$ as a solution.

Some notes:

- These sets are actually topological spaces that locally look like Euclidean space. Spaces with this property are called *manifolds*: Near each point, they look like the image of a function $\psi : V \subset \mathbb{R}^n \to \mathbb{R}^m$, which is a homeomorphism onto its image. These spaces are the analogs of the stable and unstable linear subspaces $\mathbb{E}^s$ and $\mathbb{E}^u$ in Section 4.11.

  Theorem 4.50 (The Stable Manifold Theorem) *Let* $\mathbf{p}$ *be a hyperbolic critical point of the* $C^r, r \geq 1$ *system* $\dot{\mathbf{x}} = \mathbf{F}(\mathbf{x})$*. Then* $\mathcal{W}^s_{loc}(\mathbf{p})$ *and* $\mathcal{W}^u_{loc}(\mathbf{p})$ *exist, are* C^r*, are of the same dimensions as* $\mathbb{E}^s$ *and* $\mathbb{E}^u$*, respectively, of the local linearized system around* $\mathbf{p}$*, and are tangent to* $\mathbb{E}^s$ *and* $\mathbb{E}^u$*, respectively, at* $\mathbf{p}$*.*

- One can extend $\mathcal{W}^s_{loc}(\mathbf{p})$ and $\mathcal{W}^u_{loc}(\mathbf{p})$ beyond U indefinitely as

$$\mathcal{W}^s(\mathbf{p}) = \bigcup_{t \leq 0} \varphi^t \left(\mathcal{W}^s_{loc}(\mathbf{p}) \right) \quad \text{and} \quad \mathcal{W}^u(\mathbf{p}) = \bigcup_{t \geq 0} \varphi^t \left(\mathcal{W}^u_{loc}(\mathbf{p}) \right),$$

  the *stable and unstable manifolds* of $\mathbf{p}$ in $\dot{\mathbf{x}} = \mathbf{F}(\mathbf{x})$. These spaces comprise orbits that are attracted to the critical point in either forward or backward time.

- The entirely similar concepts for discrete dynamical systems given by functions $f : X \to X$ are the subject of Definition 7.19 in Section 7.1.2, where they are generalized to periodic as well as general points in X.

4.6 Isolated Periodic Orbit Stability

Now, let $\mathbf{x} \in \mathbb{R}^n$ and $\dot{\mathbf{x}} = \mathbf{F}(\mathbf{x})$ be an ODE system. Recall that, for fixed $t \in \mathbb{R}$, $\phi^t : \mathbb{R}^n \to \mathbb{R}^n$ is the time-t map; a "slice" of the flow in trajectory space given by $\phi : \mathbb{R}^{n+1} \to \mathbb{R}^n$, so that

$$\phi^{t_0}(\mathbf{x}) = \phi(t, \mathbf{x})\big|_{t=t_0} = \phi(t_0, \mathbf{x}).$$

Let $\mathbf{p}$ be a T-periodic point that is not an equilibrium solution (think of the point $(1, \pi/2)$ in the $r\theta$-plane of the system in Equation 2.5.1 in Section 2.5. There, $T = 2\pi$.
Then $\mathbf{p} \in Fix(\phi^T)$, but $\mathbf{F}(\mathbf{p}) \neq \mathbf{0}$.

Lemma 4.51 1 *is an eigenvalue of the matrix* $D\phi^T_{\mathbf{p}}$*.*

The proof is quite straightforward and really a vector calculus calculation. However, the implications are what is interesting here:

- For the time-T map that matches the period of the cycle perfectly, the point $\mathbf{p}$ appears as a fixed point (every point on the cycle will share this property).
- The time-T map $\phi^T : \mathbb{R}^n \to \mathbb{R}^n$ is a transformation that takes the entire phase space to itself, and is in general nonlinear.

- Since **p** is fixed by ϕ^T, the derivative map $D\phi_{\mathbf{p}}^T : T_{\mathbf{p}}\mathbb{R}^n \to T_{\mathbf{p}}\mathbb{R}^n$ is simply a linear transformation of the tangent space to $\mathbb{R}^n$ at the point **p**.

Proof The directional derivative of ϕ^T in the direction of the cycle (the curve parameterized by t, really) is the vector field $\mathbf{F}(\mathbf{p})$, and

$$\mathbf{F}(\mathbf{p}) = \mathbf{F}\left(\phi^T(\mathbf{p})\right) = \frac{d}{ds}\left(\phi^s(\mathbf{p})\right)\Big|_{s=T} = \frac{d}{ds}\left[\phi^T \circ \phi^s(\mathbf{p})\right]\Big|_{s=0} = D\phi_{\mathbf{p}}^T\left(\mathbf{F}(\mathbf{p})\right).$$

The first equality is because **p** is T-periodic, the second is due to the definition of a vector field given by an ODE, the third is because of the autonomous nature of the ODE, and the last ..., well, work it out. The end effect is that we have constructed the standard eigenvalue/eigenvector equation $\lambda\mathbf{v} = A\mathbf{v}$, where here $\lambda = 1, \mathbf{v} = \mathbf{F}(\mathbf{p})$ and the derivative matrix is A. □

Definition 4.52 *For* **p** *a* T*-periodic point, call the other eigenvalues of* $D\phi_{\mathbf{p}}^T$ *the eigenvalues of* **p**.

Remark 4.53 *A cycle, or periodic solution of an ODE system, is a generalization of an equilibrium solution in many ways. It is another example of a closed, bounded solution that limits to itself. Solutions that start nearby may or may not stay nearby, and may even converge to it. This give cycles the property of stability, much like equilibria. Many mechanical systems do exhibit stable states that are not equilibria, like the undamped pendulum. And with just the right forcing function, the damped (but forced) pendulum of a grandfather clock is an asymptotically stable cycle. How to analyze the neighboring solutions to see if a cycle is stable or not requires watching the evolution of these nearby solutions. The time-t map, and its cousin the first return map, are ways to do this.*

We end this section with a result that goes back to the notion of a contraction map:

Proposition 4.54 *If* **p** *is a periodic point with all of its eigenvalues of absolute value strictly less than* 1, *then* $\mathcal{O}_{\mathbf{p}}$ *is an asymptotically stable limit cycle.*

Here the tangent linear map at **p** carries the infinitesimal variance in the vector field, which in turn betrays what the neighboring solutions do over time. The non-infinitesimal version theoretically plays the same role. Construct S, the closure of a small open subset

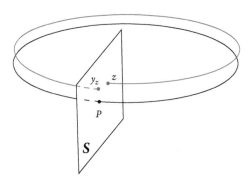

Figure 75 A Poincaré section at **p**.

of $\mathbb{R}^{n-1}$ centered at $\mathbf{p}$ and transversal to the vector field $\mathbf{F}(\mathbf{p})$ at $\mathbf{p}$ (see Figure 75). Owing to the continuity of the vector field given by the ODE system, all solutions of the ODE system that start in S sufficiently close to $\mathbf{p}$ will leave S, circle around, and again cross S. In the case where all solutions cross again at time-T (the period of $\mathbf{p}$), then the time-T map defines a discrete dynamical system on S. When this is not the case (usual in nonlinear systems), we neglect where the nearby solutions are at time T and simply look for where orbits again cross S. This latter case is the difference between a time-T map and the first return map. However, both of these constructions coalesce nicely into the infinitesimal version.

Exercise 170 Solve the first-order, autonomous nonlinear ODE system in cylindrical coordinates $\dot{r} = \frac{1}{2}r(1-r)$, $\dot{\theta} = 1$, $\dot{z} = -z$ and show that there exists an asymptotically stable limit cycle (Hint: Since the system is uncoupled, you can solve each ODE separately.) What are the eigenvalues of the 2π-periodic point at $p = (1, 0, 0)$?

Establishing that a known periodic orbit is attractive is straightforward. However, attractive periodic orbits, even in 2 dimensions, are not always easy to find. Fortunately, there are sometimes ways to detect their existence, at least in the plane, where a non-trivial periodic solution to an ODE system is a curve that separates the plane into two regions; inside and outside. This is the essence of a *Jordan curve*, defined below. Here are two ways to "see" a periodic solution.

4.6.1 The Poincaré–Bendixson Theorem

One way to detect the existence of a periodic orbit in a flow in the plane is to, in essence, trap an orbit in a bounded region. For a C^1 vector field, where the trajectories exist and are uniquely defined periodic orbits or regions whose boundaries consist of unions of orbits, you separate the plane into regions in such a way that orbits cannot move from region to region. Once an orbit is trapped in a bounded region, you would then know that its long-term behavior, encoded in its ω-limit set, also exists in that region. And since ω-limit sets always consist of entire orbits (by Proposition 2.13), one can form existence theorems based on this analysis.

First, let us set up some background objects for this discussion:

Definition 4.55 *The image of a continuous planar map* $\mathbf{c} : [0, 1] \to \mathbb{R}^2$ *is called a* Jordan arc *or* Jordan curve *if* $\mathbf{c}(t)$ *is injective, and a* closed Jordan curve *if*

(1) $\mathbf{c}(0) = \mathbf{c}(1)$, *and*

(2) $\mathbf{c}(t)$ *is injective on* $[0, 1)$.

Some notes:

- One can also define a closed Jordan curve via an injective circle map $c : S^1 \to \mathbb{R}^2$.
- Closed Jordan curves are also called *simple closed curves*.
- The complement of a closed Jordan curve $\mathbb{R}^2 \setminus \mathbf{c}$, or $\mathbb{R}^2 - \mathbf{c}$ consists of two disjoint, open regions: the "interior" of $\mathbf{c}$, or $\text{Int}(\mathbf{c})$, which is bounded by $\mathbf{c}$; and the "exterior" of $\mathbf{c}$, or $\text{Ext}(\mathbf{c})$, which is unbounded but has boundary $\mathbf{c}$.

Figure 76 D_1 is simply connected. D_2 is not.

Definition 4.56 *A region* $D \subset \mathbb{R}^2$ *is called* simply connected *if for every closed Jordan curve* $\mathbf{c} \subset D$, *we have* $\text{Int}(\mathbf{c}) \subset D$.

The basic idea of a region being simply connected is that it has no "holes." In Figure 76, D_1 is simply connected, while D_2 is not. Note that closed Jordan curves are planar objects specifically (at least limited to surfaces), but the notion of a region being simply connected is not dimension-specific. However, we will only use this notion in the plane.

Another property of a region, as a topological space, that we will employ presently is connectedness, that the space comes in "one piece." There is an interesting and powerful characterization:

Definition 4.57 *A set* $X \in \mathbb{R}^n$ *is* connected *if it cannot be written as* $X = U \cup V$, *where* U *and* V *are disjoint, open, non-empty subsets of* X.

Note that it is equivalent to say (though we will not prove this here) that a space is connected if it cannot be written as the union of two closed sets that are disjoint and non-empty. For example, break the real line (a connected set) into two pieces, with the point at the break residing on only one side. This renders that side closed. But then the other side, missing the break point, is open. Furthermore, open disjoint connected sets can be zero distance apart, as in $(-1,0)$ and $(0,1)$. Alas, so can closed, disjoint connected sets. As an example, each of the planar set of solutions to $y = 0$ and $y = 1/x$ are closed, as subsets of $\mathbb{R}^2$. And they are zero distance apart. But compact, disjoint connected sets must be separated by a positive distance. Think about this.

Proposition 4.58 (Bendixson's Criterion) *Suppose* f *and* g *are* C^1 *real-valued functions on a simply connected region* $D \in \mathbb{R}^2$. *Then the ODE system*

$$(4.6.1) \qquad \dot{\mathbf{x}} = \begin{bmatrix} \dot{x} \\ \dot{y} \end{bmatrix} = \begin{bmatrix} f(x,y) \\ g(x,y) \end{bmatrix} = \mathbf{F}(x,y)$$

has no periodic orbits $\mathbf{c}$ *lying entirely in* D *if* $\nabla \cdot \mathbf{F}$ *does not change sign in* D *or if* $\nabla \cdot \mathbf{F}$ *is not identically 0 in* $\text{Int}(\mathbf{c}) \subset D$.

Proof Suppose $\mathbf{c}$ is a closed Jordan curve that corresponds to a periodic solution of the system in Equation 4.6.1 lying entirely in D. Then, by Green's Theorem (really the planar version of the Divergence Theorem from your vector calculus course)

$$\oint_{\mathbf{c}} \mathbf{F} \cdot \mathbf{n} \, ds = \iint_{\text{Int}(\mathbf{c})} \nabla \cdot \mathbf{F} \, dx \, dy,$$

where s is the arc-length parameter on $\mathbf{c}$ and $\mathbf{n}$ is the outward-pointing unit normal to $\mathbf{c}$. Since $\mathbf{c}$ is a solution to the ODE system (it is tangent to the vector field at all points,) then the left-side integral is identically 0. But if the divergence of $\mathbf{F}$ never changes sign in D nor is identically 0 on the closed, bounded region $\mathbf{c} \cup \mathrm{Int}(\mathbf{c}) \subset D$, then this quantity cannot be 0. Hence there is no such $\mathbf{c}$. $\qquad \square$

Example 4.59 The damped linear oscillator is an example of an application modeled on a second-order, linear, homogeneous ODE with constant coefficients

$$\ddot{x} + p\dot{x} + qx = 0,$$

where $p, q > 0$. Rewriting this as a system, we get

$$\dot{\mathbf{x}} = \begin{bmatrix} \dot{x} \\ \dot{y} \end{bmatrix} = \begin{bmatrix} 0 & 1 \\ -q & -p \end{bmatrix}\begin{bmatrix} x \\ y \end{bmatrix} = \begin{bmatrix} y \\ -qx - py \end{bmatrix} = \mathbf{F}(x,y).$$

Here $\nabla \cdot \mathbf{F} = -p < 0$ on all of $\mathbb{R}^2$. Hence the damped linear oscillator has no non-trivial periodic solutions. However, notice that when $p = 0$, effectively undamping the oscillator, there may be (and, in fact, are) periodic solutions.

Exercise 171 Show that the system $\dot{x} = x(2y+1)$, $\dot{y} = x^2 - y^2$ has no periodic solutions.

Example 4.60 The van der Pol equation is $x'' - a(1 - x^2)x' + x = 0$, where a is a positive constant. Written as a first-order system, it is

$$\dot{x} = y,$$
$$\dot{y} = a(1 - x^2)y - x.$$

Here, it is easy to see immediately, that $\nabla \cdot \mathbf{F} = a(1 - x^2)$, and thus, if a periodic orbit does exist, it must non-tangentally intersect at least one of the vertical lines $x = \pm 1$. See Figure 77. For more detail on this model, also see Section 4.6.4.

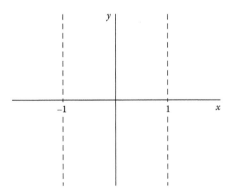

Figure 77 Phase plane for van der Pol equation.

Example 4.61 Let a vector field in $\mathbb{R}^n$ be a C^1-gradient field, defined by a C^2-potential $V : \mathbb{R}^n \to \mathbb{R}$, so that

$$\dot{\mathbf{x}} = -\nabla V(\mathbf{x}).$$

Then it is also the case that no periodic orbits can exist in this system. To see this, suppose that there exists a non-trivial periodic point $\mathbf{x}_0 \in \mathbb{R}^n$ of period $T > 0$, so that $\mathbf{x}_0 = \mathbf{x}(0) = \mathbf{x}(T)$. Then $V(\mathbf{x}(0)) - V(\mathbf{x}(T)) = 0$. But then

$$0 = V(\mathbf{x}(0)) - V(\mathbf{x}(T)) = \int_0^T \frac{dV}{dt}\, dt = \int_0^T \nabla V \cdot \dot{\mathbf{x}}\, dt = -\int_0^T ||\dot{\mathbf{x}}||^2\, dt < 0.$$

Hence there cannot exist a periodic orbit.

Bendixson's Criterion is very useful (in the cases where it can be used) to establish when periodic solutions cannot exist in a planar flow. However, it doesn't actually ensure the existence of a periodic solution (the contrapositive simply says that in those instances where a periodic solution actually exists, then the divergence of the planar flow must behave a certain way). Fortunately, at least in the plane, there are alternate criteria to employ. To understand this better, let's start with an example:

Consider the system

(4.6.2)
$$\begin{aligned}\dot{x} &= x(2 - \sqrt{x^2 + y^2}) - y \\ \dot{y} &= y(2 - \sqrt{x^2 + y^2}) + x.\end{aligned}$$

Since this is nonlinear, one may attempt a qualitative analysis to understand the dynamical behavior of the continuous dynamical system in the plane.

Exercise 172 Show that the system in Equation 4.6.2 has a unique equilibrium at the origin and use local linear theory to classify the origin as a spiral source.

Given this, one would expect that the vector field "near" the origin would have an outward component (from the origin) to it at every point distinct from the origin. Indeed, let $r = \sqrt{x^2 + y^2}$ denote the distance from the origin to the point (x,y). Then

$$\dot{r} = \frac{1}{r}(x^2(2 - r) + y^2(2 - r)) = r(2 - r).$$

So, for $\mathbf{x} = \begin{bmatrix} x \\ y \end{bmatrix}$ such that $||\mathbf{x}|| = 1$, we have $\dot{r} > 0$, and $\mathcal{O}_\mathbf{x}$ is moving away from the origin. In contrast, for $\mathbf{x}$ where $||\mathbf{x}|| = 3$, we have $\dot{r} < 0$, and $\mathcal{O}_\mathbf{x}$ is moving toward the origin. One could say that $\mathcal{O}_\mathbf{x}$ in this case is spiraling but losing altitude, perhaps. One can visualize this situation by drawing part of the vector field along these two curves that trap the annulus in Figure 78. Call this annular region $A = \left\{ (x,y) \in \mathbb{R}^4 \mid 1 \le \sqrt{x^2 + y^2} \le 3 \right\}$.

Since the vector field of this system always points inward all along the two boundary components of A, one can readily see that A is forward invariant under the flow. (It is certainly not backward invariant!) Thus A traps forward orbits. In fact, in this particular system, A traps all nontrivial forward orbits.

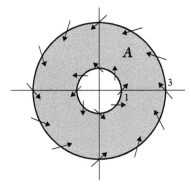

Figure 78 The vector field along the boundary of A.

Exercise 173 Convert this system to polar coordinates (indeed, one equation of the coordinate change is already present), and prove this last assertion. That is, for every $\mathbf{x} \in \mathbb{R}^2 - \mathbf{0}$, there exists a $t_x \in \mathbb{R}$ such that for all $t > t_x$, $\mathcal{O}_x(t) \in A$.

So the final questions to ask are: If all orbits are trapped in A eventually, then what happens to the orbits once inside A? Where do they go? What is the orbit structure of this system? To answer these questions, recall that the long-term behavior of orbits of a dynamical system is encapsulated in its limit sets, both in forward time, and, when defined, in backward time. So, before fully answering these questions, let take a closer look at the α- and ω-limit sets of points in a dynamical system.

4.6.2 Limit Sets of Flows

To understand the Poincaré–Bendixson Theorem, one needs to well-understand the structure of the limit sets of orbits in a continuous dynamical system as attractors. We have already seen in Proposition 2.13 (and Corollary 2.55) that ω-limit sets (and α-limit sets) are invariant sets and contain only full orbits. Here, we will explore some more of the properties of these limit sets. Our focus will be on flows in this section, as preparation for the Poincaré–Bendixson Theorem. Let X be a subset of Euclidean space with a flow on it.

Proposition 4.62 *ω-limit sets and α-limit sets are closed.*

Proof Let $x \in X$ and suppose $y \in \overline{\omega(x)}$, but $y \notin \omega(x)$ (assuming $\omega(x)$ is not closed, there must be a limit point of $\omega(x)$ that is not in $\omega(x)$). Then there exists a sequence $\{z_i\}_{i \in \mathbb{N}} \longrightarrow y$ where, for each i, $z_i \in \omega(x)$. Now, for each $i \in \mathbb{N}$, there exists a sequence $\{t_{ij}\}_{j \in \mathbb{N}} \longrightarrow \infty$ such that $\{\mathcal{O}_x(t_{ij})\}_{j \in \mathbb{N}} \longrightarrow z_i$. We use this data to construct the following sequence $\{t_{nj_n}\}_{n \in \mathbb{N}} \longrightarrow \infty$: For $n = 1$, let $t_{1j_1} = t_{11}$. For $n = 2$, choose the smallest natural number j so that $t_{2j} > t_{11}$ (every subset of the naturals has a smallest element and the sequences are monotonic). Call this j_2, so that this element is t_{2j_2}. At the nth stage, choose the smallest j such that $t_{nj} > t_{(n-1)j_{n-1}}$, and declare this element to be $t_{n j_n}$. Then the sequence $\left\{t_{n j_n}\right\}_{n \in \mathbb{N}} \longrightarrow \infty$, and it is easily seen that $\left\{\mathcal{O}_x\left(t_{n j_n}\right)\right\}_{n \in \mathbb{N}} \longrightarrow y$. Thus $y \in \omega(x)$ and thus $\omega(x)$ contains all of its limit points and is closed. □

Do keep in mind, though, that even the empty set is closed. Thus the ω-limit set of each initial point x_0 of the system $\dot{x} = c \neq 0$ (the translational flow on $\mathbb{R}^n$) is still a closed set, even if $\omega(x_0) = \alpha(x_0) = \emptyset, \forall x_0 \in \mathbb{R}^n$. However, in our introductory example above, the annular region A is a compact region in the plane (closed and bounded). It turns out that this is sufficient to ensure at least that these limit sets are not empty:

Proposition 4.63 Let $\mathcal{O}_x^+$ (resp. $\mathcal{O}_x^-$) be bounded. Then $\omega(x) \neq \emptyset$ (resp. $\alpha(x) \neq \emptyset$).

Proof Let $x \in X$, and, with a flow $\varphi(t, x)$, assume that $\mathcal{O}_x^+$ is bounded. This means that there exists a closed, bounded (hence compact) set $\mathcal{K}$ such that $\mathcal{O}_x^+ \subset \mathcal{K}$. But then, by Proposition 4.62, $\omega(x)$ is also compact, being a closed, bounded (possibly empty) subset of a compact space. Now choose a sequence $\{t_i\} \to \infty$. Then, $\left\{\mathcal{O}_x^+(t_i)\right\} \subset \mathcal{K}$ is an infinite bounded sequence in $\mathbb{R}^n$. By the celebrated Bolzano–Weierstrass Theorem, we know that any bounded, infinite sequence in $\mathbb{R}^n$ has a convergent subsequence. Limited to that convergent subsequence, the limit point satisfies the definition of an element of $\omega(x)$. Hence $\omega(x) \neq \emptyset$. □

So ω-limit sets of bounded orbits are compact and non-empty. Another feature of these limit sets of bounded orbits is that they always "come in one piece"; they are connected:

Proposition 4.64 Let $\mathcal{O}_x^+$, the (positive) semi-orbit of x, be bounded. Then $\omega(x)$ is connected.

Proof Suppose that $\omega(x)$ is not connected for some $x \in X$ with a bounded forward orbit. Then there exist two disjoint open sets $U, V \in X$ such that $U \cap \omega(x)$ and $V \cap \omega(x)$ are non-empty and $\omega(x) \subset U \cup V$. (Note here that, since $\omega(x)$ is closed, even if $d(U, V) = 0, d(U \cap \omega(x), V \cap \omega(x)) > 0$.) Now choose $u \in U$ and $v \in V$ and two sequences $\{t_i\}_{i \in \mathbb{N}} \longrightarrow \infty$ and $\{s_j\}_{j \in \mathbb{N}} \longrightarrow \infty$ such that $\{\mathcal{O}_x(t_i)\}_{i \in \mathbb{N}} \longrightarrow u$ and $\left\{\mathcal{O}_x(s_j)\right\}_{j \in \mathbb{N}} \longrightarrow v$. For each t_i, there exists k such that $s_{k-1} \leq t_i < s_k$. Then, on the interval (t_i, s_k), there exists a τ_i such that $\mathcal{O}_x(\tau_i) \notin U \cup V$. Here $\{\tau_i\}_{i \in \mathbb{N}} \longrightarrow \infty$ (it is bounded below by $\{t_i\}$), and the set $\{\mathcal{O}_x(t_i)\}$ is a bounded, infinite set of points. Hence there is a subsequence that converges to a point z. But this means that $z \in \omega(x)$ but $z \notin U \cup V$. But this is impossible, and this contradiction establishes the result. □

A beautiful corollary to this is the following exercise:

Exercise 174 Show that if $\omega(x)$ contains only equilibria, then it consists entirely of one equilibrium.

One might gather that ω-limit sets are always connected. However, this is not true, even in the plane.

Example 4.65 Let $\dot{x} = y - x(y^2 - 1), \dot{y} = x(y^2 - 1)$ be a planar system. It is then easily shown that the origin is the only equilibrium, and it is a spiral source. Not so easily shown is that the lines

$$\ell_\pm = \left\{ (x,y) \in \mathbb{R}^2 \mid y = \pm 1 \right\}$$

are invariant in this system, and for every $\mathbf{x} = (x,y) \in \mathbb{R}^2$ such that $-1 < y < 1$, $\omega(\mathbf{x}) = \ell_+ \cup \ell_-$.

And one last interesting little fact, which we shall present as a lemma for use later instead of a general result:

Lemma 4.66 *Let $x, y, z \in \mathbb{R}^n$. If, given a flow, $z \in \omega(y)$ and $y \in \omega(x)$, then $z \in \omega(x)$.*

Proof Since $y \in \omega(x)$, we know $\mathcal{O}_y \subset \omega(x)$. So if $z \in \omega(y)$, then there exists $\{t_i\} \to \infty$ where $\{y_i = \mathcal{O}_y(t_i)\} \to z$. Now, each $y_i \in \omega(x)$, so for each $i \in \mathbb{N}$, there exists $\{t_{ij}\}_{j \in \mathbb{N}} \to \infty$ such that $\{\mathcal{O}_x(t_{ij})\} \to y_i$. Create a new sequence $\{\mathcal{O}_x(t_{ii})\}_{i \in \mathbb{N}}$. Here $\{t_{ii}\} \to \infty$ and $\{\mathcal{O}_x(t_{ii})\} \to z$. Hence $z \in \omega(x)$. □

4.6.3 Flows in the Plane

Limiting our discussion now to differentiable flows in the plane, the uniqueness of solutions creates ideal conditions to limit the behavior of bounded orbits. So, let $\varphi^t : \mathbb{R}^2 \to \mathbb{R}^2$ be a C^1-flow in the plane. Recall from Section 2.5 that, at a *regular point* of the flow (where the vector field is non-zero), one can create a Poincaré section, a small co-dimension-1 subspace that is non-tangent to the flow. When first described, it was used to study the flow near a periodic-orbit since all nearby orbits that cross this Poincaré section will again cross the section at a later time, thus creating the Poincaré map of the flow, local to that periodic point. Here, though, at any regular point, one can create such a section, which is also often called a (local) *transversal* to the flow. For a planar flow, this local transversal is a curve, the image of a subinterval of $\mathbb{R}$. And while we do not explicitly need to parameterize such a curve, there is an automatic sense of direction (an orientation to the curve) compatible with the flow and the orientation of the plane. This is important, as we will presently see:

Lemma 4.67 *Given a parameterized transversal L_x to a trajectory $\mathcal{O}_x$ of a C^1-planar flow at a regular point $x \in \mathbb{R}^2$, the set of all intersections*

$$\left\{ x_i = \mathcal{O}(t_i) \mid x_i \in \mathcal{O}_x \cap L_x \right\}$$

is monotonic in time.

Proof If $\mathcal{O}_x$ intersects L_x uniquely, there is nothing to prove. So, suppose that x_1 and x_2 are two adjacent intersections of $\mathcal{O}_x$ with L_x; this means that, for $t_2 > t_1$, on the open interval (t_1, t_2), $\mathcal{O}_x \cap L_x = \emptyset$. Then we are in one of the situations shown in Figure 79:
   Denote the orbit segment from x_1 to x_2 by $\mathcal{O}_x([t_1, t_2])$ and the curve segment along L_x from x_1 to x_2 by $L_x([x_1, x_2])$. Then, in either case, the set

$$J_x = \mathcal{O}_x([t_1, t_2]) \cup L_x([x_1, x_2])$$

is a closed, Jordan curve. On the left side of Figure 79, for all $t > t_2$, $\mathcal{O}_x(t) \in \text{Int}(J_x)$, and hence any future intersections of the orbit with L_x will occur on the side opposite to x_1. On the right side of Figure 79, for all $t > t_2$, $\mathcal{O}_x(t) \in \text{Ext}(J_x)$, and hence

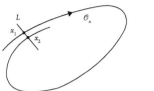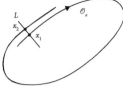

Figure 79 Jordan curves created by an orbit segment and a transversal.

any future intersections of the orbit with L_x will again occur on the side opposite to x_1. Since this is true for any two adjacent crossings, it is true for the entire intersection set. □

This last statement is true whether the number of intersections of an orbit with a transversal is finite or not. But it is even more stringent when an orbit serves as part of an ω-limit set of points in the system:

Lemma 4.68 *Let* $y \in \omega(x)$ *be a regular point. Then, for a transversal* L_y *of the flow at* y,
$$\mathcal{O}_y \cap L_y = \{y\}.$$

Proof Suppose that there does exist more than one intersection of $\mathcal{O}_y$ with L_y. Choose two such adjacent intersections, $y = y_1$ and y_2, where we will assume without any loss of generality that y_2 is later along the orbit that y_1. Then, as in the two cases of Figure 79, any future crossings of the orbit with L_y will be on the opposite side of y_2 from y_1. Hence there can be no sequence $\{t_i\} \to \infty$ where $\{\mathcal{O}_x(t_i)\} \to y_1 = y$. Hence $y \notin \omega(x)$, counter to the supposition. □

So limit sets can only cross a given transversal at one point. This limiting behavior (excuse the pun) tell us a lot about what sort of orbits can actually serve as limiting. This is especially true when the orbit, and hence its ω-limit set, is bounded. All of this leads directly to a remarkably beautiful existance theorem, named after Henri Poincaré and Ivar Bendixson, as stated in Bendixson's paper [6] in *Acta Mathematica* in 1901:

Theorem 4.69 (The Poincaré–Bendixson Theorem) *For* $x \in \mathbb{R}^2$, *if* $\mathcal{O}_x^+$ *is bounded and* $\omega(x)$ *contains no critical points, then either* x *is periodic or* $\omega(x)$ *is periodic.*

Proof $\omega(x)$ is non-empty by Proposition 4.63, so let $y \in \omega(x)$ and consider $z \in \omega(y)$. Since $y \in \omega(x)$, $\mathcal{O}_y \subset \omega(x)$ and is a bounded orbit. Hence $\omega(y)$ is also non-empty. By Lemma 4.66, then $z \in \omega(x)$. Form a transversal L_z to the flow at z. Then, there exists $\{t_i\} \to \infty$ such that $\{y_i = \mathcal{O}_y(t_i)\} \to z$, chosen so that each $y_i \in L_z$. But $y \in \omega(x)$, and is not a critical point. Hence, by Lemma 4.68, $\mathcal{O}_y \cap L_z = \{y_*\}$. Thus $y_i = y_*$ for all $i \in \mathbb{N}$, so that $\mathcal{O}_y$ is periodic and $y_* = z$. Thus $\omega(x)$, without critical points, contains a periodic orbit.

Now suppose that $u \in \omega(x)$ but $u \notin \mathcal{O}_y$. Then $\mathcal{O}_u \subset \omega(x)$, and $\mathcal{O}_u$ cannot be closed, since $\omega(x)$ is connected by Proposition 4.64.) But then $\mathcal{O}_y \subset \overline{\mathcal{O}_u}$. But, in this case, $\mathcal{O}_u \cap L_z \neq \emptyset$ and must consist of a single point that is not equal to z and hence a finite positive distance along L_z away from z. But then $\mathcal{O}_y \not\subset \overline{\mathcal{O}_u}$. This contradiction establishes that $\omega(x)$ can consist of a single periodic orbit. Hence either $\mathcal{O}_x = \omega(x)$, or tends toward the periodic $\omega(x)$. □

Some notes:

- Summarily speaking, then, at least in the plane, if a bounded, non-empty $\omega(x)$ contains no equilibria, then it consists entirely of one periodic orbit.
- In contrast, however, when $\omega(x)$ (for $\mathcal{O}_x^+$ a bounded orbit) does contain equilibria, the situation is a bit more complicated. And this sometimes leads to many alternate versions of this theorem currently in use in modern dynamical systems. If $\omega(x)$ contains a finite positive number of equilibria, then, as a connected set, it also contains orbits either heteroclinic to these fixed points, or homoclinic to the fixed point.
- But the basic idea of this theorem is the existence of a periodic orbit in any bounded region without equilibria that contains the full forward orbit of a point.

Here is an interesting consequence of the Poincaré–Bendixson Theorem:

Proposition 4.70 Let $\dot{x} = f(x)$ be a C^1-planar system with $\gamma(t)$ a non-trivial periodic solution. Then there exists an equilibrium solution within the interior of the Jordan curve $\gamma(x)$.

Note immediately that this offers a new criterion for the existence (or not) of periodic orbits in a flow in the plane: If a C^1-flow in the plane has no critical points, then it has no periodic solutions either. Something to think about.

Exercise 175 Prove Proposition 4.70.

4.6.4 Application: The van der Pol Oscillator

The van der Pol oscillator, introduced herein in Example 4.60, is modeled as a second-order, nonlinear, autonomous, and homogeneous ODE. As an example of the ODE $\ddot{x} = f(x,\dot{x})$, the equation is typically of the form

$$\ddot{x} - a(1 - x^2)\dot{x} + x = 0.$$

Written as a first-order system, with a new variable $y = \dot{x}$, we have

(4.6.3)
$$\dot{x} = y,$$
$$\dot{y} = a(1 - x^2)y - x.$$

This ODE, proposed and studied by Balthasar van der Pol [55] owing to the nature of its stable oscillatory behavior, is a model of a triode oscillator—an electronic amplifying vacuum tube used in early radios and various consumer electronics devices up to the 1970s.

Between experiments and numerical studies of the ODE, it is known that, for $a > 0$ a constant, there is a unique equilibrium at the origin and an isolated and unique asymptotically stable cycle, an attractive periodic orbit. However, the vector field is such that it is not obvious to see that this is the case without a bit of work. One way to prove the existence of such an attractive periodic orbit is via a careful construction involving the Poincaré–Bendixson Theorem. The construction involves creating a *trapping region*—a compact region in the plane bounded by one or more Jordan curves. In fact, the annulus A in Figure 78 is one such region, since the forward orbit of any point in A remains in A for all time. The trick here, though, is to properly understand the vector field in some of the more problematic regions, where it rapidly changes in small regions. The right side of Figure 80 shows the vector field of the van der Pol Equation 4.6.3, defined as a first-order system for $a = 1$. The left side of the figure shows a close-up detailing a particularly interesting region of the phase plane where the vector field is changing rapidly. A quick inspection of these figures reveals that standard choices for curves used to trap orbits, like ellipses and quadrilaterals, are not effective in creating the compact region needed to apply the theorem. Here, we construct two Jordan curves, which together enclose an annular region and properly trap orbits inside of it. The trick will be to construct the outer boundary of this annular region.

Stage 1, of course, is to locate and classify all equilibria of the system in Equation 4.6.3. It should be easy to see that the only equilibrium is at the origin, since for the vector field $\mathbf{V} = \begin{bmatrix} y \\ (1 - x^2)y - x \end{bmatrix} = \mathbf{0}$, we need $y = 0$, which leads immediately to also $x = 0$. And the Jacobian of the vector field, evaluated at the origin (constructed to create a corresponding linear system at the origin whose behavior mimics the original system), is

$$Jac(\mathbf{V})(0,0) = \begin{bmatrix} 0 & 1 \\ -2xy - 1 & (1 - x^2) \end{bmatrix}\Bigg|_{x=y=0} = \begin{bmatrix} 0 & 1 \\ -1 & 1 \end{bmatrix}.$$

The eigenvalues of this matrix satisfy the characteristic equation $r^2 - r + 1 = 0$, and are $r = \frac{1}{2} \pm \frac{\sqrt{3}}{2}i$. By directly appealing to the Hartman–Grobman Theorem 4.47, we can classify the equilibrium at the origin as a source (a spiral source.)

In Stage 2, we construct the trapping region. First, to easily construct the inner curve of what will be our annular region, we take the system in Equation 4.6.3 and convert it to polar coordinates:

(4.6.4)
$$\begin{aligned} \dot{r} &= -a(r^2 \cos^2 \theta - 1)r\sin^2 \theta, \\ \dot{\theta} &= -1 - a(r^2 \cos^2 \theta - 1)\sin\theta \cos\theta. \end{aligned}$$

Exercise 176 Using the coordinate change $x = r\cos\theta$, $y = r\sin\theta$, convert the system in Equation 4.6.3 to that of Equation 4.6.4.

Then, choosing $(r_0, \theta_0) \in \mathbb{R}^2$ such that $r_0 < 1$, for $a > 0$, the r-component of the vector field is

$$\dot{r}\big|_{(r,\theta)=(r_0,\theta_0)} = -a(r_0^2\cos^2\theta_0 - 1)r_0\sin^2\theta_0.$$

But since $r_0\sin^2\theta_0 \geq 0$ and $r_0^2\cos^2\theta_0 - 1 < 0$, we have $\dot{r} \geq 0$ along the entire circle defined by the equation $r = r_0$. So the Jordan curve

$$J_i = \left\{(r,\theta) \in \mathbb{R}^2 \mid r = r_0\right\}$$

is such that no orbit of the system can cross J_i from the outside to the inside.

Remark 4.71 *The next step is to construct an outer Jordan curve and create an annular region as our trapping region. But since the vector field depends on the value of the parameter $a > 0$, so will this outer curve. Hence, to simplify the construction, we will set the value of the parameter $a = 1$. For different values for a, one may have to alter the following construction.*

Second, to construct the outer curve of what will be our annular trapping region, we set $a = 1$ for convenience, and create the following Jordan curve $J_o = \bigcup\limits_{i=1}^{6}\ell_i$, using the following six lines, as in Figure 80 on the right:

$$\ell_1: \quad y = -7x + 28, \quad x \in \left[\tfrac{101}{25}, 3\right]$$

$$\ell_2: \quad y = 7, \quad x \in [3,1]$$

$$\ell_3: \quad y = \frac{1}{3}(4x + 17), \quad x \in \left[1, -\tfrac{101}{25}\right]$$

$$\ell_4: \quad y = -7x - 28, \quad x \in \left[-\tfrac{101}{25}, -3\right]$$

$$\ell_5: \quad y = -7, \quad x \in [-3, -1]$$

$$\ell_6: \quad y = \tfrac{1}{3}(4x - 17), \quad x \in \left[-1, \tfrac{101}{25}\right].$$

We claim that the vector field along J_o always points to the inside. Indeed, we employ two methods to establish this:

- compare the slope of the vector field along the line to the actual slope of the line; or
- show that the head of the vector field element along the line satisfies the inequality defining the half-plane region defined by the line that contains the origin.

To this end:

Case 1: The line ℓ_2 is defined by $y = 7$ on the interval $I_2 = [3,1]$ (we parameterize I_2 in this way to reflect counterclockwise movement along all of J_2). For $\mathbf{p} \in \ell_2$, we have

$$\mathbf{V_p} = \begin{bmatrix} 7 \\ (1 - x^2)7 - x \end{bmatrix}.$$

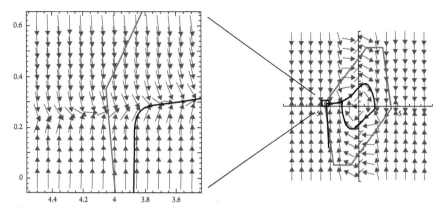

Figure 80 A van der Pol equation trapping region (right) and a vector field close-up (left).

Here, the y-component, $v_y(x) = 7 - 7x^2 - x$, is a decreasing function on I_2 ($v_y'(x) = -(14x - 1) < 0$), whose maximum is at the left endpoint $v_y(1) = -1$. Hence the vector field points downward at every point along I_2.

Case 2: The line ℓ_1 is defined by the equation $y = -7x + 28$ on the interval $I_1 = \left[\frac{101}{25}, 3\right]$. To show that the vector field elements along I_1 point to the same side of ℓ_1 as the origin, we show that the head of the vector field element satisfies the inequality $y < -7x + 28$. Indeed, for $\mathbf{w} = \mathbf{p} + \mathbf{V_p}$, we have

$$w_y = p_y + V_{p_y} = \left((1 - x^2)y - x + y\right)\big|_{y=-7x+28}$$
$$= (1 - x^2)(-7x + 28) - x + (-7x + 28)$$
$$= 7x^3 - 28x^2 - 15x + 56,$$
$$w_x = p_x + V_{p_x} = \left(y + x\right)\big|_{y=-7x+28} = -6x + 28.$$

Then the inequality is

$$7x^3 - 28x^2 - 15x + 56 < -7(-6x + 28) + 28,$$

which reduces to $7x^3 - 28x^2 - 57x + 224 < 0$ (see Figure 81). The following exercise establishes that the inequality is satisfied for all $x \in I_1$:

Exercise 177 For $g(x) = 7x^3 - 28x^2 - 57x + 224$, show that $g(x) < 0$ on $I_1 = \left[\frac{101}{25}, 3\right]$ by showing that g is less than 0 at the endpoints and g has a local minimum on I_1. Conclude that this is sufficient.

Case 3: The line ℓ_3 is defined by the equation $y = \frac{1}{3}(4x + 17)$ on the interval $I_3 = \left[-\frac{101}{25}, 1\right]$. Here again, we show that $\mathbf{w} = \mathbf{p} + \mathbf{V_p}$ satisfies the inequality $y < \frac{1}{3}(4x + 17)$ on I_3. We have

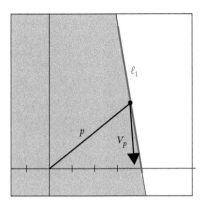
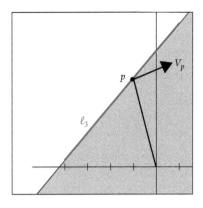

Figure 81 Vector field along trapping region elements ℓ_1 and ℓ_3.

$$w_y = p_y + V_{p_y} = \left((1-x^2)y - x + y\right)\big|_{3y=4x+17}$$

$$= (1-x^2)\left(\tfrac{1}{3}(4x+17)\right) - x + \frac{1}{3}(4x+17)$$

$$= \frac{1}{3}\left(-4x^3 - 17x^2 + 5x + 34\right),$$

$$w_x = p_x + V_{p_x} = (y+x)\big|_{3y=4x+17} = \frac{1}{9}(28x+119).$$

In this case, the inequality is

$$-4x^3 - 17x^2 + 5x + 34 < \frac{28}{3}x + \frac{119}{3},$$

which reduces to $-12x^3 - 51x^2 - 13x - 17 < 0$. Here, let $g(x) = -12x^3 - 51x^2 - 13x - 17$. We know the following:

- $\lim\limits_{x \to \infty} g(x) = -\infty,$
- $g\left(-\frac{101}{25}\right) < 0$, and
- $g(x)$ has a unique local maximum at $x_M = \frac{1}{12}(-17 + \sqrt{237}) \in I_3$, with $g(x_M) < 0$. (You should show this.)

Then we can conclude that $g(x) < 0$ on all of I_3. Again, see Figure 81.

Cases 4–6: As a function $\mathbf{V} : \mathbb{R}^2 \to \mathbb{R}^2$, $\mathbf{V}$ is odd, so that $\mathbf{V}(-\mathbf{p}) = -\mathbf{V}(\mathbf{p})$. Indeed,

$$\mathbf{V}(-\mathbf{p}) = \begin{bmatrix} (1-(-x)^2)(-y) - (-x) \\ -y \end{bmatrix} = \begin{bmatrix} -((1-x^2)y - x) \\ -y \end{bmatrix} = -\mathbf{V}(\mathbf{p}).$$

Hence $\mathbf{V}$ is symmetric with respect to the origin, so that the vector field in compatible with the rotation of the plane through π radians around the origin. Thus the vector field conditions for the lines ℓ_4, ℓ_5, and ℓ_6 will all satisfy the corresponding inequalities to those of ℓ_1, ℓ_2, and ℓ_3 above, suitably altered to reflect the new y-intercepts defining the lines and with the sense of each inequality reversed.

Exercise 178 Explicitly show that this last statement is true.

Exercise 179 For $g(x) = -12x^3 - 51x^2 - 13x - 17$, show that $g(x) < 0$ on $I_1 = \left[\frac{101}{25}, 3\right]$ by showing that g is less than 0 at the endpoints and g has a unique local extremum I_1, which is a local minimum. Conclude that this is sufficient.

Now the annular region in between the red Jordan curves in Figure 80 is a trapping region. With a simple application of the Poincaré–Bendixson Theorem 4.69, we see that there must exist a periodic orbit in this region. See the figure for a numerical graph of this orbit.

4.6.5 The Poincaré–Andronov–Hopf Bifurcation

Consider the following 2-dimensional autonomous ODE system in polar coordinates:

(4.6.5)
$$\dot{r} = ar - r^3 = f_a(r),$$
$$\dot{\theta} = 1.$$

This uncoupled system with a parameter $a \in \mathbb{R}$ has a differentiable vector field and is certainly solvable by standard integration techniques (for example, each equation is separable). Instead, we qualitatively study the system:

(a) The only equilibrium is at the origin for all values $a \in \mathbb{R}$. Notice that the second equation has no critical points. Hence the only place where a system equilibrium can exist is at a singular point of the coordinate system itself.

(b) If we graph $f_a(r)$ for various values of a, we find a qualitative change in the shape of the graph and its number of roots as a passes though 0, as can be seen in Figure 82.

(c) Neglecting $r < 0$, we see a new root of $f_a(r)$ appear for $a > 0$, corresponding to the critical point of the first equation of Equation 4.6.5, corresponding to $r = \sqrt{a}$. However, owing due to the second equation, this is an isolated periodic solution (of what period?)

(d) Looking at phase lines, we can easily classify the resulting interesting orbits as follows: For $a \leq 0$, the origin is an asymptotically stable spiral. For $a > 0$, the origin is now an unstable spiral point, and the new periodic orbit, given by $r(t) = \sqrt{a}$, is asymptotically stable.

Note that if we were to convert this system to rectilinear coordinates, we would have

(4.6.6)
$$\dot{x} = x\left(a - \left(x^2 + y^2\right)\right) - y,$$
$$\dot{y} = y\left(a - x\left(x^2 + y^2\right)\right) + x.$$

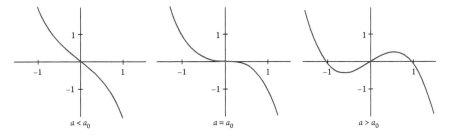

Figure 82 Qualitative change in $f_a(r)$ as a passes through 0.

Exercise 180 Perform the change of coordinates from polar to rectilinear, changing Equation 4.6.5 to Equation 4.6.6, and show that the origin is the only equilibrium in the rectilinear coordinates.

Linearizing this almost linear system at the equilibrium at the origin, we get the linear system $\dot{\mathbf{x}} = A_a\mathbf{x}$, where $A_a = \begin{bmatrix} a & -1 \\ 1 & a \end{bmatrix}$, and with eigenvalues $r = a \pm i$. It should be readily apparent that this family of linear systems has a linear Poincaré–Andronov–Hopf bifurcation at $a = 0$, studied in Section 4.2.1. In fact, it is precisely the linear system offered for study in Exercise 139.

This type of bifurcation in a system, where an isolated equilibrium switches its stability and a periodic solution is born, is called a Poincaré–Andronov–Hopf bifurcation (or, more familiarly, a Hopf bifurcation).

So what are the analytical criteria according to which such a bifurcation occurs? How can one ensure such a bifurcation has occurred without resulting to a phase portrait or actually producing solutions? For this, we turn to Andronov et al. [1], who characterized the conditions for this kind of situation:

Theorem 4.72 *Consider the locally linear system* $\dot{x} = F(x,y,a)$, $\dot{y} = G(x,y,a)$ *with an equilibrium at* $\mathbf{x}^0$ *and for which the eigenvalues of the corresponding linearized system matrix are*

$$r(a) = \alpha(a) \pm \beta(a)i.$$

Suppose that at $a = a_0$, *we have*

(i) $\alpha(a_0) = 0$, *but* $\beta(a_0) = \omega \neq 0$;

(ii) $\left.\dfrac{d\alpha(a)}{da}\right|_{a=a_0} = d \neq 0$; *and*

(iii) $b = \dfrac{1}{16}\left(F_{xxx} + F_{xyy} + G_{xxy} + G_{yyy}\right) + \dfrac{1}{16\omega}\left[F_{xy}(F_{xx} + F_{yy})\right.$
$\left. - G_{xy}(G_{xx} + G_{yy}) - F_{xx}G_{xx} + F_{xy}G_{xy}\right],$

with all derivatives evaluated at a_0. Then a unique curve of periodic solutions emanates from $\mathbf{x}^0$ into the region $a > a_0$ (for $bd < 0$), or for $a < a_0$ (for $bd > 0$), and the critical point changes stability as the parameter value passes through a_0, with the newly formed periodic orbits having the same stability as that of the equilibrium on the other side of a_0.

Notes: We call the bifurcation *supercritical* if the equilibrium changes from stable to unstable, and the new periodic orbits are stable. In contrast, if the equilibrium goes form unstable to stable, and the periodic orbits are unstable, we call this a *subcritical* Hopf bifurcation.

Example 4.73 As mentioned briefly in Exercise 140, a *brusselator* is a model for a chemical oscillator, first named after the European city by Ilya Prigogine and R. Lefever [46] in 1968. One basic version of the model is

(4.6.7)
$$\dot{x} = 1 - (a+1)x + bx^2 y,$$
$$\dot{y} = ax - bx^2 y,$$

where x and y are concentrations of two reactants (chemicals) and a and b are positive constants. Evidently, $(1, a/b)$ is the only equilibrium (you should establish this), and the parameterized matrix

$$B_{a,b} = \begin{bmatrix} a-1 & b \\ -a & -b \end{bmatrix}$$

is a 2-parameter family representing the Jacobian of the system at the equilibrium. For the purpose of this example, fix $b = 1$. Then the resulting 1-parameter families of matrices $B_{a,1} = B_a$ is the family studied in Exercise 140 in Section 4.2.1. Here, the characteristic equation of B_a is $r^2 - (a-2)r + 1 = 0$, with solutions

$$r(a) = \frac{a-2}{2} \pm \frac{\sqrt{4a - a^2}}{2} = \alpha(a) \pm \beta(a)i.$$

From this, we see that for $a = 2$, we have $\alpha(2) = 0$, and $\beta(2) = \omega = 1$, satisfying criterion (i). Also $\alpha'(a)\big|_{a=2} = \frac{1}{2} = d$, satisfying criterion (ii). And finally, with $F_x(x,y,2) = -3 + 2xy$ and $F_y(x,y,2) = x^2 = -G_x(x,y,2)$, and $G_x(x,y,2) = 2 - 2xy$, we have

$$F_{xx}(x,y,2) = -G_{xx}(x,y,2) = 2y,$$
$$F_{xxy}(x,y,2) = -G_{xxy}(x,y,2) = 2,$$
$$F_{xy}(x,y,2) = -G_{xy}(x,y,2) = 2x,$$

with all other third derivatives begin 0. Thus

$$b = \frac{1}{16}(-2) + \frac{1}{16(1)}\left[4xy - 4xy + 4y^2 - 4x^2\right]$$

which, when evaluated at the equilibrium $(x,y) = (1,2)$, gives us $b = \frac{5}{8}$, satisfying criterion (iii). Thus, by Theorem 4.72, a Hopf bifurcation occura at $a = 2$ for $b = 1$. Is this bifurcation supercritical or subcritical?

Exercise 181 Reformulate Example 4.73 leaving the parameter b as an undetermined constant and writing the Hopf bifurcation as a function of b. What is the range for b where the parameter a experiences a Hopf bifurcation? Then switch the roles of a and b, fixing the value for a and allowing b to vary. Again, develop criteria for when b experiences a bifurcation given values for a.

Exercise 182 Classify the example in Equation 4.6.5 as supercritical or subcritical, and then adapt it to provide an example of the other kind of Hopf bifurcation. Then find criteria for the constants b, d, and ω in Theorem 4.72 for which the bifurcation is supercritical and subcritical.

Exercise 183 Show that the first-order system corresponding to the *Liénard equation*

$$\ddot{x} - (a - x^2)\dot{x} + x = 0$$

has a supercritical Hopf bifurcation at $a = 0$.

4.7 Application: Competing Species

A common biological model for understanding the possible interaction between two species in a closed environment that interact only in their competition for resources is often referred to as the *Competing Species Model*. This model is different from a similar model where one species sees the other as food (called the Predator–Prey Model; detailed in Exercise 225 of Section 6.2.5 on first integrals of motion). Rather, in this model, two herbivores, say, both compete for limited food supplies. If the two species did not interact at all, their respective population equations would fit the Verhulst equation, the continuous ODE version of the logistic map from Section 2.6 and be uncoupled:

$$\dot{x} = x(\alpha_1 - \beta_1 x),$$
$$\dot{y} = y(\alpha_2 - \beta_2 y),$$

where $\alpha_1, \alpha_2, \beta_1, \beta_2 > 0$ are positive constants. We can model a simple interaction between these two species by adding in a cross-term, negative in sign (why?) and scaled by yet another parameter. We then have

(4.7.1)
$$\dot{x} = x(\alpha_1 - \beta_1 x - \gamma_1 y),$$
$$\dot{y} = y(\alpha_2 - \beta_2 y - \gamma_2 x),$$

where now all $\alpha_1, \alpha_2, \beta_1, \beta_2, \gamma_1, \gamma_2 > 0$ are again positive constants. What is the effect of these added terms? Are there steady-state solutions (equilibria, essentially)? If so, are there stable ones? To address these questions and more, let's look at these models for a few parameter assignments. Do not worry too much about just how modelers came up with this idea of simply adding terms. Let's focus on the nature of the solutions for now.

Here are two sets of parameter values, chosen solely for illustrative purposes:

(1) Let $\alpha_1 = \beta_1 = \gamma_1 = 1, \alpha_2 = \frac{3}{4}, \beta_2 = 1$ and $\gamma_2 = \frac{1}{2}$. The system is then

$$\dot{x} = x(1 - x - y),$$
$$\dot{y} = y\left(\frac{3}{4} - y - \frac{1}{2}x\right).$$

(2) Let $\alpha_1 = \beta_1 = \gamma_1 = 1, \alpha_2 = \frac{1}{2}, \beta_2 = \frac{1}{4}$ and $\gamma_2 = \frac{3}{4}$. The system is then

$$\dot{x} = x(1 - x - y),$$
$$\dot{y} = y\left(\frac{1}{2} - \frac{1}{4}y - \frac{3}{4}x\right).$$

Look at the slope fields for these two systems on the following pages and try to identify the differences between these two.

Some questions:

Question 1. Where are the critical points in these systems?

(1) Here, we solve the system

$$0 = x(1 - x - y),$$
$$0 = y\left(\frac{3}{4} - y - \frac{1}{2}x\right).$$

Of course, the origin $\mathbf{a} = (0,0)$ is one solution. But so are $\mathbf{b} = \left(0, \frac{3}{4}\right)$, $\mathbf{c} = (1,0)$, and $\mathbf{d} = \left(\frac{1}{2}, \frac{1}{2}\right)$. (Verify that you know how to find these!)

(2) In this case, the system is

$$0 = x(1 - x - y),$$
$$0 = y\left(\frac{1}{2} - \frac{1}{4}y - \frac{3}{4}x\right).$$

Again, we have $\mathbf{e} = (0,0)$. The others are $\mathbf{f} = (0,2)$, $\mathbf{g} = (1,0)$, and $\mathbf{h} = \left(\frac{1}{2}, \frac{1}{2}\right)$. See these?

Question 2. What are the type and stability of each of these equilibria?

- First note that, for any values of the parameters, the functions $F(x,y) = x(\alpha_1 - \beta_1 x - \gamma_1 y)$ and $G(x,y) = y(\alpha_2 - \beta_2 y - \gamma_2 x)$ are simply polynomials in x and y, and hence these systems are almost linear everywhere. Thus the Jacobian matrix serves as the associated linear system at each of the equilibria, at least as long as the determinant of the matrix

(4.7.2)

$$A = \begin{bmatrix} F_x\big|_{x^0} & F_y\big|_{x^0} \\ G_x\big|_{x^0} & G_y\big|_{x^0} \end{bmatrix} = \begin{bmatrix} \alpha_1 - 2\beta_1 x_0 - \gamma_1 y_0 & -\gamma_1 x_0 \\ -\gamma_2 y_0 & \alpha_2 - 2\beta_2 y_0 - \gamma_2 x_0 \end{bmatrix},$$

with $x^0 = (x_0, y_0)$ as the fixed point, is not 0.

1. In our first case, we have

$$A = \begin{bmatrix} 1 - 2x_0 - y_0 & -x_0 \\ -\tfrac{1}{2}y_0 & \tfrac{3}{4} - 2y_0 - \tfrac{1}{2}x_0 \end{bmatrix},$$

and at the four critical points we have

$$A_a = \begin{bmatrix} 1 & 0 \\ 0 & \tfrac{3}{4} \end{bmatrix}, \quad A_b = \begin{bmatrix} \tfrac{1}{4} & 0 \\ -\tfrac{3}{8} & -\tfrac{3}{4} \end{bmatrix}$$

$$A_c = \begin{bmatrix} -1 & -1 \\ 0 & \tfrac{1}{4} \end{bmatrix}, \quad A_d = \begin{bmatrix} -\tfrac{1}{2} & -\tfrac{1}{2} \\ -\tfrac{1}{4} & 0 \end{bmatrix}.$$

None of these has determinant 0, so the system is almost linear at each of these equilibria. The eigenvalues tell us that the corresponding linear systems have a source at **a**, saddles at both **b** and **c**, and a sink at **d**. (You should verify this!) So, by the Hartman–Grobman Theorem 4.47, the nonlinear equilibria will also have these types and their corresponding stability. The only one of these that is stable is the asymptotically stable sink at **d**.

2. Contrast these fixed points with those at **e**, **f**, **g**, and **h**. In the same fashion, we get the matrix

$$A = \begin{bmatrix} 1 - 2x_0 - y_0 & -x_0 \\ -\tfrac{3}{4}y_0 & \tfrac{1}{2} - \tfrac{1}{2}y_0 - \tfrac{3}{4}x_0 \end{bmatrix}.$$

Thus we have the four linear systems determined by the matrices

$$A_e = \begin{bmatrix} 1 & 0 \\ 0 & \tfrac{1}{2} \end{bmatrix}, \quad A_f = \begin{bmatrix} -1 & 0 \\ -\tfrac{3}{2} & -\tfrac{1}{2} \end{bmatrix},$$

$$A_g = \begin{bmatrix} -1 & -1 \\ 0 & -\tfrac{1}{4} \end{bmatrix}, \quad A_h = \begin{bmatrix} -\tfrac{1}{2} & -\tfrac{1}{2} \\ -\tfrac{3}{8} & -\tfrac{1}{8} \end{bmatrix}.$$

Again, calculate the eigenvalues, and verify that there is an unstable node (a source) at the origin (**e**), sinks at **f** and **g**, and a saddle at **h**.

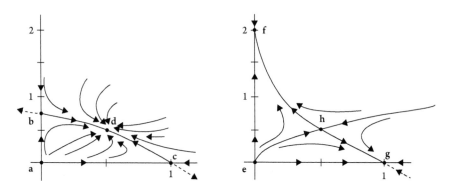

Figure 83 Competing Species phase portraits for $\delta = 0$ and $\delta = 1$.

Given all of this data, think about how the stability of each equilibrium would affect the nearby solutions. At the saddles, we would have to find the approximate directions of linear motion. The nonlinear saddles will locally defined stable and unstable dimension-1 manifolds emanating from them; curves comprising solutions all of which are asymptotic to the equilibrium in either forward or backward time. All other nearby solutions eventually veer away from the equilibrium. The curves of forward and backward asymptotic solutions wind up being tangent at the equilibrium to the directions of linear travel from the associated linear saddle at that equilibrium. This give an idea of how the nonlinear saddle is oriented. See Figure 83 for rough phase portraits of these two systems.

Last question: Suppose we construct a system using sliders for each of the parameters so that we can watch how the phase portrait changes as we alter the parameters continuously. Notice in the two examples above that in both we had $\alpha_1 = \beta_1 = \beta_2 = \gamma_1 = 1$. But in the first example, we had $\alpha_2 = \frac{3}{4}$, $\beta_2 = 1$ and $\gamma_2 = \frac{1}{2}$, and in the second, $\alpha_2 = \frac{1}{2}$, $\beta_2 = \frac{1}{4}$ and $\gamma_2 = \frac{3}{4}$. One way to link the two portraits in Figure 83 is to construct a single slider that simultaneously moves the three parameters from their values in the left phase portrait to those in the right.

So, we create a new variable δ, ranging from 0 to 1. We can use a linear parameterization from any vector $\mathbf{a} \in \mathbb{R}^3$ to any other vector $\mathbf{b} \in \mathbb{R}^3$ to "slide" these parameter values, via $\mathbf{x} = \mathbf{a} + \delta(\mathbf{b} - \mathbf{a})$. Then when $\delta = 0$, $\mathbf{x} = \mathbf{a}$, and when $\delta = 1$, $\mathbf{x} = \mathbf{b}$. Here, we do this with the three parameters above simultaneously:

$$\alpha_2 = \tfrac{3}{4} + \delta\left(\tfrac{1}{2} - \tfrac{3}{4}\right) = \tfrac{3}{4} - \tfrac{1}{4}\delta = \tfrac{3-\delta}{4},$$

$$\beta_2 = 1 + \delta\left(\tfrac{1}{4} - 1\right) = 1 - \tfrac{3}{4}\delta = \tfrac{4-3\delta}{4},$$

$$\gamma_2 = \tfrac{1}{2} + \delta\left(\tfrac{3}{4} - \tfrac{1}{2}\right) = \tfrac{1}{2} + \tfrac{1}{4}\delta = \tfrac{2+\delta}{4}.$$

Given this, we can continuously change the phase portrait and look for places where the number, type, and/or stability of any of the equilibria change. Given the profound differences between the two phase portraits, we will find something for some intermediate value of δ.

4.7.1 The Fixed Points

First in this analysis, understand that since we are changing the values of the parameters continuously, the fixed points (equilibria) will either stay where they are or move continuously also. So we can track them. We will do this by rewriting the vector field functions $F(x,y)$ and $G(x,y)$ in terms of δ instead of the parameters: here, equilibria are the solutions to the equations

$$F(x,y) = 0 = x(1 - x - y),$$
$$G(x,y) = 0 = y(\alpha_2 - \beta_2 y - \gamma_2 x) = y\left(\frac{3-\delta}{4} - \frac{4-3\delta}{4}y - \frac{2+\delta}{4}x\right).$$

By inspection, we find the following:

(1) Whenever $y = 0$, $G(x,y) = 0$. Hence any equilibria along the x-axis will not depend on δ for position at all. Hence the equilibria at $(0,0)$ and $(1,0)$ do not move for $\delta \in [0,1]$.

(2) For the non-trivial equilibrium along the y-axis, where $x = 0$ but $y \neq 0$, $F(x,y) = 0$ but $G(x,y) = 0$ only when $\alpha_2 - \beta_2 y = 0$, so $y = \alpha_2/\beta_2$. In terms of δ, there will be a critical point for the system when $x = 0$, and

$$y = \frac{\frac{3-\delta}{4}}{\frac{4-3\delta}{4}} = \frac{3-\delta}{4-3\delta}.$$

(3) Lastly, there seems to persist a critical point in the open first quadrant $x,y > 0$. This equilibrium will satisfy both

$$\left.\begin{array}{rcl} 1 - x - y & = & 0, \\ \alpha_2 - \beta_2 y - \gamma_2 x & = & 0 \end{array}\right\} \Rightarrow \left\{\begin{array}{rcl} x + y & = & 1, \\ \gamma_2 x + \beta_2 y & = & \alpha_2. \end{array}\right.$$

Combining these via $y = 1 - x$, we get

$$\gamma_2 x + \beta_2(1 - x) = \alpha_2, \quad \text{or} \quad x = \frac{\alpha_2 - \beta_2}{\gamma_2 - \beta_2}.$$

In terms of δ, we get

$$x = \frac{\frac{3-\delta}{4} - \frac{4-3\delta}{4}}{\frac{2+\delta}{4} - \frac{4-3\delta}{4}} = \frac{\frac{-1+2\delta}{4}}{\frac{-2+4\delta}{4}} = \frac{1}{2}.$$

Thus $y = 1 - x = \frac{1}{2}$ and we see that the equilibrium strictly in the first quadrant does not move for $\delta \in [0,1]$ and is at $\left(\frac{1}{2}, \frac{1}{2}\right)$.

Thus the four critical points for this model, in terms of δ are

$$(0,0), \quad (1,0), \quad \left(0, \frac{3-\delta}{4-3\delta}\right), \quad \text{and} \quad \left(\frac{1}{2}, \frac{1}{2}\right).$$

4.7.2 Type and Stability

Now the analysis moves toward a classification of these equilibria for the various values of $\delta \in [0,1]$. Recall that at any critical point $\mathbf{x}^0 = (x_0, y_0)$ of an almost linear system, the Jacobian of the system at $\mathbf{x}^0$ is the matrix A in Equation 4.7.2. Here, then

$$A_{(0,0)}(\delta) = \begin{bmatrix} 1 & 0 \\ 0 & \alpha_2 \end{bmatrix} = \begin{bmatrix} 1 & 0 \\ 0 & \frac{3-\delta}{4} \end{bmatrix}.$$

Thus, in this case, $r_1 = 1 > 0$ and $r_2 = r_2(\delta) = \frac{3-\delta}{4} > 0, \forall \delta \in [0,1]$. By the Hartman–Grobman Theorem 4.47, the equilibrium at the origin is a source for all $\delta \in [0,1]$.

At the static fixed point at $(1,0)$, we have

$$A_{(1,0)}(\delta) = \begin{bmatrix} -1 & -1 \\ 0 & \alpha_2 - \gamma_2 \end{bmatrix} = \begin{bmatrix} -1 & -1 \\ 0 & \frac{1-2\delta}{4} \end{bmatrix}.$$

Eigenvalues of $A_{(1,0)}(\delta)$ are immediately available to us, since the matrix is upper triangular, so the eigenvalues are the entries on the main diagonal:

$$r_1 = -1, \quad \text{and} \quad r_2 = \frac{1-2\delta}{4}.$$

One can readily show, via the eigenvector equation, that an eigenvector for $r_1 = -1$ is $\mathbf{v}_1 = \begin{bmatrix} 1 \\ 0 \end{bmatrix}$. Along the "other" direction, we have an eigenvalue/eigenvector pair

$$r_2 = \frac{1-2\delta}{4}, \quad \mathbf{v}_2 = \begin{bmatrix} 1 \\ \frac{2\delta-5}{4} \end{bmatrix}.$$

The interesting effect is at $\delta = \frac{1}{2}$, where the non-horizontal eigendirection is seen to slow to a stop, creating a curve of equilibria emanating from $(1,0)$. As δ passes through $\frac{1}{2}$, the eigenvalue r_2 goes from positive to negative, and the saddle bifurcates to a sink, passing through the value where the node is not isolated. This is a planar bifurcation where an unstable node can become stable.

Now, for the critical point $\left(0, \frac{3-\delta}{4-3\delta}\right)$, we get

$$A_{\left(0, \frac{3-\delta}{4-3\delta}\right)}(\delta) = \begin{bmatrix} 1 - y_0 & 0 \\ \gamma_2 y_0 & \alpha_2 - \beta_2 y_0 \end{bmatrix}$$

$$= \begin{bmatrix} 1 - \frac{3-\delta}{4-3\delta} & 0 \\ -\left(\frac{2+\delta}{4}\right)\left(\frac{3-\delta}{4-3\delta}\right) & \frac{3-\delta}{4} - 2\left(\frac{4-3\delta}{4}\right)\left(\frac{3-\delta}{4-3\delta}\right) \end{bmatrix}$$

$$= \begin{bmatrix} \frac{1-2\delta}{4-3\delta} & 0 \\ \left(\frac{2+\delta}{4-3\delta}\right)\left(\frac{\delta-3}{4}\right) & \frac{\delta-3}{4} \end{bmatrix}.$$

Here, the eigenvalues are $r_1 = \frac{1-2\delta}{4-3\delta}$ and $r_2 = \frac{\delta-3}{4}$. For the eigenvector $\mathbf{v}_2 = [v_1, v_2]^T$ corresponding to r_2, we have the eigenvector system

$$\left(\tfrac{1-2\delta}{4-3\delta}\right) v_1 = \left(\tfrac{\delta-3}{4}\right) v_1$$

$$\left(\tfrac{2+\delta}{4}\right)\left(\tfrac{\delta-3}{4}\right) v_1 + \left(\tfrac{\delta-3}{4}\right) v_2 = \left(\tfrac{\delta-3}{4}\right) v_2.$$

This system is solved by $v_1 = 0$ and v_2 anything non-trivial, so that the vector $\mathbf{v}_2$ is along the y-axis $\forall \delta \in [0,1]$.

For r_1, the eigenvector system is

$$\tfrac{1-2\delta}{4-3\delta} v_1 = \tfrac{1-2\delta}{4-3\delta} v_1,$$

$$\left(\tfrac{2+\delta}{4}\right)\left(\tfrac{\delta-3}{4}\right) v_1 + \left(\tfrac{\delta-3}{4}\right) v_2 = \left(\tfrac{1-2\delta}{4-3\delta}\right) v_2.$$

While this is a fairly messy calculation, it boils down to

$$v_1 = \tfrac{3\delta^2 - 21\delta + 16}{(2+\delta)(\delta-3)} v_2.$$

Upon inspection, one can readily see that both components will be non-zero for every $\delta \in [0,1]$, except for at one value: $\delta \sim 0.87$. At this point, one can show that $r_1 = r_2$, and there is only one eigendirection.

Exercise 184 Establish what is happening for this value of δ at the non-trivial critical point along the vertical axis.

Lastly, for this case, notice again, that one of the eigenvalues $r_1 = 0$, when $\delta = \frac{1}{2}$. This is precisely an instance of a (nonlinear) saddle-node bifurcation from a saddle to a sink, where one of the eigendirections slows down its repellent motion, stops, and then reverses direction. Compare this nonlinear model with the linear version in Section 4.2.2 and, in particular, in Figure 64.

And finally, let's analyze the stability, type and structure of the phase space at the point $\left(\frac{1}{2}, \frac{1}{2}\right)$. We have

$$A_{\left(\frac{1}{2},\frac{1}{2}\right)}(\delta) = \begin{bmatrix} -\frac{1}{2} & -\frac{1}{2} \\ -\frac{2+\delta}{8} & \frac{3-\delta}{4} - \frac{4-3\delta}{4} - \frac{2+\delta}{8} \end{bmatrix} = \begin{bmatrix} -\frac{1}{2} & -\frac{1}{2} \\ -\frac{2+\delta}{8} & \frac{-4+3\delta}{8} \end{bmatrix},$$

with eigenvalues

$$r = -\frac{(-8+3\delta) \pm \sqrt{(-8+3\delta)^2 - (32-64\delta)}}{16} = \frac{(8-3\delta) \pm \sqrt{9\delta^2 + 16\delta + 32}}{16}.$$

By inspection, both of the eigenvalues are real, with one of them remaining negative for all $\delta \in [0,1]$. The other one, however, is negative on $\delta \in \left[0, \frac{1}{2}\right)$, positive on $\delta \in \left(\frac{1}{2}, 1\right]$, and 0 when $\delta = \frac{1}{2}$. This, again, indicated a bifurcation value for δ, with the equilibrium going from a sink to a saddle.

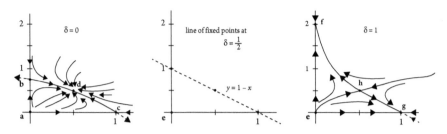

Figure 84 Can you draw the phase portrait for $\delta = \frac{1}{2}$?

In fact, at $\delta = \frac{1}{2}$, we have that strange situation where the three non-trivial critical points all have 0 as an eigenvalue of their linearization. This suggests a curve of critical points in the phase plane. We can appeal directly to the original ODE system to find this curve: Let $\delta = \frac{1}{2}$. Then the critical points are all at

$$F(x,y) = x(1 - x - y) = 0,$$

$$G(x,y) = y\left(\frac{3-\frac{1}{2}}{4}\right) - \left(\frac{4-\frac{3}{2}}{4}\right)y - \left(\frac{2+\frac{1}{2}}{4}\right)x\right) = 0$$

$$= y\left(\frac{5}{8} - \frac{5}{8}y - \frac{5}{8}x\right) = 0$$

$$= \frac{5}{8}y(1 - x - y) = 0.$$

At $\delta = \frac{1}{2}$, there will be a line of critical points, ranging within the first quadrant along the line $y = 1 - x$, from the equilibrium at $(1,0)$, through the equilibrium at $\left(\frac{1}{2}, \frac{1}{2}\right)$ to the the fixed point at $(1,0)$. Can you envision this?

Exercise 185 Figure 84 shows the equilibria of the nonlinear ODE Competing Species Model for $\delta = \frac{1}{2}$. The equilibria are at the origin and on the line $y = 1 - x$. Draw in the rest of the phase portrait as accurately as possible by linearizing the four persistent equilibria and extrapolating what happens off the line.

5 Recurrence

So far, we have explored many systems and contexts where dynamical systems have exhibited simple, or fairly simple, behavior. We will now begin to explore more complicated behavior. However, to start, we will stay with maps of the type we have already introduced. But we will change the place on which they are acting. This, in and of itself, changes the nature of the orbits. It turns out that when the space is Euclidean, orbits can converge to something or wander away toward the edge of the space. However, when a space is compact—roughly speaking, when its edges are not infinitely far away (if, in fact, they exist at all)—then an orbit that does not converge to any particular thing must go somewhere within the space. How to describe the long-term behavior of orbits in this context will introduce us to behavior that is more complicated than what we have already seen. To begin, consider the following definition:

Definition 5.1 *For $f : X \to X$ a continuous map of a metric space, a point $x \in X$ is called*

- positively recurrent *with respect to f if $\exists$ a sequence $\{n_k\} \longrightarrow \infty$ such that $\{f^{n_k}(x)\} \longrightarrow x$;*
- *if f is invertible,* negatively recurrent *if $\exists$ a sequence $\{n_k\} \longrightarrow -\infty$ such that $\{f^{n_k}(x)\} \longrightarrow x$;*
- recurrent *if it is both positively and negatively recurrent.*

Roughly speaking, the orbit of a point x will visit every open set containing x an infinite number of times. Another way to say this is that x is one of the limit points of its own orbit. It should be obvious that fixed and periodic points are recurrent; there is a subsequence of the orbit consisting of the constant orbit x. But in the simple dynamical systems we have studied so far, the only recurrent points have been fixed and periodic points. However, in the right context, non-periodic points can also be recurrent. This chapter begins a study of relatively simple maps that exhibit this more complicated behavior. And this behavior is captured in the notion of recurrence.

5.1 Rotations of the circle

Recall, from Section 3.3.2, that we can think of S^1 either as the set of unit-modulus numbers of the complex plane

$$S^1 = \left\{ z \in \mathbb{C} \mid z = e^{2\pi i \theta}, \theta \in \mathbb{R} \right\}$$

or as the quotient space of the real line modulo the integers, $S^1 = \mathbb{R}/\mathbb{Z}$. Under this last interpretation, one can imagine S^1 to be the unit interval $[0,1] \subset \mathbb{R}$, but one agrees to identify the endpoints (hence the notation $0 = 1$ we used in Remark 3.49). For clarity and to distinguish points between the two spaces, we denote by $\bar{x}, \bar{y} \in S^1$ their respective points $x, y \in \mathbb{R}$ under the exponential map $\rho : \mathbb{R} \to S^1, \rho(\theta) = e^{2\pi i \theta}$.

- Here $\bar{x} = \bar{y}$ iff $x - y \in \mathbb{Z}$, or $x \equiv y \,(\bmod\ 1)$.
- $\bar{x}, \bar{y}$ are the equivalence classes of points in $\mathbb{R}$ under the equivalence relation imposed on $\mathbb{R}$ by the map ρ.
- We can add and subtract points in S^1 in a way compatible with addition and subtraction in $\mathbb{R}$ under this equivalence relation,

$$\bar{x} + \bar{y} = \overline{x+y}, \text{ and } \bar{x} - \bar{y} = \overline{x-y}$$

by, in essence, declaring $\overline{x+y}$ to be "to the right" (in the sense of increasing $\theta \in \mathbb{R}$) of $\bar{x}$ at a distance $\bar{y}$. This immediately gives us a well-defined notion of multiplication of the equivalence classes on S^1 by integers, $n\bar{x} = \overline{nx}$, for $n \in \mathbb{Z}$. However, in general, one cannot multiply $\bar{x}$ and $\bar{y}$ in a way that is compatible with multiplication in $\mathbb{R}$.

Example 5.2 To understand that multiplication of elements in S^1 as equivalence classes is not compatible with the corresponding product on $\mathbb{R}$, let $x_1 = \frac{1}{2}$ and $x_2 = \frac{3}{2}$. Then $\bar{x}_1 = \bar{x}_2 = \overline{\frac{1}{2}}$, but $x_1 x_2 = \frac{3}{4}$. So

$$\overline{\tfrac{1}{4}} = \bar{x}_1 \bar{x}_2 \neq \overline{x_1 x_2} = \overline{\tfrac{3}{4}}.$$

Now define a metric on S^1 by simply inheriting the one it has as it sits in $\mathbb{C}$ (or, if you will, $\mathbb{R}^2$). This is essentially the Euclidean metric and measures the straight-line distance in the plane between two points. Really, this is the length of the chord, or secant line, joining the points. See Figure 85.

But we can also define a distance between points by the arc length between them. In some ways, this is preferable, since, in the abstract, S^1 doesn't really sit anywhere. There is no interior and exterior of S^1, unless you call the actual points on the curved line making the circle the interior points. The problem with using arc length to determine the distance between points is that there are two distinct paths going from one point to another. We must determinate which one to choose. Choosing the minimal path is a nice choice, but how does one do this mathematically? The answer comes from the view that

S^1 is actually the real line $\mathbb{R}$ infinitely coiled up like a slinky by the exponential map ρ given above, and length in $\mathbb{R}$ is easy to describe, and passes through this map, at least locally. Define

(5.1.1) $$d\left(\bar{x},\bar{y}\right) = \min\left\{|x-y| \mid x,y \in \mathbb{R}, x \in \bar{x}, y \in \bar{y}\right\}.$$

Figure 86 shows the equivalence classes of the points $\bar{x} = \frac{1}{3}$ and $\bar{y} = \frac{3}{4}$. Choosing arbitrary representatives x and y and calculating their distance in $\mathbb{R}$ will lead to many different results. However, the minimum distance between representatives of these two classes is well-defined and, in this case, $d\left(\bar{x},\bar{y}\right) = \frac{5}{12}$. Notice that, in fact, the closest two distinct distances between two equivalence classes in $\mathbb{R}$ correspond precisely to the arc lengths in S^1 along the two distinct paths joining $\bar{x}$ and $\bar{y}$.

Lemma 5.3 *These two metrics are uniformly equivalent.*

Proof We leave the proof as Exercise 186. $\qquad\square$

Exercise 186 Show that the inherited Euclidean metric on $S^1 \subset \mathbb{R}^2$ is uniformly equivalent to the arc-length metric in Equation 5.1.1.

Denote by R_α the translation of the points in S^1 by the real number α. We could say that R_α denotes the rigid rotation of S^1 by the angle α, but, as we will see, the parameterization

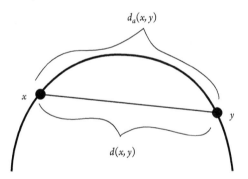

Figure 85 Equivalent metrics on S^1.

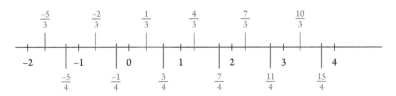

Figure 86 The equivalence classes in $\mathbb{R}$ of $\bar{x} = \frac{1}{3}$ (blue) and $\bar{y} = \frac{3}{4}$ (red) in S^1.

of S^1 becomes vitally important here. We have currently parameterized S^1 as the unit interval in $\mathbb{R}$, with $0 = 1$. So, even though $\alpha \in \mathbb{R}$, $R_{\alpha+n} = R_\alpha$ for all $n \in \mathbb{Z}$. (Note that this is definitely not the case for a continuous dynamical system given by $\dot{x} = \alpha$, $x \in S^1$. Can you see why?) Here $R_\alpha(\bar{x}) = \bar{x} + \alpha \pmod 1 = \bar{x} + \bar{\alpha} = \overline{x + \alpha}$. In complex notation, we view rotations as linear maps, with multiplication by the factor $z_\alpha = e^{2\pi i\alpha}$, so that $R_\alpha(z) = z_\alpha z$. In each case, then, $R_\alpha : S^1 \to S^1$, with either

$$R_\alpha^n(\bar{x}) = R_{n\alpha}(\bar{x}) = \bar{x} + n\alpha \pmod 1 = \bar{x} + \overline{n\alpha} = \overline{x + n\alpha},$$

or $R_\alpha^n(z) = z_\alpha^n z$.

Q. What can we say about the dynamics of a circle rotation?

Q. What if $\alpha \in \mathbb{Q}$?

Q. What if $\alpha \notin \mathbb{Q}$?

The quick answers are that, when α is rational, all orbits are periodic, and all of the same period. Indeed, for $\alpha \in \mathbb{Q}$, $\alpha = p/q$ for $p, q \in \mathbb{Z}$. Then, for $n = q$, we have

$$R_\alpha^q(\bar{x}) = \overline{x + q\alpha} = x + q\left(\frac{p}{q}\right) = \overline{x + p} = \bar{x}, \quad \forall \bar{x} \in S^1.$$

It doesn't even matter if the fractional representation of α is in lowest terms. When α is not rational, however, there are no periodic orbits at all. The trick is to understand what R_α^n looks like for each n, and what it means for the orbit of a point under a rotation to be non-periodic in the circle.

Exercise 187 Let $R_\alpha : S^1 \to S^1$, be the rotation $r_\alpha(\bar{x}) = \bar{x} + \alpha = \overline{x + \alpha}$. Show that no orbit is periodic when $\alpha \notin \mathbb{Q}$.

Exercise 187 leads to a deeper concern: Without fixed or periodic points in S^1 for what is called an *irrational rotation*, the question is, where do the orbits go? They cannot converge to a point in the circle, since in many cases (and, in fact, in general), if they converged to a point in S^1, then that point would have to be a fixed point (if orbits converge, they must converge to another orbit). The answer is that they go everywhere! And that tells one a lot about the dynamics.

Remark 5.4 *The above notion of an irrational rotation was based on the parameterization of S^1 given by the interval $[0, 1)$. There, the rotation R_α was irrational as a rotation when α wasn't rational as a number. However, the parameterization is critical here, and the rationality is of the rotation with respect to the integer 1, the maximum value of the parameter (more precisely, the length of the interval, or the maximum value of the parameter minus the minimum value). To see this, suppose instead we parameterized S^1 via the interval $[0, 2\pi)$, another rather common parameterization given by the map $\rho : \mathbb{R} \to S^1$, where $\rho(x) = e^{ix}$. Here, a rotation halfway around the circle is given by R_π, where $\alpha = \pi$ is irrational (as a number!). Thus the rotation R_π is not irrational at all, since every point* is 2-periodic. However, the rotation by 1, R_1, on S^1 parameterized by the interval $[0, 2\pi)$ would have no periodic orbits. And what would be the orbit structure for a rotation by the number e on either version of S^1?

Exercise 188 Show there are no periodic orbits for the rotation R_1 on S^1 parameterized via the map $\rho : \mathbb{R} \to S^1$, where $\rho(x) = e^{ix}$.

Exercise 189 It is known that Gelfond's number e^{π} is transcendental (cannot be a root of a polynomial with rational coefficients). Use this fact to show that $\frac{\ln 2}{\pi}$ is irrational. Then show that $R_{\ln 2}$ has no periodic points on S^1 under either parameterization via the maps $\rho_1(x) = e^{2\pi i x}$ or $\rho_{2\pi}(x) = e^{ix}$.

The correct conclusion to draw here is that the rationality of the rotation R_α depends on the parameterization. We offer a definition to be clear:

Definition 5.5 A rotation $R_\alpha : S^1 \to S^1$, where S^1 is parameterized by the interval $[0, T)$ for $T > 0$, is called irrational if $\alpha/T \notin \mathbb{Q}$. Otherwise, the rotation is called rational.

Proposition 5.6 For R_α an irrational rotation of S^1, all orbits are dense in S^1.

We will defer the proof of this fact for a short while to better understand the structure of orbits under an irrational rotation. But the basic idea is that once we know that irrational rotation orbits are not periodic (Exercise 187), we can show that a particular point like $\bar{0} \in S^1$ is recurrent for R_α. This will imply that the orbit of $\bar{0}$ is dense in the circle. And, finally, it will become clear that there is nothing special about $\bar{0}$.

Note: All rotations are invertible, right? In fact, they are all homeomorphisms (actually isometries). So, define $R_\alpha^{-1}(\bar{x}) = R_{-\alpha}(\bar{x})$. To show density, we have to show that the orbit of $\bar{x}$ will visit any size open neighborhood of $\bar{x}$. Here is a nice technique for helping to show this:

5.1.1 Continued Fraction Representation

The continued fraction representation (CFR) of a real number is a representation of real numbers as a sequence of integers in a way that essentially determines the rationality of the number. This is very much like the standard decimal representation of real numbers, in that it also provides a ready way to represent all real numbers. However, the sequences of integers that represent real numbers in a base-10 decimal expansion represent some rational numbers as finite-length sequences (think $\frac{11}{8} = 1.375$) and others as infinite-length sequences (think $\frac{4}{9} = 0.44444\ldots$). The CFR, instead, is a base-free representation in which all and only rational number representations are the finite length sequences. Plus, the CFR is another nice way to approximate a real number by either truncating its sequence or simply not calculating the entire sequence.

Exercise 190 For $i : (0, 1) \to \mathbb{R}, i(x) = \frac{1}{x}$, show that $i(x)$ is a homeomorphism onto its image $(1, \infty)$. (Note that $i(x)$, as a function, is a restriction of a more general function $i : \mathbb{R}^+ \to \mathbb{R}^+$. And, on the positive real numbers, it is an example of an *involution*: a function $f : X \to X$ that is its own inverse. This means that $f^{-1}(x) = f(x)$, or $f^2(x) = x$ for all $x \in X$.)

Indeed, given Exercise 190, any real number in $(0,1)$ can be written as $\frac{1}{s}$, where $s \in (1,\infty)$. More generally, then, any real number r can be written as an integer and a real number in $(0,1)$, as

$$r = n + \frac{1}{s}, \text{ where } n \in \mathbb{Z}, \text{ and } s \in (1,\infty).$$

If $s \in \mathbb{N}$, then this expression is considered the CFR of r, and is often written in the form $r = [m:s]$ to accentuate the integers. As an example, $\frac{5}{2} = 2 + \frac{1}{2} = [2:2]$.

Now suppose $s \notin \mathbb{N}$. Then since $s \in (1,\infty)$, $s = m + \frac{1}{t}$, for $m \in \mathbb{N}$, and $t \in (1,\infty)$. Thus,

$$r = n + \frac{1}{m + \frac{1}{t}}, \text{ where } n \in \mathbb{Z}, \ m \in \mathbb{N}, \text{ and } t \in (1,\infty).$$

Again, if $t \in \mathbb{N}$, then we stop and $r = [n:m,t]$ is the CFR of r. If it is not, we again let $t = p + \frac{1}{u}$, for $p \in \mathbb{N}$ and $u \in (1,\infty)$, so

$$r = n + \frac{1}{m + \frac{1}{p + \frac{1}{u}}}, \quad \text{where } n \in \mathbb{Z}, \ m, p \in \mathbb{N}, \text{ and } u \in (1,\infty).$$

Again, if $u \in \mathbb{N}$, we stop, and the CFR of r is $[n:m,p,u]$. If not, then we continue indefinitely. The CFR is a finite sequence iff $r \in \mathbb{Q}$.

Exercise 191 Compute the CFR of $-\frac{33}{13}$.

Exercise 192 Calculate the fraction whose CFR is $[0:3,5,7]$.

Exercise 193 Calculate the number whose CFR is $[1:1,1,1,\ldots]$. Then calculate the number whose CFR is $[1:2,1,2,1,\ldots]$. (Hat tip to Alec Farid, a Johns Hopkins University math and physics double major, who took the course AS.110.421 Dynamical Systems in Spring 2016 and suggested this exercise.)

Remark 5.7 *As mentioned, rational numbers in the CFR have a finite representation. This means that the representation $[n : m_1, m_2, m_3, \ldots]$ has a position $k \in \mathbb{N}$, such that $m_i = 0$, for all $i > k$. One can also talk about quadratic irrationals, real numbers of the form $\frac{a + b\sqrt{c}}{d}$, for $a, b, c, d \in \mathbb{Z}$. These are basically roots of quadratic polynomials with integer coefficients. Quadratic irrationals have CFRs in which the sequences are periodic, such as the two CFRs given in Exercise 193. Two famous numbers that are quadratic irrationals are the Golden Ratio $\varphi = \frac{1+\sqrt{5}}{2}$, and the Silver Ratio $\delta_S = 1 + \sqrt{2}$, both of which have been studied for their interesting properties since the ancient Greeks. Two numbers $a > b > 0$ are said to be in a Golden Ratio if $\frac{a+b}{a} = \frac{a}{b} = \varphi$, and in a silver ratio if $\frac{2a+b}{a} = \frac{a}{b} = \delta_S$.*

Exercise 194 Calculate the CFRs of the Golden and Silver Ratios.

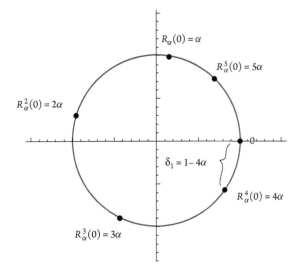

Figure 87 The first five iterates of $R_\alpha(0)$.

Here is a nice example of the use of the CFR to show the recurrence of $\overline{0}$ under an irrational rotation. Let R_α be a rotation of S^1 for $\alpha = \dfrac{1}{4 + \frac{1}{2 + \frac{1}{\pi}}}$. Since π is not rational and $S^1 = \mathbb{R}/\mathbb{Z}, \alpha \notin \mathbb{Q}$.

To start, observe that $\frac{1}{5} < \alpha < \frac{1}{4}$. (Why? This is the strength of the CFR.) In Figure 87, we graph the first five iterates of R_α, evaluated at 0 (we will drop the bar notation at this point). $R_\alpha(0)$, $R_\alpha^2(0)$, $R_\alpha^3(0)$, $R_\alpha^4(0)$, and $R_\alpha^5(0)$. One of the latter two winds up being the early, closest approach to 0 of the orbit $\mathcal{O}_0^+$. A quick calculation yields

$$1 - 4\alpha = 1 - \frac{1}{4 + \frac{1}{2 + \frac{1}{\pi}}} = \frac{\frac{1}{2 + \frac{1}{\pi}}}{4 + \frac{1}{2 + \frac{1}{\pi}}} = \frac{1}{9 + \frac{4}{\pi}},$$

$$5\alpha - 1 = \frac{5}{4 + \frac{1}{2 + \frac{1}{\pi}}} - 1 = \frac{1 - \frac{1}{2 + \frac{1}{\pi}}}{4 + \frac{1}{2 + \frac{1}{\pi}}} = \frac{1 + \frac{1}{\pi}}{9 + \frac{4}{\pi}}.$$

Hence our closest approach is $R_\alpha^4(0)$, and we set $\delta_1 = \frac{1}{9 + \frac{4}{\pi}}$.

So the fourth iterate is the first closest return of $\mathcal{O}_0^+$ to 0 at this point. The next questions are: Will the orbit ever get closer to 0? And if it does, then which iterate? Probing these questions will ultimately lead us to a method to show that the orbit will eventually get arbitrarily close to 0. One strategy would be to simply hunt for the next return. However, we could instead be clever and calculate it. Here is the idea: It took four steps to get within δ_1 of the initial point 0. (We could say it took four steps to get δ_1-close to 0.) Construct an open δ_1-neighborhood of 0, $N_{\delta_1}(0)$, and notice that $R_\alpha^4 = R_{4\alpha}$ is a map that takes $x \mapsto x - \delta_1$. Hence $R_{4\alpha}(\alpha) = \alpha - \delta_1$ and $R_{4\alpha}^n(\alpha) = \alpha - n\delta_1$, for $n \in \mathbb{N}$. Is there an n where

$$-\delta_1 < \alpha - n\delta_1 < 0?$$

That is, is there an iterate of $R_{4\alpha}$ that takes α to something inside the lower half of $N_{\delta_1}(0)$. Convince yourself that there is such an n. Indeed, this is true when

$$n - 1 < \frac{\alpha}{\delta_1} < n,$$

where

$$\frac{\alpha}{\delta_1} = \frac{4 + \frac{1}{2 + \frac{1}{\pi}}}{\frac{1}{9 + \frac{4}{\pi}}} = \frac{9 + \frac{4}{\pi}}{4 + \frac{1}{2 + \frac{1}{\pi}}} = 2 + \frac{1}{\pi}.$$

Hence $n = 3$ and $-\delta_1 < R_{4\alpha}^3(\alpha) < 0$, and $R_{4\alpha}^3(\alpha) = R_\alpha^{12}(\alpha) = R_\alpha^{13}(0)$. Set

$$\delta_2 = 1 - (13\alpha \mod 1) = 3 - \frac{13}{4 + \frac{1}{2 + \frac{1}{\pi}}}.$$

Then $R_\alpha^{13} = R_{13\alpha}$ is a map that takes $x \mapsto x - \delta_2$, as in Figure 88 on the left. Again, $R_{13\alpha}^n(\alpha) = \alpha - n\delta_2$, and, via the same calculation, we solve $-\delta_2 < R_{13\alpha}^n(\alpha) < 0$, or $-\delta_2 < \alpha - n\delta_2 < 0$. or $n - 1 < \frac{\alpha}{\delta_2} < 0$. Here, we see that $3 < \frac{\alpha}{\delta_2} = \frac{2 + \frac{1}{\pi}}{1 - \frac{1}{\pi}} < 4$. With $n = 4$, we conclude that $-\delta_2 < R_{13\alpha}^4(\alpha) = R_\alpha^{53}(0) < 0$.

Finally, set $\delta_3 = 1 - (53\alpha \mod 1) = 13 - \frac{53}{4 + \frac{1}{2 + \frac{1}{\pi}}}$. We find that $n = 6$, and that

$$-\delta_3 < R_\alpha^{319}(0) < 0,$$

continuing the process. On the right of Figure 88 is an idea of the nesting of the neighborhoods $N_{\delta_i}(0)$ resulting from this process.

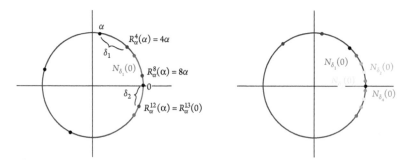

Figure 88 A few iterates of $R_\alpha^4(\alpha)$ (left) and the nesting of closest return neighborhoods of 0 (right).

Using this construction as a guide, we can now prove the recurrence of the point $\overline{0}$ for any irrational rotation:

Lemma 5.8 $\overline{0}$ *is recurrent for* R_α *on* $S^1 = \mathbb{R}/\mathbb{Z}$ *for* $\alpha \notin \mathbb{Q}$.

Proof Given $\alpha \in (0,1)$ irrational, let $n_1 = \lfloor \frac{1}{\alpha} \rfloor$. Then $n_1 \alpha < 1$, but $(n_1 + 1)\alpha > 1$. Define $\delta_1 = 1 - n_1\alpha$ and create $N_{\delta_1}(\overline{0})$. To find a future iterate of $\overline{0}$ under R_α that is in $N_{\delta_1}(\overline{0})$, notice that $R_\alpha^{n_1}(x) = R_{n_1\alpha}(x) = x - \delta_1$. Iterating, we have $R_{n_1\alpha}^m(\alpha) = \alpha - m\delta_1$ and $R_{n_1\alpha}^m(\alpha) \in N_{\delta_1}(0)$ AND in the lower half precisely when

$$-\delta_1 < \alpha - m\delta_1 < 0, \quad \text{or} \quad m - 1 < \frac{\alpha}{\delta_1} < m.$$

Here, then, call this number $m_1 = \lceil \frac{\alpha}{\delta_1} \rceil$. And since $R_{n_1\alpha}^{m_1}(\alpha) = R_\alpha^{n_1 m_1}(\alpha) = R_\alpha^{n_1 m_1 + 1}(\overline{0})$, let $n_2 = n_1 m_1 + 1$ and $\delta_2 = 1 - (n_2\alpha \mod 1) = 1 - R_\alpha^{n_2}(\overline{0})$, and construct $N_{\delta_2}(\overline{0})$.

Continue in this fashion to create the sequences $\{n_i\}_{i\in\mathbb{N}}$ and $\{\delta_i\}_{i\in\mathbb{N}}$. Here $\{n_i\}_{i\in\mathbb{N}}$ is an unbounded, strictly increasing sequence, while $\{\delta_i\}_{i\in\mathbb{N}}$ is a strictly decreasing sequence bounded below by 0. Hence it converges.

Suppose $\{\delta_i\}_{i\in\mathbb{N}}$ converges to $\delta_* > 0$. Then, $N_{\delta_*}(\overline{0}) \cap \mathcal{O}_{\overline{0}} = \{\overline{0}\}$ only. Choose $\epsilon = \frac{\delta_*}{2}$ and form $N_\epsilon(\delta_*)$. Then, $\exists j, k \in \mathbb{N}$ such that δ_j, δ_k satisfy $|\delta_j - \delta_k| < \epsilon$ (δ_* is an accumulation point). But since $\delta_k = R_\alpha^{n_k}$, we have $\left| R_\alpha^{n_j}(\overline{0}) - R_\alpha^{n_k}(\overline{0}) \right| = \left| R_\alpha^{|n_j - n_k|}(\overline{0}) - \overline{0} \right| < \epsilon$.

Hence $R_\alpha^{|n_j - n_k|}(\overline{0}) \in N_{\delta_*}(\overline{0})$, contradicting the assertion that $N_{\delta_*}(\overline{0}) \cap \mathcal{O}_{\overline{0}} = \{\overline{0}\}$ only $(R_\alpha^{|n_j - n_k|}(\overline{0}) \neq \overline{0})$. Thus $\{\delta_i\}_{i\in\mathbb{N}} \longrightarrow 0$ and we are done. $\square$

Using this construction, we are now in a position to prove the proposition:

Proof of Proposition 5.6 We use the construction in Lemma 5.8 directly. Since $\overline{0}$ is recurrent for $R_\alpha(\overline{x})$, $\alpha \notin \mathbb{Q}$, given $\epsilon > 0$, we can choose an $i \in \mathbb{N}$ such that $R_\alpha^{n_i}(\overline{0})$ is δ_i-close to $\overline{0}$ and $\delta_i < \epsilon$. But then, for $r_i = \lfloor \frac{1}{\delta_i} \rfloor$ (recall this means that $r_i \delta_i < 1$, but $(r_i + 1)\delta_i > 1$), the sequence

$$\left\{ \overline{0}, R_\alpha^{n_i}(\overline{0}), R_\alpha^{2n_i}(\overline{0}), \dots, R_\alpha^{r_i n_i}(\overline{0}) \right\}$$

partitions S^1 into arcs where the largest arc length is at most δ_i. Hence $\mathcal{O}_{\overline{0}}$ is δ_i-close to every $\overline{x} \in S^1$ (actually $\frac{\delta_i}{2}$-close, but this is not an important distinction). As $\overline{0}$ is recurrent, we can make this partition size arbitrarily small. so that $\mathcal{O}_{\overline{0}}$ is arbitrarily close to every point in S^1. Thus $\mathcal{O}_{\overline{0}}$ is dense in S^1. And finally, as $R_\alpha^n(\overline{x}) = R_\alpha^n(\overline{0}) + \overline{x}$ (we will leave this also as an exercise), it then follows that $\mathcal{O}_{\overline{x}}$ is dense in S^1 for every $\overline{x} \in S^1$. $\square$

Exercise 195 Show the sequence $\{\sin n\}_{n=1}^\infty$ is dense in $[-1,1]$. The first few iterates are graphed in Figure 89. (Hint: Relate this sequence to an irrational rotation of S^1.)

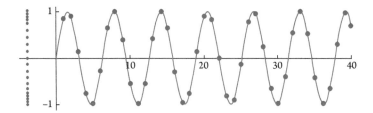

Figure 89 The sequence $\{\sin n\}_{n \in \mathbb{N}}$.

Visually, one can also view an irrational rotation $R_\alpha(\bar{x})$ by its action as a simple translation $x \mapsto x + \alpha$ back up on the real line. Approaching and getting closer to $\bar{0} \in S^1$ means that our translation orbit in $\mathbb{R}$ will at some point come close to an integer value (any integer will do, since they all represent 0 in the circle!). See Figure 90. We will return to this idea of maps on S^1 corresponding to compatible maps on $\mathbb{R}$ shortly, since it will become an important tool in our study. But first, there is another way to understand the density of irrational rotation orbits in the circle, namely, via the frequency with which an orbit visits a region in S^1. The *dynamical frequency* is a measure of how often an orbit visits a small arc in S^1 relative to how much time it is outside of the interval.

5.2 Equidistribution and Weyl's Theorem

To start, fix $\Delta \subset S^1$ an arc (in this section, we will follow the notation and development of Hasselblatt and Katok [26] and drop the overbar from the notation for points of S^1.) Then, for $x \in S^1$ and $n \in \mathbb{N}$, define

$$F_\Delta(x, n) = \#\{ k \in \mathbb{Z} | 0 \le k < n, R_\alpha^k(x) \in \Delta \}.$$

Here, the number sign # denotes the cardinality of the set. For example, in Figure 88 (see also Figure 90) with our choice of α, and $\Delta = N_{\delta_1}(0)$, we have

$$F_\Delta(0, 18) = F_{N_{\delta_1}(0)}(0, 18) = \#\{0, 8, 13\} = 3.$$

Note that for Δ small, and any $x \in S^1$, $F_\Delta(0, n)$ will be small relative to n. And, for Δ large, $F_\Delta(0, n)$ will be bigger, or closer to n, but always a non-negative integer less than (or equal to) n. So we can say that $0 \le F_\Delta(x, n) \le n$, for every x and Δ. In fact, for irrational

$$1 - R_\alpha^4(0) = \delta_1 \qquad\qquad 3 - R_\alpha^{13}(0) = \delta_2 < \delta_1$$

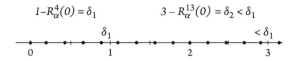

Figure 90 R_α as a translation on $\mathbb{R}$.

rotations, F has a nice property, detailed in the following lemma. First, some introductory notes:

- Define $\ell(\Delta) = $ length of Δ (under some metric).
- For an irrational rotation, the relative frequency really does not depend on whether Δ is open, closed, or neither (why not?). Is this still true also for rational rotations?
- The convention is to take representatives for arcs to be of the "half-closed" form $[\cdot,\cdot)$. Then it is easy to see whether unions of arcs are connected or not.

Lemma 5.9 For $\alpha \notin \mathbb{Q}$, let $\Delta, \Delta' \subset S^1$ be arcs, with $\ell(\Delta) \leq \ell(\Delta')$. Then $\exists N \geq 1$, such that, $\forall n \geq N$ and $\forall x \in S^1$, we have

$$F_{\Delta'}(x, n+N) \geq F_\Delta(x, n).$$

Proof Without any real loss of generality, let's assume that both $\Delta = [x_1, x_2)$, and $\Delta' = [y_1, y_2)$ are in $[0, 1)$. Then $y_2 - y_1 > x_2 - x_1$. Let $\delta = \frac{1}{2}[(y_2 - y_1) - (x_2 - x_1)]$. Now every orbit of R_α is dense in S^1. Hence, for any choice of $x \in S^1$, $\exists N \in \mathbb{N}$ such that $R_\alpha^N(x) \in (y_1, y_1 + \delta)$. But then $R_\alpha^N(\Delta) \subset \Delta'$, since

$$R_\alpha^N(x_2) = x_2 - x_1 + R_\alpha^N(x_1) < x_2 - x_1 + y_1 + \delta$$

$$= x_2 - x_1 + y_1 + \frac{(y_2 - y_1) - (x_2 - x_1)}{2} < y_2.$$

Thus

$$F_\Delta(x, n) = \#\left\{ k \in \mathbb{Z} \mid 0 \leq k < n, R_\alpha^k(x) \in \Delta \right\}$$

$$= \#\left\{ k \in \mathbb{Z} \mid 0 \leq k < n, R_\alpha^{k+N}(x) \in R_\alpha^N(\Delta) \right\}$$

$$= \#\left\{ k \in \mathbb{Z} \mid 0 \leq k < n, R_\alpha^{k+N}(x) \in \Delta' \right\}$$

$$= \#\left\{ k \in \mathbb{Z} \mid 0 \leq k < n+N, R_\alpha^k(x) \in \Delta' \right\} = F_{\Delta'}(x, n+N). \qquad \square$$

Hence, eventually, $F_\Delta(x, n)$ on larger arcs dominates that on smaller arcs. Of course, for any choice of x and Δ, as n grows, $F_\Delta(x, n)$ is also monotonically increasing. However, for $\alpha \notin \mathbb{Q}$, $\lim_{n\to\infty} F_\Delta(x, n) = \infty$.

Exercise 196 For any positive length arc Δ and any $x \in S^1$, show that $\lim_{n\to\infty} F_\Delta(x,n) = \infty$ for an irrational rotation.

Hence, instead of documenting the frequency with which the orbit of a point visits an arc, we study the *relative frequency* of visits as n gets large, or the quantity

$$\frac{F_\Delta(x, n)}{n}.$$

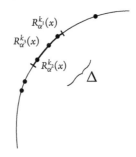

Figure 91 Orbit segment visits to $\Delta \in S^1$.

Suppose on the orbit segment of a point x under the irrational rotation by α given by $\{R_\alpha^i(x)\}_{i=0}^m$, we found that, given the arc Δ, we had $R_\alpha^{k_1}(x), R_\alpha^{k_2}(x), R_\alpha^{k_3}(x) \subset \Delta$ and that these were the only three (see Figure 91). Then we know that the frequency $F_\Delta(x, m) = 3$, and the relative frequency $F_\Delta(x, m)/m = 3/m$. In our example from above in the figures, the relative frequency of hits on the interval $N_\delta(0)$ on the orbit segment $\{R_\alpha^i(x)\}_{i=0}^{18}$ is $\frac{1}{18}F_{N_\delta(0)}(0) = \frac{3}{18} = \frac{1}{6}$. The goal is to study the relative frequency of a rotation on any arc of any length and be able to say something meaningful about how often, on average, the *entire orbit* visits the arc. In terms of this concept of relatively frequency, this translates to a study of

$$\lim_{n \to \infty} \frac{F_\Delta(x, n)}{n}.$$

However, it is yet not entirely clear that this limit actually exists. We first address this point.

Definition 5.10 *Given a sequence* $s = \{a_n\}_{n \in \mathbb{N}}$ *of points in* $\mathbb{R}$, *the extended number* $\ell \in \mathbb{R}^* = \mathbb{R} \cup \{-\infty, \infty\}$ *is a limit point of* s *if there exists a subsequence* $\{a_{n_k}\}_{k \in \mathbb{N}}$ *of* s *where*

$$\lim_{k \to \infty} a_{n_k} = \ell.$$

It is a classical result in analysis that every bounded sequence in $\mathbb{R}^n$ has limit points (this again is the celebrated Bolzano–Weierstrass Theorem.) Using this extended notion of $\mathbb{R}$, we can say that every sequence in $\mathbb{R}$ has limit points in $\mathbb{R}^*$. We note here that in Definition 5.1 above, a point is recurrent if it is a limit point of its own orbit. Categorizing the limit points of a sequence provides important information about the extent of a sequence. One way to categorize them is to find their bounds:

Definition 5.11 *For* $\{a_n\}_{n \in \mathbb{N}}$ *a sequence of points in* $\mathbb{R}$, *the limit inferior is*

$$\liminf_{n \to \infty} a_n = \varliminf_{n \to \infty} a_n = \lim_{n \to \infty} \left(\inf_{m \geq n} a_m \right),$$

where $\inf\limits_{m \geq n} a_m$ *is the infimum of the sequence* $\{a_m\}_{m=n}^{\infty}$.

In fact, it is the largest number smaller than or equal to all of the remaining elements of the sequence. It is the largest number smaller than all but a finite number of the elements of the sequence. One can define the *limit superior* similarly as the smallest of the numbers larger than all but a finite number of elements of the sequence. We use the notation $\limsup_{n\to\infty}$ or $\overline{\lim}_{n\to\infty}$ for the limit superior.

It should be intuitively obvious that

(1) the limits inferior and superior always exist in $\mathbb{R}_*$;

(2) for any sequence s, $\liminf s \leq \limsup s$;

(3) for any sequence s, $\liminf s = \limsup s$ iff $\lim s$ exists.

In essence, if $\liminf s = a$ and $\limsup s = b$, then the interval $[a,b]$ will contain all possible limit points of s. Note that this does not mean that every point in $[a,b]$ may be a limit point of the sequence s. As an example, the set of limit points of the sequence $\{x_n\} = \{(-1)^n\}$ are all contained in the interval $[-1,1]$ defined by the limits inferior and superior. However, the sequence has no limit points in $(-1,1)$.

Example 5.12 Consider the function $f : \mathbb{N} \to \mathbb{R}$ defined by $f(n) = (1 + 1/n)\sin n$. A priori, we do not know whether $\lim_{n\to\infty} f(n)$ exists or not (Think of the continuous version of the function in calculus.) It should be clear, though, that there will be only a finite number of values of f greater than or equal to any value $c > 1$, outside the interval $[-1,1]$ (the same is true for values less than or equal to $c < -1$). But, from Exercise 195, every point of the interval $[-1,1]$ serves as a limit point of f. Hence $\varliminf_{n\to\infty} f = -1$ and $\varlimsup_{n\to\infty} f = 1$.

In our case, let A be a disjoint union of arcs. Then define

$$\overline{f}_x(A) = \limsup_{n\to\infty} \frac{F_A(x,n)}{n}, \quad \underline{f}_{-x}(A) = \liminf_{n\to\infty} \frac{F_A(x,n)}{n}.$$

It turns out that these two quantities not only exist—they also are equal, so the actual limit does exist. The value of this limit is a remarkable result of Hermann Weyl in 1914 [60]:

Theorem 5.13 (Weyl Equidistribution Theorem) *For any arc $\Delta \subset S^1$ and every $x \in S^1$, and any irrational rotation R_α, $\alpha \notin \mathbb{Q}$ on S^1, we have*

$$f(\Delta) := \lim_{n\to\infty} \frac{F_\Delta(x,n)}{n} = \ell(\Delta).$$

Remark 5.14 *The concept of equidistribution, or of a sequence being equidistributed in an interval or in S^1, is based on the notion of a uniform distribution. A sequence is equidistributed if the number of its terms falling inside a subinterval vis-á-vis outside is proportional to (actually equal to) the size of the interval. Thus Weyl's Theorem actually defines equidistribution. There are many proofs of this theorem. Here, we will follow an elementary proof found in Hasselblatt and Katok [26], using similar notation.*

Proof We will proceed in steps: Our first claim is that, for $\ell(\Delta) \leq 1/k$, $\bar{f}_x(\Delta) \leq 1/(k-1)$. Indeed, partition S^1 into $k-1$ equal-length arcs Δ_i, so that for $i = 1, \ldots, k-1$, $\ell(\Delta_i) = 1/(k-1)$. By Lemma 5.9, we have

$$F_{\Delta_i}(x, n+N) \geq F_\Delta(x, n).$$

So, for $i = 1, \ldots, k-1$, we have $F_\Delta(x, n) \leq F_{\Delta_i}(x, n+N_i) \leq F_{\Delta_i}(x, n+N)$, where $N = \max_i N_i$. But since this is true for each i, it is true for the average, and

$$F_\Delta(x, n) \leq \frac{1}{k-1} \sum_{j=1}^{k-1} F_{\delta_j}(x, n+N) = \frac{1}{k-1} F_{\Delta_1 \cup \cdots \cup \Delta_{k-1}}(x, n+N)$$

$$= \frac{1}{k-1} F_{S^1}(x, n+N) = \frac{n+N}{k-1}.$$

Thus

$$\frac{F_\Delta(x, n)}{n} \leq \frac{1}{k-1} \cdot \frac{n+N}{n},$$

and

$$\bar{f}_x(\Delta) = \limsup_{n \to \infty} \frac{F_\Delta(x, n)}{n} \leq \limsup_{n \to \infty} \frac{1}{k-1} \cdot \frac{n+N}{n} = \frac{1}{k-1}.$$

Hence the claim holds.

The next claim is that for any arc $\Delta \subset S^1$, $\bar{f}_x(\Delta) \leq \ell(\Delta)$. To see this, first note that it is true for $\Delta = S^1$. More generally, consider for any $\epsilon > 0$ and an arc $\Delta \in S^1$, we can construct a strictly larger arc $\Delta' \supset \Delta$ such that

$$\ell(\Delta) < \ell(\Delta') = \frac{l}{k} < \ell(\Delta) + \epsilon$$

for some rational l/k. Then, cutting up Δ' into l equal-length arcs means partitioning Δ into l arcs Δ_i, each of length at most $1/k$. Then, by Claim 1 above, on each subarc, $\bar{f}_x(\Delta_i) \leq 1/(k-1)$. But then, by additivity,

$$\bar{f}_x(\Delta) \leq \frac{l}{k-1} < (\ell(\Delta) + \epsilon) \frac{k}{k-1}.$$

Pushing ϵ to 0 (and thereby pushing k to ∞) means $\bar{f}_x(\Delta) \leq \ell(\Delta)$, establishing the new claim.

We note here that, for any arc $\Delta \subset S^1$, we have $F_\Delta(x, n) = n - F_{\Delta^c}(x, n)$, where Δ^c is the complement of Δ in S^1. Thus $\frac{F_\Delta(x,n)}{n} = 1 - \frac{F_{\Delta^c}(x,n)}{n}$ and hence $\bar{f}_x(\Delta) = 1 - \underline{f}_x(\Delta^c)$.

We can use this now to show that $\underline{f}_{-x}(\Delta) \geq \ell(\Delta)$. Indeed, we can apply the construction of Claim 2 above to show that $\overline{f}_x(\Delta^c) \leq \ell(\Delta^c)$. But then

$$\underline{f}_{-x}(\Delta) = 1 - \overline{f}_x(\Delta^c) \geq 1 - \ell(\Delta^c) = \ell(\Delta).$$

But now we have, for any arc $\Delta \in S^1$,

$$\overline{f}_x(\Delta) \leq \ell(\Delta) \leq \underline{f}_{-x}(\Delta),$$

from which, since, by definition, $\underline{f}_{-x}(\Delta) \leq \overline{f}_x(\Delta)$, we must have, for any $x \in S^1$,

$$\overline{f}_x(\Delta) = \ell(\Delta) = \underline{f}_{-x}(\Delta).$$

The result follows. ☐

Hence, we say that any orbit of an irrational rotation of S^1 is uniformly distributed on S^1. This is our notion of a set being dense in another set, and for these orbits, one can actually "see" the notion of recurrence. To further understand this new type of dynamical behavior, we look at an application. But first, let's continue with a little more nomenclature.

Definition 5.15 *A continuous map $f : X \to X$ is called* topologically transitive *if $\exists x \in X$ such that $\mathcal{O}_x$ is dense in X. A non-invertible map is called topologically transitive if $\exists x \in X$ such that $\mathcal{O}_x^+$ is dense in X.*

Definition 5.16 *A continuous map $f : X \to X$ is* minimal *if $\forall x \in X$, $\mathcal{O}_x$ is dense in X (the forward orbit is dense for a non-invertible map).*

Definition 5.17 *A closed invariant set is* minimal *is there does not exist a proper closed invariant subset.*

More notes:

- Recall Proposition 2.21 that a set is closed if it contains all of its limit points. So the closure $\overline{X}$ of a set X is the set obtained by adding to X all of its boundary points. In the case of a minimal map $f : X \to X$, for any $x \in X$, we have $\overline{\mathcal{O}}_x = X$.
- The same is true for a topologically transitive map f, if one takes any point on the dense orbit.
- Irrational rotations of the circle are minimal!.

5.2.1 Application: Periodic Function Reconstruction via Sampling

Consider the two functions in Figure 92:

- Each is periodic and of the same period as the other.
- Each can be viewed as a real-valued smooth function on S^1. And each takes values in the interval $I = [-1, 1]$.

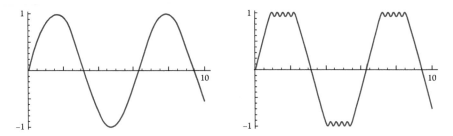

Figure 92 Two functions of equal period and range, but with values distributed differently.

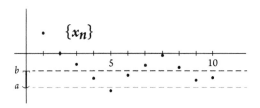

Figure 93 Value frequency of a sequence.

- Question: Are the values of these two functions distributed equally (or even evenly) on I?

- Question: If we knew the period and range of some unknown function, and needed to sample the function (create a sequence of function values) to see which of the above two functions was the one we are seeking, how can we design our sampling to ensure we can differentiate between these two?

Dynamics attempts to answer these questions. Let $\{x_n\}$ be a sequence (think of this sequence as a sampling of the function) and $a < b$ two real numbers. Define

$$F_{a,b}(n) = \#\{k \in \mathbb{Z} \mid 1 \le k \le n, a < x_k \le b\}$$

as the number of times the sequence up to element n visits the interval $(a, b) \subset \mathbb{R}$. Actually, this is the same definition of F as before on the arc $\Delta \subset S^1$. The only change in this case is that we are defining F in this context as an interval in $\mathbb{R}$. Then the relative frequency is defined as before. In Figure 93, the relative frequency of $\{x_n\}$ on the interval $(a, b]$ shown is

$$\left. \frac{F_{a,b}(n)}{n} \right|_{n=6} = \frac{2}{6} = \frac{1}{3}.$$

We say that $\{x_n\}$ has an *asymptotic distribution* if $\forall a, b$, where $-\infty \le a < b \le \infty$, the quantity

$$\lim_{n \to \infty} \frac{F_{a,b}(n)}{n}$$

exists. In a sense, we are defining the percentage of the time that a sequence visits a particular interval.

In the case where the sequence has an asymptotic distribution, the function

$$\Phi_{\{x_n\}}(t) = \lim_{n \to \infty} \frac{F_{-\infty,t}(n)}{n}$$

is called the *distribution function* of the sequence $\{x_n\}$. Here Φ is monotonic and measures how often the values of a sequence visit regions of the real line as one varies the height of an interval $(-\infty, t]$.

Definition 5.18 *A real-valued function φ on a closed, bounded interval is called* piecewise monotonic *if the domain can be partitioned (in the sense of Remark 3.43) into finite many subintervals on which φ is monotonic. A real-valued function on $\mathbb{R}$ is* piecewise monotonic *if it is piecewise monotonic on every closed bounded subinterval of $\mathbb{R}$.*

Remark 5.19 *Monotonic means strictly monotonic here. In fact, this means that there are no flat (purely horizontal on an open interval) regions of the graph of φ. Think of functions like $f(x) = \sin x$, and polynomials of degree larger than 1, which are piecewise monotonic, and functions like*

$$g(x) = \begin{cases} -(x+2)^2, & -4 \le x < -2, \\ 0, & -2 \le x \le 0, \\ x^2, & 0 < x \le 2, \end{cases}$$

which is not piecewise monotonic (see the graph of $g(x)$ on the left side of Figure 94).

When φ is piecewise monotonic on a closed, bounded interval, the pre-image of any codomain interval I is a finite union of intervals in the domain.

Definition 5.20 *The φ-length of an interval I is*

$$\ell_\varphi(I) := \ell\left(\varphi^{-1}(I)\right).$$

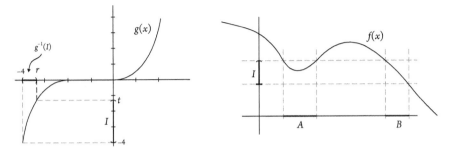

Figure 94 A non–piecewise monotonic $g(x)$, and a piecewise monotonic $f(x)$.

- This is the total length of all pieces of the domain that map into I. On the right side of Figure 94, $\ell_f(I) = \ell(A) + \ell(B)$.
- For piecewise monotonic functions φ, the φ-length is a continuous function of the end points of I (vary one endpoint of I continuously, and the φ-length of I also varies continuously. This doesn't work with flat regions, since the φ-length ℓ_φ would then jump as one hit the value of the flat region.

Indeed, let's look at $g(x)$ in Figure 94 more closely. Here, one can calculate the φ-length. Indeed, choose the interval $I = [-4, t]$. Here, t is the function value, and there is only a single interval mapped onto i for any value of t.

For $t < 0$, this interval is given in Figure 95 as the interval of the domain $g^{-1}(I) = [-4, r]$, where $g(r) = t$. Solving the equation $g(r) = t$ for r yields

$$-(r+2)^2 = t \iff r = -\sqrt{-t} - 2,$$

where we chose the negative branch of the square root function in the middle step to account for the domain restrictions. Here, the g-length of I,

$$\ell_g(I) = \ell\left(g^{-1}\left([-4, t]\right)\right)$$
$$= -2 - \sqrt{-t} - (-4) = 2 - \sqrt{-t}.$$

Now, for $t > 0$, the same calculation yields $\ell_g(I) = 4 + \sqrt{t}$ for $I = [-4, t]$. Putting these two pieces of the g-length function together yields the graph of

$$\ell_g(I) = \begin{cases} 2 - \sqrt{-t}, & -4 \le t < 0, \\ 4 + \sqrt{t}, & 0 < t \le 4, \end{cases}$$

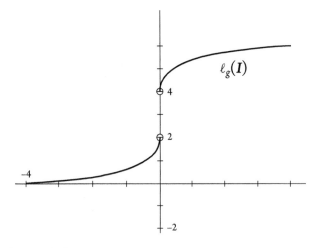

Figure 95 A discontinuity for ℓ_g.

which has a jump discontinuity at $t = 0$. In fact, the only way to change $g(x)$ to make the g-length function continuous is to remove the middle piece of the function $g(x)$ and translate one or the other pieces right or left to again make $g(x)$ continuous. But that would have the effect of moving the two pieces of the graph of $\ell_g(I)$ together. The jump discontinuity becomes a hole in the graph, easily filled. But, in this case, the changed $g(x)$ has been made piecewise monotonic!

One can show that for a piecewise monotonic function φ, a distribution function for φ is

$$\Psi : \mathbb{R} \to \mathbb{R}, \quad \Psi_\varphi(t) = \ell_\varphi((-\infty, t)).$$

We can use this for the following theorem:

Theorem 5.21 [26] *Let φ be a T-periodic function of $\mathbb{R}$ such that $\varphi_T = \varphi|_{[0,T]}$ is piecewise monotonic. If $\alpha \notin \mathbb{Q}$ and $t_0 \in \mathbb{R}$, then the sequence $x_n = \varphi(t_0 + n\alpha T)$ has an asymptotic distribution with distribution function*

$$\Phi_{\{x_n\}}(t) = \frac{1}{T}\Psi_\varphi(t) = \frac{\ell\left(\varphi^{-1}((-\infty, t))\right)}{T}.$$

We won't prove this or study it in any more detail. But there is an interesting conclusion to be drawn from it. In the theorem, the sequence of samples of the T-periodic function φ has the same distribution function as the actual function φ (defined over the period, that is) precisely when the sampling is taken at a rate that is an irrational multiple of the period T. In this way, the sequence, over the long term, will fill out the values of φ over the period in a dense way. In a way, one can recover the function φ from a sequence of regular samples of it only if the sampling is done in a way that ultimately allows for all regions of the period to be visited evenly. This is a very interesting result.

Example 5.22 Here is an example very similar in construction to that in [26]. We calculate the asymptotic distribution of the sequence $\{2 + \cos(2n)\}$. The function

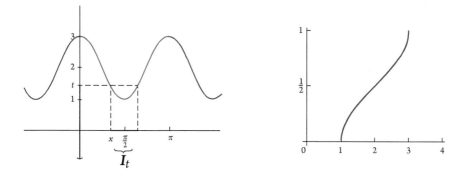

Figure 96 $f(x) = 2 + \cos 2x$ (left) and the asymptotic distribution function (right).

$\varphi(x) = 2 + \cos 2x$ is π-periodic, where $\pi \notin \mathbb{Q}$. So, for $t_0 = 0$ and $\alpha = 1/\pi$, we can use φ to generate our sequence. The length of the interval I_t in Figure 96, as a function of t, is given as follows: Let $t = 2 + \cos 2x$, so that $x = \frac{1}{2} \arccos(t - 2)$. Then the length of I_t is $2 \left(\frac{1}{2}\pi - \frac{1}{2}\arccos(t - 2) \right) = \pi - \arccos(t - 2)$. Thus

$$\Phi_{\{x_n\}}(t) = \Psi_{y_\pi}(t) = \begin{cases} 0, & t < 1, \\ 1 - \pi^{-1}\arccos(t - 2), & 1 \le t \le 3, \\ 1, & t > 3. \end{cases}$$

Exercise 197 Calculate the distribution function Ψ for the triangular-wave function

$$f(x) = 4 * \left| \frac{t+1}{4} - \left\lfloor \frac{t+3}{4} \right\rfloor \right| - 1.$$

Here $\lfloor x \rfloor$ is the *floor* function: the value of the floor function on any real number x is the greatest integer less than or equal to x. We will use the floor function in an important way in Section 5.5 on circle homeomorphisms.

Exercise 198 Calculate the distribution function for the sequence $\cos n$.

5.3 Linear Flows on the Torus

Here is another type of dynamical system where circle rotations and their implications play a vital role. This one involves generalizing circle maps via a corresponding circle flow into more than one dimension.

To start, recall the definition of a flow from Chapter 1: a flow for an IVP system $\dot{\mathbf{x}} = \mathbf{f}(\mathbf{x}), \mathbf{x}(0) = \mathbf{x}_0 \in \mathbb{R}^n$ is a map $\varphi : I \times \mathbb{R}^n \to \mathbb{R}^n$ defined on a interval $I \subset \mathbb{R}$ that satisfies the following:

- $\forall t \in I, \varphi^t = \varphi(t, \cdot) : \mathbb{R}^n \to \mathbb{R}^n$ is the time-t map of the IVP; and
- $\forall s, t \in I$, where $s + t \in I$,

$$\varphi^s \circ \varphi^t(\mathbf{x}) = \varphi^{s+t}(\mathbf{x}).$$

Now define a flow on $S^1 = \mathbb{R}/\mathbb{Z}$ by the differential equation $dx/dt = \alpha$, and choose an initial value $x(0) = x_0$ to create an IVP. This IVP is solved by $x(t) = \alpha t + x_0$, or, written in flow form, $\varphi_\alpha^t(x) = \alpha t + x$. Notice that in this last expression, we have included the subscript α to denote the dependence of the flow on the value of the parameter α. Here the time-1 map is just

$$\varphi_\alpha^1(x) = \alpha + x = R_\alpha(x), \quad x \in S^1.$$

The time-1 map is just a rotation map of the circle by α. Keep in mind, however, that the IVP will share the same time-1 map as the new IVP given by $dx/dt = \alpha + n, x(0) = x_0$ for n any positive integer. However, the flows will all be very different! (Do you see this?)

Linear flows on S^1 are not very interesting (are you starting to get used to this term in mathematics yet?). They differ only by speed (and possibly direction), and ultimately all look like continuous rotations of the circle, whether α is rational or not. However, we can generalize this flow to a situation that does produce somewhat interesting dynamics.

Consider now a flow given by the pair of uncoupled circle ODEs:

(5.3.1)
$$\frac{dx_1}{dt} = \omega_1, \quad \frac{dx_2}{dt} = \omega_2.$$

This system, which can be written as the uncoupled vector ODE $\dot{\mathbf{x}} = \boldsymbol{\omega}$, or $\begin{bmatrix} \dot{x}_1 \\ \dot{x}_2 \end{bmatrix} = \begin{bmatrix} \omega_1 \\ \omega_2 \end{bmatrix}$, can be viewed as defining a flow on the two-torus $\mathbb{T}^2 = S^1 \times S^1$, and has the solution $\mathbf{x}(t) = \begin{bmatrix} x_1 + \omega_1 t \\ x_2 + \omega_2 t \end{bmatrix}$.

In flow notation, we can write either

$$T^t_{\boldsymbol{\omega}}(x_1, x_2) = (x_1 + \omega_1 t, x_2 + \omega_2 t) \qquad \text{or} \qquad \varphi^t_{\boldsymbol{\omega}}(\mathbf{x}) = \mathbf{x} + \boldsymbol{\omega} t.$$

Graphically, solutions are simply translations along $\mathbb{R}$ (Figure 97) or straight-lines motions in $\mathbb{R}^2$ (Figure 98). Note that in this last interpretation, the slope of the solution line is $\gamma = \omega_2/\omega_1$.

However, each of these uncoupled ODEs can also be considered as a flow on S^1, and hence the system can be considered a flow on $S^1 \times S^1 = \mathbb{T}^2$. Suppose, for example, that $1 < \omega_1 < 2$, while $0 < \omega_2 < 1$. The flow from time $t = 0$ to time $t = 1$ would take the

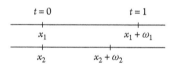

Figure 97 A pair of translation flows on $\mathbb{R}$.

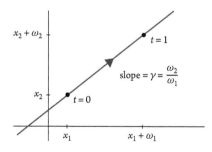

Figure 98 A translational flow in the plane.

origin on one circle to the point $\omega_1 - 1$, and the flow line would start at $x_1 = 0$ and travel once around the circle before stopping to ω_1. The flow on the other circle would take $x_2 = 0$ partway around the circle to ω_2. Viewed via the two periodic coordinates of $\mathbb{T}^2$, we have the flow line in Figure 99.

Another way to see this is to go back to the plane and consider the equivalence relation given by the exponential map on each coordinate. The set of equivalence classes is given by the unit square in the plane, under the idea that the left side of the square (the side on the $x_1 = 0$ line) and the right side (the $x_1 = 1$ side) are considered the same points (this is the $0 = 1$ idea of the circle identification). Similarly, the top and bottom of the square are to be identified. This version of a 2-dimensional torus, viewed as the unit square in the plane with the opposite sides identified, is called a *square torus* or a *Euclidean torus* since objects drawn on it obey Euclidean geometry (see Figure 100). Here, the flow line at the origin under the ODE system 5.3.1 is a straight line of slope γ emanating from the origin and meeting the right edge of the unit square at the point $(1, \gamma)$. But, by the identification, we can restart the graph of the line at the same height on the left side of the square (at the point $(0, \gamma)$). Continuing to do this, we will eventually reach the top of the square. But, by the identification again, we will drop to the bottom point and continue the line as before, as depicted in Figure 100. In essence, we are graphing the flow line as it would appear on the unit square. When we pull this square out of the plane and bend it to create our torus $\mathbb{T}^2$, the flow line will come with it. Suppose $\gamma \notin \mathbb{Q}$. What can we say about the positive flow line?

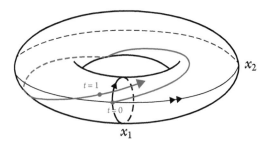

Figure 99 The flow line of the origin in $\mathbb{T}^2$.

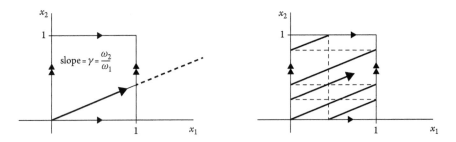

Figure 100 Wrapping the flow onto $\mathbb{T}^2$, the unit square with opposite sides identified.

Proposition 5.23 *if $\gamma = \omega_2/\omega_1$ is irrational, then the flow is minimal. If $\gamma \in \mathbb{Q}$, then every orbit is closed.*

Proof Choose a point $\bar{\mathbf{x}} = (\bar{x}_1, \bar{x}_2)$ and let

$$S_{\bar{\mathbf{x}}} = \left\{ (\bar{x}_1, \bar{y}) \in \mathbb{T}^2 \mid \bar{y} \in S^1 \right\}$$

be a Poincaré section (a first return map for the flow). Then, along $S_{\bar{\mathbf{x}}}$, the Poincaré map is precisely R_γ (see the left side of Figure 101). Since $\gamma \notin \mathbb{Q}$, $\mathcal{O}_{\bar{\mathbf{x}}} \cap S_{\bar{\mathbf{x}}}$ is dense in $S_{\bar{\mathbf{x}}}$.

So let $\bar{\mathbf{y}}$ be some arbitrary point in $\mathbb{T}^2$. We claim that arbitrarily close to $\bar{\mathbf{y}}$ is a point in $\mathcal{O}_{\bar{\mathbf{x}}}$. Indeed, choose $\epsilon > 0$ and construct $N_\epsilon(\bar{\mathbf{y}})$, an open ϵ-neighborhood of $\bar{\mathbf{y}}$ in $\mathbb{T}^2$. Then $\exists \bar{\mathbf{z}} \in S_{\bar{\mathbf{x}}}$ and a small value $t > 0$, such that $\bar{\mathbf{y}} = \omega t + \bar{\mathbf{z}}$. But then $\exists \bar{\mathbf{u}} \in \mathcal{O}_{\bar{\mathbf{x}}} \cap S_{\bar{\mathbf{x}}}$, where $d(\bar{\mathbf{u}}, \bar{\mathbf{z}}) < \epsilon$. And then $\bar{\mathbf{v}} = \omega t + \bar{\mathbf{u}} \in N_\epsilon(\bar{\mathbf{y}})$ and $\bar{\mathbf{v}} \in \mathcal{O}_{\bar{\mathbf{x}}}$. See the right side of Figure 101. $\qquad\square$

Remark 5.24 *For a Poincaré first return map, if every orbit intersects the Poincaré section, we call the section a* global section. *Otherwise, it is a* local section.

Now let $\gamma = \omega_2/\omega_1 \in \mathbb{Q}$. Then R_γ is a rational rotation on $S_{\bar{\mathbf{x}}}$. Then $\exists\, t_0 > 0$ such that $\bar{\mathbf{x}} = \omega t_0 + \bar{\mathbf{x}} \mod 1$. Thus $\omega t_0 = 0 \mod 1$ and $\mathcal{O}_{\bar{\mathbf{x}}}$ is closed. But then, for any other point $\bar{\mathbf{y}} \in \mathbb{T}^2$, we have $\bar{\mathbf{y}} - \bar{\mathbf{x}} = \bar{\mathbf{z}} \mod 1$. Thus $\bar{\mathbf{x}} = \bar{\mathbf{y}} - \bar{\mathbf{z}} \mod 1$ and $\bar{\mathbf{y}} - \bar{\mathbf{z}} = \omega t_0 + \bar{\mathbf{y}} - \bar{\mathbf{z}}$, so that $\bar{\mathbf{y}} = \omega t_0 + \bar{\mathbf{y}} \mod 1$. Hence $\mathcal{O}_{\bar{\mathbf{y}}}$ is closed. $\qquad\square$

Exercise 199 Draw enough of the flow lines passing through the origin to indicate general behavior for the following values of γ:

(a) $\gamma = 1$, **(b)** $\gamma = 2$, **(c)** $\gamma = \dfrac{1}{2}$, **(d)** $\gamma = \dfrac{1 + \sqrt{5}}{2}$, **(e)** $\gamma = \dfrac{1 - \sqrt{5}}{2}$.

Exercise 200 Show that a linear map on the real line $f(x) = kx$ corresponds to a continuous map on S^1 iff k is an integer. Graph the circle map on the unit interval (with the endpoints identified) that corresponds to $f(x) = 3x$. Identify all fixed points. Can you find a period-3 orbit? Hint: Remember that two points on the real line

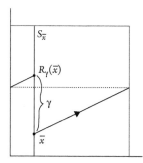
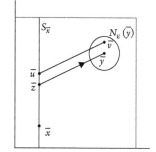

Figure 101 The orbit of $\bar{\mathbf{x}}$ is dense in $\mathbb{T}^2$.

correspond to the same point on the circle if their distance apart is an integer. This is the parameterization that we will always refer to by default.

Exercise 201 Let $h : \mathbb{R}^2 \to \mathbb{R}^2$ be a linear map so that $h(\mathbf{x}) = A\mathbf{x}$, where $A = \begin{bmatrix} a & b \\ c & d \end{bmatrix}$ and $a, b, c, d \in \mathbb{Z}$. Do the following:

(a) Show that h induces a map on $\mathbb{T}^2$. Hint: Two vectors in the plane are in the same equivalence class on the torus (correspond to the same point on $\mathbb{T}^2$) if they differ by a vector with integer entries.

(b) Describe, as best as you can, the dynamics of the linear maps on $\mathbb{T}^2$ corresponding to

$$\text{i. } A = \begin{bmatrix} 0 & 1 \\ -1 & 0 \end{bmatrix}, \quad \text{ii. } A = \begin{bmatrix} 1 & 1 \\ 0 & 1 \end{bmatrix}, \quad \text{iii. } A = \begin{bmatrix} 2 & 1 \\ 1 & 1 \end{bmatrix}.$$

5.3.1 Application: Lissajous Figures

We can look at toral flows in another way: via an embedding of a torus into $\mathbb{R}^4$ with simultaneous rotations in the two orthogonal coordinate planes. To see this, think of $S^1 \in \mathbb{R}^2$ as a circle of radius r centered at the origin. Then we can represent $\mathbb{T}^2$ as the set

$$\mathbb{T} = \left\{ (x_1, x_2, x_3, x_4) \in \mathbb{R}^4 \mid x_1^2 + x_2^2 = r_1^2, \ x_3^2 + x_4^2 = r_2^2 \right\}.$$

Remark 5.25 *This embedding of $\mathbb{T}^2$ into $\mathbb{R}^4$ is related to what is called the Clifford torus, where the radii of the two circles that define the torus are both $r = \frac{1}{\sqrt{2}}$. It is an example of what is called a "flat" or Euclidean torus, since it can be flattened out into the plane without stretching, and geometric onjects defined on it obey Euclidean geometry. Actually, it is isometric to the square torus above.*

Now recall that a continuous rotation in $\mathbb{R}^2$ is given by the linear ODE system $\dot{\mathbf{x}} = B_\alpha \mathbf{x}$, where B is the matrix $\begin{bmatrix} 0 & \alpha \\ -\alpha & 0 \end{bmatrix}$ whose eigenvalues $\pm \alpha i$ are purely imaginary. Do this for each pair of coordinates (each of two copies of $\mathbb{R}^2$ in the Clifford torus) to get the partially uncoupled system of ODEs on $\mathbb{R}^4$,

$$\dot{\mathbf{x}} = A\mathbf{x}, \qquad \begin{bmatrix} \dot{x}_1 \\ \dot{x}_2 \\ \dot{x}_3 \\ \dot{x}_4 \end{bmatrix} = \begin{bmatrix} 0 & \alpha_1 & 0 & 0 \\ -\alpha_1 & 0 & 0 & 0 \\ 0 & 0 & 0 & \alpha_2 \\ 0 & 0 & -\alpha_2 & 0 \end{bmatrix} \begin{bmatrix} x_1 \\ x_2 \\ x_3 \\ x_4 \end{bmatrix}.$$

We will eventually see that this is the model for the spherical pendulum. For now, Figure 102 shows two planes, which, under uncoupled rotations, would leave concentric circles invariant. Joining these two planes only at the origin gives $\mathbb{R}^4$ (hard to visualize, no?) The 2-torus $\mathbb{T}^2$ is a subspace of $\mathbb{R}^4$ invariant under this flow.

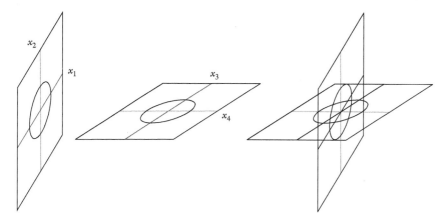

Figure 102 The x_1x_2- and x_3x_4-planes as subspaces of $\mathbb{R}^4$.

Some notes:

- The two circles $x_1^2 + x_2^2 = r_1^2$ and $x_3^2 + x_4^2 = r_2^2$ are invariant under this flow.
- We can define angular coordinates on $\mathbb{T}^2$ via the equations

$$x_1 = r_1 \cos 2\pi \varphi_1, \qquad x_2 = r_1 \sin 2\pi \varphi_1,$$
$$x_3 = r_2 \cos 2\pi \varphi_2, \qquad x_4 = r_2 \sin 2\pi \varphi_2.$$

Then, restricted to these angular coordinates and with $\omega_i = \alpha_i/2\pi$, $i = 1,2$, we recover

$$\dot{\varphi}_1 = -\omega_1, \qquad \dot{\varphi}_2 = -\omega_2.$$

Motion is independent along each circle, and the solutions are $\varphi_i(t) = \omega_i(t - t_0)$.

- If $\alpha_2/\alpha_1 = \omega_2/\omega_1 \notin \mathbb{Q}$, then the flow is minimal.

Exercise 202 Do the change of coordinates explicitly to show that these two interpretations of linear toral flows are the same.

Now, for a choice of ω and $r_1 = r_2 = 1$, project a solution onto either the (x_1, x_3)- or the (x_2, x_4)-plane. The resulting figure is a plot of a parameterized curve whose two coordinate functions are cosine (respectively sine) functions of periods that are rationally dependent iff ω is rational. In this case, the figure is closed, and is called a *Lissajous figure*. See Figure 103 for the case of two sine functions (projection onto the (x_2, x_4)-plane, in this case), where ω_1 and ω_2 have the values shown. Notice that the last one is not yet a closed figure in this partial rendering. Will it ever be?

Exercise 203 Draw the Lissajous figure corresponding to the x_2x_4-planar projection of the toral flow when $\omega_1 = 2$ and $\omega_2 = 5$. For these same values, draw the orbit

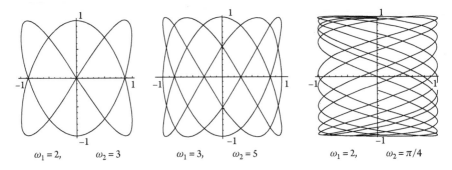

Figure 103 Some Lissajous figures.

also on the torus using the unit square in $\mathbb{R}^2$ representation (with sides identified appropriately), and as well on the "surface of a doughnut" representation in $\mathbb{R}^3$.

A nice physical interpretation of this curve is as the trajectory of a pair of uncoupled harmonic oscillators, given by

$$\ddot{x}_1 = -\omega_1 x_1,$$
$$\ddot{x}_2 = -\omega_2 x_2.$$

Toral flows also show up as a means to study a completely different class of dynamical system; a billiard. We will eventually focus on some more general features of this class of dynamical systems. For now, we will introduce a particularly interesting example, where a linear flow on a torus allows us to answer a rather deep question.

5.3.2 Application: A Polygonal Billiard

Recall from Section 1.1.4 our construction of a system of two point-beads moving at constant velocity along a finite-length wire. Presently, we will explore this system in detail. Consider the unit interval $I = [0, 1]$ with two point masses x_1 and x_2, with respective masses m_1 and m_2, free to move along I but confined to stay on I. Without outside influence, these point masses will move at a constant initial velocity. Eventually, they will collide with each other and with the walls (see the left side of Figure 104). Assume also that these collisions are elastic, with no energy absorption or loss due to friction. Here, elastic means that when a point mass collides with a wall, the velocity of the point mass will only switch sign. And when two point masses collide, they will exchange velocities. For now, assume that $m_1 = m_2 = 1$.

The state space in $\mathbb{R}^2$ is

$$T = \left\{ (x_1, x_2) \in \mathbb{R}^2 \,\middle|\, 0 \le x_1 \le x_2 \le 1 \right\}.$$

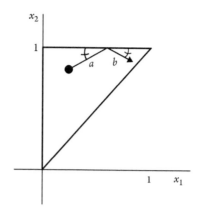

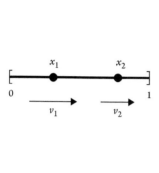

Figure 104 The state space (right) of two point masses in the unit interval (left).

Here T is the region in the unit square above the diagonal line, which is the graph of the identity map on I (the right side of Figure 104). The edges of the region T are included; since the point masses have no size, they can occupy the same position at the point of contact. An interesting question to ask yourself is: How would the state space change if the point masses were extended objects?

Now, given an initial set of data, with initial positions and velocities v_1 and v_2, respectively, what is the evolution of the system? The answer lies in the study of these types of dynamical systems called billiards. Evolution will look like movement in T. A point in T comprises the simultaneous positions of the two particles, and movement in T will consist of a curve parameterized by time t. The idea is that this curve will be a line since the two velocities are constants. In Figure 104, the slope of line a (the trajectory before any collisions), is v_2/v_1. As can be seen, its value is positive and less than 1 here. (Why?) Once a collision happens, though, this changes. There are two types of collisions: Assuming that v_2/v_1 is the ratio of the velocities of the two point masses before collision, we have the following:

- When a point mass hits a wall, it "bounces off," traveling back into I with equal velocity and of opposite sign. Thus the new velocity is $-v_2/v_1$ (this is the slope of line b in Figure 104).

- When the two point masses collide, they exchange their velocities (think of real billiard balls here). Thus the new velocity is v_1/v_2. Caution: This reciprocal velocity is not the slope of a perpendicular line, which would be the negative reciprocal.

Envision these collisions in the diagram and the resulting trajectory curves before and after each type of collision, as in the figure. What you see are perfect rebounds off of each of the three walls, where the angle of reflection equals the angle of incidence. This type of collision is sometimes also called *specular*. This is an ideal billiard table, although one with no pockets.—which leads to the obvious question: What happens if a trajectory heads straight into a corner? For now, we will accept the stipulation that

- When the two point masses collide with a wall simultaneously, either at separate ends of I or at the same end, both velocities switch sign. While this will not change the slope of the trajectory, it will change the direction of travel along that piece of trajectory line.

Some questions to ask:

Q. Can there exist closed trajectories?

Q. Can there exist a dense orbit?

Q. The orbits of points in T will very much intersect each other, and many trajectories will intersect themselves also. The phase space will get quite messy. Is there a way to better "see" the orbits of points more clearly?

The answer to the last question is yes, although this table is fairly special. Here, one can "unfold" the table:

- Think of the walls of T as mirrors. Do you see why we use the term "specular" now? When a trajectory hits a wall, it rebounds off in a different direction. However, its reflection in the mirror simply continues its straight-line motion. Think of a reflected region T across this wall. The trajectory looks to simply pass through the wall and continue on, as in Figure 105.

- Envision each collision that follows also via its reflection. Motion continues in a straight-line fashion through each mirrored wall. By continuing this procedure, the motion will look linear for all forward time, no?

- This idea works because this particular triangle, under reflections, will eventually cover the plane in such a way that only its edges overlap and all points in $\mathbb{R}^2$ are covered by at least one triangle. This is called a (regular) *tessellation* of the plane by T, and works only because T has some very special properties. See below.

- The unfolded trajectory is called a linear flow on the billiard table $\mathbb{R}^2$.

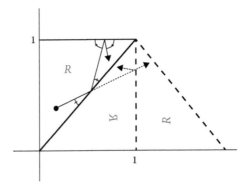

Figure 105 Starting to unfold the triangular billiard table.

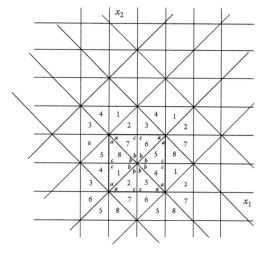

Figure 106 A fundamental domain.

So what does a billiard flow in $\mathbb{R}^2$ look like? Obviously, it is just straight-line motion at a slope v_2/v_1 forever, since there are no collisions. The better question to ask is: What does this tell us about the original flow on the triangle T?

By continually unfolding (reflecting) the table T, one starts to notice that there are only eight different configurations: the four orientations of T given by rotations by multiples of $\pi/2$ radians, and the reflection of each. If you collect up a representative of each of these configurations into a connected region, you wind up with enough information to characterize the entire flow in $\mathbb{R}^2$: each time your $\mathbb{R}^2$ linear flow re-enters a region of a particular configuration of T, you can simply note the trajectory in your representative of that region. This region of representative configurations is called a *fundamental domain* for the flow. One such fundamental domain of this flow is the square of side length 2 in Figure 106.

Noting the configurations and their arrangement, we can see that as the trajectory leaves the square, it enters a triangle with the same configuration as one on the other side of the square. We can view the trajectory then as re-entering the square from the other side. Similarly, when we leave the square at the top, it enters a configuration represented at the bottom of the square. Thus we can continue the trajectory as if it had re-entered the square at the bottom.

Remark 5.26 *There was a famous arcade video game called Asteroids, an Atari, Inc., game released in 1979, where a space ship was planted in the middle of a square screen. It could turn but not move. See Figure 107. Various boulders (asteroids) would float in and out of the screen. Should an asteroid hit the ship, the game would be over. The ship could fire a weapon at an asteroid, and, if hit. the asteroid would break into two smaller ones, which would go off in different directions. The asteroids (or pieces of asteroids) always traveled in straight lines. And as an asteroid left the screen, it would always reappear on the opposite side and travel in*

Figure 107 A toral universe?

the same direction. In fact, the asteroids were only exhibiting a linear toral flow. Who would have though that in playing this game, one was actually playing in a universe that was not the plane at all but rather the torus $\mathbb{T}^2$?

Hence linear flows on $\mathbb{R}^2$ again look like toral flows on this fundamental domain, which comprises the space of configurations of T as one uses T to tile $\mathbb{R}^2$. So what do linear toral flows say about the trajectories on T?

Proposition 5.27 *If the ratio of initial velocities $v_2/v_1 \in \mathbb{Q}$, then the orbit is closed (on $\mathbb{T}^2$ and thus also on T). If $v_2/v_1 \notin \mathbb{Q}$, then the orbit is dense in T.*

Note: Now it is easier to see what a collision in a corner of T will look like. Unfortunately, this billiard table T is quite special in many ways, and these methods of analysis do not generalize well. However, with this choice of T, we can say much more:

Proposition 5.28 *For any starting set of data (point mass positions and velocities), the trajectory will assume at most eight different velocity ratios.*

Count them: There are two possible ratio magnitudes, each with two signs. That makes four. But travel along the lines of each of these slopes can be in each of the two directions owing to the reflected configurations in the fundamental domain.

Here are some final comments before moving on: First, how can one generalize these results to other tables:

- **Unequal masses.**

  - An elastic collision between unequal masses will not result in what would look like a reflection off of the diagonal wall in T. One could certainly accurately chart the collision as a change in direction off of the wall. However, when unfolding the table, the resulting flow in $\mathbb{R}^2$ will only be piecewise linear. While this is workable, it is not such an easy leap to a conclusion.

  - One can also actually change the table. Use momenta to define the collision between the point masses, and alter the diagonal wall so that a collision remains specular. The resulting flow will not be linear. And the new table will no longer tile the plane, although in many cases the unfolded table will cover the plane with many holes (the reflecting curve will be concave, so will fit into the original T). The unfolded flow will look linear until it hits a hole, where it will reflect through the hole perpendicularly through its center axis and appear on the other side to continue at the same slope.

- **Other tessellations of $\mathbb{R}^2$.** It is easy to see that some shapes tessellate the plane, while others do not. For regular polygons, only triangles, squares, or hexagons tessellate the plane. Rectangles, and a few non-regular triangles, also work fine. But examples are fairly rare. And, in each case, one would need to find a fundamental domain and then interpret the resulting flow on that domain in terms of the original flow as well as that on the plane. We will not pursue this further in this text.

We will return to more general billiards in Section 6.4. For the present, we will stay with the idea of dynamical systems on a torus, but will switch from continuous ODE systems to maps. Here, some surprising relationships occur between flows on a torus and the corresponding time-t maps.

5.4 Toral Translations

As with Euclidean space $\mathbb{R}^n$, one can generalize the construction of the 2-torus $\mathbb{T}^2 = S^1 \times S^1$ by considering a system of equations involving more than two angular coordinates: The n-dimensional torus, or n-torus, denoted by $\mathbb{T}^n$ is simply the n-fold product of n circles

$$\mathbb{T}^n = \overbrace{S^1 \times \cdots \times S^1}^{n \text{ times}}.$$

Then

$$\mathbb{T}^n = \mathbb{R}^n / \mathbb{Z}^n = \mathbb{R}/\mathbb{Z} \times \cdots \times \mathbb{R}/\mathbb{Z}.$$

One could imagine a linear flow on $\mathbb{T}^n$ by simply adding the appropriate number of uncoupled equations to the system in Equation 5.3.1; with n-independent rotational flows in each copy of $\mathbb{S}^1$, the resulting flow may be seen as linear motion on $\mathbb{T}^n$.

Indeed, view the n-torus via an identification within $\mathbb{R}^n$. Recall that the unit square with it opposite sides identified acts as a good model for $\mathbb{T}^2$. And the unit interval $[0,1]$ with its opposite boundary points 0 and 1 identified yields us the circle $S^1 = \mathbb{T}^1$. This construction generalizes for all the natural numbers: For $n = 3$, take the unit cube in $\mathbb{R}^3$ (see the general shape of Figure 108). Identify each of the opposite pairs of faces, squares in this case (think of a die, and identify two sides if their numbers add up to 7). The resulting model is precisely the (square or flat) 3-torus $\mathbb{T}^3$. This works well if one wants to watch a linear flow on $\mathbb{T}^3$. Simply allow the flow to progress in the unit cube, and whenever it hits a wall, simply let it vanish and reappear on the opposite wall, entering back into the cube, as in Figure 108. Note that here the origin is at the lower left corner, intersections with $\mathcal{O}_{(0,0,0)}$ at the top and bottom are in red, while intersections at the back and front walls are in blue. This represents a periodic orbit. Can you see this?

Now, the vectorized exponential map $(\theta_1, \ldots, \theta_n) \xmapsto{\exp} \left(e^{2\pi i \theta_1}, \ldots, e^{2\pi i \theta_n} \right)$ maps $\mathbb{R}^n$ onto $\mathbb{T}^n$. We can define a (vector) rotation on $\mathbb{T}^n$ by the vector $\boldsymbol{\alpha} = (\alpha_1, \ldots, \alpha_n)$, where $R_{\boldsymbol{\alpha}}(\mathbf{x}) = (x_1 + \alpha_1, \ldots, x_n + \alpha_n) = \mathbf{x} + \boldsymbol{\alpha}$. Normally, this is called a *translation* (by $\boldsymbol{\alpha}$) on the torus. In Figure 108, $\boldsymbol{\alpha} = (1, 2, 3)$, and only the orbit of the origin is shown for the toral flow corresponding to this $\boldsymbol{\alpha}$. Note that it should be obvious that if all of the α_i's are rational, then the resulting flow on $\mathbb{T}^n$ will have closed orbits. Questions to ask are: Are these the only maps of this type that are linear? If one or more α_i's are not rational, can there still be periodic orbits? And if there cannot, are the non-periodic orbits dense in the torus?

Consider the linear flow in $\mathbb{T}^n$ whose time-1 map is $R_{\boldsymbol{\alpha}}$. This would be the flow whose ith-coordinate solution is $x_i(t) = x_i + \alpha_i t$. Again, with ALL of the α_i's rational, the flow would have all closed orbits. Now allow one of the coordinate rotation numbers to be irrational. We saw how it was the ratio of the two flow rates that determined whether the flow had closed orbits on $\mathbb{T}^2$. Does this hold in higher dimensions? Do the properties of the time-1 map still reflect accurately the properties of the flow? Does the irrationality

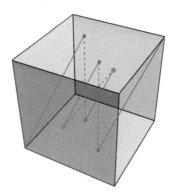

Figure 108 A periodic orbit of a translation flow on $\mathbb{T}^3$.

of some or all of the coordinate rotations imply minimality of the map? Of the flow? All of these questions will rely on a good notion of measuring the relative ratios of the individual pairs of map rotations and flow rates. And how do we define these ratios in higher dimensions? By a notion of the rational independence of sets of numbers:

Definition 5.29 *A set of n real numbers $\{\alpha_i\}_{i=1}^n$ is said to be* rationally independent *if, given $k_1, \ldots, k_n \in \mathbb{Z}$, the only solution to*

$$k_1 \alpha_1 + \ldots + k_n \alpha_n = 0$$

is for $k_1 = \cdots = k_n = 0$.

Another way to say this is the following: For all non-trivial integer vectors $\mathbf{k} = (k_1, \ldots, k_n) \in \mathbb{Z}^n - \{\mathbf{0}\}$,

$$\sum_{n=1}^{n} k_i \alpha_i = \mathbf{k} \cdot \boldsymbol{\alpha} \neq 0.$$

We have the following:

Proposition 5.30 *A toral translation on $\mathbb{T}^n$, given by R_α, is minimal iff the numbers $\alpha_1, \ldots, \alpha_n, 1$ are rationally independent.*

Proposition 5.31 *The flow on $\mathbb{T}^n$ whose time-1 map is the translation R_α is minimal iff the numbers $\alpha_1, \ldots, \alpha_n$ are rationally independent.*

Do you see the difference? One way to view this is to restrict to the case of a 2-torus. Here, the second proposition says that the flow will be minimal if, in essence, $k_1 \alpha_1 + k_2 \alpha_2 = 0$ is only satisfied when $k_1 = k_2 = 0$. In fact, if there were another solution, then it would be the case that $\alpha_2/\alpha_1 = k_1/k_2 \in \mathbb{Q}$.

On the other hand, the first proposition indicates that both α_1 and α_2 need to be rationally independent and also both rationally independent from 1! That means that not only do the two α's need to be rationally independent from each other, but neither α_1 nor α_2 can be rational (then it would be a rational multiple of 1). Hence a flow can be minimal on a torus, while the time-1 map isn't. Why is this so? Let's study the situation via an example.

Example 5.32 On the 2-torus, let $\alpha_1 = \frac{1}{4}$ and $\alpha_2 = \pi/16$. The flow will be minimal here, since $\alpha_2/\alpha_1 = \pi/4 \notin \mathbb{Q}$ (α_1 and α_2 are rationally independent). However, the time-1 map of this flow, for $\alpha = \begin{bmatrix} \alpha_1 \\ \alpha_2 \end{bmatrix}$, is R_α, and since

$$k_1 \cdot \frac{1}{4} + k_2 \cdot \frac{\pi}{16} + k_3 \cdot 1 = 0 \text{ is solved by } \begin{bmatrix} k_1 & k_2 & k_3 \end{bmatrix} = \begin{bmatrix} 4 & 0 & -1 \end{bmatrix},$$

the translation will not be minimal (the orbits are not dense in the torus). The fact that α_1 is already rational is the problem. See Figure 109. Essentially, the orbit coordinates of the translation in the x_1 direction will only take the values $0, \frac{1}{4}, \frac{1}{2}, \frac{3}{4}$, while the x_2-

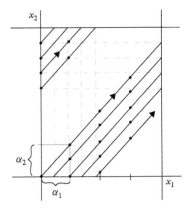

Figure 109 The orbit of the origin from Example 5.32.

coordinates will "fill out" the vertical direction. The result is that the orbit of the translation will only be dense on the vertical lines corresponding to the x_1-coordinates of the orbit. This sits in contrast to the flow, in which every orbit will "fill" the torus.

Exercise 204 Given the flow in Example 5.32 show that on the global Poincaré section corresponding to the set $S_{\frac{1}{2}} = \left\{ \left(\frac{1}{2}, y\right) \in \mathbb{T}^2 \mid y \in S^1 \right\}$, the first return map is minimal.

Exercise 205 Rework Example 5.32 with the following new values of α:

(a) $\alpha = \left(\frac{3}{4}, \frac{1}{5}\right)$;   (b) $\alpha = \left(\frac{1}{5}, e\right)$;   (c) $\alpha = (\pi, e)$;

(d) $\alpha = \left(\sqrt{2}, \sqrt{3}\right)$;   (e) $\alpha = \left(\frac{1}{\sqrt{2}}, \sqrt{2}\right)$.

5.5 Invertible Circle Maps

Let's return to maps on the circle, but now relax the condition that the map be a rigid rotation. Again, think of S^1 as the identification space $S^1 = \mathbb{R}/\mathbb{Z}$, given by the map $x \mapsto e^{2\pi i x}$, but let's rewrite the map a bit. Consider the map

$$\pi : \mathbb{R} \to S^1, \quad \pi(x) = [x] = x - \lfloor x \rfloor,$$

where $[x]$ is simply what is left over by disregarding the integer part. Here again, $\lfloor x \rfloor$ is called the *floor* function, introduced in Exercise 197 of Section 5.2.1. The function $\pi(x) = [x]$, as a function from $\mathbb{R}$ to $\mathbb{R}$ is sometimes also called the *sawtooth function*, due to the shape of its graph. The sawtooth function is a projection from $\mathbb{R}$ to $\mathbb{R}$, and is only piecewise continuous on $\mathbb{R}$. However, the image is precisely the set of equivalence classes forming the points of $S^1 = \mathbb{R}/\mathbb{Z}$, and is continuous as a function from $\mathbb{R}$ to S_1. This new notation, however, will be easier to utilize in the discussion.

Definition 5.33 *A map $f : X \to X$ is called a* projection *if $\forall x \in X, f(x) = f^2(x)$, that is, if f equals its square (composition with itself).*

A couple of notes here:

- Technically speaking, S^1 is not a subset of $\mathbb{R}$, so the map π is only a projection if we think of the equivalence classes comprising S^1 as the half-open interval $[0, 1) \subset \mathbb{R}$.
- A map or an operation that doesn't change the effect on inputs upon repeated application after the initial one is called *idempotent*. Projections are idempotent. Another example is the absolute value function, defined on $\mathbb{R}$.
- The use of the symbol $\pi = \pi(x)$ for this function is common and has no relation to the number π. It is common in mathematics to denote a projection by π.

Proposition 5.34 *For any continuous map $f : S^1 \to S^1$, there exists an associated map $F : \mathbb{R} \to \mathbb{R}$, called a* lift *of f to $\mathbb{R}$, where*

$$f \circ \pi = \pi \circ F, \quad equivalently f([x]) = [F(x)].$$

Some notes:

- One way to see this is via the commutative diagram

$$
\begin{array}{ccc}
\mathbb{R} & \xrightarrow{\ F\ } & \mathbb{R} \\
{\scriptstyle \pi}\downarrow & & \downarrow{\scriptstyle \pi} \\
S^1 & \xrightarrow{\ f\ } & S^1
\end{array}
$$

- The lift F is unique up to an additive constant (sort of like how the anti-derivative of a function is unique only up to an additive constant).
- The quantity

$$\deg(f) = F(x+1) - F(x)$$

is well-defined for all $x \in \mathbb{R}$ and is called the *degree* of f.
- If f is a homeomorphism, then $\left|\deg(f)\right| = 1$.
- The structure of F is quite special. It looks like the sum of a periodic function with the line $y = \left(\deg(f)\right)x$. This is due to the structure of the projection π.
- The degree of a map is well-defined with regard to composition, and the degree of a composition of maps is the product of the degrees of the individual maps.

So just how much information about f can we learn by the study of the lifts of f? Certainly, maps on $\mathbb{R}$ are fairly easy to study. (This is what your first calculus class was really all about! The full title of a first semester calculus course could be "the calculus of functions of one real, independent variable".) And maps with the structure of these lifts F may be easier still. If we can use these lifts to say fairly general things about how an f may behave, this would be quite important. For example, this quantity $\deg(f)$ is defined solely by a choice of lift F of f. But since $\deg f$ is a property of f, the definition must be independent of the choice of lift. We will see presently just what information $\deg(f)$ conveys. For a moment, let's first take a look at why some of the assertions we just made are true.

- *Lifts always exist.* Simply construct one using the definition. This will become clear from the examples and exercises below.

- *F is unique up to a constant.*

  Proof Suppose $\overline{F}$ is another lift. Then

  $$\left[\overline{F}(x)\right] = f\left([x]\right) = [F(x)] \quad \forall x \in \mathbb{R}.$$

  This is just another way of saying that $\pi \circ \overline{F} = f \circ \pi = \pi \circ F$, $\forall x \in \mathbb{R}$. But then $\overline{F} - F$ is always an integer! Indeed, $\overline{F} - F$ is the difference between two continuous functions, and hence is continuous. But a continuous function on $\mathbb{R}$ that takes values in the integers is necessarily constant. □

- deg(*f*) *is well-defined.*

  Proof Here $\deg(f) = F(x+1) - F(x)$ is a continuous function on $\mathbb{R}$ that takes values in the integers (it must, owing to the projection π and the commutativity of the diagram defining F). Thus it also is a constant for all $x \in \mathbb{R}$. □

- *If f is a homeomorphism, then* $\left|\deg(f)\right| = 1$.

  Proof Suppose that $\left|\deg(f)\right| > 1$. Then $|F(x+1) - F(x)| > 1$. And since $F(x+1) - F(x)$ is continuous, by the Intermediate Value Theorem, $\exists y \in (x, x+1)$ where $\left|F(y) - F(x)\right| = 1$. But then $f\left([y]\right) = f\left([x]\right)$ for some $y \neq x$. Thus f cannot be one-to-one, and hence cannot be a homeomorphism.

  Now suppose that $\left|\deg(f)\right| = 0$. Then $F(x+1) = F(x)$ $\forall x$, and hence F is not one-to-one on the interval $(x, x+1)$. But then neither is f, and, again, f cannot be a homeomorphism. □

- $F(x) - x\deg(f)$ *is periodic.*

  Proof It is certainly continuous. (Why?) To see that it is periodic (of period-1), simply evaluate this function at $x+1$:

  $$F(x+1) - (x+1)\deg(f) = \left(F(x) + \deg(f)\right) - (x+1)\deg(f)$$
  $$= F(x) - x\deg(f).$$ □

- *For f, g : $S^1 \to S^1$ continuous,* $\deg(g \circ f) = \deg(g)\deg(f)$.

  Proof Given lifts $F, G : \mathbb{R} \to \mathbb{R}$ of f and g, for $s \in S^1$ and $k \in \mathbb{Z}$,

  $$G(s+k) = G(s+k-1) + \deg(g)$$
  $$= G(s+k-2) + 2\deg(g) = \cdots = G(s) + k\deg(g).$$

  But this means

  $$G(F(s+1)) = G\left(F(s) + \deg(f)\right) = G(F(s)) + \deg(f)\deg(g). \quad □$$

Hence, for instance, $\deg(f^n) = \left(\deg(f)\right)^n$. So, given these properties, what do lifts of circle maps look like in practice? Here are some examples:

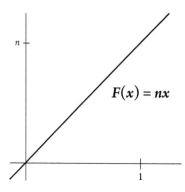

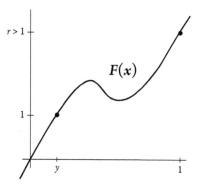

Figure 110 A lift of $f(x) = x^n$ (left) and a lift of some degree-r map (right).

Example 5.35 Let $f(x) = x$. This is the "identity" map on S^1, since all points are taken to themselves. A suitable lift for f is the map $F(x) = x$ on $\mathbb{R}$ (see Figure 110 for $n = 1$). To see this, make sure the definition works. Question: Are there any other lifts for f? What about the map $\overline{F}(x) = x + a$ for a a constant? Are there any restrictions on the constant a? The answer is yes. For a to be an acceptable constant, we would need the definition of a lift of be satisfied. Thus

$$[\overline{F}(x)] = [x + a] = f([x]) = [x].$$

So the condition that a must satisfy is $[x + a] = [x]$ on S^1. Hence $a \in \mathbb{Z}$.

Exercise 206 For a real number $a \notin \mathbb{Z}$, can $\overline{F}(x) = x + a$ serve as a lift of a circle map? What sort of circle map?

Exercise 207 Find a suitable lift $F : \mathbb{R} \to \mathbb{R}$ for the rotation map $R_\alpha : S^1 \to S^1$ where $\alpha = 2\pi$ and verify that is works. Graph both F and R_α. Keep in mind that we are using $S^1 = \mathbb{R}/\mathbb{Z}$ as our model of the circle.

Example 5.36 Let $f(x) = x^n$. Thinking of x as the complex number $x = e^{2\pi i\theta}$, for $\theta \in \mathbb{R}$, then

$$f(x) = f\left(e^{2\pi i\theta}\right) = \left(e^{2\pi i\theta}\right)^n = e^{2\pi i(n\theta)}.$$

Hence a suitable lift is obviously $F(x) = nx$, as in Figure 110 on the left (I say obviously, since the variable in the exponent <u>is</u> the lifted variable!) Question: This is a degree-n map. For which values of n does the map f have an inverse" And what does the map f actually do for different values of n?

Example 5.37 Let f be a general degree-r map. Then $F(1) - F(0) = r = \deg(f)$. Suppose that $F(0) = 0$. Then $F(1) = r$, and if, for example, $r > 1$, it is now easy to see that there will certainly be a $y \in (0, 1)$, where $F(y) = 1$. This was a fact that we used in the proof above to show that f cannot be a homeomorphism (see Figure 110 on the right). In this case, where $r > 1$, at every point in $y \in (0, 1)$ where $F(y) \in \mathbb{Z}$, we will

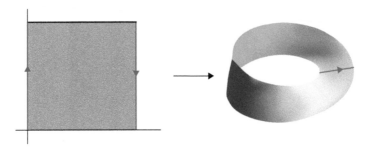

Figure 111 The unit square and, with its identifications, the Möbius band.

have $\pi \circ F(y) = [F(y)] = 0$ on S^1. This won't happen when $r = 1$. When $r = 0$, the map F will be periodic, which is definitely not one-to-one. Question: What happens when $r < 0$? Draw some representative examples to see.

Definition 5.38 *Suppose that* $f : S^1 \to S^1$ *is invertible. Then*

(1) *if* $\deg(f) = 1, f$ *is orientation-preserving;*

(2) *if* $\deg(f) = -1, f$ *is orientation-reversing.*

Recall from your study of vector calculus that *orientation* is, in a sense, a choice of direction in the parameterization of a space. (Actually, it exists outside of any choice of coordinates on a space, but once you put coordinates on a space, you have essentially chosen an orientation for that space.) This is true at least for those spaces that actually are orientable. For example, one way to construct a *Möbius band* is to take the unit square in the plane and identify the left wall with the right wall, but with a twist—one creates a kind of "twisted" cylinder. This would be similar to the construction of a torus in Section 3.3.4, except that we will leave the roof and floor alone. See Figure 111. What results is a geometric figure with one edge and one side. Can you see this?

On $\mathbb{R}$, an orientation is simply a choice of direction for the symbol ">" in the standard strict total ordering on the reals. More generally, for an orientable curve in $\mathbb{R}^n$, it is again simply a choice of direction along the curve. On an orientable surface, it is a little more complicated. However, if that surface is C^1 and a subset of $\mathbb{R}^3$, one can simply make a choice of side, which amounts to choosing a unit normal vector to the surface. Then, any basis for one of the tangent spaces to the surface in $\mathbb{R}^3$, along with a choice of one of the two unit normal vectors, is either compatible with the *right-hand rule* or not. On S^1, a map is orientation-preserving if, in the image, points to the right of a small neighborhood of a designated point remain on the right. An orientation-reversing map will flip a very small neighborhood of a point. And we note that there are ways of defining orientation of a topological space regardless of whether it is a subspace, whether it "lives" inside some other space. However, we will not go into this general detail in this text.

Dynamically, circle maps may or may not have periodic points. And, given an arbitrary homeomorphism, without regard to any other specific properties of the map, we would expect to be able to construct maps with lots of periodic points of any period. However, it turns out that circle homeomorphisms are quite restricted, since they are one-to-one and onto, or bijective. To explain, we will need to develop another property of circle homeomorphisms.

Proposition 5.39 *Let $f : S^1 \to S^1$ be an orientation-preserving homeomorphism, with $F : \mathbb{R} \to \mathbb{R}$ a lift. Then the quantity*

(5.5.1)
$$\rho(F) := \lim_{|n| \to \infty} \frac{F^n(x) - x}{n}$$

(1) *exists $\forall x \in \mathbb{R}$;*

(2) *is independent of the choice of x and is defined up to an additive integer; and*

(3) *is rational iff f has a periodic point.*

Given these qualities, the additional quantity $\rho(f) = [\rho(F)]$ is well-defined and is called the *rotation number* of f.

Some notes:

- This quantity and this proposition were proposed and proved by Henri Poincaré [42] back in 1885.

- $\rho(f)$ is also sometimes called the *map winding number,* although it is different from the winding number used in algebraic topology or complex analysis. Be careful here.

- $\rho(R_\alpha) = [\alpha]$.

  Exercise 208 Use Definition 5.5.1 to calculate $\rho(R_\alpha)$ for the circle rotation R_α, $\alpha \in (0, 1)$.

- ρ represents in a way the *average rotation* of points in a circle homeomorphism.

It turns out that the rotation number is a very telling property of a circle homeomorphism. And like interval maps, there is a certain rigidity to the properties of circle homeomorphisms:

Proposition 5.40 *If $\rho(f) = 0$, then f has a fixed point.*

Another way of stating this proposition is that if there is no average rotation of the circle map, then somewhere a point doesn't move under f. This is like the Intermediate Value Theorem in first semester calculus: if the points whose image is to the right are averaged out by points whose image is to the left, then somewhere a point must not move at all.

Proof We prove the contrapositive. Assume that a circle homeomorphism $f : S^1 \to S^1$ has no fixed points. Then, given a lift $F : \mathbb{R} \to \mathbb{R}$, where $F(0) \in (0, 1)$, if $F(x) - x \in \mathbb{Z}$ for any $x \in \mathbb{R}$, then f must have a fixed point, by the definition of a lift. Hence $\forall x \in \mathbb{R}$, $F(x) - x \notin \mathbb{Z}$. But as a function on $\mathbb{R}$, the function $F(x) - x$ is continuous, and hence, by the Intermediate Value Theorem,

$$0 < F(x) - x < 1, \quad \forall x \in \mathbb{R},$$

and on the interval $[0,1]$, $F(x) - x$ must achieve its maximum and minimum by the Extreme Value Theorem. So there must exist a constant m, where

$$0 < m \leq F(x) - x \leq 1 - m < 1.$$

But, by the above, $F(x) - x$ is periodic, and hence this last inequality holds on all of $\mathbb{R}$. Choosing $x = 0$, we get that $m \leq F(0) \leq 1 - m$ implies $nm \leq F^n(0) \leq n(1 - m)$ by additivity, so that

$$m \leq \frac{F^n(0)}{n} \leq 1 - m.$$

Since this holds for all $n \in \mathbb{N}$, it holds in the limit. But this limit is the rotation number $\rho(f)$ and is independent of the choice of initial point. Hence $\rho(f) \neq 0$. □

Exercise 209 Show that if a continuous $f : S^1 \to S^1$ is not surjective, then $\rho(f) = 0$.

One can generalize quite readily to q-periodic points by looking at the fixed points of f^q, so that we get the following:

Proposition 5.41 *If f is an orientation-preserving homeomorphism of S^1, then $\rho(f)$ is rational if and only if f has a periodic point.*

Exercise 210 Prove this.

But it gets even more restrictive. If f has a q-periodic point, then, for a lift F, we have $F^q(x) = x + p$ for some $p \in \mathbb{Z}$. For example, let $f = R_{\frac{6}{7}}$. Then a suitable lift for f would necessarily satisfy $F^7(x) = x + 6$, $\forall x \in \mathbb{R}$. Notice that there would be no room for any other periodic points in this case. But this is true in general.

Proposition 5.42 *Let $f : S^1 \to S^1$ be an orientation-preserving homeomorphism. Then all periodic points must have the same period.*

This last point is quite restrictive. Essentially, if an orientation-preserving homeomorphism has a fixed point, it cannot have periodic points of any other period, say. Note that this is not true of a orientation-reversing map. For example, the map that flips the unit circle in $\mathbb{R}^2$ across the y-axis will fix the two points $(0, 1)$ and $(0, -1)$, while every other point is of order two.

Proof Suppose $f^p(x_*) = x_*$. Then $S^1 - \{x_*\}$ is simply an interval, and f^p, as a homeomorphism, extends to a bijective increasing map on $I = \overline{S^1 - \{x_*\}}$ that fixes both endpoints. By Proposition 2.59, every point of I is either fixed by f^p (and therefore a period-p point) or asymptotic to a fixed point. Thus, there can be no other m-periodic points, for $m \neq p$. □

It is quite typical that degree-1 circle homeomorphisms, when they have n-periodic points, have an even number of distinct n-orbits, weaving attractive orbits with repelling ones. Can you see this? Do you recognize this behavior from other contexts? Go back to Figure 23 in Section 2.3.4 on homoclinic and heteroclinic points as you ponder these questions.

Example 5.43 Let $f : S^1 \to S^1$ be the orientation-preserving homeomorphism given by the expression

$$f(x) = x + \frac{1}{2} + \frac{1}{4\pi}\sin(4\pi x).$$

It is not easy to verify that f is indeed a homeomorphism analytically. But from the plot of f in Figure 112 on the left, one can verify its properties readily. Notice that f has no fixed points, but does have two easily verified period-2 orbits (one can "see" them in the argument of the sine), namely, at $\mathcal{O}_0 = \{0, \frac{1}{2}\}$ and at $\mathcal{O}_{\frac{1}{4}} = \{\frac{1}{4}, \frac{3}{4}\}$. And the stability of these period-2 orbits? Take a look at the graph of $f^2 = f \circ f$ in Figure 112 on the right. The period-2 orbits appear as fixed points here, and their stability is readily readable here from the derivative information. Just cobweb a bit to verify.

Exercise 211 Show that any lift of the rotation map $R_{\frac{6}{7}}$ must satisfy $F^7(x) = x + 6, \forall x \in \mathbb{R}$, and explicitly construct two such lifts.

Exercise 212 Find the rotation number for the following invertible circle map:

$$f(x) = \begin{cases} \frac{1}{3}(x_3 + 1), & 0 \leq x < \frac{1}{3}, \\ 6x - \frac{44}{27}, & \frac{1}{3} \leq x < \frac{10}{27}, \\ 2x - \frac{4}{27}, & \frac{10}{27} \leq x < \frac{1}{2}, \\ \frac{8}{5}\left(x - \frac{1}{2}\right) + \frac{23}{27}, & \frac{1}{2} \leq x < \frac{21}{27}, \\ \frac{1}{6}\left(x - \frac{21}{27}\right) + \frac{35}{27}, & \frac{21}{27} \leq x < 1. \end{cases}$$

Circle homeomorphisms with rational rotation number have another interesting feature: they tend to trap orbits in between periodic orbits in a certain precise way. Indeed, the power of the map that has fixed points will trap orbits between adjacent fixed points.

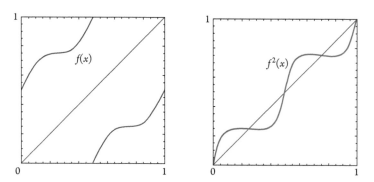

Figure 112 The circle homeomorphism f and its square $f^2 = f \circ f$.

Proposition 5.44 *Let* $f : S^1 \to S^1$ *be an orientation-preserving homeomorphism such that* $\rho(f) = p/q \in \mathbb{Q}$. *Then, if* $\bar{x}$ *is not periodic, it is heteroclinic under* f^q *to two fixed points of* f^q *(two periodic points of* f*).*

Remark 5.45 *It is possible that the set of all periodic points consists of a single fixed point. In this case, all of the non-fixed points will be homoclinic to the fixed point. Periodic orbits of maps also can have stability properties, as we will see in Section 7.5. So orbits of non-periodic points here can be heteroclinic to two periodic orbits (of* f*), or homoclinic to one, possibly.*

Proof For clarity, assume that p/q is in lowest terms, although this is not strictly necessary. Then, by Proposition 5.42, f has only q-periodic points in $\mathrm{Per}(f)$. Thus, f^q, still an orientation preserving homeomorphism, has only fixed points in $\mathrm{Per}(f^q)$.

Let $\bar{x}$ be non-periodic for f and hence for f^q. Then there exists a smallest closed arc $\bar{I}_{\bar{x}} = \left[\bar{a}, \bar{b}\right] \subset S^1$ such that $\bar{x} \in \bar{I}_{\bar{x}}, f^q(\bar{a}) = \bar{a}$ and $f^q(\bar{b}) = \bar{b}$, and $\forall \bar{y} \in \bar{I}_{\bar{x}}, \bar{y}$ is not periodic. Lift f^q to a map F^q, where $F^q(0) \in [0, 1)$. Then, for $\pi(a) = \bar{a}$ and $\pi(b) = \bar{b}$, the closed, bounded $I_x = [a, b]$ satisfies $0 \le a < b \le 1$, $F^q(a) = a$, $F^q(b) = b$, and $F^q : I \to I$ is an increasing interval map with no fixed points on the interior. By Proposition 2.59, $\mathcal{O}_x$ is forward asymptotic to one of a or b and backward asymptotic to the other of the two endpoints. Now project $\mathcal{O}_x$ back down to S^1 via π to $\mathcal{O}_{\bar{x}}$. □

For now, this will be enough for circle homeomorphisms. We will return very briefly to this topic in Section 8.1 with a famous classification theorem of Arnauld Denjoy for circle homeomorphisms.

6 Phase Volume Preservation

In Chapter 5, we first looked at what was considered "recurrent" behavior at a point in a dynamical system, which roughly means that the orbit of a point passes arbitrarily close to the point. This worked well in the classification of rigid circle rotations, since either the orbit of a point was closed (the orbit was periodic; the rotation was rational) or the orbit was dense (for an irrational rotation). In either case, every point was recurrent. The same was true for the linear toral flows and their time-t maps.

Contrast this with the dynamical systems that we studied in Chapters 2 and 4. Here, with examples like contracting maps and interval maps with sinks and sources, as well as planar maps and flows with sinks, sources, and saddles, the only recurrent points were the fixed and periodic points, and there were very few of those in each system (except for centers, that is). More generally, maps can exhibit much more complicated behavior. To understand this behavior, we will have to broaden our idea of how to study such systems. This chapter begins this study.

To start, let's change our perspective. Given a dynamical system, let's not focus on the behavior of an individual orbit so much as on how whole families of nearby orbits evolve. This will be more like following all of the orbits that start in a small open subset of the state space over the evolution of the map. For a contraction, this would be easy and not very insightful. (Why is that again?) But for a general map, this idea can be quite interesting.

6.1 Incompressibility

The notion of *incompressibility* in a dynamical system means that positive-volume domains in the state space do not change their volume as the orbits of their points evolve. This notion is also called *phase volume preservation*. Suppose we have a dynamical system where this property holds; as we evolve via a flow, or iterate via a map, the volume of a small domain does not change. When this happens, the volume is said to be preserved by the flow (respectively, map), or is said to be invariant under the flow (respectively, map). Obvious examples include translational flows in $\mathbb{R}^n$, rotation maps on S^1 (remember that volume in a space like $\mathbb{R}$ or S^1 is just length, and in dimension 2 is just area), and linear toral flows. Examples of systems that do not preserve volume include contraction maps, and flows (defined by ODEs) that include sinks and sources.

In fact, if the map is an isometry, or the flow has all of its time-t maps given by isometries, then the volume will be preserved in phase space. This should be obvious, since if all of the distances between the points of a small domain are preserved, the volume cannot change. The converse is not true, however. Lots of maps and flows preserve volume but are not isometries.

Example 6.1 The linear twist map on the cylinder

$$T : S^1 \times [0,1] \longrightarrow S^1 \times [0,1], \quad T(x,y) = (x+y,y)$$

is an area-preserving map that is not an isometry. See Figure 113.

Exercise 213 Show that (1) T is not an isometry, but (2) T preserves area on the cylinder.

Exercise 214 Show that the linear flow on the 2-torus in Equation 5.3.1 is an isometry. (Hint: Build the proper metric on the torus.)

Now, let's consider a linear map on $\mathbb{R}^n$:

$$f : \mathbb{R}^n \to \mathbb{R}^n, \quad f(\mathbf{x}) = A\mathbf{x},$$

where A is a $n \times n$ matrix. Choose an orthonormal basis for $\mathbb{R}^n$. Then the standard cube C whose sides are these orthonormal basis vectors (this is the "unit cube" relative to the chosen basis) will be mapped by f to a parallelepiped. The volume of C is 1. The volume of its image $f(C)$, however, will in general not be 1:

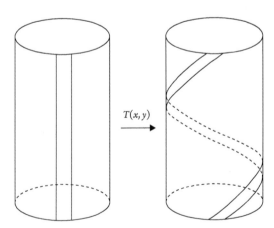

$T(x,y)$

Figure 113 A cylinder twist.

$$\text{vol}\big(f(C)\big) = |\det A|.$$

This is standard linear algebra, where, according to f, each standard basis vector $\mathbf{e}_i$ is mapped to the ith column of A. Hence the image of the unit cube is the parallelepiped with edges the columns of A. Hence the determinant of A (in absolute value) is the volume of the image of the unit cube under f. Hence volume is preserved by f if $|\det A| = 1$. What conclusion can one draw from this? Read f as the linear model for the infinitesimal version of any smooth map on $\mathbb{R}^n$. Then we have:

Proposition 6.2 *Let $U \in \mathbb{R}^n$ be an open domain. A differentiable map $f : U \to \mathbb{R}^n$ preserves volume iff $\big|\det(Df_x)\big| = 1, \forall x \in U$.*

Recall that the *Jacobian matrix*, or simply the *Jacobian*, of a function $f : U \subset \mathbb{R}^m \to \mathbb{R}^n$ at a point $x \in U$ is just the derivative matrix, as a matrix of numbers. We can extend this to an $n \times m$ matrix of partial derivative functions of f, sometimes denoted $\text{Jac}(f)$, or

$$\frac{\partial \big(f_1, \ldots, f_n\big)}{\partial (x_1, \ldots, x_m)},$$

or simply Df as the derivative matrix. Then the ijth element of $\text{Jac}(f)$ is the function $(\partial f_i / \partial x_j)(x_1, \ldots, x_n)$. Evaluated at a point $x \in U$, $Df(x) = Df_x$ is then just the matrix of real numbers, called the derivative of f at x. When $n = m$, the derivative matrix is square, and its determinant becomes an important property. In the square matrix case, it is common to refer to the determinant of this derivative matrix as simply the *Jacobian determinant* of f, $\text{Jac}(f)$.

Remark 6.3 *Be careful here, since the Jacobian may refer to the matrix of partial derivatives as functions, or the matrix of numbers when those partial derivatives are evaluated at a point in the domain, or the determinant of either of these without explicit reference to which. Context and care will usually determine the meaning, however.*

Definition 6.4 *A map $f : U \to \mathbb{R}^n$, where $U \subset \mathbb{R}^n$ is a domain, preserves orientation if $\forall x \in U, \text{Jac}(f) > 0$.*

"Nice" ODEs (where solutions exist and are unique everywhere, for example), are always orientation-preserving. Recall the relationship between the time-1 map of any linear ODE system on $\mathbb{R}^2$ and its corresponding flow. The time-1 map is a linear transformation on the plane, and its matrix always has eigenvalues that are related to those of the original flow by the exponential map. Under the exponential map, the time-1 map will always have a positive Jacobian. (Can you see why?)

More generally, let $\dot{\mathbf{x}} = \mathbf{f}(\mathbf{x})$ be an ODE on $\mathbb{R}^n$, where $\mathbf{f}$ is a C^1 vector field on $\mathbb{R}^n$. Let $\varphi : \Omega \times I \to \mathbb{R}^n$ be its corresponding flow, where $0 \in I \subset \mathbb{R}$ is open and $\Omega \subset \mathbb{R}^n$ is a domain (recall we often write $\varphi(t, \mathbf{x}) = \varphi^t(\mathbf{x})$ to accentuate the one-parameter group of transformations of phase space as well as the exponent notation of function iteration compatible with the additive group). Here, for a fixed $t \in I$, $\varphi^t : \Omega \to \mathbb{R}^n$ is a diffeomorphism onto its image, and $\varphi^0(\mathbf{x}) = \varphi(0, \mathbf{x}) = \mathbf{x}$.

Denote by $\text{Jac}(\varphi^t)$ the Jacobian matrix of the flow map φ^t. Here $\text{Jac}(\varphi^t)$ is just the derivative of the time-t map of the flow and $\text{Jac}(\varphi^0) = I_n$. This way, $\det\left(\text{Jac}(\varphi^t)\right)$ measures the relative volume of bounded domains at time t in Ω relative to their original volumes at $t = 0$. Then $(d/dt)\big|_{t=0} \det\left(\text{Jac}(\varphi^t)\right)$ measures the instantaneous rate of change of volume along the flow. Recall that this is precisely one of the geometric interpretations of the divergence of a vector field:

$$\text{div}(\mathbf{f}) = \frac{d}{dt}\bigg|_{t=0} \det\left(\text{Jac}(\varphi^t)\right),$$

when φ^t is the flow corresponding to the vector field $\mathbf{f}$.

Indeed, we know the following:

(1) Flows solve their corresponding ODE systems:

$$\frac{d}{dt}\varphi^t = \mathbf{f}(\varphi^t) \qquad \text{and} \qquad \frac{d}{dt}\bigg|_{t=0} \varphi^t = \mathbf{f}(\mathbf{x});$$

(2)
$$\frac{d}{dt}\bigg|_{t=0} \text{Jac}(\varphi^t) = \text{Jac}\left(\frac{d}{dt}\bigg|_{t=0} \varphi^t\right),$$

since time and phase space coordinates are independent of each other.

Thus we have

$$\frac{d}{dt}\bigg|_{t=0} \text{Jac}(\varphi^t) = \text{Jac}\left(\frac{d}{dt}\bigg|_{t=0} \varphi^t\right) = \text{Jac}(\mathbf{f}).$$

We now have a beautiful result from (multi-)linear algebra (actually matrix calculus), whose proof we leave as an exercise, since it is purely constructive.

Proposition 6.5 *Let $A(t)$ be a C^1 family of $n \times n$-matrices (whose entries are C^1 functions of t) defined on some $I \subset \mathbb{R}$ containing 0, where $A(0) = I_n$. Then*

$$\frac{d}{dt}\bigg|_{t=0} \det A(t) = \text{trace}\left(\frac{d}{dt}\bigg|_{t=0} A(t)\right).$$

Exercise 215 Prove this, noting the special form for the derivative of a determinant of a matrix of functions:

$$\frac{d}{dt}\det A(t) = \sum_i \text{Row}_i'(A(t)),$$

where $\text{Row}_i'(A(t))$ is the matrix $A(t)$ with the ith row's entries replaced with their corresponding derivatives.

Then, since $\text{Jac}(\varphi^t)$ is just a 1-parameter family of $n \times n$ matrices, we have

(6.1.1) $$\frac{d}{dt}\Big|_{t=0} \det\left(\text{Jac}(\varphi^t)\right) = \text{trace}\left(\frac{d}{dt}\Big|_{t=0}\text{Jac}(\varphi^t)\right) = \text{trace}\left(\text{Jac}(\mathbf{f})\right) = \text{div}(\mathbf{f}).$$

This leads immediately to the following conclusion:

Proposition 6.6 *If the divergence of the vector field $\mathbf{f}$ vanishes (that is, if $\text{div}(\mathbf{f}) = 0$), then $\mathbf{f}$ preserves volume.*

Proof Via Equation 6.1.1, we have $(d/dt)|_{t=0}\det\left(\text{Jac}(\varphi^t)\right) = 0$. And since $\text{Jac}(\varphi^0) = I_n$, $\det\left(\text{Jac}(\varphi^t)\right) = 1, \forall t \in I$. $\qquad\square$

6.2 Newtonian Systems of Classical Mechanics

Your previous work in ODEs suggested a general premise about systems of differential equations. If they are defined "nicely," then the present state of a mechanical system determines its future evolution through other states uniquely. One can place this in the language of dynamical systems to say that if a mathematical construction accurately models a mechanical system, than the construction determines a dynamical system on the space of all possible states of the system. The trick in many cases is to have a good understanding of what constitutes a state of a mechanical system. To start, given a mechanical system, the *configuration space* of the system is the set of all possible positions (value combinations of all of its variables) of the system. The *state space*, on the other hand, is the set of all possible states the system can be in. This is usually much broader a description.

For example, consider the pendulum: a mass is attached to the free end of a massless rigid rod, while the other end of the rod is fixed. The set of all possible configurations of the pendulum is simply a copy of S^1. However, for each configuration, the pendulum is in a different state depending on what the mass's velocity is when it resides in a configuration. Visually, see Figure 2 in Section 1.1.1. One can think of all possible states as the space $S^1 \times \mathbb{R}$. This reflects the data necessary to completely determine the future evolution of the pendulum from knowledge of its position and velocity at a single moment, and from the fixed rule, which is a second-order, possibly nonlinear and non-autonomous, ODE in the general form

$$\ddot{x} = f(t, x, \dot{x}).$$

In the case of a pendulum, time is not explicit on the right-hand side, and the equation is autonomous. Under the standard practice of converting this ODE into a system of two first-order ODEs, we can interpret the evolution as giving a vector field on the state space $S^1 \times \mathbb{R}$ (a state space cylinder, Figure 118 in the upcoming Section 6.2.4 on the planar mathematical pendulum), with coordinates x and $\dot{x}$. This vector field determines a flow, which solves the ODE and determines the future evolution of the system based on knowledge of the state of the system at a particular moment in time.

Many systems behave in such a way that their future states are deterministic, completely determined by their present position and velocity, along with a notion of how they are changing. In classical (Newtonian) mechanics, Newton's Second Law of Motion states roughly that the force acting on an object is proportional to how the velocity of the object is changing. This is the famous equation $f = ma$, where f is the total force acting on the object and a is its acceleration. As the velocity depends on the current position of an object, a good notion of how an object moves through a space under the influence of a force is that this is completely determined by how its position and velocity are changing, at least when the force is static (that is, not changing with respect to time):

$$f(x) = ma = m\ddot{x} = m\frac{d^2 x}{dt^2}.$$

This is a special case of the general second-order ODE mentioned above.

Example 6.7 An object under the influence of only gravity satisfies Newton's Second Law, and the differential equation is $\ddot{x} = -g$, where g is the acceleration due to gravity. This is solved by integrating the "pure-time" ODE twice, to give

$$x(t) = -g\frac{t^2}{2} + v_0 t + s_0,$$

where v_0 and s_0 are the initial velocity and initial position, respectively (the two constants of the integrations).

Remark 6.8 *A pure-time ODE is a separable, first-order ordinary differential equation of the form $\dot{x} = f(t)$. One can think of a separable ODE $\dot{x} = f(t,x) = g(t)h(x)$ as autonomous iff $g(t)$ is a constant function, and pure-time iff $h(x)$ is a constant. The general solution to a pure-time ODE is simply the anti-derivative*

$$x(t) = \int f(t) \, dt = F(t) + C$$

one would calculate in any standard calculus course.

Example 6.9 A harmonic oscillator is a mechanical system that behaves in the following manner; when displaced from its equilibrium position, it experiences a restoring force proportional to the displacement. This concept is embodied in Hooke's Law: the amount an object is deformed is linearly related to the force causing the deformation. The mathematical equation corresponding to this notion is then $\ddot{x} = -kx$, which is autonomous. Solutions are given by

$$x(t) = a\sin\sqrt{k}t + b\cos\sqrt{k}t,$$

where a and b are related to the initial starting position and velocity of the mass.

As stated above, note that any ODE of the form $\ddot{x} = f(t, x, \dot{x})$ can be converted to a system of two first-order (generally coupled) ODEs of the form

$$\dot{x} = v,$$
$$\dot{v} = f(t, x, v),$$

which together define a vector field (a static vector field if t does not appear explicitly in the equations) on the (x, v) state space. In this autonomous case, we get for Newton's equation, $\dot{x} = v$ and $\dot{v} = m^{-1} f(x, v)$. Often, the model neglects the dependence of the vector field on the velocity component, as in the case where friction is ignored. In this case, Newton's equation(s) reduce to $\dot{x} = v$ and $\dot{v} = m^{-1} f(x)$. This is the case in the two examples above. We will treat this case presently.

Note: For the system defined by n coordinates and their velocities, we get the $2n$-dimensional system of first order equations defined as

$$\dot{\mathbf{x}} = \mathbf{v},$$
$$\dot{\mathbf{v}} = \mathbf{f}(t, \mathbf{x}, \mathbf{v}).$$

The state space consists of the $2n$-vectors $\begin{bmatrix} \mathbf{x} \\ \mathbf{v} \end{bmatrix}$. Restricting to the case where time is not explicit in the ODEs and velocity-dependent effects are ignored, the vector field of this $2n$-system attaches the vector $V = \begin{bmatrix} \mathbf{v} \\ m^{-1}\mathbf{f}(\mathbf{x}) \end{bmatrix}$ to each point $\begin{bmatrix} \mathbf{x} \\ \mathbf{v} \end{bmatrix}$.

Proposition 6.10 *Let $\ddot{\mathbf{x}} = m^{-1}\mathbf{f}(\mathbf{x})$ be an autonomous Newtonian system where the total force $\mathbf{f}$ is a function of position $\mathbf{x}$ alone. Then the corresponding flow in phase space preserves volume.*

Proof The divergence of the vector field $V = \begin{bmatrix} \mathbf{v} \\ m^{-1}\mathbf{f}(\mathbf{x}) \end{bmatrix}$ is

$$\nabla \cdot V = \sum_{i=1}^{n} \frac{\partial}{\partial x_i}(v_i) + \sum_{i=1}^{n} \frac{\partial}{\partial v_i}\left(\frac{1}{m}f_i(\mathbf{x})\right) = 0.$$

Hence, by Proposition 6.6, the flow preserves volume. □

6.2.1 Generating Flows from Functions: Lagrange

One way to understand Newtonian mechanics is via a formulation developed by Joseph Louis Lagrange in the late 1700s. This approach is essentially a variational approach that says, roughly, that the path of a particle though a force field can be described not only via the equations of motion in the standard Cartesian coordinates by forces determining the various constraints of the motion, but also via a set of independent *generalized coordinates* that completely characterize the motion of the particle. Here, generalized coordinates characterize motion restrictions via new, parameterization variables that, in turn, eliminate the need for the force constraints. This approach usually reduces the number of coordinates needed to completely describe the motion (by parameterizing a subspace of Euclidean space on which motion is constrained) and in some cases greatly simplifies the process of solving the equations of motion. And this formulation works for conservative and non-conservative systems.

This idea of motion being described via parameterizations of subspaces of Euclidean spaces acts in concert with *Newton's First Law of Motion*: An object not under the influence of an external force will move linearly and with constant (maybe zero) velocity. The question is, how does this notion of linear (constant velocity) motion appear in spaces like Euclidean space? On $\mathbb{T}^n$ or S^2? Or on an arbitrary metric space? To understand this, we must get a better understanding of just what a straight line is in a possibly curved space. and here we can use the metric on a (sub)space to define a straight line as the path that is the shortest distance between two points. This path, which obviously depends on the metric, is called a *geodesic* (of the metric).

For example, on a smooth surface $S \subset \mathbb{R}^3$, the Euclidean metric on $\mathbb{R}^3$ induces a metric on S. Choose a point $x \in S$, and the surface has a well-defined tangent plane to S at x. With this tangent plane, we can choose a normal $\mathbf{n}$ to the surface at x, as well as a desired direction $\mathbf{v}$ in the tangent plane. Now, for a particle moving freely along the surface S in the direction of $\mathbf{v}$ at x, the only force acting on the particle is the force keeping it on S. Thus, the acceleration vector of the particle is in the direction of $\mathbf{n}$. With no component of the force in the direction of motion, the speed $||\mathbf{v}||$ is constant along this intersection line. See Figure 114.

Indeed, if the motion in $\mathbb{R}^n$ is constrained to a subspace in a way that can be described via a set of transformational equations

$$\mathbf{x} = \mathbf{x}(q_1, \ldots, q_k, t),$$

then we can use this parameterization as a way to rewrite the system in terms of the generalized coordinates q_i, $i = 1, \ldots, k$. Here $k \in \mathbb{N}$ is called the *number of degrees of freedom* of the system. And motion, constrained to the subspace, will still satisfy Newton's Second Law without regard to the force keeping the object in the subspace.

Example 6.11 For the (mathematical) pendulum in $\mathbb{R}^2$, with $\mathbf{x} = \begin{bmatrix} x \\ y \end{bmatrix}$, motion is constrained to the circle of radius ℓ about the origin. where ℓ is the length of the rigid rod and we have placed the origin of the plane at the fixed end of the rod. Then, a choice of generalized coordinate is the angular coordinate θ (of the standard polar

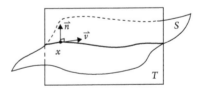

Figure 114 A geodesic on a surface $S \in \mathbb{R}^3$.

coordinate system in the plane centered at the fixed end of the rod) along this circle, where $\theta \in S^1$. For this θ-coordinate, we place the origin on the circle at the lowest point of the vertically oriented plane $\begin{bmatrix} 0 \\ -\ell \end{bmatrix}$ for physical reasons (why?) Then

$$\mathbf{x}(\theta(t)) = \begin{bmatrix} \ell \sin\theta \\ -\ell \cos\theta \end{bmatrix} \quad \text{and} \quad \dot{\mathbf{x}}(\theta,\dot\theta) = \begin{bmatrix} \ell\dot\theta \cos\theta \\ \ell\dot\theta \sin\theta \end{bmatrix}.$$

With this choice of coordinate, the pendulum has only one degree of freedom.

For a system of n particles with constraints, we can list the transformation equations $\mathbf{x}_i = \mathbf{x}_i(q_1,\ldots,q_k,t)$, $i = 1,\ldots,n$. Then the total kinetic energy of the system is

$$T = \sum_i \tfrac{1}{2} m_i \dot{\mathbf{x}}_i \cdot \dot{\mathbf{x}}_i,$$

and if the force field depends on position alone, then one can often define the total potential energy as the energy of position $V = V(\mathbf{x}_i) = V(\mathbf{q}_i)$.

The Lagrangian (function) L is defined as the difference between the kinetic and potential energies: $L = T - V$. This function contains the core dynamical information of the system. We will bypass much of the detail of the origin and derivation of this function here, and state that the equations of motion can be calculated from L via the Euler–Lagrange equations (sometimes called the *Lagrange equations of the second kind*):

(6.2.1)
$$\frac{d}{dt}\left(\frac{\partial L}{\partial \dot q_j}\right) = \frac{\partial L}{\partial q_j}, \quad j = 1,\ldots,k.$$

This is in general a system of k second-order ODEs, one for each generalized coordinate or degree of freedom.

Example 6.12 For the pendulum, we have

$$T(\mathbf{x}) = \tfrac{1}{2} m \dot{\mathbf{x}}_i \cdot \dot{\mathbf{x}}_i = \tfrac{1}{2} m\left(\ell^2\dot\theta^2 \cos^2\theta + \ell^2\dot\theta^2 \sin^2\theta\right) = \tfrac{1}{2} m\ell^2\dot\theta^2 = \tfrac{1}{2} m\left(\ell\dot\theta\right)^2,$$

and

$$V(\mathbf{x}) = mg(\ell + y) = mg(\ell - \ell\cos\theta),$$

written this way to place the lowest potential energy at the low point on the circle and at 0. Then, with

$$L = T - V = \tfrac{1}{2} m\left(\ell\dot\theta\right)^2 - mg(\ell - \ell\cos\theta),$$

we can derive the equations of motion as

$$\frac{d}{dt}\left(\frac{\partial L}{\partial \dot\theta}\right) = m\ell^2\ddot\theta = -mg\ell \sin\theta = \frac{\partial L}{\partial \theta},$$

or $m\ell^2\ddot\theta + mg\ell \sin\theta = 0$.

The advantages of using the Lagrangian formulation to study Newtonian physics are many:

- The equations of motion can all be derived from a single function through simple calculus. That all of the dynamic information about a system is contained within a single function on the phase space is a very powerful idea.
- The Lagrangian is simply a difference of energies, which as scalar fields are easier to calculate than forces (vector fields).
- This formulation works well in all coordinate systems.

However, there are also some shortcomings to this approach: Often, the Lagrangian does not have a physical interpretation as some measurable quantity. And it can be quite difficult or impossible to actually solve the resulting second-order differential equation system.

6.2.2 Generating Flows from Functions: Hamilton

There is another formulation of classical mechanical systems, similar to the Lagrangian approach, that has been shown to be much more robust in generalizing to other areas of physics: the Hamiltonian formulation. Here, instead of using generalized coordinates and their velocities, one replaces the velocities with the corresponding coordinates' conjugate momenta. This was developed by William Rowan Hamilton in the 1830s. To start, for each choice of generalized coordinate q_i in the Lagrangian formulation, define a corresponding conjugate momentum p_i via

(6.2.2)
$$p_i = \frac{\partial L}{\partial \dot{q}_i}.$$

Then, in the new coordinate system $(q_1, \ldots, q_n, p_1, \ldots, p_n)$, define the function $H(\mathbf{q}, \mathbf{p}, t) = \sum_i \dot{q}_i p_i - L(\mathbf{q}, \dot{\mathbf{q}}, t)$, the Legendre transform of the Lagrangian. The function H is called the Hamiltonian of the system and often measures the *total energy* of the system—see Exercise 216.

Exercise 216 Show that for $L = T - V$, under the Legendre transform, the Hamiltonian $H = T + V$.

We can calculate the infinitesimal change in H via the differential

$$dH = \sum_i \left(\dot{q}_i \, dp_i + p_i \, d\dot{q}_i - \frac{\partial L}{\partial q_i} \, dq_i - \frac{\partial L}{\partial \dot{q}_i} \, d\dot{q}_i \right) - \frac{\partial L}{\partial t} \, dt$$

$$= \sum_i \left(\frac{\partial H}{\partial q_i} \, dq_i + \frac{\partial H}{\partial p_i} \, dp_i \right) + \frac{\partial H}{\partial t} \, dt.$$

But, with the definition of conjugate momenta in Equation 6.2.2 and the Euler–Lagrange equations 6.2.1, as well as the conclusion that

$$\sum_i \left(-\frac{\partial L}{\partial q_i} \right) dq_i = \sum_i \frac{\partial H}{\partial q_i} dq_i,$$

we can match coefficients to obtain Hamilton's equations of motion:

(6.2.3)
$$\dot{p}_i = -\frac{\partial H}{\partial q_i}, \quad \dot{q}_i = \frac{\partial H}{\partial p_i}, \quad \frac{\partial H}{\partial t} = -\frac{\partial L}{\partial t}.$$

One big advantage of the Hamiltonian formulation is in the symmetries of these last equations:

- The value of H along a solution does not vary if H is not an explicit function of t:

$$\frac{dH}{dt} = \sum_i \left(\frac{\partial H}{\partial q_i} \dot{q}_i + \frac{\partial H}{\partial p_i} \dot{p}_i \right) + \frac{\partial H}{\partial t} = \sum_i \left(-\dot{p}_i \dot{q}_i + \dot{q}_i \dot{p}_i \right) + \frac{\partial H}{\partial t} = \frac{\partial H}{\partial t}.$$

Hence if $\partial H/\partial t = 0$, then H is a constant of motion.

- Suppose one coordinate q_j does not appear in the Hamiltonian (or the Lagrangian for that matter). Then the corresponding conjugate momentum p_j is also a conserved quantity, since

$$\dot{p}_j = -\frac{\partial H}{\partial q_j} = 0.$$

Effectively, then, the $2n$-system is now only a system of $2n - 2$ first-order equations. This stands in contrast to the Lagrangian formulation, where the corresponding velocity $\dot{q}_j$ may still appear in L and hence there would still be n second-order ODEs.

Example 6.13 The classic two-body problem in mechanics is to determine the motion of two point particles of masses m_1 and m_2 that interact only with each other, like a planet and a star, or a satellite and a planet. The formulation is somewhat tricky in general, but for the purpose of illustration, we will simplify the system: Once converted to generalized coordinates to fix the center of mass $m = m_1 m_2/(m_1 + m_2)$ and expressed via spherical coordinates in 3-space, we can reduce the motion to a one-body problem in a plane with a central potential $U(r)$ using polar coordinates (this becomes the configuration space). The Lagrangian is then

(6.2.4)
$$L(r,\theta,\dot{r},\dot{\theta},t) = \tfrac{1}{2} \left(\dot{r}^2 + r^2 \dot{\theta}^2 \right) - U(r),$$

By inspection, we see that L is not a function of θ. Hence the momentum corresponding to this angular coordinate is conserved, so

$$p_\theta = \frac{\partial L}{\partial \dot{\theta}} = mr^2 \dot{\theta} = \ell,$$

a constant. Thus

(6.2.5)
$$\dot{\theta} = \frac{\ell}{mr^2}.$$

Now, if we substitute Equation 6.2.5 back into the Lagrangian before calculating the equations of motion, we lose the angular equation of motion completely $(\partial L/\partial\theta = (d/dt)(\partial L/\partial\dot{\theta}))$. Indeed,

$$L(r,\dot{r},t) = \frac{1}{2}\left(\dot{r}^2 + r^2\left(\frac{\ell}{mr^2}\right)^2\right) - U(r)$$

$$= \frac{1}{2}\left(\dot{r}^2 + \frac{\ell^2}{m^2r^2}\right) - U(r),$$

so that the sole equation of motion is in the r-direction and yields

$$\frac{\partial L}{\partial r} = -\frac{\ell^2}{mr^3} - \frac{\partial U}{\partial r} = m\ddot{r} = \frac{d}{dt}\left(\frac{\partial L}{\partial \dot{r}}\right),$$

or

$$\ddot{r} = -\frac{1}{m}\left(\frac{\ell^2}{mr^3} + \frac{\partial U}{\partial r}\right).$$

This is unfortunate, since it neglects the fact that Equation 6.2.5 is just a related rates problem (from calculus), and we can recover the second equation of motion by simple differentiation, so that the full equations of motion are

$$\ddot{r} = r\dot{\theta}^2 - \frac{1}{m}\frac{\partial U}{\partial r},$$

(6.2.6)

$$\ddot{\theta} = -\frac{2\dot{r}\dot{\theta}}{r}.$$

Now, using Equation 6.2.5 as a substitution, we can generate the partially uncoupled system

$$\ddot{r} = \frac{\ell^2}{mr^3} - \frac{1}{m}\frac{\partial U}{\partial r},$$

(6.2.7)

$$\ddot{\theta} = -\frac{2\ell\dot{r}}{mr^3}.$$

Exercise 217 Construct the system 6.2.6 from the Euler–Lagrange equations and the Lagrangian in Equation 6.2.4 directly. Then, also differentiate Equation 6.2.5 to obtain the second equation of the system 6.2.6.

Now consider the same system in its Hamilton formulation. Here $p_r = m\dot{r}$ is linear and the angular momentum is again $p_\theta = mr^2\dot{\theta}$. And, with $\dot{p}_\theta = 0$, the Hamiltonian is

$$H(r,\theta,p_r,p_\theta,t) = \dot{r}p_r + \dot{\theta}p_\theta - L(r,\theta,\dot{r},\dot{\theta},t)$$

$$= m\dot{r}^2 + mr^2\dot{\theta}^2 - \tfrac{1}{2}m\left(\dot{r}^2 + r^2\dot{\theta}^2\right) + U(r)$$

$$= \frac{1}{2m}\left(p_r^2 + \frac{p_\theta^2}{r^2}\right) + U(r).$$

The equations of motion are then

$$\dot{r} = \frac{\partial H}{\partial p_r} = \frac{1}{m}p_r,$$

$$\dot{p}_r = -\frac{\partial H}{\partial r} = \frac{p_\theta^2}{mr^3} - \frac{\partial U}{\partial r},$$

(6.2.8)

$$\dot{\theta} = \frac{\partial H}{\partial p_\theta} = \frac{p_\theta}{mr^2},$$

$$\dot{p}_\theta = -\frac{\partial H}{\partial \theta} = 0.$$

Exercise 218 Show that the system 6.2.8 is equivalent to either system 6.2.6 or system 6.2.7.

Remark 6.14 *Notice that in the two-body problem, motion is confined to a plane. In the Hamiltonian formulation, this is particularly easy to see:*

$$H = \frac{1}{2m}\left(p_r^2 + \frac{p_\theta^2}{r^2}\right) + U(r) = \frac{p_r^2}{2m} + \left(\frac{p_\theta^2}{2mr^2} + U(r)\right) = \frac{p_r^2}{2m} + U_{\text{eff}}(r),$$

where p_θ is a constant. Hence everything in the large parentheses is a function only of r. This quantity is sometimes called the effective potential, which we denote here by $U_{\text{eff}}(r)$. So motion is relegated to the 2-dimensional rp_r-space with a modified potential.

Dynamically speaking, we can also play the game the other way: Let $n \in \mathbb{N}$ and endow $\mathbb{R}^{2n}$ with the coordinates $(q_1, \ldots, q_n, p_1, \ldots, p_n)$. Then any C^1 function $H : \mathbb{R}^{2n} \to \mathbb{R}$ can play the role of a total energy function (read: Hamiltonian) of some Newtonian system. The vector field can be given by $X_H = \begin{bmatrix} \partial H/\partial p_i \\ -\partial H/\partial q_i \end{bmatrix}$, so that the equations of motion are

$$\dot{q}_i = \frac{\partial H}{\partial p_i} \quad \text{and} \quad \dot{p}_i = -\frac{\partial H}{\partial q_i}.$$

Notice that this vector field is automatically conservative, since

$$\nabla \cdot X_H = \sum_i \frac{\partial}{\partial q_i}\left(\frac{\partial H}{\partial p_i}\right) + \frac{\partial}{\partial p_i}\left(-\frac{\partial H}{\partial q_i}\right) = 0.$$

Exercise 219 Solve the planar system given by the function $H(x, y) = 3x^2 + 2xy + y^2$ and classify the equilibrium at the origin.

Remark 6.15 *We can generalize this Hamiltonian construction to many even-dimensional non-Euclidean spaces, as long as they satisfy certain conditions. We will not go into this in this text, but the spaces where this can be done are called* symplectic, *or, more generally,* Poisson *spaces. This leads to a mathematical version of Hamiltonian dynamics called symplectic geometry, a synthesis of differential geometry, differential topology, and sometimes smooth dynamics.*

Now let's go back to a simple system and assume that the force $f(x)$ is a gradient field (this means that the force is the gradient of a function of position alone, or $f = -\nabla V$ for some $V(x)$). Then

$$f(x) = ma = m\dot{v} = -\nabla V.$$

Here, the function V is called the potential energy (energy of position), and the energy of motion, the kinetic energy is

$$K = \tfrac{1}{2} m \|v\|^2 = \tfrac{1}{2} m (v \cdot v).$$

The total energy $H = K + V$ satisfies

$$\frac{d}{dt}(H) = \frac{dK}{dt} + \frac{dV}{dt} = m\dot{v} \cdot v + \sum_{i=1}^{n} \frac{\partial V}{\partial x_i} \cdot \frac{\partial x_i}{\partial t} = m\dot{v} \cdot v + \nabla V \cdot v = (\nabla V + m\dot{v}, v) = 0.$$

The conclusion is that the total energy H is conserved as a system like this evolves. As H is a function defined on the state space given by the vectors x and mv, the solutions to the system of ODEs are confined to the level sets of this function. Recall that a system like this, whose vector field is generated by the gradient of a function, is called *conservative*, and the function that generates the vector field is called a *potential*, regardless of whether the system has a physical interpretation. In a sense, you have seen this before in the guise of exact ODEs.

6.2.3 Exact Differential Equations

Start with $H(x,y) = 3x^2 + 2xy + y^2$ and consider both x and y as dependent on time. If motion is confined to the level sets of H, then

$$(6.2.9) \qquad \frac{\partial H}{\partial t} = 0 = (6x + 2y)\frac{dx}{dt} + (2x + 2y)\frac{dy}{dt}.$$

But this is simply an exact ODE $M\,dx + N\,dy = 0$, with $M = 6x + 2y$ and $N = 2x + 2y$, and $M_y = N_x = 2$. The connection is due to the fact that the integral curves of the ODE are forced to "live" on the level sets of the function. Hence the value of the function is constant along the solution curves of the vector field generated by the ODE. This is no accident, and persists in higher dimensions with some subtleties.

Exercise 220 Write the exact ODE in Equation 6.2.9 as a single ODE for y as a function of x (x as the independent variable) and then as the linear system

$$\dot{\mathbf{x}} = \begin{bmatrix} 2 & 2 \\ -6 & -2 \end{bmatrix} \mathbf{x}, \quad \mathbf{x}(t) = \begin{bmatrix} x(t) \\ y(t) \end{bmatrix}.$$

Solve the system and show that the trajectories are indeed the same ellipses as the level sets of $H(x,y)$.

Consider the nonlinear system of two first-order, linear, autonomous differential equations in two variables

(6.2.10)
$$\dot{x} = 4 - 2y,$$
$$\dot{y} = 12 - 3x^2.$$

This system can also be written as a single differential equation by "factoring out the parameter t" and considering y as a function of x

(6.2.11)
$$(12 - 3x^2)\,dx - (4 - 2y)\,dy = 0.$$

Note that this equation is exact, and separable, and upon integration one obtains

$$4y - y^2 = 12x - x^3 + C.$$

This defines our solutions implicitly. In fact, we can use it directly.
Define a function $\varphi(x,y) = 4y - y^2 - 12x + x^3$. Then φ is conserved by the flow, and the flow must live along the constant level sets of φ (the sets that satisfy $\varphi(x,y) = C$). Some of these level sets are shown in Figure 115.
   Now recall from Section 2.5 that an ODE $M\,dx + N\,dy = 0$ is exact if $M_y = N_x$ (this notation again refers to the first partial derivatives of the functions with respect to the subscripts). The reason is vector-calculus in nature: The solution is a function $\varphi(x,y) = C$ satisfying $\partial\varphi/\partial x = M$ and $\partial\varphi/\partial y = N$. The condition for exactness is simply the statement that for any C^2 function $\varphi(x,y)$, the mixed partials are equal:

$$M_y = \frac{\partial M}{\partial y} = \frac{\partial}{\partial y}\frac{\partial\varphi}{\partial x} = \frac{\partial^2\varphi}{\partial y\partial x} = \frac{\partial^2\varphi}{\partial x\partial y} = \frac{\partial}{\partial x}\frac{\partial\varphi}{\partial y} = \frac{\partial N}{\partial x} = N_x.$$

But we can go even further: The vector field $\mathbf{F}(x,y) = (4 - 2y, 12 - 3x^2)$ of the system in Equation 6.2.10 corresponds to the exact ODE given by Equation 6.2.11 when $M = 12 - 3x^2$ and $N = -(4 - 2y)$, or $\mathbf{F}(x,y) = (-N, M)$. With this,

$$\mathrm{div}(\mathbf{F}) = \frac{\partial}{\partial x}(-N) + \frac{\partial}{\partial y}(M) = -N_x + M_y = 0.$$

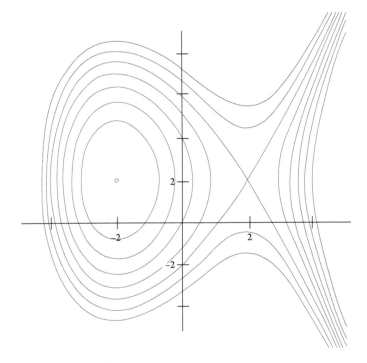

Figure 115 Some level sets of $\varphi(x,y)$.

The vector field **F** is conservative, and the flow will preserve volume (area) in the plane. This is a general fact for vector fields of exact ODEs and leads directly to the following:

Proposition 6.16 *The flow of an exact ODE in $\mathbb{R}^2$ preserves volume in phase space.*

Corollary 6.17 *Equilibria of exact ODEs in $\mathbb{R}^2$ can only be saddles or centers.*

Exercise 221 Complete the phase diagram for this ODE by noting directions of motion along the level curves. Also, note the values of the level sets corresponding to the two equilibrium solutions. Finally, show analytically that the equilibrium at $(2,2)$ is unstable, while the equilibrium at $(-2,2)$ is stable.

The repercussions of these facts are quite important: For instance, $\mathbf{p} = (-2,2)$ is an equilibrium solution of Equation 6.2.10. What is its type and stability (forgetting the figure for a moment, that is)? Of course, this system is almost linear everywhere (see Section 4.5). Hence we can linearize the system at $\mathbf{p}$:

$$
\begin{bmatrix} \dot{x} \\ \dot{y} \end{bmatrix} = \begin{bmatrix} \dfrac{\partial(-N)}{\partial x}(-2,2) & \dfrac{\partial(-N)}{\partial y}(-2,2) \\[2mm] \dfrac{\partial M}{\partial x}(-2,2) & \dfrac{\partial M}{\partial y}(-2,2) \end{bmatrix} \begin{bmatrix} x \\ y \end{bmatrix} = \begin{bmatrix} 0 & -2 \\ 12 & 0 \end{bmatrix} \begin{bmatrix} x \\ y \end{bmatrix}.
$$

The eigenvalues $r = \pm\sqrt{24}i$ are purely imaginary. Hence the linearlized equilibrium at the origin is a center. But centers are NOT *structurally* stable, in that a small perturbation in a center may result in a sink or a source, as well as a center (the eigenvalues may take on small real parts, either negative or positive). And centers are not hyperbolic, so the Hartman–Grobman Theorem 4.47 does not apply. Hence, by itself, we cannot declare that **p** is in fact a center via the linearized system. However, we know that the system has a conservative vector field, and in the plane this means, by Corollary 6.17, that a center cannot perturb into a sink or source and remain conservative; that is, within the space of conservative vector fields, centers are structurally stable, and only systems whose linearization is the matrix $\pm I_2$ are not. This leads directly to a beautiful restatement of the classification theorem for the isolated equilibria of vector fields generated by functions:

Theorem 6.18 (Hartman–Grobman Theorem for conservative systems in $\mathbb{R}^2$)

Suppose $\mathbf{x}^0$ is a critical point of a C^1, conservative ODE system in the plane whose local linearization matrix is not $\pm I_2$. Then the phase portrait of the system at $\mathbf{x}^0$ and the phase portrait of the associated linear system are locally homeomorphic in a neighborhood of $\mathbf{x}^0$.

Thus, for a planar system $\dot{\mathbf{x}} = \mathbf{F}(\mathbf{x})$, when **F** is conservative, an equilibrium of the nonlinear system whose local linearization is a center must also be a center. Such is an import of phase volume preservation.

6.2.4 Application: The Planar Pendulum

One can model the planar pendulum (Figure 116 by the autonomous second-order differential equation derived in Example 6.11:

$$(6.2.12) \qquad\qquad 2\pi m\ell\ddot{x} + mg\sin(2\pi x) = 0.$$

Some notes:

- This is the undamped pendulum as stated. If we were to consider damping, we could model this by adding a term involving $\dot{x}$. A common such term is $c\ell\dot{x}$.

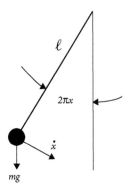

Figure 116 A pendulum.

- This equation can be rewritten as

$$\frac{2\pi \ell}{g}\ddot{x} + \sin(2\pi x) = 0.$$

- To simplify even further, we can scale time by $\tau = t/T$, where $T = \sqrt{g/2\pi L}$. Then we get

$$\ddot{x} + \sin(2\pi x) = 0.$$

So the model becomes

(6.2.13) $\dot{x} = v,$

(6.2.14) $\dot{v} = -\sin 2\pi x,$

which is Newtonian with $f(x) = -\sin 2\pi x$. Here the kinetic energy is $K = \frac{1}{2}v^2$ and V is the potential energy, where

$$f(x) = -\nabla V \quad \text{and} \quad V = \int \sin 2\pi x\, dx = -\frac{1}{2\pi}\cos 2\pi x.$$

The total energy is $H = K + V = \frac{1}{2}v^2 - \frac{1}{2\pi}\cos 2\pi x$ and is conserved. Hence motion is along the level sets of H.

Some dynamical notes:

- For low energy values $H \in \left(-\frac{1}{2\pi}, \frac{1}{2\pi}\right)$, motion is periodic and all orbits are closed.
- For high energy values $H > \frac{1}{2\pi}$, motion looks unbounded. But is it really? What is the pendulum actually doing for these energy values? Recall the idea from before that the actual phase space is a cylinder (the horizontal coordinate—the position of the pendulum—is angular and hence periodic). Hence motion is still periodic and orbits are still closed.
- What happens for the energy value $H = \frac{1}{2\pi}$? What is the lowest energy value? What are the energy values of the equilibria solutions?

Exercise 222 On the phase plane in Figure 117, complete the diagram by orienting the solution curves (choose your directions carefully and justify your choice by showing that it is compatible with your choice of coordinates. Then create a Poincaré section along the vertical axis of the phase diagram as an open interval running from the top separatrix to the bottom separatrix. Compare the first return map with any time-t map within the region bounded by the two separatrices.

The preceding two examples—the exact ODE and the planar pendulum—illustrate some very important phenomena. One important facet is that they are examples of nonlinear, autonomous, first-order ODEs in the plane. And although the exact ODE

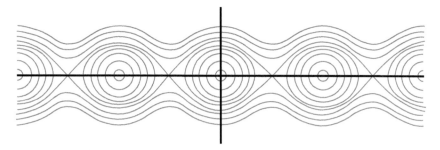

Figure 117 The phase plane of the planar pendulum.

system can be solved (the pendulum cannot), both can be effectively studied via an analysis of a few particular orbits. We take a moment here to expound on this.

Let's linearize the pendulum around the equilibrium at $(x, v) = (0,0)$. Here, in Equation 6.2.13, $f(x, v) = v$ and $g(x, v) = -\sin 2\pi x$, so the linearized system is

$$
\begin{bmatrix} \dot{x} \\ \dot{y} \end{bmatrix} =
\begin{bmatrix}
\dfrac{\partial f}{\partial x}(0,0) & \dfrac{\partial f}{\partial y}(0,0) \\[2mm]
\dfrac{\partial g}{\partial x}(0,0) & \dfrac{\partial g}{\partial y}(0,0)
\end{bmatrix}
=
\begin{bmatrix} 0 & 1 \\ -2\pi & 0 \end{bmatrix}
\begin{bmatrix} x \\ y \end{bmatrix}.
$$

Note that this linear system is simply that of a harmonic oscillator $\ddot{x} - kx = 0$, with $k = -2\pi$.

One can see immediately the following:

- The eigenvalues of the matrix $\begin{bmatrix} 0 & 1 \\ -2\pi & 0 \end{bmatrix}$ are $\lambda = \pm\sqrt{2\pi}\,i$. Hence the linear system has a center at the origin. But, according to the the the Poincaré–Lypaunov theorem, we cannot automatically use this to classify the origin of the nonlinear system. This is true even though we do know that, in this case at least, solutions of the undamped pendulum are in fact periodic. However, this is a conservative system in 2 dimensions. Hence we can use Theorem 6.18 to state that the original nonlinear system does, in fact, have a center at the origin, given that the linearized system does.

- The total energy of this "classical" system is $H = \frac{1}{2}v^2 + \pi x^2$ and is conserved. Hence motion is along the level sets of H in the plane, which are concentric ellipses.

- If one solves the linear system, all of the periods of motion along the ellipses are the same. The question is, are the periods the same for the undamped pensulum?

Exercise 223 Solve the associated linear systems at both equilibria explicitly and compare the linear system solutions directly with the nonlinear phase portraits.

Now linearize around the other equilibrium solution at $\left(\frac{1}{2},0\right)$. We get the linear system

$$\begin{bmatrix} \dot{x} \\ \dot{y} \end{bmatrix} = \begin{bmatrix} 0 & 1 \\ 2\pi & 0 \end{bmatrix} \begin{bmatrix} x \\ y \end{bmatrix}.$$

Here, the eigenvalues are $\lambda = \pm\sqrt{2\pi}$, real and distinct, and the Poincaré–Lyapunov Theorem classifies this equilibrium as a saddle, both for the linearized system and for the original nonlinear system. Questions: What do the level sets of total energy look like for the linear system here? Which level set corresponds to the solutions that limit to the equilibrium? Notice that in the phase cylinder (figure 118), these are the homoclinic points of the pendulum. Can you describe just what a solution here looks like in terms of the actual mechanical device pendulum? In detail, what does a homoclinic solution look like physically? See Figure 119.

6.2.5 First Integrals

In any conservative system, the total energy H is an example of a first integral of the equations of motion. More generally, we have the following:

Definition 6.19 *Given a first-order system*

(6.2.15) $$\mathbf{F}(t,\mathbf{x},\dot{\mathbf{x}}) = \mathbf{0},$$

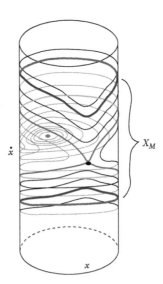

Figure 118 Phase cylinder of the planar pendulum.

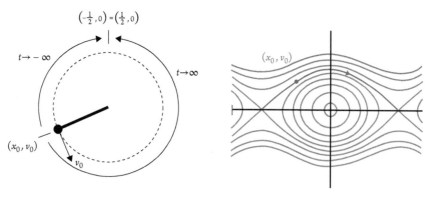

Figure 119 The separatrix orbit of the pendulum. It never stops advancing toward the vertical position, nor does it ever reach it.

with solutions $\mathbf{x} : I \in \mathbb{R} \to \mathbb{R}^n$, a non-constant function $\varphi : \mathbb{R} \times \mathbb{R}^n \to \mathbb{R}$ that is locally constant along any solution (that is, $\frac{d\varphi}{dt}(t, \mathbf{x}(t)) = 0$, where $\mathbf{x} : J \to \mathbb{R}^n$ solves 6.2.15) is called a first integral of the system.

Some notes:

- For $n = 1$, $\dot{x} = f(t, x)$, a first integral $\varphi(t, x)$ would satisfy

$$\frac{d\varphi}{dt}(t, x(t)) = 0 = \frac{\partial \varphi}{\partial t} + \frac{\partial \varphi}{\partial x}\frac{dx}{dt} = \frac{\partial \varphi}{\partial t} + \frac{\partial \varphi}{\partial x}f(t, x(t)).$$

  Thus solutions to the ODE can be defined, at least implicitly, by this partial differential equation.

- Solutions are thus confined to "live" on the level sets of $\varphi(t, x(t))$. Hence the value of φ is constant along the solutions and we sometimes call first integrals *constants of motion*.

- Again, for $n = 1$, a single first integral effectively solves the system, at least implicitly. (Why is this?) And, if $\dot{x} = f(t, x)$ is non-autonomous, we can readily create an equivalent autonomous 2-dimensional system by letting $x_1(t) = t$ and $x_2(t) = x(t)$, so that

(6.2.16)
$$\dot{\mathbf{x}} = \begin{bmatrix} \dot{x}_1(t) \\ \dot{x}_2(t) \end{bmatrix} = \begin{bmatrix} 1 \\ f(x_1, x_2) \end{bmatrix}.$$

  Here, solutions are parameterized curves in the $x_1 x_2$-plane, and, again, a single first integral, defined on phase space, solves the system inplicitly.

- If we go back to our exact ODE in Equation 6.2.11, and consider the solution function as a kind of total energy of the system, $\varphi(x, y) = H(x, y) = 4y - y^2 - 12x + x^3$, then, as x and y vary with respect to t, so does H, and

$$\frac{\partial H}{\partial t} = \frac{\partial H}{\partial x}\frac{\partial x}{\partial t} + \frac{\partial H}{\partial y}\frac{\partial y}{\partial t} = (-12 + 3x^2)(4 - 2y) + (4 - 2y)(12 - 3x^2)$$

$$= 0.$$

So H is constant with respect to t along the flow.

So, what happens in higher dimensions? For an ODE system with phase space $\mathbb{R}^n$ (or some subset), a first integral, or constant of the motion, is a function $H : \mathbb{R}^n \to \mathbb{R}$. A regular level set of H (recall that the notion of regular here means that the $1 \times n$-matrix DH_x is not the 0-matrix) is an $(n-1)$-dimensional subset of $\mathbb{R}^n$, called a *hypersurface*, given by the set of solutions to the equation $H(\mathbf{x}) = c \in \mathbb{R}$. Note that the hypersurface is regular as long as c is not a critical value of H, by the Implicit Function Theorem. The hypersurface is also called the inverse image of c in $\mathbb{R}^n$, and denoted by $H^{-1}(c) \subset \mathbb{R}^n$ even though with $n > 1$ there is no possibility of the function H actually having an inverse.

Now, if one can find two such non-constant functions $G : \mathbb{R}^n \to \mathbb{R}$ and $H : \mathbb{R}^n \to \mathbb{R}$, and these two functions are "sufficiently different" from each other, then one can view solutions to the ODE system as living on the level sets of each of G and H, simultaneously. This means that solutions will live on the intersections of the two $(n-1)$-dimensional hypersurfaces. When the level sets not are tangent to each other, motion is restricted to live on a lower-dimensional surface. This constrains the solutions further and is an effective tool for solving systems of ODEs in more than two variables. First, we need a good notion of what "sufficiently different" means here:

Definition 6.20 *For $n \geq m > 1$, a collection of C^1 functions $H_i : \mathbb{R}^n \to \mathbb{R}$, $i = 1, \ldots, m$, are said to be* functionally independent *on some domain $U \subset \mathbb{R}^n$ if ∇H_1, ∇H_2, ..., ∇H_m are linearly independent as vectors at each $x \in U$.*

Note that, equivalently, one can create the function $H : \mathbb{R}^n \to \mathbb{R}^m$, $H(\mathbf{x}) = (H_1, H_2, \ldots, H_m)$. Then the Jacobian of H is just the derivative of H, denoted by $DH(\mathbf{x})$. This is the $m \times n$ matrix whose ith row is the derivative of H_i, or the transpose of the gradient of H_i. The functions are functionally independent iff the Jacobian has full rank on U.

For example, suppose that $G : \mathbb{R}^3 \to \mathbb{R}$ and $H : \mathbb{R}^3 \to \mathbb{R}$ are two functionally independent first integrals of a system of first-order autonomous ODEs in $\mathbb{R}^3$. Then the solution passing through $\mathbf{x}_0 \in \mathbb{R}^3$ must live on the intersection of the two level sets, or $\mathbf{x}(t) \subset G^{-1}(c) \cap H^{-1}(d)$ for all t where the solution is defined, and where $G(\mathbf{x}_0) = c$ and $H(\mathbf{x}_0) = d$. Figure 120 offers some examples of what this situation may look like. And, in Example 6.13, both the total energy H and the angular momentum p_θ, as functions, are constants of motion. Hence solutions are confined to live on the intersections of the level sets of these two functions. Can you envision what the intersections of the level sets of these two functions look like?

Note that if one can find a sufficient number of functionally independent first integrals for a system, then solutions will be confined to the 1-dimensional common intersection of all of them. In this case, the system is considered solved, at least implicitly. We call such a system *completely integrable*. It is also then the case that any other first integral would have to be functionally dependent on these. We won't spend more time in this area for now, except for an example.

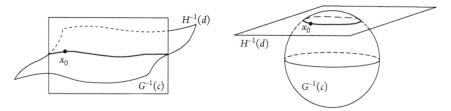

Figure 120 A solution in 3-space constrained by two first integrals.

Example 6.21 Let $\ddot{\mathbf{x}}(t) + \nabla U(\mathbf{x}(t)) = \mathbf{0}$ be a second-order autonomous system for $U:$ $\mathbb{R}^n \to \mathbb{R}$ a C^1 potential function. Then the $2n$-system

$$\dot{\mathbf{x}} = \mathbf{v},$$
$$\dot{\mathbf{v}} = -\nabla U(\mathbf{x})$$

is Hamiltonian. By setting $\dot{x}_i = \partial H / \partial v_i$ and $\dot{v}_i = -\partial H / \partial x_i$, we can recover the Hamiltonian, a first integral

$$H(\mathbf{x}, \mathbf{v}) = \frac{||\mathbf{v}||^2}{2} + U(\mathbf{x}).$$

If $n > 1$, then to solve the system, one would have to find others. For those who know something about classical mechanics, think about angular momentum.

Exercise 224 For Example 6.21 and for $n = 3$, show that the components of the angular momentum $\mathbf{x} \times \mathbf{v}$ are also preserved.

Exercise 225 The Lotka–Volterra equations, commonly referred to as the Predator–Prey Model, can be taken as a model of two interacting species in a closed ecosystem where one species forages for food while the other hunts the forager. The nonlinear system is given by

$$\dot{x} = ax - bxy, \quad \dot{y} = -cy + dxy,$$

where $a, b, c, d > 0$ are positive real constants and $x(t)$ and $y(t)$ are the population sizes of the two species. (Can you identify within the equations which is the predator and which is the prey?) Show that the function $F(x, y) = dx + by - \ln\left(x^c y^a\right)$ is a first integral of the motion.

6.2.6 Application: The Spherical Pendulum

Let's take the pendulum in Section 6.2.4 and allow it to move freely in 3-space, still under the influence of gravity alone, and place its anchor at the origin. Here, the set of all configurations will include any point at a distance ℓ from the origin: the 2-sphere of radius ℓ. Using spherical coordinates and acknowledging that the range coordinate is constant, we can parameterize the configuration space via two angular coordinates: φ, the rotation angle about the z-axis (the longitude angle), and θ, the distance along the sphere from the south pole (the latitude angle). In terms of these two angles, we can write the original coordinates via the parameterization

$$x = \ell \sin\theta \cos\varphi,$$
$$y = \ell \sin\theta \sin\varphi,$$
$$z = -\ell \cos\theta.$$

Note that here the pendulum has a stable equilibrium at the south pole (rectilinear coordinates $(x,y,z) = (0,0,-\ell)$). The potential energy is solely given by position in height above the south pole, and is $U(\theta,\varphi) = -mg\ell \cos\theta$. The kinetic energy is

$T = \frac{1}{2}m||\mathbf{v}||^2$, where $\mathbf{v} = \begin{bmatrix} \dot{x} \\ \dot{y} \\ \dot{z} \end{bmatrix}$, so that

$$T = \frac{1}{2}m||\mathbf{v}||^2 = \frac{1}{2}m(\dot{x}^2 + \dot{y}^2 + \dot{z}^2) = \frac{1}{2}m\ell^2\left(\dot{\theta}^2 + \sin^2\theta\,\dot{\varphi}^2\right).$$

Exercise 226 Verify this calculation.

Immediately, we see that the Lagrangian is

$$L = T - U = \frac{1}{2}m\ell^2\left(\dot{\theta}^2 + \sin^2\theta\,\dot{\varphi}^2\right) + mg\ell\cos\theta.$$

We can use L directly to form the equations of motion:

(1) $\dfrac{d}{dt}\left(\dfrac{\partial L}{\partial \dot{\varphi}}\right) - \dfrac{\partial L}{\partial \varphi} = 0.$

Since φ does not appear in the expression for L, we see immediately that $\partial L/\partial\varphi = 0$, so that $\partial L/\partial\dot{\varphi} = $ constant, or

$$m\ell^2 \sin^2\theta\,\dot{\varphi} = \text{constant}.$$

(2) $\dfrac{d}{dt}\left(\dfrac{\partial L}{\partial \dot{\theta}}\right) - \dfrac{\partial L}{\partial \theta} = 0.$

We have

$$\ddot{\theta} + \frac{g}{\ell}\sin\theta - \sin\theta\cos\theta\,\dot{\varphi} = 0 = \ddot{\theta} + \frac{g}{\ell}\sin\theta - \frac{\cos\theta}{\sin^3\theta}k^2,$$

for $k = \sin^2\dot{\varphi}$ a constant because of (1).

These are uncoupled equations of motion and can be solved separately. In fact, here are two interesting cases:

- $\varphi = \varphi_0 = $ constant. This implies that $\dot{\varphi} = 0$, and the second equation of motion reduces to

$$\ddot{\theta} + \frac{g}{\ell}\sin\theta = 0,$$

or the planar pendulum. Movement is confined to the intersection of the 2-sphere with the plane through the vertical axis that contains all points satisfying $\varphi = \varphi_0$.

- $\theta = \theta_0 = $ constant. Her $\dot{\varphi} = $ constant. Motion here is confined to the intersection of the 2-sphere and the horizontal plane at $z = -\ell\cos\theta_0$, and is sometimes called the *conical pendulum* since it traces out a right circular cone.

Alternatively, we can form the Hamiltonian. This involves first calculating the conjugate momenta for the two coordinates:

$$p_\varphi = \frac{\partial L}{\partial \dot{\varphi}} = m\ell^2 \sin^2\theta \dot{\varphi} = \text{constant},$$

$$p_\theta = \frac{\partial L}{\partial \dot{\theta}} = m\ell^2 \dot{\theta}.$$

Then

$$H = p_\varphi\dot{\varphi} + p_\theta\dot{\theta} - L$$
$$= \frac{1}{2}m\ell^2\left(\dot{\theta}^2 + \sin^2\theta\dot{\varphi}^2\right) - mg\ell\cos\theta$$
$$= \frac{1}{2}m\ell^2\left(\left(\frac{p_\theta}{m\ell^2}\right)^2 + \sin^2\theta\left(\frac{p_\varphi}{m\ell^2\sin^2\theta}\right)^2\right) - mg\ell\cos\theta$$
$$= \frac{1}{2m\ell^2}\left(p_\theta^2 + \frac{p_\varphi^2}{\sin^2\theta}\right) - mg\ell\cos\theta.$$

Here, both H and p_φ are first integrals, where p_φ is the angular momentum about the vertical axis.

Exercise 227 Show that, in rectilinear coordinates, the third coordinate of angular momentum $\mathbf{x} \times \dot{\mathbf{x}}$ is equal to p_φ.

6.3 Poincaré Recurrence

Back to the mathematical pendulum in Section 6.2.4, we ask a new question: What can we say about the recurrent points in the phase space cylinder? First, let's bound the total energy so that $H < M$, for some $M > -\frac{1}{2\pi}$, so

(6.3.1) $$X_M = \left\{(x,v) \in S^1 \times \mathbb{R} \mid H(x,v) < M\right\}.$$

What can we say about which points are recurrent? Here, except for the level set of H corresponding to the separatrices (the level set containing the saddle points), every point is periodic, even though the periods are not all the same. Hence one can say that "almost every" point on X_M is recurrent.

Remark 6.22 *One must be careful here, as the term* almost every *has a very precise meaning in different contexts. In this context, roughly speaking, "almost every" means all points except possibly those on a set that has zero volume. In a more general sense, we would say "except on a set of measure 0", where measure is a generalization of volume that gives size to sets. We have avoided the motion of measure in this text, even though it is a very important concept in dynamical systems. For our purposes, volume is a measure and is sufficient for our purposes.*

So, how common is this idea that, for a volume-preserving flow or map defined on a compact space, almost every point is recurrent? It turns out that it is very common. But, to state the result, let's first look a little closer at the orbits of points in these situations, and include a point's neighbors in the discussion.

6.3.1 Non-Wandering Points

Notice that, for the non-recurrent points on a separatrix in the phase space of a pendulum, arbitrarily close (in any neighborhood) are points that are recurrent. In a sense, this captures a form of complicated dynamical behavior not seen in circle rotations or toral translations, where every point was recurrent, or increasing interval maps, where only fixed points were recurrent. To understand better the repercussions on this kind of behavior, we first study the nature of neighborhoods of points under volume-preserving transformations.

Proposition 6.23 *let X be a finite-volume domain in $\mathbb{R}^n$ or $\mathbb{T}^n$, and let $f : X \to X$ be an invertible, volume-preserving C^1 map. Then, for every $x \in X$ and every $\epsilon > 0$, there exists $n \in \mathbb{N}$ such that*

$$f^n(B_\epsilon(x)) \cap B_\epsilon(x) \neq \emptyset.$$

Proof This can be easily seen as follows: Suppose $\exists x \in X$, and $\exists \epsilon > 0$ such that $\forall n \in \mathbb{N}$,

$$f^n(B_\epsilon(x)) \cap B_\epsilon(x) = \emptyset.$$

Since f is volume-preserving, we must have at the nth iterate,

$$\infty > \text{vol}(X) > \sum_{i=1}^{n} \text{vol}\left(f^i(B_\epsilon(x))\right) = n \cdot \text{vol}(B_\epsilon(x)).$$

But, for all choices of $\epsilon > 0$,

$$\lim_{n \to \infty} n \cdot \text{vol}(B_\epsilon(x)) = (\text{vol}(B_\epsilon(x))) \lim_{n \to \infty} n = \infty,$$

since $\text{vol}(B_\epsilon(x)) > 0$. This contradiction establishes the proof. $\qquad\square$

An immediate consequence of Proposition 6.23 is that, arbitrarily close to any point $x \in X$, there are points whose orbits are close to x. Indeed, using Figure 121 as a guide, given any small neighborhood of $B_\epsilon(x) \subset X$, there will be an iterate of f in the forward

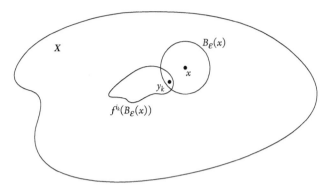

Figure 121 Volume preserving map behavior.

orbit of $B_\epsilon(x)$ that will intersect $B_\epsilon(x)$. A point y in this intersection has the property that both it and $f^i(y)$ lie within the intersection. Now choose a new $\epsilon > 0$ where $\epsilon < d(x,y)$, and repeat the procedure. Play this game for a decreasing sequence of ϵ_k's going to 0. At each stage, you produce a y_k close to x that has a forward iterate $f^{i_k}(y_k)$ that is at least as close. Of course, the i_k can be chosen—or must be chosen—to approach infinity. In the limit, you produce a sequence $\{y_k\} \to x$ of points converging to x, all of which have iterates $\{f^{i_k}(y_k)\} \to x$, where $\{i_k\} \to \infty$. And this is true even for points that are not recurrent.

Exercise 228 Produce this sequence.

Recall Definition 5.1 of a point being recurrent. Also recall that, for a point $x \in X$ under a map f, the ω-limit set is the set of all accumulation points of $\mathcal{O}_x$ as a sequence. So, if x is positively recurrent, then $x \in \omega(x)$ (x is in its own ω-limit set). In fact, we can use this as a definition of the set of all recurrent points:

Definition 6.24 For $f : X \to X$ a continuous map of a metric space, call

$$\mathcal{R}_f(X) = \left\{ x \in X \mid x \in \omega(x) \right\}$$

the set of all recurrent points of f on X, and its complement $\mathcal{N}\mathcal{R}_f(X) = \mathcal{R}_f(X)^c$ the set of all non-recurrent points.

Definition 6.25 For $f : X \to X$ a continuous map of a metric space, a point $x \in X$ is called wandering if there exist $\epsilon > 0$ and an $N \in \mathbb{N}$ such that for every $n > N$, $f^n(B_\epsilon(x)) \cap B_\epsilon(x) = \emptyset$. Points that are not wandering are called non-wandering.

Exercise 229 Develop a precise definition of a non-wandering point.

One can collect up all of the wandering and non-wandering points of a map $f : X \to X$ as $\mathcal{W}_f(X)$ and $\mathcal{N}\mathcal{W}_f(X)$, respectively, and note immediately that $\mathcal{R}_f(X) \subset \mathcal{N}\mathcal{W}_f(X)$ (recurrent points are non-wandering.) However, these sets are not the same: being non-wandering is a property of a point that is based on what happens to orbits near x.

But non-recurrence is a property only of the point x. Indeed, there exist non-wandering points that are not recurrent: For example, let $\ddot{x} = x^2 - x$ be a second-order, autonomous ODE (Is the vector field conservative)? Or, if you prefer, the system $\dot{x} = y, \dot{y} = x^2 - x$.

Exercise 230 Solve the system by constructing an equivalent exact ODE.

Here, as in Figure 122, the shaded region consists of the non-wandering points. But this set is a closed region, and $\mathcal{NW}$ also contains the separatrix forming the orbit line containing $\mathcal{O}_{\left(-\frac{1}{2},0\right)}$. The points on this orbit line are all homoclinic to the unstable equilibrium at $(1,0)$. They are not recurrent. But they are also non-wandering, since any neighborhood of an initial point on this orbit line will contain pieces of periodic orbits.

Exercise 231 Let $T_2 : [0,1] \to [0,1]$ and $T_2(x) = 1 - 2\left|x - \frac{1}{2}\right|$. ($T_2$ is an example of what is called a *tent map*. We will define this map as a parameterized family of maps later in Example 7.55 in Section 7.1.6 on Markov partitions. But you can also see its graph on the left of Figure 176 in Section 8.1). For T_2, show that any point x that is a dyadic rational is non-recurrent and non-wandering. Note that a *dyadic rational* is a rational number whose denominator, when in lowest terms, is a power of 2.

Exercise 232 Show, by construction, that the rotation map $R_\alpha : S^1 \to S^1$ has the property that $\mathcal{R}_{R_\alpha}(S^1) = \mathcal{NW}_{R_\alpha}(S^1) = S^1, \forall \alpha \in \mathbb{R}$.

Exercise 233 Show the same by construction for a translation on $\mathbb{T}^2$.

And, finally, back to the pendulum, one can say that for a given finite region $H \le M$, for some $M > -\frac{1}{2\pi}$, in the region X_M defined by Equation 6.3.1, all points are non-wandering, even though only almost all points are non-recurrent.

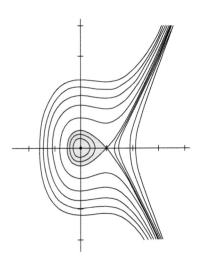

Figure 122 Non-wandering points.

6.3.2 The Poincaré Recurrence Theorem

Back in 1890, Henri Poincaré [43] recognized the prevalence of recurrent points in certain systems. He saw that many systems of this nature will, after an amount of time, return to a state very close to the initial state. The time involved is now referred to as the *Poincaré recurrence time*, although this quantity is highly dependent on what it means to be close to an initial state and also the initial state itself. He discussed the phenomenon, but the subsequent fact wasn't fully proved until 1919 by Constantin Cathéodory [15]. The result still carries Poincaré's name, however:

Theorem 6.26 (Poincaré Recurrence Theorem) *Let X be a closed finite-volume domain in $\mathbb{R}^n$ or $\mathbb{T}^n$ and $f : X \to X$ an invertible volume-preserving map. Then the set of recurrent points for f is dense in X.*

Remark 6.27 *This certainly does not mean that all points are recurrent. In fact, there may be tons of points whose ω-limit sets do not include the original point. It does mean, however, that every point either is recurrent or has a recurrent point arbitrarily close to it. So it will be non-wandering.*

The actual proof of Theorem 6.26 is a bit too advanced for this text. However, we detail here a rough idea of why this result holds. Let $x \in \mathcal{N}\mathcal{R}_f(X)$ be a non-recurrent point. Choose $\epsilon > 0$ small enough that $B_\epsilon(x)$ satisfies $(B_\epsilon(x) - x) \cap \mathcal{O}_x = \emptyset$. Now denote by

$$E_x = B_\epsilon(x) \cap \mathcal{N}\mathcal{R}_f(X)$$

the set of all non-recurrent points in $B_\epsilon(x)$. We claim that E_x contains no interior points. Indeed, if E_x contained an interior point, then it would contain a small open set U about that point, lying entirely inside E_x. But then, by Proposition 6.23, there would be a future iterate of U that would intersect U, contradicting the fact that $U \subset E_x$. But subsets in $\mathbb{R}^n$ or $\mathbb{T}^n$ without interior points have zero volume (recall that the volume of a set in Euclidean space is defined by filling the set with open cubes and taking the limit). Now, since every point in E_x is not an interior point, any neighborhood of a point in E_x will contain points in the complement of E_x in $B_\epsilon(x)$, which are points $\mathcal{R}_f(X) \cap B_\epsilon(x)$.

6.4 Billiards

We return to billiard maps now and present a more general situation. The two-particle model that set up the triangular billiard table in Section 5.3.2 is part of an entire field of study called convex billiards. To start, let D be a bounded, closed, convex domain in the plane, where $B = \partial D$ is the boundary of D. Orbits of motion are line segments in D with endpoints in B, and adjacent line segments meet in B. When B, as a curve in the plane, is C^1, the angle which a line segment makes with the tangent to B at the end point is the same as the angle the adjacent line segment makes. This is what is meant by "angle of incidence equals angle of reflection." Should B also contain corners (points where B is not C^1), declare that an orbit entering the corner end there; this is sometimes referred to as "pocket" billiards. Motion is always considered with constant velocity on line segments, and collisions with B are specular (elastic).

Some dynamical criteria:

- Every orbit is completely determined by its starting point and direction.
- Recall that for polygonal billiards, a billiard flow is a continuous flow per unit time. It is certainly not a differentiable flow, since it fails at the collisions with B, (Note: One can certainly define a smooth flow whose trajectory has corners. All that is necessary is for the flow to slow up and momentarily stop at the corner, to allow it to change direction smoothly. This is quite common for parameterized curves. Here, though, the flow does not slow up before a boundary collision.)
- In the billiard flow on the triangle, we cured the non-differentiable flow points by "unfolding" the table. Here, instead, we will analyze this situation by creating a completely different state space that collects only the relevant information necessary to help distinguish orbits.

First, ignore the time between collisions of line segments with B, and consider an orbit as simply a sequence of points on B, along with their angle of incidence. For each collision of an orbit with B, the point and the angle completely determine the next point and angle of collision. In the "space" of points of B and possible angles of collision, we get an assignment of the next point of collision and angle for each previous one. It turns out that this assignment is quite well-defined. Call this assignment Φ, where $(x_1,\theta_1) \mapsto (x_2,\theta_2) \mapsto \cdots \mapsto (x_n,\theta_n) \mapsto \cdots$. For now, let B be C^1. Collect up all of the points of B, and you get a copy of S^1, although we parameterize this copy by arclength, representing an actual table, as we will see. Collect up all possible angles of incidence, and you get the open interval $(0,\pi)$. But notice that, with a very small positive angle (near 0) and a convex table, the orbit segment will again meet B in a nearby point, and as the angle goes to 0, so does the length of the orbit segment. Hence we can extend the billiard map to $[0,\pi)$ by setting $\Phi(x,0) = (x,0)$, for all $x \in B$. At the other end, as the angle goes to π, we again see that the orbit segment's length goes to 0. However, at this end, it looks like the billiard map actually makes a full revolution, even though all points are again mapped to themselves. We will return to this later, but it implies that we can easily extend the billiard map to the closed interval $[0,\pi]$. The state space is all of the points in B along with all of the incidence angles, and is a copy of $C = S^1 \times [0,\pi]$, the cylinder. The assignment takes $(x_1,\theta_1) \mapsto (x_2,\theta_2) \mapsto \cdots \mapsto (x_n,\theta_n)$. The resulting cylinder, along with the evolution map Φ is called the billiard map.

6.4.1 Circular Billiards

Let

$$D = \left\{ (x,y) \in \mathbb{R}^2 \mid x^2 + y^2 \le 1 \right\}$$

be the unit disk in the plane. Here $B = \partial D = S^1$ is the unit circle, parameterized by the standard angular coordinate θ from polar coordinates in the plane. Note that this parameter takes values in $[0,2\pi)$, corresponding precisely to the arclength of the unit circle, and is quite different from the parameterization we have been using for S^1 given by the exponential map $x \mapsto e^{2\pi i x}$. The state space is then $C = S^1 \times I$, where $I = [0,\pi]$.

What are the dynamics? Draw a chord between two points in S^1. You will see that the acute angle between the tangent line and the chord at each of its two endpoints is the same. Hence, if this chord were an orbit of a billiard, the next segment of the orbit would consist of a similar chord, starting at the endpoint of the previous one: the initial angle of incidence would never change across the orbit, and the evolution map would be constant on the second coordinate.

Exercise 234 Show that for S^1 the unit circle, $\Phi(s,\theta) = (s+2\theta,\theta)$.

Exercise 235 Show that this is not quite true for a billiard table whose radius in not 1 by computing the billiard map $\Phi : C \to C$ for a circular table of radius $r > 0$.

Now do you recognize the evolution map on the state space in this dynamical system? This is the twist map on the cylinder, albeit with twice the rotation angle of the map in Example 6.1, a map that you already showed was area-preserving in Exercise 213. And you can already well-understand the dynamics of this map. To continue our study, we can say more about the orbit structure within each invariant cross-section (constant-θ section) of the cylinder: To each $\theta = \theta_0$ is associated a *caustic*:

- In optics, a caustic is the envelope of light rays reflected or refracted by a curved surface or object, or the projection of that envelope of rays on another surface.
- Alternatively, a caustic is a curve or surface to which each of the light rays is tangent, defining a boundary of an envelope of rays as a curve of concentrated light.
- In differential geometry and geometric optics (mathematics, in general), a caustic is the envelope of rays (directed line segments) either reflected or refracted by a manifold.

Exercise 236 Shine a light from a small hole horizontally into a circular mirrored room. Try to pass the light beam directly through the center of the room (force $\theta_0 = \pi/2$. What happens as you "focus" the light? How does the light fill the room as you approach $\pi/2$, and when you reach $\pi/2$? For the circle billiard, let $\theta/2\pi \notin \mathbb{Q}$. Then the caustic is the edge of the region basically filled with light. Write the equation for this caustic as a function of the angle θ_0.

Exercise 237 For any starting point x on the boundary of a circular billiard, construct an n-periodic orbit for any natural number $n > 1$. Then calculate the number of distinct n-orbits starting at x, for $n = 2,\ldots,8$. You will notice that regular polygons constitute some of these orbits. But there are others.

Exercise 238 Show that there is no dense orbit of a circular billiard.

6.4.2 Elliptic Billiards

Let

$$D = \left\{ (x,y) \in \mathbb{R}^2 \ \middle|\ \frac{x^2}{a^2} + \frac{y^2}{b^2} \leq 1 \right\}$$

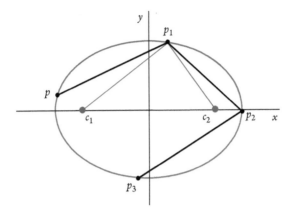

Figure 123 An elliptic billiard.

be an ellipse, where the *diameter* is $2a$ and the *width* is $2b$. Recall that an alternate definition of an ellipse is the set of points in the plane whose combined distance from two reference points is a constant. Figure 123 details an example, with c_1 and c_2 the two reference points, called the foci of the ellipse. The sum of the lengths of the two red line segments $\overline{c_1 p_1}$ and $\overline{p_1 c_2}$ is constant as one moves along the ellipse.

This distance, which is the diameter $2a$, is called the *focal sum*. Written as above, the ellipse is centered at the origin in the plane, and the diameter and the width are along the horizontal and vertical axes, respectively. Also in the figure is a partial orbit $\mathcal{O}_p^3 = \{p, p_1, p_2, p_3\}$. Upon inspection, one can see that it will not be the case, as with the circle, that there is a period-2 orbit passing through each point. There are four such points, though, and all four of these lie on one of the axes. Why is this true? We will see.

This billiard table has notable differences from the circular one, beyond the relative lack of period-2 orbits. To understand these differences better, we introduce a tool to analytically study billiards: the generating function.

Parameterize the boundary by arc length s and let p and p_1 be two points on $B = \partial D$. Now define a real-valued function on $B \times B$ by

$$H(s, s_1) = -d(p, p_1),$$

where d is the standard Euclidean metric in the plane. This function H is called the *generating function* for the billiard.
Some notes:

- This function helps to identify points on the same orbit.
- Critical points of H determine period-2 orbits (think about what this means for the ellipse.)
- Rarely can we find a good working expression for H in terms of s and s_1. But we can discuss its properties and use them effectively.

Example 6.28 Let $a = b = 1$, and we are back at the circular billiard. Here $H(s, s_1) = -2 \sin \frac{1}{2}(s - s_1)$.

Exercise 239 Derive the expression for H in Example 6.28 using the geometry of the unit circle.

Exercise 240 For $a > b$, we do not have a good expression for H. However, we can surmise that the diameter boundary points are at a minimum for H (remember the minus sign), and the width boundary points are a saddle point for H. Why is this? Can you see it?

As in circular billiards, one way to discuss the orbit structure for an elliptic billiard is to try to describe any possible caustics (curves tangent to orbits), which help to define edges of envelopes of orbit regions. To this end, let E denote an elliptical billiard table with c_1 and c_2 its two foci. Denote by $\overline{c_1 c_2}$ the line segment joining the foci, which we will call the *focal line*. In Figure 123, the focal line is on the horizontal axis. We begin with a lemma:

Lemma 6.29 *For $p \in E$, an orbit segment of $\mathcal{O}_p$ crosses the focal line iff every orbit segment of $\mathcal{O}_p$ crosses the focal line.*

Proof Suppose there exist a point $p \in E$ and a partial orbit $\mathcal{O}_p^2 = \{p, p_1, p_2\}$ where $\overline{pp_1} \cap \overline{c_1 c_2} = \emptyset$, but $\overline{p_1 p_2} \cap \overline{c_1 c_2} \neq \emptyset$. The initial angle θ_0 of the orbit at p is either close to 0 or close to π, since $\overline{pp_1}$ doesn't cross the focal line by supposition. We will consider only the former case, since the latter case is entirely symmetrical. Now vary the orbit by holding p fixed and reducing the value of θ from θ_0 to 0. It should be easy to see that both p_1 and p_2 are continuous functions of θ, and that

$$\lim_{\theta \to 0} p_1(\theta) = \lim_{\theta \to 0} p_2(\theta) = p,$$

and that for every $\theta \in [0, \theta_0]$, $\overline{pp_1(\theta)}$ does not cross the focal line. By continuity, however, for values of θ near θ_0, $\overline{p_1(\theta)p_2(\theta)}$ does intersect the focal line. In contrast, for $\theta \approx 0$, $p_2(\theta)$ will be inside the original arc from p to p_1. Thus, for values of θ near 0,

$$\overline{pp_1(\theta)} \cap \overline{c_1 c_2} = \overline{p_1(\theta)p_2(\theta)} \cap \overline{c_1 c_2} = \emptyset.$$

But then, again by continuity, there must be a θ_*, where $\overline{p_1(\theta_*)p_2(\theta_*)}$ intersects one of the foci. But then it must be the case by the geometry of E that $\overline{pp_1(\theta_*)}$ must intersect the other focus. But this is impossible. The result follows. $\qquad \square$

This result gives us two cases here: orbits whose segments all fall between the foci and orbits whose segments never do. The "in between" case we will detail later. Note that two distinct ellipses are confocal if they share the same foci, and when they do, as Jordan curves, one will always be in the interior of the other. We will call the one on the inside the smaller of the two (since it will have a smaller perimeter—can you calculate this?). The following proof is based on a construction first detailed in a Russian text by Victor Gutenmacher and N. B. Vasilyev, translated into English as *Lines and Curves* [23]:

Proposition 6.30 *Every smaller confocal ellipse is a caustic.*

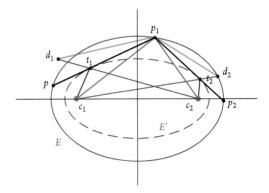

Figure 124 An elliptical caustic.

Proof Choose a point $p \in E$ and an angle such that the partial orbit $\mathcal{O}_p^2 = \{p, p_1, p_2\}$ does not cross the focal line. (By Lemma 6.29, all we need is for the orbit segment $\overline{pp_1}$ to not cross $\overline{c_1 c_2}$.) Connect each focus to p_1 via a line segment (the two red lines in Figure 124). Reflect these two lines across the orbit segments $\overline{pp_1}$ and $\overline{p_1 p_2}$, respectively, creating the two new endpoints d_1 and d_2 as the reflected foci (these are the green lines in the figure). Now connect the two original foci to these new endpoints to create two new line segments $\overline{d_1 c_2}$ and $\overline{c_1 d_2}$ (the thin black lines in the figure). These two new lines intersect $\mathcal{O}_p^2$ in the points t_1 and t_2, respectively. Reflect back the two line segments $\overline{d_1 t_1}$ and $\overline{t_2 d_2}$ across the orbit segments $\overline{pp_1}$ and $\overline{p_1 p_2}$, respectively (the purple lines in the figure).

Now we claim that t_1 and t_2 determine a unique ellipse confocal to E and so that $\mathcal{O}_p^2$ is tangent to this new ellipse at t_1 and t_2. Indeed, the line segments $\overline{d_1 c_2}$ and $\overline{c_1 d_2}$ are the same size, since the triangles formed by connecting their respective endpoints to p_1 are congruent.

Exercise 241 Show that the two triangles $\triangle(d_1 p_1 c_2)$ and $\triangle(c_1 p_1 d_2)$ are congruent.

Hence, for $\ell(\cdot)$ Euclidean length, we have

$$\ell\left(\overline{d_1 c_2}\right) = \ell\left(\overline{c_1 t_1}\right) + \ell\left(\overline{t_1 c_2}\right) = \ell\left(\overline{c_1 t_2}\right) + \ell\left(\overline{t_2 c_2}\right) = \ell\left(\overline{c_1 d_2}\right) = L_{E'},$$

where $L_{E'}$ is the focal sum of the new ellipse E', confocal to E, containing t_1 and t_2. That the orbit segments are tangent to this new ellipse comes from the fact that $\angle(c_1 t_1 p) = \angle(pt_1 d_1)$ by reflection and that $\angle(pt_1 d_1) = \angle(c_2 t_1 p_1)$ as vertical angles. Thus $\overline{pp_1}$ is tangent to this new ellipse. The other side is similar. $\qquad\square$

And for orbits whose segments intersect the focal line, it turns out that these orbits also have caustics and these caustics are also conic sections. We will leave the proof of this as an exercise, noting that it is almost exactly the same as the proof for the confocal ellipse case.

Proposition 6.31 *Every hyperbola confocal to the ellipse is a caustic.*

Exercise 242 Prove Proposition 6.31.

Remark 6.32 *A playful way to construct an ellipse "by hand" on a piece of paper is to place two pins a set distance apart and wrap a non-stretchable loop of string around both pins and a pencil, and then, keeping the string taught, move the pencil around the two pins. This verifies the idea of the constant sum of lengths between the pencil and the two pins. However, from the above discussion, there is another way to draw a confocal ellipse when one is given: simply wrap a non-stretchable loop of string (of length larger than the given ellipse) around the given ellipse and a pencil and, again keeping the string taught, draw the pencil around the ellipse. This result is known as Graves's Theorem, given in 1841 by Robert Graves, a mathematician and Bishop of Limerick, Ireland. This result is a special case of a theorem by Gottfried Leibnitz on caustics, and also a special case of a type of problem posed by Daniel Bernoulli called coëvolution, the transformation of a curve in the plane to a family of other curves of the same. One can read more about this interaction and the details in a 1965 historical note by Samuel Roberts [48]. The general idea is that, for any two tangents to an ellipse that intersect, the sum of the line segments from the tangencies to the common intersection is constant over the entire ellipse. And this sum is the focal sum of the new, larger ellipse.*

As is well known, ellipses and hyperbolas are conic sections, and related via their *eccentricity*, a non-negative number that parameterizes conic sections via a ratio of their data. Indeed, along the major axis (the diameter) of a conic section, one can measure the distance from the curve to the origin (let's keep all conic sections centered at the origin for now). Call this the radius a. One can also measure the distance from the center to one of the foci, denoted c. Then eccentricity e is the ratio of these two numbers:

- For $e = c/a = 0$ (implying that $c = 0$), the section is a circle.
- For $0 < e = c/a < 1$, the section is an ellipse.
- For $e = c/a = 1$, the section is a parabola.
- For $e = c/a > 1$, the section is a hyperbola.

For the case of a circle, the equation is elliptical, with $a = b$, and we have $x^2/a^2 + y^2/a^2 = 1$, or $x^2 + y^2 = a^2$. For the case of a hyperbola, we have $x^2/a^2 - y^2/b^2 = 1$. In fact, we can classify all confocal conic sections (ellipses and hyperbolas) via one parameterized equation:

(6.4.1)
$$\frac{x^2}{a^2 + \gamma} + \frac{y^2}{b^2 + \gamma} = 1.$$

We will call this a *confocal family*.

Exercise 243 Find the ranges of the parameter γ in Equation 6.4.1 corresponding to ellipses, hyperbolas, and neither. Rectify the range of values corresponding to each with the equations for ellipses and hyperbolas given separately above.

Exercise 244 Fix a confocal family by choosing $a > 0$, so that the two foci are at $c_1 = (-a,0)$ and $c_2 = (a,0)$ in the plane. Then show that for any point $p \in \mathbb{R}^2$ not on the two axes, there exists a unique ellipse and a unique hyperbola passing through p. Also show that this hyperbola and ellipse are perpendicular at p.

And as a couple of final dynamic notes:

- An orbit that passes through one focus must pass through the other. What are the implications of this for the resulting orbit?

- There are tons of periodic orbits in elliptic billiards, of all periods. Can you draw some? Period 4 should be easy to see, as should period 2. How about period 3?

Exercise 245 Construct a period-4 orbit for an elliptic billiard and show analytically that it exists.

Exercise 246 Describe the long-term behavior of any orbit that has a orbit segment that pass through one of the foci.

6.4.3 General Convex Billiards

Going back to the generating functions, we can say more about orbits in general. Recall Definition 2.40 in Section 2.2.2, according to which a region is convex if for any two points in the region, the entire line segment joining them is in the region. So, let $D \subset \mathbb{R}^2$ be a bounded convex billiard table (either with a C^1 boundary, or at most with a finite number of pockets). Here are some more general properties. Define the generating function $H(s,s_1) = -d(p,p_1)$, where d is the standard Euclidean metric in $\mathbb{R}^2$. Choose a point p along the boundary where the tangent line is defined and an angle θ with respect to the tangent. Then the next collision p_1 is well-defined, and, if not a pocket, the next angle θ_1 is also. Then H is differentiable at the corresponding s and s_1, and we have the following:

Lemma 6.33 $\dfrac{\partial}{\partial s_1}H(s,s_1) = -\cos\theta_1$, and $\dfrac{\partial}{\partial s}H(s,s_1) = \cos\theta$.

Proof This really is simply calculus. For the first result, fix and parameterize a small arc in the ellipse centered at s_1, $c(t)$, where $c(t_0) = p_1$. Choose a parameterization such that the tangent vector is unit length. Then, noting that $d(p,p_1) = ||p - p_1||$, we have

$$\frac{\partial}{\partial s_1}H(s,s_1) = \frac{d}{dt}\bigg|_{t=t_0} -d\big(p,c(t)\big) = \frac{-1}{2d(p,c(t))}\left(2\left(c'(t)\cdot(p-c(t))\right)\right)\bigg|_{t=t_0}$$
$$= \frac{-||c'(t_0)||\,||p-c(t_0)||\cos\theta_1}{||p-c(t_0)||} = -\cos\theta_1,$$

by the cosine formula for the dot product of two vectors and since $||c'(t_0)|| = 1$ by the parameterization. Hence $\partial H/\partial s_1 = -\cos\theta_1$. The other result is similar. □

Now apply this idea to any three points s_{-1}, s_0, and s_1 on the ellipse, and ask the question: Can these three points lie successively on an orbit? The answer is yes, as long as the points are chosen so that the infinitesimal condition holds at the middle point, so that the incoming angle at s_0 equals the outgoing angle. One can state this as a variational problem by searching for a critical point s_0 to the assignment

$$s \longmapsto H(s_{-1}, s) + H(s, s_1).$$

We will not pursue this idea directly in this text, but it will reappear in the following discussion.

Experiment 1 *Consider a strictly convex billiard with one pocket (corner) p, as in Figure 125. Can one sink a ball via a single back shot from anywhere on the table? Is there more than one shot from each point? An infinite number at some points?*

Recall the notion of a strictly convex domain from Definition 2.40, where $B = \partial D$ has non-zero curvature (where B is C^2 and where the second derivative is non-zero). Visually, this means that there are no straight-line segments on B, and certainly no inflection points (changes in concavity). It also means that we must be careful in how we define the angle associated with the start of an orbit. An angle of 0 effectively means that there is no movement, and the billiard map takes $(p, 0)$ to $(p, 0)$; it is the identity map. However, pushing θ to π, is this still true? We will return to this later, but the transformation here is more that each point is mapped to itself after a full revolution around the border. Think about this. It implies that the billiard map, even for a convex table that is not a circle, is still a twist map, like that in Example 6.1, except that is will be nonlinear. And, like the twist map in the example, the billiard map for a general convex billiard table preserves area! The quick way to see this is to study the generating function a bit more closely.

Indeed, we already know that for the generating function $H(s, s_1)$, we have

$$dH = \frac{\partial H}{\partial s}\, ds + \frac{\partial H}{\partial s_1}\, ds_1 = -\cos\theta\, ds + \cos\theta_1\, ds_1.$$

And since, by exterior differentiation,

$$0 = d(dH) = d(-\cos\theta) \wedge ds + d(\cos\theta_1) \wedge ds_1 = \sin\theta\, d\theta \wedge ds - \sin\theta_1\, d\theta_1 \wedge ds_1,$$

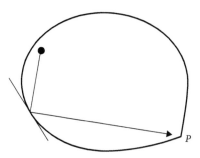

Figure 125 A bank shot.

we see immediately that $\sin\theta\, d\theta \wedge ds = \sin\theta_1\, d\theta_1 \wedge ds_1$, implying that the area form $\sin\theta\, d\theta \wedge ds$ is invariant under the billiard map. That this is an actual area form is ensured, given that $\sin\theta > 0$ on $(0,\pi)$. However, George D. Birkhoff [10] saw the invariance of this area under the billiard map in an interesting geometric (and quite elementary) fashion. To prove area invariance, he studied the invariance of the integral $\iint_\sigma \sin\theta\, d\theta\, ds$ evaluated on small open sets and their images. We detail his proof presently:

Proposition 6.34 *For a convex billiard, the billiard map*

$$\Phi(s,\theta) = (\mathcal{S}(s,\theta),\Theta(s,\theta)) : C \to C$$

is area- and orientation-preserving.

Proof With the notation $\Phi(s,\theta) = (s_1,\theta_1)$, let $\sigma_1 = \Phi(\sigma)$ be the image of a small open region in the phase space cylinder. Then, for $M(s,\theta)$ a function, we want

$$\iint_{\sigma_1} M(s_1,\theta_1)\, d\theta_1\, ds_1 = \iint_\sigma M(s,\theta)\, d\theta\, ds.$$

By the Change of Variables Theorem in calculus,

$$\iint_{\sigma_1} M(s_1,\theta_1)\, d\theta_1\, ds_1 = \iint_\sigma M(s_1,\theta_1)J\, d\theta\, ds,$$

so that $M(s_1,\theta_1)J = M(s,\theta)$ over the arbitrarily chosen region σ (since σ is chosen arbitrarily, the integrals will be invariant iff this equality holds). Hence $\iint_\sigma \sin\theta\, d\theta\, ds$ will be inveriant under Φ iff

$$J = \frac{\sin\theta}{\sin\theta_1}.$$

To show this, let

$$x = F(s), \quad y = G(s)$$

be a convex curve in the plane, knowing that the tangent vectors are of unit length (s is arc length). Then, at s and s_1, we have (as in Figure 126 at left) $\tan\tau = G'(s)/F'(s)$ and $\tan\tau_1 = \frac{G'(s_1)}{F'(s_1)}$, where τ is the angle with respect to the positive horizontal x-axis of the tangent vector to the parameterized curve. Letting α be the angle that the orbit segment from s to s_1 makes with respect also to the positive x-axis direction, we immediately see that $\theta = \alpha - \tau$ and $\theta_1 = \tau_1 - \alpha$. Indeed, since the tangent vector has unit modulus, we also get $\sin\tau = G'(s)$ and $\cos\tau = F'(s)$, yielding

$$\tau = \tan^{-1}\frac{G'(s)}{F'(s)}.$$

Similarly, $\tau_1 = \tan^{-1}\left(G'(s_1)/F'(s_1)\right)$. Also, notice (by Figure 126 at right) that

$$\tan(\alpha - \pi) = \tan\alpha = \frac{G(s_1) - G(s)}{F(s_1) - F(s)},$$

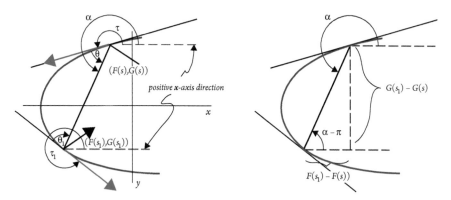

Figure 126 Angles at the points $(f(s), G(s))$ and $(f(s_1), G(s_1))$.

so that $\sin\alpha$ and $\cos\alpha$ are proportional (by the same constant) to $G(s_1) - G(s)$ and $F(s_1) - F(s)$, respectively. With this, we obtain

$$
\Phi: \quad
\begin{aligned}
\theta &= \tan^{-1}\frac{G(s_1)-G(s)}{F(s_1)-F(s)} - \tan^{-1}\frac{G'(s)}{F'(s)} = L(s,s_1), \\[2mm]
\theta_1 &= \tan^{-1}\frac{G'(s_1)}{F'(s_1)} - \tan^{-1}\frac{G(s_1)-G(s)}{F(s_1)-F(s)} = M(s,s_1).
\end{aligned}
$$

This is sufficient to completely characterize the transformation Φ. In fact, infinitesimally speaking, $d\theta = L_s\,ds + L_{s_1}\,ds_1$ and $d\theta_1 = M_s\,ds + M_{s_1}\,ds_1$ and we can use these to solve for ds_1 and get

$$
ds_1 = \frac{1}{L_{s_1}}\,d\theta - \frac{L_s}{L_{s_1}}\,ds,
$$

$$
\begin{aligned}
d\theta_1 &= M_{s_1}\,ds_1 + M_s\,ds = M_{s_1}\left(\frac{1}{L_{s_1}}\,d\theta - \frac{L_s}{L_{s_1}}\,ds\right) + M_s\,ds \\[2mm]
&= \frac{M_{s_1}}{L_{s_1}}\,d\theta + \left(M_s - \frac{L_s}{L_{s_1}}M_{s_1}\right)ds.
\end{aligned}
$$

The area form is then

$$
\begin{aligned}
d\theta_1 \wedge ds_1 &= J\,d\theta \wedge ds \\[2mm]
&= -\frac{M_{s_1}L_s}{(L_{s_1})^2}\,d\theta \wedge ds + \left(\frac{M_s}{L_{s_1}} - \frac{M_{s_1}L_s}{(L_{s_1})^2}\right)ds \wedge d\theta \\[2mm]
&= -\frac{M_s}{L_{s_1}}\,d\theta \wedge ds,
\end{aligned}
$$

so that $J = -M_s/L_{s_1}$. Now,

$$
M_s = \frac{\partial}{\partial s}M(s,s_1) = \frac{\partial}{\partial s}\left(-\tan^{-1}\frac{G(s_1)-G(s)}{F(s_1)-F(s)}\right).
$$

Hence we can show the following:

Exercise 247 Show that

$$M_s = \frac{G'(s)\left(f(s_1) - F(s)\right) - F'(s)\left(G(S_1) - G(s)\right)}{\left(f(s_1) - F(s)\right)^2 + \left(G(s_1) - G(s)\right)^2}$$

and

$$L_{s_1} = \frac{G'(s_1)\left(f(s_1) - F(s)\right) - F'(s_1)\left(G(S_1) - G(s)\right)}{\left(f(s_1) - F(s)\right)^2 + \left(G(s_1) - G(s)\right)^2}.$$

Then

$$J = -\frac{M_s}{L_{s_1}} = -\frac{G'(s)\left(f(s_1) - F(s)\right) - F'(s)\left(G(s_1) - G(s)\right)}{G'(s_1)\left(f(s_1) - F(s)\right) - F'(s_1)\left(G(s_1) - G(s)\right)}$$

$$= -\frac{\sin\tau\cos\alpha + \cos\tau\sin\alpha}{\sin\tau_1\cos\alpha - \cos\tau_1\sin\alpha} = \frac{\sin(\alpha - \tau)}{\sin(\tau_1 - \alpha)} = \frac{\sin\theta}{\sin\theta_1}. \qquad \square$$

Proposition 6.35 *If $B = \partial D$ is C^k (which means that the Euclidean coordinates are C^k functions of the length parameter), then both S and R are C^{k-1}.*

Proof This is the Implicit Function Theorem. $\qquad\square$

Proposition 6.36 *For D strictly convex, the billiard map has at least two period-2 orbits: at the diameter and at the width.*

For a compact, convex region in the plane, one can define the *width* in the following way. Take two distinct vertical parallel lines that intersect the region and maximize the distance between them. For a region with C^1 boundary, these lines will be tangent to the boundary. As one slowly rotates the region, maintaining the intersections and maximum distances causes the distance between these lines to change (unless the table is circular). When one reaches the *diameter* of the table (the largest possible Euclidean distance between two points in the region), the two points will lie along a common line perpendicular to the vertical lines. This perpendicular line segment represents one of the period-2 orbits. The other comes at the point when the two vertical lines reach a local minimum distance (which is the minimum distance for a strictly convex table). At this point again, the line segment joining the two tangencies will be perpendicular to the vertical lines and represent another period-2 orbit. This is the *width* of the convex region. Note that if the region were an ellipse, the diameter and width defined here would correspond to the major and minor axes of the ellipse, respectively.

Finding these period-2 orbits using this method involves finding where the vertical lines reach a minimum and a maximum distance from each other. But this is what the generating function H is also doing, and why the generating function is particularly good at finding period-2 orbits. Indeed, think of the generating function as a C^1 function on the direct product of two copies of the boundary of the billiard table, so

$$H : S^1 \times S^1 \to \mathbb{R},$$

with coordinates s and s_1. As the domain is compact (it is basically a 2-torus), and H is continuous, it will achieve its maximum and minimum by the Extreme Value Theorem. The maximum value of H is 0, and will occur at least over the entire *diagonal* $\{(s,s) \mid s \in S^1\}$. $H < 0$ otherwise, and will achieve its minimum (maximum negative) value at a critical point, a place where the gradient

$$\nabla H = \begin{bmatrix} \dfrac{\partial H}{\partial s} \\ \dfrac{\partial H}{\partial s_1} \end{bmatrix} = \mathbf{0} = \begin{bmatrix} \cos\theta \\ -\cos\theta_1 \end{bmatrix}.$$

Of course, this occurs when $\theta = \theta_1 = \pi/2$. At the diameter, this obviously occurs at a period-2 orbit of maximum total length. But can you see what is happening at the width?

For higher-order periodic points, we can generalize this argument directly: First, consider any three points s, s_1, s_2 on the boundary and connect them with chords to form a triangle. Does this triangle coincide with a period-3 orbit of the billiard map? The answer is yes, if the angles at each vertex of the triangle are compatible with the angle conditions with respect to the boundary tangent lines. Create a new function

$$\mathcal{H} : S^1 \times S^1 \times S^1 \to \mathbb{R}, \quad \mathcal{H}(s, s_1, s_2) = H(s, s_1) + H(s_1, s_2) + H(s_2, s).$$

Again, this is a continuous function on a compact space, with maxima along the diagonal and a minimum at some critical point.

$$\nabla H = \mathbf{0} = \begin{bmatrix} \dfrac{\partial H}{\partial s} \\ \dfrac{\partial H}{\partial s_1} \\ \dfrac{\partial H}{\partial s_2} \end{bmatrix} = \begin{bmatrix} \dfrac{\partial H}{\partial s}(s, s_1) + \dfrac{\partial H}{\partial s}(s_2, s) \\ \dfrac{\partial H}{\partial s_1}(s_1, s_2) + \dfrac{\partial H}{\partial s_1}(s, s_1) \\ \dfrac{\partial H}{\partial s_2}(s_2, s) + \dfrac{\partial H}{\partial s_2}(s_1, s_2) \end{bmatrix}.$$

These are simply the same individual variational conditions as before, and one of these critical points will correspond to the triangle of maximum perimeter inscribed in the table. Generalizing in the obvious way, we can easily establish the following:

Proposition 6.37 *Given a C^1 strictly convex billiard table, there exists an n-periodic orbit of prime period n for all natural numbers $n > 1$.*

Example 6.38 Figure 127 shows some examples of low-period orbits of a elliptical billiard table. You will notice that the symmetries of the table are reflected in corresponding symmetries in the orbits. These orbits were created numerically by minimizing the function $\mathcal{H}$ and being careful to keep the rotational order of the points along the orbit.

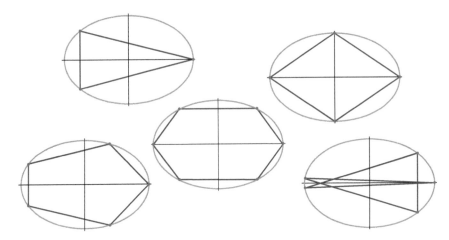

Figure 127 Some periodic orbits of an elliptical billiard table.

Notice also that there are two kinds of period-5 orbits: pentagons as well as penta-grams. Once the ordering of vertices is fixed around the border, any relatively prime pair (p,q), where $p > q$, has a maximum perimeter-inscribed polygon.

6.4.4 Poincaré's Last Geometric Theorem

Note that if we delete a single point p from the interior of a compact, convex region D in the plane, then it is possible to create a fixed-point-free transformation $D - \{p\}$. For example, the map $r \to r$, $\theta \to \theta + a$, is fixed-point-free on $D = B^2 - \{(0,0)\}$, for $a \in (0, 2\pi)$. Hence, fixed-point-free transformations of the annulus

$$\mathcal{A} = \left\{ (x,y) \in \mathbb{R}^2 \mid 0 < a^2 \leq x^2 + y^2 \leq b^2 \right\}$$

are easy to construct (simple rotations will suffice.) But Poincaré, in his studies of the interactions of many point masses under gravity (the celebrated N-body problems), devised an interesting simple problem: Must a transformation of an annulus that preserves area and rotates the invariant boundary circles in opposite directions have a fixed point? In 1912 [45], he conjectured that this is so, but only proved it in limited cases. This result, proven the following year by George Birkhoff [9], (extending the number of distinct fixed points to at least two) eventually became known as the Poincaré–Birkhoff Theorem, and is also known as the Poincaré Fixed-Point Theorem, although it is often referred to by Birkhoff's name for it: Poincaré's Last Geometric Theorem. Henri Poincaré died in July of 1912 of an embolism, possibly as a complication of surgery, at 58 years old.

Theorem 6.39 (Poincaré's Last Geometric Theorem) *An area-preserving continuous transformation of an annulus that rotates the boundaries in opposite directions necessarily contains a fixed point.*

Poincaré's proof involved showing that any fixed-point-free such transformation would necessary map some Jordan curve to another that lies either completely inside the first or completely outside of it. Either way, this would violate the area-preserving aspect of the map, establishing the result. But his assumptions were too restrictive to prove the general case, and Birkhoff not only fully proved the result but established that there are at least two such fixed points, of different characteristics, which we will not go into here.

How does this relate to billiards? Recognize that the phase cylinder of the billiard map has many of the same properties as an area-preserving annulus map:

- The compact cylinder is homeomorphic to the compact annulus. See Exercise 93.
- The billiard map is continuous (indeed, it is differentiable if the boundary is also).
- The boundary circles of the billiard map are invariant (indeed fixed), although we interpret the upper boundary circle to have rotated one full rotation.

So, under a suitable homeomorphism like

$$h : C \to \mathcal{A}, \quad h(\theta, \varphi) = ((\varphi + \pi) \cos \theta, (\varphi + \pi) \sin \theta),$$

we can view the billiard map as an annular map by (topological) conjugation with h on

$$\mathcal{A} = \left\{ (r, \theta) \in \mathbb{R}^2 \mid \pi \le r \le 2\pi \right\}.$$

Note that this map on the annulus does not satisfy the suppositions of Poincaré's Last Geometric Theorem. But Birkhoff understood how to use this map effectively. Noticing that, say, two iterations of an annular map T constructed via h and a billiard map fixed the inside boundary and rotated the outside boundary by 4π, he composed T^2 with the rigid rotation map $R_{-2\pi}$. By itself, $R_{-2\pi}$ fixes all points of $\mathcal{A}$, but $R_{-2\pi} \circ T^2$ has the effect of mapping the annulus to the annulus and rotating the two boundaries in opposite directions by 2π. Hence, by Theorem 6.39, there must be an interior fixed point. As this point cannot be fixed (there are no period-1 points of a billiard), this fixed point must correspond to a period-2 orbit. One can easily compose $R_{-2\pi}$ with T^n to create the conditions for the existence of a period-n orbit.

6.4.5 Application: Pitcher Problems

In the movie "Die Hard with a Vengeance" (20th Century Fox, produced by John McTiernan and Michael Tadross and directed by John McTiernan), John McClain (Bruce Willis) and Zeus (Samuel L. Jackson) are confronted with a puzzle in their ordeal to stop a terrorist's plot run by Peter Krieg (Jeremy Irons). The puzzle is part of a family of brain-teasers called the Pitcher Problems: Given two pitchers of a certain size, in integer gallons, how can one measure out an intermediate integer gallon amount through a combination of events that include a complete filling of a pitcher, a complete emptying, and/or a complete transfer of fluid from one pitcher to another? In the movie, near a fountain stand two empty jugs, one a 5-gallon jug and the other a 3-gallon jug. There is also a bomb here, which can be disabled by weighing out exactly 4 gallons of fluid. The heroes figure out the process as a 6-stage one: Fill the 5-gallon jug; use it to fill the 3-gallon jug; empty the 3-gallon jug; put the remaining 2 gallons into the 3-gallon jug; fill the 5-gallon jug; and use it to refill the 3-gallon jug. What remains in the 5-gallon jug is 4 gallons. How did they come up with this procedure? Watch the movie. Is there a systematic way to find the shortest procedure for doing this? Yes, and billiards is one way to work it out.

Construct a parallelogram billiard table with an acute angle of $\pi/6$ radians, whose side lengths are, in this case, 5 and 3 units long, as in Figure 128. The integer points in this parallelogram form a lattice. Define the corner collisions in the following way:

- If a point meets an obtuse corner coming in along an edge, then the collision looks like a collision with the other wall, with reflection angle $\pi/6$.

- Any other corner collision ends the orbit. (This is all that we will need.)

Now create an orbit that begins at the lattice point $(0,0)$ and runs along one of the edges. You have two choices. What is its fate? Well, one result is that the orbit will always end at the obtuse corner opposite the first one encountered in the orbit. But, even more interestingly, the orbit will meet every lattice point along the boundary, except for the opposite acute corner. In fact, for a table like this with side lengths $p \leq q$, this is true iff p and q are relatively prime (have no factors in common except for 1). Why is this important?

Follow the orbit. Every horizontal segment corresponds to either filling (left to right) or emptying (right to left) one of the jugs, and every vertical segment (defined along the other edge) corresponds to either filling (up) or emptying (down) the other jug. The diagonal segments (diagonal with respect to the sides) corresponds to transferring fluid from one jug to the other. So each lattice point (a, b) corresponds to the current state of the fluid in each jug, and at least one of these jugs in each edge collision is either empty or full. The result is that each lattice point in an orbit corresponds to a "next move." If and when the desired entry k appears as a part of an orbit, you can count how many moves it takes to create that desired amount of fluid. There are only two possible starting moves, so there is a shortest path (possibly two of them?). There is a beautiful number theory result concerning just how long these shortest orbits may be to reach any desired intermediate measure.

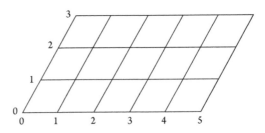

Figure 128 A 5 × 3-parallelogram billiard table.

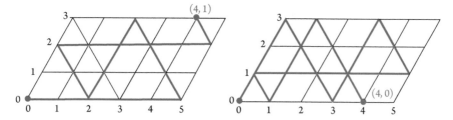

Figure 129 The two orbits from $(0,0)$ to the first instance of a boundary lattice point with a 4.

In general, for any two pitchers of size $p < q \in \mathbb{N}$ units, the number of fillings, emptyings, and transfers needs to measure out any $r < p \in \mathbb{N}$ is at most $q + 1$. Finding it, however, by simply exploring possibilities, may take a while. With a parallelogram billiard, just take a shot!

Exercise 248 Find a way to measure out 1 gallon of fluid with only two jugs of sizes 8 and 11 gallons. What is the shortest number of fillings, emptyings, and transfers needed to do so.

Exercise 249 Given two jugs of sizes 6 gallons and 9 gallons, determine precisely which intermediate measurements are NOT possible.

Exercise 250 Find a way to cook a perfectly timed 11-minute boiled egg, using only a 5-minute egg-timer and a 9-minute egg-timer.

7 Complicated Orbit Structure

7.1 Counting Periodic Orbits

For an adequate study of more general metric space maps $f : X \to X$, we will need to broaden our analysis from the identification and classification of just the possible fixed points. One way to do this is to look for points that may be fixed under an iterate of a map but are not fixed under the map itself: non-trivial periodic points. This makes sense since:

- for $n > 1$, n-periodic points, as fixed points of f^n, may also have stability features;
- there are existence theorems for periodic points;
- often, we can "solve" for them without actually solving the dynamical systems; and
- the number of n-periodic points, as a function of n, is an important feature of a map.

Recall Definition 2.8 of the set of n-periodic points of a map $f : X \to X$: $\mathrm{Per}_n(f) := \{x \in X \mid f^n(x) = x\}$. Here, we are interested in the cardinality of this set.

Definition 7.1 *For $f : X \to X$ a map, let*

$$P_n(f) := \#\{x \in X \mid f^n(x) = x\}$$

be the number of all n-periodic points of f.

Note that $P_n(f)$ also counts all m-periodic points when $m \mid n$. In particular, the 1-periodic points are the fixed points, and these are counted in $P_n(f)$ for all $n \in \mathbb{N}$. As a sequence, $\{P_n(f)\}_{n \in \mathbb{N}}$ can say a lot about f.

As a motivational example, consider the map

(7.1.1) $$E_2 : S^1 \to S^1, \quad E_2(z) = z^2,$$

where $z = e^{2\pi i x} \in \mathbb{C}$ is a complex number of unit modulus. Alternatively, this is the map $E_2(s) = (2s \mod 1)$, for $s \in S^1 = \mathbb{R}/\mathbb{Z}$. This is an example of what is called an *expanding map*: a map that locally increases distances throughout the domain.

Definition 7.2 *A map* $f : X \to X$ *on a metric space is called* expanding *if, for every* $x \in X$, *there exists* $\epsilon > 0$ *such that for all* $y \in B_\epsilon(x)$, $d(f(x), f(y)) > d(x,y)$.

While this may not be very interesting on a domain like $\mathbb{R}$ (think $f(x) = 2x$ on all real numbers), a map that expands distances on a closed, bounded (compact) domain like S^1 has significant consequences. We will discuss the more general properties of expanding maps shortly. For now, though, one consequence of expanding maps on a closed, bounded space is that they can have lots of periodic points:

Proposition 7.3 $P_n(E_2) = 2^n - 1$, *and all periodic points are dense in* S^1. *That is,* $\overline{Per(E_2)} = S^1$.

Proof Using the model $E_2(z) = z^2$, we find that z is an n-periodic point if

$$E_2^n(z) = \left(\cdots \left(\left(z^2 \right)^2 \right)^2 \cdots \right)^2 = z \quad \text{or} \quad z^{2^n} = z \quad \text{or} \quad z^{2^n - 1} = 1.$$

Thus every periodic point is an order-$(2^n - 1)$ root of unity (and vice versa). And there are exactly $2^n - 1$ of these, uniformly spaced around the circle. In fact, to any

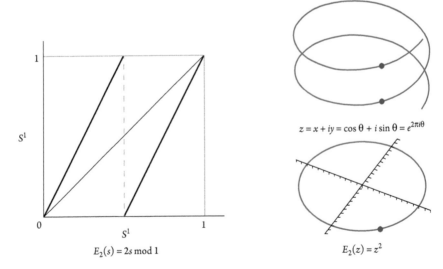

$z = x + iy = \cos\theta + i\sin\theta = e^{2\pi i \theta}$

$E_2(s) = 2s \mod 1$

$E_2(z) = z^2$

Figure 130 The map E_2 on the unit circle.

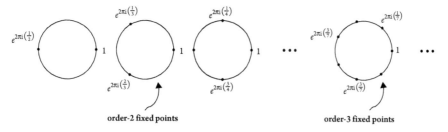

order-2 fixed points   order-3 fixed points

Figure 131 Order-n fixed points of E_2 spaced evenly in S^1.

rational $p/q \in \mathbb{Q}$, the point $e^{2\pi i(p/q)}$ is a qth root of unity. If $q = 2^n - 1$, for $n \in \mathbb{N}$, then $e^{2\pi i(p/q)}$ is an order-n fixed point. Now, as n goes to ∞, the spacing between order-$(2^n - 1)$ roots of unity goes to 0. Hence any point $x \in S^1$ can be written as the limit of a sequence of these points. Hence it will be in the closure of $Per(E_2)$. □

Example 7.4 z is 2-periodic if $z^{2^2-1} = z^3 = 1$. These are the points $z = e^{2\pi i\left(\frac{k}{3}\right)}$, for $k = 1, 2$. Indeed,

$$E_2\left(e^{2\pi i\left(\frac{1}{3}\right)}\right) = \left(e^{2\pi i\left(\frac{1}{3}\right)}\right)^2 = e^{2\pi i\left(\frac{1}{3}\right)*2} = e^{2\pi i\left(\frac{2}{3}\right)}, \text{ while}$$

$$E_2\left(e^{2\pi i\left(\frac{2}{3}\right)}\right) = \left(e^{2\pi i\left(\frac{2}{3}\right)}\right)^2 = e^{2\pi i\left(\frac{2}{3}\right)*2} = e^{2\pi i\left(\frac{4}{3}\right)} = e^{2\pi i}e^{2\pi i\left(\frac{1}{3}\right)} = e^{2\pi i\left(\frac{1}{3}\right)}.$$

We can calculate the growth rate of $P_n(E_2)$ in the obvious way: Define the *truncated natural logarithm*

$$\ln_+ x = \begin{cases} \ln x & x \geq 1, \\ 0 & \text{otherwise.} \end{cases}$$

Then define

$$p(f) = \varlimsup_{n\to\infty} \frac{\ln_+ P_n(f)}{n}$$

as the relative logarithmic growth of the number of n-periodic points of f with respect to n.

For our case, then, where $E_2(z) = z^2$,

$$p(E_2) = \varlimsup_{n\to\infty} \frac{\ln_+ (2^n - 1)}{n} = \varlimsup_{n\to\infty} \frac{\ln_+ 2^n(1 - 2^{-n})}{n}$$

$$= \varlimsup_{n\to\infty} \frac{\ln_+ 2^n + \ln_+(1 - 2^{-n})}{n} = \ln 2.$$

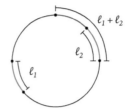

Figure 132 The inverse images under E_2 of the blue interval are the two red intervals.

This is the exponential growth rate of the periodic points of the map E_2. Note that the growth factor is 2 at each stage, and the exponential growth rate is the exponent of e that corresponds to the growth factor. Here $2 = e^{\ln 2}$.

Proposition 7.5 *For $E_m : S^1 \to S^1$, $E_m(z) = z^m$, where $m \in \mathbb{Z}$ and $|m| > 1$,*

$$P_n(E_m) = \left| m^n - 1 \right|,$$

the set of all periodic points is dense in S^1, and $p(E_m) = \ln |m|$.

Exercise 251 Show this for $m = -3$.

Here is an interesting fact about this family of maps: The image of any small arc under the map E_2 on S^1 is twice as long as the original arc. However, the map is a 2–1 map, meaning that each point has two distinct pre-images. Hence there are actually two disjoint pre-images of each small arc, and each is exactly half the size. Combined, the sum of the lengths of these two pre-images exactly matches the total length of the image. See Figure 132, where the pre-images of the blue arc are the two red arcs, each of length $\frac{1}{8}$, under the model of $S^1 = \mathbb{R}/\mathbb{Z}$, with the metric induced by the standard metric on $\mathbb{R}$ under the quotient map. Thus this expanding map on S^1 actually preserves length! Some notes about this:

- This is true for all of the expanding maps $E_m : S^1 \to S^1$, $E_m(z) = z^m$, where $m \in \mathbb{Z}$, and $|m| > 1$.

- This is a somewhat of a broadening of the idea of area preservation for a map. When the map is onto but not 1–1 (in this case, the map is 2–1), the relationship between pre-image and image is more intricate, and care is needed to understand the relationship properly.

But even without such a rigid structure given by maps like E_m, counting fixed and periodic points for expanding maps on S^1 is still a straightforward process, especially if the map is differentiable. Then, as with the use of the derivative in defining a contraction on an interval map in Proposition 2.31 in Section 2.1.4 on Lipschitz continuity, one can also use the derivative to define an expanding map:

Definition 7.6 *A C^1 map $f : S^1 \to S^1$ is expanding if $|f'(x)| > 1$, $\forall x \in S^1$.*

It should be obvious from this definition that the map E_m, where $m \in \mathbb{Z}$ and $|m| > 1$, is expanding, since $E_m(x) = mx \mod 1$ is differentiable and $\left| E'_m(x) \right| = |m| > 1$ for all

$x \in S^1$. But more general expanding maps of S^1 won't have a constant derivative and hence such a well-defined way to locate periodic points. Nevertheless, counting periodic points is still dependent solely on the degree of the map:

Proposition 7.7 *If $f : S^1 \to S^1$ is expanding, then $|\deg(f)| > 1$ and $P_n(f) = |(\deg(f))^n - 1|$.*

Proof Here, the first assertion is very straightforward. If $|f'(x)| > 1$ on all of S^1, then this is true also for any lift F, so that $|F'(x)| > 1$. Now recall from the definition of degree in Section 5.5 on invertible circle maps that

$$\deg(f) = F(x+1) - f(x).$$

But, if $|F'(x)| > 1$, then $|F(x+1) - F(x)| = |\deg(f)| > 1$.

As for the second assertion, note two claims:

- If f is expanding, then so is f^n, for $n \in \mathbb{N}$. This is due to the Chain Rule. And
- $\deg(f^n) = (\deg(f))^n$, since the degree of a composition is the product of the degrees.

Because of this, we only need to show that the proposition is true for the case $n = 1$, the fixed points.

Now, by the definition of a lift, we know that $f([x]) = [F(x)]$, so, at a fixed point $f([x]) = [x]$, we have $[F(x)] = [x]$, so the fixed points will satisfy $F(x) - x \in \mathbb{Z}$. For the function $i(x) = F(x) - x$, we also know two things:

- $i(1) - i(0) = F(1) - 1 - F(0) = \deg(f) - 1$, meaning that there will be at least $|\deg(f) - 1|$ fixed points. And
- $i'(x) > 0$, since $|F'(x)| > 1$.

Hence $i(x)$ is strictly monotonic, implying that $\mathrm{Fix}(f) = |\deg(f) - 1|$. $\qquad\square$

7.1.1 The Quadratic Map: Beyond 4

Let's return to the logistic map, first introduced in Section 2.6: For $\lambda \in \mathbb{R}$, let $f_\lambda : \mathbb{R} \to \mathbb{R}$, where $f_\lambda(x) = \lambda x(1 - x)$. For $\lambda \in [0,4]$, we can restrict to $I = [0,1]$, and $f_\lambda : I \to I$ is the 1-parameter family of interval maps we have already introduced and partially studied. To summarize our results so far: For $\lambda \in [0,3]$, the dynamics are quite simple. There are only fixed points, and no non-trivial periodic points, and all other points are asymptotic to them. The fixed points are at $x = 0$ and $x = 1 - 1/\lambda$.

Some new facts:

1. For $\lambda \in [3,4]$, a lot happens! We will defer a discussion of this map for parameter values in this interval until Section 7.5.

2. For $\lambda > 4$, I is not invariant.

3. Since f_λ is quadratic, f_λ^n is at most of degree 2^n. Thus the set of n-periodic points must be solutions to the equation $f_\lambda^n(x) = x$. Bringing x to the other side of

the equation, the set $P_n(f_\lambda)$ must consist of the roots of an (at most) 2^n-degree polynomial. Hence

$$P_n(f_\lambda) \leq 2^n \text{ for all } \lambda \in \mathbb{R}.$$

4. For $\lambda > 4$, many points escape the interval I. However, as we will see, many points have orbits that do not. Hence we can still study the map on a new domain: the set of all of points in I whose entire orbits remain in I.

Remark 7.8 *The particular logistic map $f_4(x) = 4x(1-x)$ is sometimes referred to as the Ulam map or the Ulam–von Neumann map, because of Stanislaw Ulam's and John von Neumann's early interest in its dynamics. Apparently, von Neumann proposed the map as a form of random number generator in 1947. And early studies by Ulam and Paul Stein popularized the map as an example of very complicated behavior arising from an interval map.*

Let $\lambda > 4$, and consider the first iterate of f_λ. Using Figure 133 as a guide, notice (top) that the intervals I_0 and I_1 are both mapped onto $[0, 1]$, that the map is strictly monotonic on each of these intervals, and that each of these intervals contains exactly one fixed point. Under the second iterate of the map, f_λ^2 (the middle plot of Figure 133), only points in the four intervals J_i, $i = 0, 1, 2, 3$, remain in $[0, 1]$, and the map restricted to these intervals is again surjective and strictly monotonic (hence injective). Here there are four fixed points (again one in each interval). But notice that only two of them are new: y_1 and y_2. These two new ones are period-2 points that are not fixed points. See the period-2 orbit in the cobwebbed figure at the bottom of Figure 133.

Continuing to iterate in this fashion, one can see the following:

- There will be 2^n intervals of points that remain in I after n iterates, all disjoint and all of positive length;
- The next iterate of f_λ maps each of these 2^n intervals onto $[0, 1]$, creating a single fixed point in each interval (of f_λ^n);
- It will follow (by induction) that $P_n(f_\lambda) = 2^n$, when $\lambda > 4$.

Exercise 252 Show that for f_λ where $\lambda > 4$, if $\mathcal{O}_x^+ \not\subseteq I$, then $\mathcal{O}_x \longrightarrow -\infty$. Also show that once an iterate of x under f_λ leaves I, it never returns.

So what about the points whose orbits stay in I? We can construct the set of all points whose orbits are in $[0, 1]$ as follows: For $x \in I$, call $\mathcal{O}_x^n$ the nth partial (positive) orbit of x, where

$$\mathcal{O}_x^n = \left\{ y \in I \mid y = f^i(x), \ i = 0, \dots, n-1 \right\}.$$

Then define

$$C_n = \left\{ x \in I \mid \mathcal{O}_x^n \in I \right\}.$$

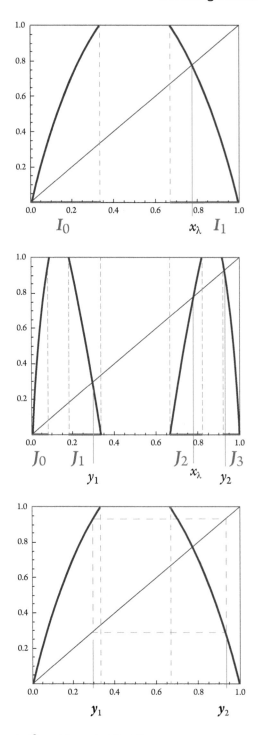

Figure 133 The map f_λ, f_λ^2, and the period-2 orbit.

Then $C_0 = I$, $C_1 = I_0 \cup I_1$, $C_2 = J_0 \cup J_1 \cup J_2 \cup J_3$, and $C_n \subset C_{n-1}$ for all $n \in \mathbb{N}$. Finally, define

$$C_\lambda = \bigcap_{n=0}^{\infty} C_n,$$

Then $f_\lambda : C_\lambda \to C_\lambda$ is a discrete dynamical system.

What does this set C_λ look like? For starters, it seems quite similar in construction to our ternary Canter set. Be careful here, though. The connected subintervals of C_n will not always be the same length in C_n (though they will all have positive length). You can see this in Figure 133, but should also check specifically for C_2. It should be certain that if $x \in I$ is n-periodic, then $x \in C_\lambda$. But are these the only points whose entire orbit lies in I? What about a point $y \in [0,1]$ that is well-approximated by periodic points? This means that there is a sequence of periodic points in I that converges to y. Is that enough to ensure that $\mathcal{O}_y \in [0,1]$? This is an important question (the answer to which should be yes, by continuity). It turns out that there are a lot of non-periodic points in C_λ. In fact, there are an uncountable number. In fact, a Cantor set's-worth! For the following discussion, recall the construction of the ternary Cantor set in Section 3.4.

Proposition 7.9 Let $f_\lambda : I \to \mathbb{R}$ be defined by $f_\lambda(x) = \lambda x(1-x)$, where $\lambda > 4$ and let

$$C_\lambda = \left\{ x \in I \mid \mathcal{O}_x \in I \right\}.$$

Then C_λ is a Cantor set and $f_\lambda|_{C_\lambda}$ is a discrete dynamical system.

Proof By Exercise 252, we can already deduce that all periodic points are in C_λ. For the moment, let's consider only the case that $\lambda > 2 + \sqrt{5} > 4$. In this case, we are assured that $\left| f_\lambda'(x) \right| > \mu(\lambda) > 1$, $\forall x \in C_1$ and some number μ dependent on λ. Hence f_λ is an expanding map on the points whose orbits remain in $[0,1]$.

Exercise 253 Verify that f_λ is expanding on points whose orbits stay in $[0,1]$ when $\lambda > 2 + \sqrt{5} > 4$.

Recall from Definition 3.50 that a subset of a closed interval is a Cantor set if it is closed, totally disconnected, and perfect. Since C_λ is an arbitrary intersection of closed sets, it is certainly closed.

As for totally discontinuous, let's assume that for $x, y \in C_\lambda$, where $x \neq y$, the interval $[x,y] \in C_\lambda$. Then the orbit of the entire interval lies completely in C_λ. But since f_λ is expanding, $\left| f(x) - f(y) \right| > \mu \left| x - y \right|$. And, for each $n \in \mathbb{N}$, $\left| f^n(x) - f^n(y) \right| > \mu^n \left| x - y \right|$. Choose $n > -\ln \left| x - y \right| / \ln \mu$. Then $\left| f^n(x) - f^n(y) \right| > 1$. But then $f^{n+1}\left([x,y]\right) \notin C_\lambda$. This contradiction means that no positive-length intervals exist in C_λ, and establishes that C_λ is totally discontinuous.

To see that C_λ is perfect, assume for a minute that there exists an isolated point $z \in C_\lambda$. Being isolated means that there is a small open interval $U(z) \subset I$, where, for all $x \in U(z)$, where $x \neq z$, we have $x \notin C_\lambda$. Now, since $z \in C_\lambda$, it is in a subinterval of every C_n. For any choice of $n \in \mathbb{N}$, call the interval $[x_n, y_n] \in C_n$ where $z \in [x_n, y_n]$. Create a sequence of nested closed intervals $\left\{[x_i, y_i]\right\}_{i\in\mathbb{N}}$, where, for every i, $z \in [x_i, y_i] \subset C_i$. Each endpoint x_i is eventually fixed, and hence $x_i \in C_\lambda$ for all $i \in \mathbb{N}$. But C_λ is totally disconnected. Hence the intersection

$$\bigcap_{i=1}^{\infty} [x_i, y_i]$$

can consist of only one point, and z is in this set. Thus, as a sequence, $\{x_i\}_{i\in\mathbb{N}} \longrightarrow z$, and z is not isolated in C_λ. Hence C_λ is perfect, and hence C_λ is a cantor set.

As a final note, we will relegate a discussion of why C_λ is still a Canter set when $4 < \lambda < 2 + \sqrt{5}$ to the following remark, noting that the proof requires a subtle bit of finesse not totally germane to the current discussion. $\square$

Remark 7.10 *When* $4 < \lambda < 2 + \sqrt{5}$, *the map* f_λ *is not expanding on* C_1. *Indeed, for* $\epsilon > 0$, *let* $\lambda = 4 + \epsilon$, *as in Figure 134. Then the first intersection of the graph of* f_λ *and the*

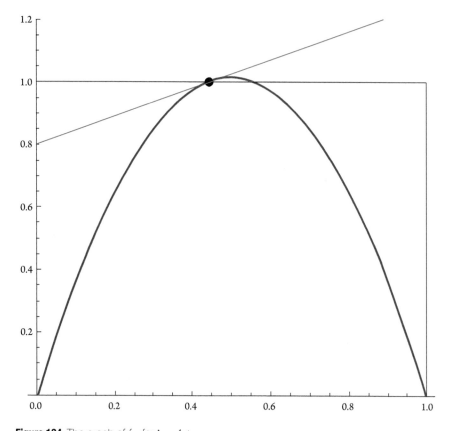

Figure 134 The graph of f_λ, for $\lambda = 4 + \epsilon$.

line $y = 1$ is at $x_1 = \frac{1}{2}\left(1 - \sqrt{1 - 4/\lambda}\right)$. The derivative of f_λ at this crossing is $f_\lambda'(x_1) = \sqrt{\lambda^2 - 4\lambda}$, which evaluates (when $\lambda = 4 + \epsilon$; see Figure 134) to

$$f_\lambda'(x_1) = \sqrt{\lambda^2 - 4\lambda} = \sqrt{4\epsilon + \epsilon^2} > 2\sqrt{\epsilon}.$$

The derivative of the square of the map at x_1 has a much larger derivative since the derivative of the image of x_1 is $-\lambda = -(4 + \epsilon)$ at the image point $f_\lambda(x_1) = 1$. Hence the derivative of the square of this map is greater, in magnitude, than $8\sqrt{\epsilon}$. This happens all through the interval, and the map can be said to be eventually expanding, in that $\exists N \in \mathbb{N}$ where for all $n > N$ the map $f_\lambda^n(x)\big|_{C_n}$ is expanding. Then the proof above holds. Thus the proposition is true for all $\lambda > 4$.

This quadratic family is an example of a *unimodal* map: a continuous map defined on an interval that has a unique turning point. In this case, this means that the map is increasing to the left of an interior point and decreasing thereafter.

Proposition 7.11 *Let $f : [0,1] \to \mathbb{R}$ be continuous with $f(0) = f(1) = 0$ and suppose there exists $c \in (0,1)$ such that $f(c) > 1$. Then $P_n(f) \geq 2^n$. If, in addition, f is unimodal and expanding, then $P_n(f) = 2^n$.*

Now recall that a map on a closed, bounded domain may preserve volume (length or area) even if it is many-to-one (not injective), once we understand what that means. The definition of expanding can incorporate this concept of looking backward in this case:

Definition 7.12 *A map $f : [0,1] \to [0,1]$ is expanding if*

$$|f(x) - f(y)| > |x - y|$$

on each interval of $f^{-1}\left([0,1]\right)$.

Examples of expanding maps include the logistic map for suitable values of $\lambda > 4$, and the circle maps E_m, where $m \in \mathbb{Z}$ and $|m| > 1$, as long as one interprets the map as either a circle map or an interval map with appropriate boundary conditions.

Exercise 254 Modify Definition 7.12 here to include maps of S^1.

Note here:

- In Proposition 7.11, the condition $f(0) = f(1) = 0$ and continuity ensure that the map will "fold" the image over the domain.
- The condition $f(c) > 1$ ensures that the folding will be complicated, with lots of points escaping, while lots of points will not.

Example 7.13 Let $g(x) = \frac{3}{128}\left(8x^4 - 16x^3 + 11x^2 - 3x\right)$. From the left graph in Figure 135, once can see that $g(0) = g(1) = 0$, and that there exists a point $c \in [0,1]$ such that $g(c) > 1$. However, $g(x)$ is not unimodal on $[0,1]$, and one can easily count that there are more than two fixed points. Also, this feature propogates through the iterates, as one can see in Figure 135 on the right, in the graph of $g^2(x)$. What is $P_2(g)$ here?

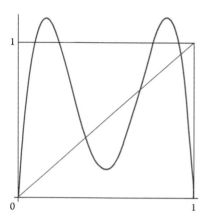 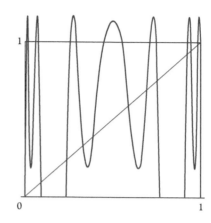

Figure 135 A non-unimodal map $g(x)$ on $[0, 1]$ (left) and its square (right).

Example 7.14 Another often studied unimodal map is a tent map. We will define this map in full generality in Example 7.55 in Section 7.1.6 on Markov partitions. For now, consider the map $T_3 : [0, 1] \to \mathbb{R}$, $T_3(x) = \frac{3}{2}\left(1 - 2\left|x - \frac{1}{2}\right|\right)$. This map does satisfy the suppositions of Proposition 7.11. Figure 136 shows the graphs of both T_3 and T_3^2. If we restrict the domain to points whose orbits stay in $[0, 1]$, as in the logistic map above, we again find that T_3 is a dynamical system on a Cantor set. Note that, in these first two iterates, only the red parts of the graph correspond to orbits that remain in $[0, 1]$ up through the second iterate. Do you recognize the points whose full forward orbits will remain in $[0, 1]$?

Exercise 255 Show that the set of points of $[0, 1]$ whose forward orbits under the tent map T_3 from Example 7.14 lie entirely within $[0, 1]$ is precisely the ternary Cantor set.

7.1.2 Hyperbolic Toral Automorphisms

One interesting way to generalize the linear expanding maps of S^1 is to consider a linear map on a product of copies of S^1 where the expressions on each copy of S^1 (on each factor) are coupled, in the sense that at least one expression involves more than one variable We start with a rather famous example: Let $L : \mathbb{R}^2 \to \mathbb{R}^2$, $L(x,y) = (2x+y, x+y)$. We can also write L as the linear vector map

$$L(\mathbf{x}) = A\mathbf{x}, \quad \text{where } A = \begin{bmatrix} 2 & 1 \\ 1 & 1 \end{bmatrix}.$$

Remark 7.15 *Note here that if vector spaces are finite-dimensional and bases are given or understood, then any linear map can be represented by the matrix defining the action. Hence we will simply call L the transformation of the plane given by the matrix A, and refer to L as either the map or the matrix $L = \begin{bmatrix} 2 & 1 \\ 1 & 1 \end{bmatrix}$. In mathematics, this is known as an abuse*

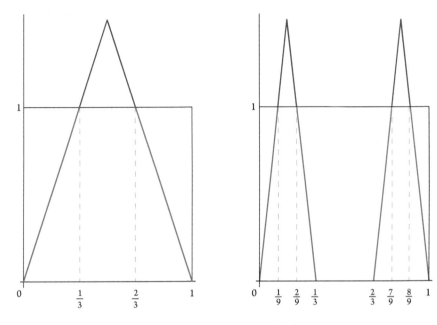

Figure 136 A tent map $T_3(x)$ on $[0, 1]$ (left) and its square (right).

of notation. *And when little added confusion results, it can be an effective way to reduce the number of objects involved in a discussion.*

We know that since A has integer entries, it takes integer vectors to integer vectors, and hence descends to a map on the 2-torus $\mathbb{T}^2$. Indeed, if $\mathbf{x}_1 = \begin{bmatrix} x_1 \\ y_1 \end{bmatrix}$ and $\mathbf{x}_2 = \begin{bmatrix} x_2 \\ y_2 \end{bmatrix}$ satisfy $\mathbf{x}_1 - \mathbf{x}_2 \in \mathbb{Z}^2$, then

$$L(\mathbf{x}_1 - \mathbf{x}_2) = L(\mathbf{x}_1) - L(\mathbf{x}_2) \in \mathbb{Z}^2.$$

But then $L(\mathbf{x}_1) - L(\mathbf{x}_2) = \mathbf{0} \mod 1$, which means $L(\mathbf{x}_1) = L(\mathbf{x}_2) \mod 1$. Hence the map L induces a map on $\mathbb{T}^2$ that assigns

$$(x,y) \longmapsto (2x + y \mod 1, x + y \mod 1) = (2x + y, x + y) \mod 1.$$

We will call this new *induced map* on the torus $f_L : \mathbb{T}^2 \to \mathbb{T}^2$, where

$$f_L(\mathbf{x}) = A\mathbf{x}, \quad A = \begin{bmatrix} 2 & 1 \\ 1 & 1 \end{bmatrix}, \quad \mathbf{x} \in \mathbb{T}^2.$$

The map f_L, for this choice of L, is sometimes called *Arnol'd's cat map* owing to Vladimir I. Arnol'd's use of an image of a cat to illustrate the dynamics of this map in a 1968 treatise [2] (see Section 7.1.3 below). It is often also referred to as the 2111-map.

Some notes:

- The map f_L is an example of an *automorphism* of $\mathbb{T}^2$: a homeomorphism that preserves also the ability of points on the torus to be added together (multiplied, if one defines the multiplication correctly).

- f_L—as a linear transformation of the torus (again we will abuse notation here)—is also invertible, since it is an integer matrix of determinant 1. The inverse map f_L^{-1} : $\mathbb{T}^2 \to \mathbb{T}^2$ is given by the matrix $A^{-1} = \begin{bmatrix} 1 & -1 \\ -1 & 2 \end{bmatrix}$.

- The eigenvalues of f_L (really the eigenvalues of A) are the solutions to the quadratic equation $\lambda^2 - 3\lambda + 1 = 0$, or

$$\lambda_1 = \frac{3 + \sqrt{5}}{2} > 1, \quad \text{and } \lambda_2 = \lambda_1^{-1} = \frac{3 - \sqrt{5}}{2} < 1,$$

so that the matrix defining f_L is a hyperbolic matrix (see Remark 4.20 in Section 4.3 on linear planar maps). Hence L here is a hyperbolic map of the plane. The map f_L is an example of a *hyperbolic toral automorphism*.

We can now generalize:

Definition 7.16 *Let* $L : \mathbb{R}^2 \to \mathbb{R}^2$, $L(x,y) = (ax + by, cx + dy)$ *be a linear map of the plane, where* $a, b, c, d \in \mathbb{Z}$ *and* $|ad - bc| = 1$. *Then* L *induces a* C^∞ *diffeomorphism* f_L : $\mathbb{T}^2 \to \mathbb{T}^2$ *called a* toral automorphism. *If, in addition,* $|a + d| = |\operatorname{tr} L| > 2$, *then* f_L *is a* hyperbolic toral automorphism.

Remark 7.17 *The classification of toral automorphisms above is a bit more general that that of Section 4.20 in that (1) we want only matrices with integer entries (sometimes called integer matrices), and (2) we want to allow matrices of negative unit determinant. These also preserve area, but not orientation. A square* $(n \times n)$ *integer matrix of determinant 1 or* -1 *is called a* unimodular matrix, *and, via matrix multiplication, the set of all such* $n \times n$ *matrices forms a group called the* unimodular group. *For* $n = 2$, *the unimodular group is*

$$GL(2, \mathbb{Z}) = \left\{ \begin{bmatrix} a & b \\ c & d \end{bmatrix} \,\middle|\, a, b, c, d \in \mathbb{Z}, \ |ad - bc| = 1 \right\}.$$

The hyperbolic ones are those whose eigenvalues are off of the unit circle in $\mathbb{C}$.

Remark 7.18 *Also keep in mind that only the general linear integer matrices of determinant* ± 1 *have inverses that are also integer matrices. Thus, for example,* $A = \begin{bmatrix} 4 & 3 \\ 2 & 1 \end{bmatrix} \notin$ $GL(2, \mathbb{Z})$, *since* $A^{-1} = -\frac{1}{2} \begin{bmatrix} 1 & -3 \\ -2 & 4 \end{bmatrix}$ *and is not an integer matrix.*

Exercise 256 Show that a determinant-1, 2×2 matrix with integer entries is hyperbolic iff the trace has magnitude greater than 2.

Exercise 257 Show, for $A \in GL(2, \mathbb{Z})$ a hyperbolic element, that

(1) the eigenvalues of A are not rational;

(2) the eigenspaces, as lines in the plane through the origin, have irrational slope; and

(3) the two eigenlines are orthogonal if A is symmetric as a matrix.

This last exercise is quite important. Recall from Proposition 5.23 how an linear flow on the torus with irrational slope is minimal, so that all orbits are dense in the torus. For a hyperbolic element of $GL(2, \mathbb{Z})$, the eigenspaces are invariant and have irrational slope. By Definition 4.11, they correspond to $\mathbb{E}^s$ and $\mathbb{E}^u$ of the transformation of the plane, and, since the transformation is linear, they will correspond to the stable and unstable manifolds of the origin under L:

Definition 7.19 *For $f : X \to X$ a homeomorphism of a metric space, the sets*

$$W^s(x) = \left\{ y \in X \mid d(f^n(x) - f^n(y)) \to 0, \text{ as } n \to \infty \right\}$$

and

$$W^u(x) = \left\{ y \in X \mid d(f^{-n}(x) - f^{-n}(y)) \to 0, \text{ as } n \to \infty \right\}$$

are the stable and unstable sets of $x \in X$. That is, they are the sets of points in X whose orbits are forward (respectively backward) asymptotic to the orbit of x in X.

Naturally, these sets are analogous to the flow case in Definition 4.49 and the discussion around it. We will not develop this too deeply here, but if X is also a C^k-diffeomorphism ($k \geq 1$), then these sets also look locally like Euclidean space, as a subspace of X, and are usually called *manifolds*. These linear transformations are smooth (C^∞), and hence we can talk of the stable and unstable manifolds of points like the origin in $\mathbb{R}^2$ under the hyperbolic map L. They are respectively the eigenspaces corresponding to the eigenvalue whose modulus is less than one and the eigenvalue whose modulus is greater than one.

Now, given the map $\rho : \mathbb{R}^2 \to \mathbb{T}^2$, the images of $W^s(0)$ and $W^u(0)$ wind densely around the torus. But this will also be true for all stable sets of all points:

Exercise 258 Let $x \in X$ and $f_L : \mathbb{T}^2 \to \mathbb{T}^2$ be a hyperbolic toral automorphism. Show that

$$W^s(x) = x + W^s(0) \quad \text{and} \quad W^u(x) = x + W^u(0).$$

Question 7.20 *For L a hyperbolic element of $GL(2, \mathbb{Z})$, like the 2111-map, how does f_L act on $\mathbb{T}^2$?*

In fact, the answer to this question relies on how L acts on $\mathbb{R}^2$. Using the model of $\mathbb{T}^2$ as the square torus in the plane, the unit square in $\mathbb{R}^2$ with the opposite sides identified, we can study the action of f_L on $\mathbb{T}^2$ through the action of L on $\mathbb{R}^2$. Note that this is the 2-dimensional version of studying a lift of a circle map on $\mathbb{R}$ as a means of studying the circle map, with L acting as the lift map of f_L.

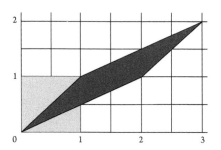 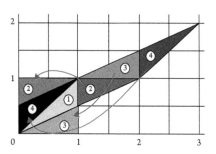

Figure 137 The 2111-map L, and how f_L acts on $\mathbb{T}^2$.

Linear maps of the plane take lines to lines. Hence they take polygons to polygons, and, in this case, they take parallelograms to parallelograms. The image of the unit square can be found by simply finding the images of the four corners of the square and constructing the parallelogram determined by those points by connecting corresponding adjacent points via lines. See the left side of Figure 137. But there is more: L, as a linear transformation, preserves area in the plane. Hence the image of the parallelogram that is the unit square also has area 1 (remember the discussion around volume-preserving linear maps of $\mathbb{R}^n$ in Section 6.1 on incompressibility). And, because of the equivalence relation given by the exponential map on $\mathbb{R}^2$, every point in the image of the unit square has a representative within its equivalence class inside the unit square. We can reconstruct the unit square by translating all of these outside points back into the square. This places the image of points on the torus back into the torus, as seen in Figure 137 on the right. Notice in the figure that

$$L\left(\begin{bmatrix}1\\1\end{bmatrix}\right)=\begin{bmatrix}3\\2\end{bmatrix}, \quad L\left(\begin{bmatrix}1\\0\end{bmatrix}\right)=\begin{bmatrix}2\\1\end{bmatrix}, \quad \text{and} \quad L\left(\begin{bmatrix}0\\1\end{bmatrix}\right)=\begin{bmatrix}1\\1\end{bmatrix}.$$

Exercise 259 Draw the image of the square torus in the plane under the following hyperbolic toral automorphisms. Then, for each, draw the torus in its representation in $\mathbb{R}^3$ (as the surface of a doughnut using the parameterization in Equation 3.3.2), with its two canonical loops that correspond to the edges of the unit square in $\mathbb{R}^2$, viewed as a fundamental domain. Then carefully draw the images of these two curves under the toral maps given by the unimodular matrices.

(a) $\begin{bmatrix}2 & 1\\1 & 1\end{bmatrix}$;

(b) $\begin{bmatrix}3 & 2\\1 & 1\end{bmatrix}$;

(c) $\begin{bmatrix}0 & 1\\1 & 1\end{bmatrix}$.

Question 7.21 *How are any periodic points distributed?*

We have the following proposition:

Proposition 7.22 *The set of all periodic points of $f_L : \mathbb{T}^2 \to \mathbb{T}^2$ is dense in $\mathbb{T}^2$, and*
$P_n(f_L) = \lambda_1^n + \lambda_1^{-n} - 2$.

Proof The first claim we will make to prove this result is the following: *Every rational point in $\mathbb{T}^2$ is periodic*. To see this, note that every rational point in $\mathbb{T}^2$ is a point in the unit square with coordinates $x = s/q$, and $y = t/q$, for some $q, s, t \in \mathbb{Z}$. For every point like this, $f_L(x,y)$ is also rational, with the <u>same</u> denominator. (Of course, we are neglecting fraction reduction—do you see why?) But there are only q^2 distinct points in $\mathbb{T}^2$ that are rational and that have q as the common denominator. Hence, at some point, the sequence $\mathcal{O}_{(x,y)}$ will repeat itself. Hence this claim is proved. Now notice that the set of all rational points in $\mathbb{T}^2$ is dense in $\mathbb{T}^2$, or

$$\overline{\mathbb{Q} \cap [0,1]} \times \overline{\mathbb{Q} \cap [0,1]} = [0,1]^2.$$

Hence the periodic points are dense in $\mathbb{T}^2$.

The next claim is: *Only rational points are periodic*. To see this, assume $f_L\left(\begin{bmatrix} x \\ y \end{bmatrix}\right) = \begin{bmatrix} x \\ y \end{bmatrix}$, where $L = \begin{bmatrix} a & b \\ c & d \end{bmatrix}$, as in Definition 7.16. Then

$$f_L\left(\begin{bmatrix} x \\ y \end{bmatrix}\right) = \begin{bmatrix} a & b \\ c & d \end{bmatrix}\begin{bmatrix} x \\ y \end{bmatrix} = \begin{bmatrix} x \\ y \end{bmatrix} \quad \text{mod 1,}$$

and this forces the system of equations

$$ax + by = x + k_1,$$
$$cx + dy = y + k_2, \quad \text{for } k_1, k_2 \in \mathbb{Z}.$$

Simply solve this system for x and y, and you find that $x, y \in \mathbb{Q}$.

Exercise 260 Solve this system for x and y.

To count the periodic points, we resort to a technique similar to how we counted periodic points of expanding maps on S^1: In the proof of Proposition 7.7, we created a function $i(x) = F(x) - x$, looking for points of the lift of an expanding map where the difference between the image and the point is an integer. These corresponded to fixed points. Here, create a parameterized family of new linear maps $g_n : \mathbb{T}^2 \to \mathbb{T}^2$, where

$$g_n\left(\begin{bmatrix} x \\ y \end{bmatrix}\right) = f_L^n\left(\begin{bmatrix} x \\ y \end{bmatrix}\right) - \begin{bmatrix} x \\ y \end{bmatrix} = (f_L^n - I_2)\begin{bmatrix} x \\ y \end{bmatrix}.$$

The n-periodic points are then precisely the kernel of this linear map:

$$Per_n(f_L) = \ker(g_n) = \left\{ \begin{bmatrix} x \\ y \end{bmatrix} \in \mathbb{T}^2 \,\middle|\, g_n\left(\begin{bmatrix} x \\ y \end{bmatrix}\right) = \begin{bmatrix} 0 \\ 0 \end{bmatrix} \right\}.$$

We can now easily see and count n-periodic points by lifting these maps to linear maps of $\mathbb{R}^2$. Let $G_n : \mathbb{R}^2 \to \mathbb{R}^2$, where $G_n = L^n - I_2$. Then the kernel of G_n consists of all pre-images of integer vectors when the domain is restricted to the square torus. And we can count these!

Claim *All pre-images of* $\begin{bmatrix} 0 \\ 0 \end{bmatrix}$ *under the map* g_n *are given by the integer vector values of* $G_n = L^n - I_2$ *evaluated on the square torus, namely,* $\mathbb{Z}^2 \cap (L^n - I_2) \big([0,1) \times [0,1) \big)$.

- Entirely analogous to the idea that f_L is simply the matrix L where images are taken modulo 1, the map g_n is simply the map $G_n = L^n - I_2$ where images are also taken modulo 1.

- To avoid over-counting points, we modify our unit square as the model for the square torus in the plane, eliminating twice-counted points on two edges (the top and right), and quadruply counted points (the adjacent corners to the eliminated edges). Consider the "half-open box" $[0,1)^2$ as our model of $\mathbb{T}^2$. In this model, every point lives in its own equivalence class.

We try a few early iterates:

Example 7.23 $g_1 = f_L - I_2$. Then the corresponding map on $\mathbb{R}^2$ is the linear map with matrix

$$G_1 = L - I_2 = \begin{bmatrix} 2 & 1 \\ 1 & 1 \end{bmatrix} - \begin{bmatrix} 1 & 0 \\ 0 & 1 \end{bmatrix} = \begin{bmatrix} 1 & 1 \\ 1 & 0 \end{bmatrix}.$$

This map is a shear on $[0,1)^2$, and we see that the only integer vector in the image is the origin: the red dot in Figure 138. Thus

$$P_1(f_L) = \lambda^1 + \lambda^{-1} - 2 = \frac{3 + \sqrt{5}}{2} + \frac{3 - \sqrt{5}}{2} - 2 = 3 - 2 = 1.$$

Example 7.24 $G_2 = L^2 - I_2$. Here

$$G_2 = L^2 - I_2 = \begin{bmatrix} 2 & 1 \\ 1 & 1 \end{bmatrix} \begin{bmatrix} 2 & 1 \\ 1 & 1 \end{bmatrix} - \begin{bmatrix} 1 & 0 \\ 0 & 1 \end{bmatrix} = \begin{bmatrix} 5 & 3 \\ 3 & 2 \end{bmatrix} - \begin{bmatrix} 1 & 0 \\ 0 & 1 \end{bmatrix} = \begin{bmatrix} 4 & 3 \\ 3 & 1 \end{bmatrix}.$$

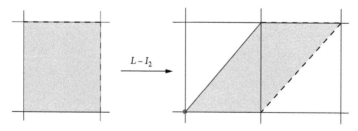

Figure 138 G_1 action on $\mathbb{T}^2$ as seen through the map $L - I_2$ on $[0,1)^2$.

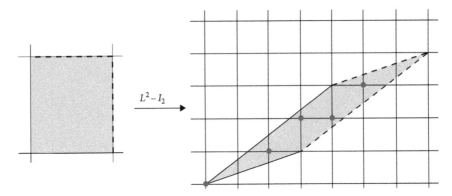

Figure 139 G_2 action on $\mathbb{T}^2$ as seen through the map $L^2 - I_2$ on $[0, 1)^2$.

This map is a little more complicated, and we see in Figure 139 that there are a few more integer vectors in the image, namely, the points $(2, 1)$, $(3, 2)$, $(4, 2)$, and $(5, 3)$. And, since

$$P_2(f_L) = \lambda^2 + \lambda^{-2} - 2 = 7 - 2 = 5,$$

we see that the formula continues to hold.

Exercise 261 What were the original points in $[0, 1)^2$ that correspond to these five integer vectors under G_2?

Exercise 262 Draw the image of $[0, 1)^2$ in $\mathbb{R}^2$ under the linear map corresponding to G_3 for f_L above. Calculate $P_3(f_L)$ via the formula and verify by marking the integer points in $(L^3 - I_2)([0, 1)^2)$. Choose two non-zero integer vectors in the image and identify the original 3-periodic points in $\mathbb{T}^2$ that correspond to them.

This proof ends by appealing to a strikingly beautiful result by Georg Alexander Pick in 1899 [40], which we now call Pick's Theorem:

Theorem 7.25 (Pick's Theorem) *Let $A \in \mathbb{R}^2$ be a polygon in the plane whose vertices are integer points (points with integer coordinates). Then, for I the number of integer points in the interior of A, and B the number of integer points on A, we have*

$$\text{area}(A) = I + \tfrac{1}{2}B - 1.$$

Here, the polygonal integer points are edge points and vertex points, so that $\frac{1}{2}B - 1 = (B - 2)/2$. For a parallelogram, as in our case, this amounts to collecting up all integer points on the polygon, and counting each as a half and also counting all four vertices as 1. But this is precisely the recipe for double counting edge points and quadruple counting vertex points. Hence the formula, applied to our square $[0, 1)^2$ with its only two edges and one vertex, reduces to a simple counting of interior points and the remaining edge points (see Figure 140). In sum: the area of $G_n([0, 1)^2)$ is precisely equal to the number of integer vectors in the image. And the latter is given by

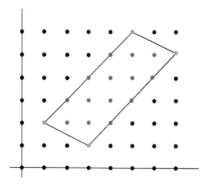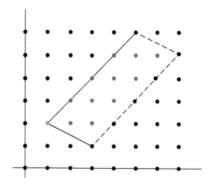

Figure 140 Pick's formula for parallelograms.

$$|\det(G_n)| = |\det L^n - I_2| = |(\lambda^n - 1)(\lambda^{-n} - 1)| = \lambda^n + \lambda^{-n} - 2,$$

where λ is the largest eigenvalue (in magnitude) of L. □

Note: G_2 on $\mathbb{R}^2$ is certainly not area-preserving, since $\det(G_2) = \begin{vmatrix} 4 & 3 \\ 3 & 1 \end{vmatrix} = 5$.

Remark 7.26 *The map f_L above is area-preserving on the torus. It is also invertible. Any determinant-1 integer matrix is invertible, and the inverse is also an integer matrix! In fact, elements of this type comprise $SL(2,\mathbb{Z})$, the group of all 2×2, determinant-1 matrices with integer entries, a subgroup of both $SL(2,\mathbb{R})$ mentioned in Remark 4.20 and $GL(2,\mathbb{Z})$ from Section 7.17.*

However, area preservation does not ensure invertibility of the map. The prime example is the circle map $E_m : S^1 \to S^1$, where $E_m(z) = z^m$. The map is area-preserving, if we sum all of the lengths of the disjoint pre-images of small sets. But it is also of degree m. And if $|m| > 1$, then the map is m-to-1. Invertibility is a very desirable property for a map, since it allows us to work both forwards and backwards in constructing orbits. Fortunately, there are ways to study non-invertible maps by encoding their information in a (different) invertible dynamical system.

7.1.3 Application: Image Restoration

We can spend a little more time with the 2111-map f_L on the torus, and display a curious effect on the discrete version of $\mathbb{T}^2$: Break up the unit square, used to construct the square torus, into an $N \times N$ grid, effectively pixelizing the square torus. Of course, the pixel locations will be rational points on the torus, and hence will all be periodic under the transformation. But the orbits of neighboring pixels will not stay nearby, and if the pixels represented an image, the image would effectively be a mess after only a few iterates. However, since all orbits would be finite (they do represent rational points on the square torus), one can conclude that eventually, we should be able to restore the image simply by

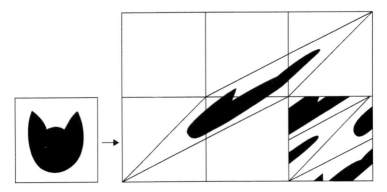

Figure 141 Arnol'd's original cat.

iterating long enough. This is what Vladimir Arnol'd [2] demonstrated with his cat map, using the image of a cat, reproduced here in Figure 141.

One way to see this is to consider an image of size $N \times N$ pixels, and form the square torus of side size N in the plane, $\mathbb{T}_N^2$, which is simply a reparameterized unit-square version of $\mathbb{T}^2$. Now discretize it by considering only the integer vectors in $\mathbb{T}_N^2$. The 2111-map will act on these integer vectors as

$$\mathbf{x}_{i+1} = \begin{bmatrix} 2 & 1 \\ 1 & 1 \end{bmatrix} \mathbf{x}_i, \qquad \mod N,$$

where $\mathbf{x} = \begin{bmatrix} x_i \\ y_i \end{bmatrix}$ has entries $x, y \in \{0, 1, \ldots, N-1\}$. These vectors are simply the pixel locations of the image. Of course, f_L takes integer vectors to integer vectors, so it takes pixels to pixels, and, in a sense, simply shuffles them around in a highly non-trivial fashion. And, of course, there are only N^2 positions, and the map is bijective on these positions, so every orbit will be finite. Hence, ultimately, a future iterate of the map should bring the image back to irs original configuration.

The question is, how long will this take, based on the size of the image? It turns out that the relationship between N and the number of iterates m_N it takes to restore the image is quite complicated.

For example, for $N = 3$, it takes only $m_3 = 4$ iterates to return the matrix to its original configuration. Indeed,

$$\begin{bmatrix} p_{0,2} & p_{1,2} & p_{2,2} \\ p_{0,1} & p_{1,1} & p_{2,1} \\ p_{0,0} & p_{1,0} & p_{2,0} \end{bmatrix} \longmapsto \begin{bmatrix} p_{2,2} & p_{1,0} & p_{0,1} \\ p_{1,1} & p_{0,2} & p_{0,1} \\ p_{0,0} & p_{2,1} & p_{1,2} \end{bmatrix} \longmapsto \begin{bmatrix} p_{0,1} & p_{2,1} & p_{1,1} \\ p_{0,2} & p_{2,2} & p_{1,1} \\ p_{0,0} & p_{2,0} & p_{1,0} \end{bmatrix}$$

$$\longmapsto \begin{bmatrix} p_{1,1} & p_{2,0} & p_{0,2} \\ p_{2,2} & p_{0,1} & p_{0,2} \\ p_{0,0} & p_{1,2} & p_{2,1} \end{bmatrix} \longmapsto \begin{bmatrix} p_{0,2} & p_{1,2} & p_{2,2} \\ p_{0,1} & p_{1,1} & p_{2,1} \\ p_{0,0} & p_{1,0} & p_{2,0} \end{bmatrix}.$$

Figure 142 The image and its iterates (from left top to bottom right, 1, 10, 30, 40, 50, 59, 60).

Here, we have associated $p_{i,j}$ with the integer vector $\begin{bmatrix} i \\ j \end{bmatrix}$, and indexed the matrix according to how the square torus $\mathbb{T}_3^2$ sits in the first quadrant.

As a more interesting example, the image in Figure 142 at top left is a square image of size 183×183. The first iterate is the second image, and one can see a similar folding of the image onto the square torus to that in Figure 137 in Section 7.1.2. The image is unrecognizable for many of its iterates. However, the image is restored at iterate $m_{183} = 60$.

Other examples include $m_{100} = 150$, $m_{150} = m_{300} = 300$, $m_{74} = 114$ and $m_{257} = 258$. To date, some calculations and bounds have been given by Freeman Dyson and Harold Falk [19], so that, for example, $m_N \leq 3N$, with equality only when N is twice an integer power of 5. And their results are based on their determination that "the period $[m_n]$ is related to the divisibility properties of Fibonacci numbers," using the fact that the 2111-map is the square of the hyperbolic toral automorphism given by the matrix $\begin{bmatrix} 1 & 1 \\ 1 & 0 \end{bmatrix}$, intimately related to the Fibonacci numbers via the linear planar map studied in Section 4.3.5 on saddle points, although that map was given by the slightly altered (reindexed) hyperbolic map $A = \begin{bmatrix} 0 & 1 \\ 1 & 1 \end{bmatrix}$, having the same eigendata.

Exercise 263 Calculate m_3 for the hyperbolic toral automorphisms given by both $\begin{bmatrix} 0 & 1 \\ 1 & 1 \end{bmatrix}$ and $\begin{bmatrix} 3 & 2 \\ 1 & 1 \end{bmatrix}$.

7.1.4 Inverse Limit Spaces

One issue with many-to-one maps like E_2 above is precisely that the map is not invertible: each point in the range of the function has two distinct pre-images. For example,

$$E_2^{-1}\left(\tfrac{1}{4}\right) = \left\{\tfrac{1}{8}, \tfrac{5}{8}\right\}.$$

That the map is volume-preserving is well hidden by this property, as we have seen. One can account for the non-injectivity via a "choice" of pre-image for a particular point, but one cannot move backward along an orbit in general. There is a way to account for all orbits' pre-origins, however, which we do now.

For X a metric space with $f : X \to X$ a surjective map, let

$$(X,f) = \left\{\underline{x} = (\ldots, x_i, \ldots, x_{-2}, x_{-1}, x_0) \mid x_i \in X, i \in -\mathbb{N}, f(x_i) = x_{i+1}\right\}$$

be the set of all sequences of points in X that serve as orbits leading up to x_0, for all choices of $x_0 \in X$. The set (X,f) may seem a bit unwieldy and complicated, but it has some nice properties. For example, one can use the topology of X to endow this set with its own topology. Indeed, (X,f) is a subset of the bigger set of all infinite sequences of X, sometimes called X^∞, a space given the (infinite) product topology. By a famous theorem of Andrey Tychanoff [53], if X is compact, then so is (X,f). And, as a topological space, (X,f) also can be made a metric space with the metric

$$\underline{d}(\underline{x}, \underline{y}) = \sum_{i \leq 0} 2^i d(x_i, y_i),$$

where d is the metric on X.

Exercise 264 Show that this is a metric. Describe an ϵ-ball around a point in (X,f).

Now, define the map $T_f : (X,f) \to (X,f)$ by

$$T_f(\ldots, x_{-3}, x_{-2}, x_{-1}, x_0) = (\ldots, x_{-2}, x_{-1}, x_0, f(x_0)).$$

It turns out that this map is a homeomorphism: It is continuous in the product topology (we will not show this), as is its inverse (drop off the final term), and we have the following exercise:

Exercise 265 Show T_f is one-to-one and onto.

In fact,

$$T_f(\underline{x}) = \underline{f(x)}.$$

Here, (X,f) is an example of an *inverse limit space* of X, given f. It catalogs every orbit that leads to x_0. Now create the space of bi-infinite sequences (or a two-sided sequence space)

$$X' = \left\{(\ldots, x_{-3}, x_{-2}, x_{-1}, x_0, O_{x_0}^+) \mid x_i \in X, i \in \mathbb{Z}, f(x_i) = x_{i+1}\right\}$$

and extend T_f to X'. Then T_f is just the *left shift map* on X'; a map on sequences that simply shifts all sequences by one element, mapping the index $i \mapsto i+1$. It remains a homeomorphism, X' is also compact when X is, and every point in X' constitutes an entire $\mathbb{Z}$-orbit. The inverse, T_f^{-1} is the *right shift map*.

Example 7.27 For $E_2(x) = 2x \mod 1$ on S^1, some of these sequences that correspond to $x_0 = 1$ look like

0th place

↓

$$\{\dots, \tfrac{1}{8}, \tfrac{1}{4}, \tfrac{1}{2},\ \mathbf{1}\ , 1, 1, \dots\}$$

$$\{\dots, \tfrac{1}{4}, \tfrac{1}{2}, 1,\ \mathbf{1}\ , 1, 1, \dots\}$$

$$\{\dots, \tfrac{3}{8}, \tfrac{3}{4}, \tfrac{1}{2},\ \mathbf{1}\ , 1, 1, \dots\}$$

$$\{\dots, \tfrac{7}{8}, \tfrac{3}{4}, \tfrac{1}{2},\ \mathbf{1}\ , 1, 1, \dots\}$$

Definition 7.28 *For X a metric space and $f : X \to X$ continuous, the inverse limit is a discrete dynamical system defined on the space of sequences*

$$X' = \left\{ \{x_n\}_{n \in \mathbb{Z}} \mid x_n \in X, f(x_n) = x_{n+1}, \forall n \in \mathbb{Z} \right\}$$

by the map $F\left(\{x_n\}_{n \in \mathbb{Z}}\right) = \{x_{n+1}\}_{n \in \mathbb{Z}}$.

Here, X' is called the *inverse limit space*. Note that since this map takes entire sequences to sequences, it is 1–1, and hence we can go backwards. On sequences, this map is invertible, since the entire history of a point is already in the "point" (read: sequence).

Example 7.29 Back to the map E_2 on S^1, the inverse limit space is

$$\mathbb{S} = \left\{ \{x_n\}_{n \in \mathbb{Z}} \mid x_n \in S^1, E_2(x_n) = x_{n+1}, \forall n \in \mathbb{Z} \right\}$$

with the map $F\left(\{x_n\}_{n \in \mathbb{Z}}\right) = \{2x_n \mod 1\}_{n \in \mathbb{Z}}$. The space $\mathbb{S}$ is called a *solenoid* or the *Vietoris solenoid*, since it was first described by Leopold Vietoris [57] in 1927. The generalization to $|n| > 1$ was described shortly thereafter by David van Dantzig [54].

Visually, what is happening here? Each $x \in S^1$ has two pre-images $w_1, w_2 \in S^1$, each of which has two pre-images v_1, v_2, v_3, v_4. These are evenly spaced along the circle and one can envision unwinding the circle to "see" the pre-images, as in Figure 143. Now place a copy of S^1 as the center curve in a space called the *solid torus*, a space homeomorphic to $M = S^1 \times D^2$, where D^2 is the 2-dimensional disk (a copy of S^1, seen as the unit circle in the plane, along with its interior). Visually, one can view the torus $\mathbb{T}^2$ in 3-space as in the parameterization given in Section 3.3.4, and then include its interior.

Create the map $f : M \to M$, where

(7.1.2) $$f(s, x, y) = \left(2s, \tfrac{1}{4}x + \tfrac{1}{2}\cos 2\pi s, \tfrac{1}{4}y + \tfrac{1}{2}\sin 2\pi s\right).$$

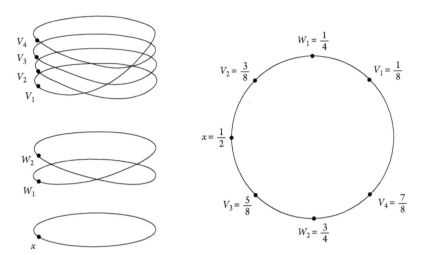

Figure 143 Pre-images of $x \in S^1$ under the map E_2.

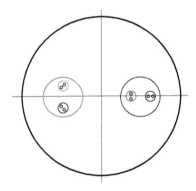

Figure 144 The invariant disk D_0 under the map $f : M \to M$.

In the directions comprising D, this map looks like a contraction, mapping the disk into itself and shifting the origin. However, in the S^1 direction, the map is E_2. Hence the vertical (the disk part) slice has image generally in some other slice.

However, the disk corresponding to $0 \in S^1$, which we will call the $s = 0$ disk or D_0, is invariant under f, so that any orbit that begins in this disk remains. However, many other orbits enter D_0, and remain thereafter. In Figure 144, the image of D_0 under f is the black-edged disk in the right semi-disk, and its image is the smaller disk on the right within this first image.

Exercise 266 The map $f : D_0 \to D_0$ is a contraction. Calculate Lip(f) and find the coordinates of the fixed point.

Figure 145 The first few images of $f : M \to M$.

Back in Figure 144, the green-edged disk in the left semidisk is the image of the disk corresponding to $s = \frac{1}{2}$, while the red-edged disk is the image of the $s = \frac{1}{2}$ under two iterations. The orange-edged disk is the image of the $s = \frac{1}{4}$ disk under two iterations, and the blue-edged disk, the smallest circle in the third quadrant closest to the horizontal axis, is the image of the $s = \frac{3}{4}$ disk after three iterations. One can see these first few iterations in Figure 145, a beautiful rendition of the solenoid produced in a paper by Olga Isaeva, Sergey Kuznetsov, and co-workers [36].

And finally, note that the two constants in Equation 7.1.2, namely, $\frac{1}{4}$ and $\frac{1}{2}$, are only chosen to keep the images of f from self-intersecting. In fact as long as the coefficient of the coordinates (the current $\frac{1}{4}$) is less than the coefficients of the trig functions (currently $\frac{1}{2}$), and as long as their sum is less than 1, the image of the solid torus remains a solid torus strictly inside the original, and winding around twice on each iterate.

7.1.5 Shift Spaces

Thinking along the lines of the inverse limit spaces we saw in Section 7.1.4, we can adapt that construction to create a new type of dynamical system that serves as a beautiful and important model for many of the concepts we will see soon. To start, recall from Section 1.1.3 our definition of $\mathcal{M}_m = \{0, 1, 2, \ldots, m-1\}$, a finite, discrete set of m symbols (we will use numbers here instead of letters to facilitate the discussion to follow, but there is no natural reason why). The elements of $\mathcal{M}$ are typically called *letters* and $\mathcal{M}$ is often called an *alphabet*. We can actually make $\mathcal{M}_m$ into a space by "topologizing" it with the discrete topology (recall Remark 3.3; essentially every subset, including a subset consisting of a single individual letter, is considered "open") so that $\mathcal{M}_m$ is a topological space. It is even a metric space. Indeed, arranging $\mathcal{M}_m$ as the first m non-negative natural numbers (including 0) in $\mathbb{R}$, we can define a metric $d_{\mathcal{M}_m}$ as the restriction of the standard Euclidean metric on $\mathbb{R}$, so

$$d_{\mathcal{M}_m}(x, y) = |x - y|, \quad x, y \in \mathcal{M}_m.$$

Note that another commonly used metric here is the discrete metric (recall Exercise 26): Let δ_{ij} be the *Kronecker delta function*, a function on pairs of natural numbers, where $\delta_{ij}(m, n)$ equals 1 iff $m = n$, and 0 otherwise. Then define $\delta_{\mathcal{M}_m}(x_i, x_j) = 1 - \delta_{ij}$ on $\mathcal{M}_m$ as a metric where every element is distance 1 from every other element.

Exercise 267 Show that $\delta_{\mathcal{M}_m}$ is a metric on $\mathcal{M}_m$.

The use of these spaces of symbols is central to the construction of *shift spaces*, which we define now.

Given a finite symbol space $\mathcal{M}_m$ on m letters with a metric $d_{\mathcal{M}}$, construct two other sets,

$$\mathcal{M}_m^{\mathbb{N}} = \big\{ \mathbf{x} = \{x_i\} \mid i \in \mathbb{N}, x_i \in \mathcal{M}_m \big\},$$
$$\mathcal{M}_m^{\mathbb{Z}} = \big\{ \mathbf{x} = \{x_i\} \mid i \in \mathbb{Z}, x_i \in \mathcal{M}_m \big\}.$$

We can think of these new sets in two ways:

- the set of all infinite and bi-infinite sequences of elements of $\mathcal{M}_m$, respectively; and
- the set of all functions $\mathbb{N} \to \mathcal{M}_m$ and $\mathbb{Z} \to \mathcal{M}_m$, respectively, assigning an element of $\mathcal{M}_m$ to each index value.

It is easy to see that these sets can also be made into topological spaces via the product topology. In this topology, one can construct open sets via a metric. Let

$$(7.1.3) \qquad\qquad d(\mathbf{x}, \mathbf{y}) = \sum_{i \in \mathbb{K}} \frac{d_{\mathcal{M}_m}(x_i, y_i)}{\rho^{|i|}}$$

for $\mathbb{K}$ either $\mathbb{N}$ or $\mathbb{Z}$ and any $\rho > 1$. The requirement on ρ is necessary to ensure that the series defining the metric converges. It should be obvious here that the more two sequences agree on positions near the 0th position in the sequences, the closer they are to each other. In fact, with ρ chosen to be greater than m, one can show that the $(1/\rho^n)$-neighborhoods of a particular sequence $\mathbf{x} \in \mathcal{M}_m^{\mathbb{N}}$ coincide precisely with set of all sequences that agree with $\mathbf{x}$ on the first $n + 1$ positions. Indeed, on $\mathcal{M}_m$, the metric $d_{\mathcal{M}_m}$ is no bigger than $m - 1$. So choose $\mathbf{x}, \mathbf{y} \in \mathcal{M}_m^{\mathbb{N}}$ such that $x_i = y_i$, for $i = 0, \ldots, n$. Then

$$d(\mathbf{x}, \mathbf{y}) = \sum_{i=0}^{\infty} \frac{|x_i - y_i|}{\rho^i} = \sum_{i=n+1}^{\infty} \frac{|x_i - y_i|}{\rho^i}$$

$$\leq \frac{1}{\rho^{n+1}} \sum_{j=0}^{\infty} \frac{|x_{n+1+j} - y_{n+1+j}|}{\rho^j}$$

$$\leq \frac{1}{\rho^{n+1}} \sum_{j=0}^{\infty} \frac{m-1}{\rho^j} = \frac{m-1}{\rho^{n+1}} \frac{1}{1 - \rho^{-1}} = \frac{m-1}{\rho-1} \frac{1}{\rho^n}.$$

Hence, as long as $\rho > m$, we know that

$$(7.1.4) \qquad\qquad \mathbf{y} \in B_{1/\rho^n}(\mathbf{x}) = \left\{ y \in \mathcal{M}_m^{\mathbb{N}} \mid d(\mathbf{x}, \mathbf{y}) < \frac{1}{\rho^n} \right\}.$$

And, given a possible $\mathbf{y}$, where $\mathbf{y} \in B_{1/\rho^n}(\mathbf{x})$, if there was a position $0 \leq i \leq n$ where $x_i \neq y_i$, then it should be clear that $d(\mathbf{x}, \mathbf{y}) \geq 1/\rho^i > 1/\rho^n$. Hence, in $\mathcal{M}_m^{\mathbb{N}}$, small metric neighborhoods coincide precisely with sets of sequences that agree on all early positions. Note that, in the case of $\mathcal{M}^{\mathbb{Z}}$, where we would like $(1/\rho^n)$-neighborhoods to coincide with sets of sequences that agree on all positions from $-n$ to n, we would need ρ to be larger.

Exercise 268 Find a bound for ρ in the metric in Equation 7.1.3 such that

$$B_{1/\rho^n}(\mathbf{x}) = \left\{ \mathbf{y} \in \mathcal{M}_m^{\mathbb{Z}} \mid x_i = y_i, \ i = -n, \ldots, 0, \ldots, n \right\}.$$

Metric neighborhoods defined in this way are often called *cylinders* in the product topology.

One may think that these spaces are somewhat unwieldy. In fact, they are quite easy to work with. One of the more beautiful (and surprising), aspects of these spaces is that they are compact.

Definition 7.30 *A topological space is called* sequentially compact *if every infinite sequence in the space has a convergent subsequence.*

Example 7.31 The closed, unit interval $[0,1] \in \mathbb{R}$ is sequentially compact. This is a direct consequence of the famous *Bolzano–Weierstrass Theorem*: Every bounded sequence in $\mathbb{R}^n$ has a convergent subsequence. But this also then means that every closed and bounded subset of $\mathbb{R}^n$ is sequentially compact.

Example 7.32 $\mathbb{R}$ is not sequentially compact. In fact, the sequence $\{n\}_{n\in\mathbb{N}}$ does not have a convergent subsequence.

Since closed and bounded is equivalent to compact as far as subsets of $\mathbb{R}^n$ go (this is another famous theorem, the *Heine–Borel Theorem*), one may want to conclude that compact and sequentially compact are also equivalent. This is not true in general, owing to the deeper aspects of topologizing an infinite product of spaces. However, when the space in question possesses the additional structure of a metric space, these two properties are equivalent. We will offer this fact without the technical machinery to prove it herein, but rather refer the reader to any book on elementary analysis or topology: a metric space is sequentially compact iff it is compact. Accepting this fact, we can show that $\mathcal{M}^{\mathbb{Z}}$ (note that, for now, we will omit the subscript m) is compact by showing that it is sequentially compact.

Proposition 7.33 $\mathcal{M}^{\mathbb{Z}}$ *is sequentially compact.*

Remark 7.34 *Two points to note: (1) One needs to be careful here. To prove this proposition, we will need to construct a sequence of elements in $\mathcal{M}^{\mathbb{Z}}$ that are themselves sequences. To help mitigate this possible source of confusion, we will refer to elements of $\mathcal{M}^{\mathbb{Z}}$ as points, at least in this proof. And (2), this proof will employ a Cantor's diagonal argument. This argument is common in set theory to create a distinct element from an enumeration of all* *possible sets of a type. The most famous Cantor diagonal argument is the one commonly used to show that the set of real numbers in the unit interval is uncountable.*

Proof Let $\mathbf{x} \in \mathcal{M}^{\mathbb{Z}}$ be a point, and consider a sequence $\{\mathbf{x}_n\}_{n \in \mathbb{N}}$. Since $\mathcal{M}$ is a finite set, the sequence $\{\mathbf{x}_n\}$ must contain a(n infinite) subsequence $\{\mathbf{x}_{m_0}\}^0$, all of whose elements agree in the 0th place. We use the superscript only to keep track of our subsequences. Then, using this subsequence, we can create a new subsubsequence, all of whose elements also agree on the -1st and the 1st positions. Call this new subsubsequence $\{\mathbf{x}_{m_1}\}^1$. Continue creating new recursive infinite subsequences $\{\mathbf{x}_{m_i}\}^i$, whose elements all agree on all places from position $-i$ to i. Note here that all of the elements in $\{\mathbf{x}_{m_i}\}^i$ are within $1/2^{i-1}$ of each other. Thus

$$\cdots \subset \{\mathbf{x}_{m_i}\}^i \subset \cdots \subset \{\mathbf{x}_{m_1}\}^1 \subset \{\mathbf{x}_{m_0}\}^0 \subset \{\mathbf{x}_n\}.$$

Now create a new sequence $\{\mathbf{x}_{n_i}\}$ by choosing from each of these the element whose index m_i is the smallest one larger than the one chosen previously. That is, $n_0 = 0$, $n_1 = m_1 > 0$, $n_2 = m_2 > n_1$, and, for each $i \in \mathbb{N}$, choose $n_i = m_i > n_{i-1}$. Then $\{\mathbf{x}_{n_i}\}$ converges exponentially. $\qquad\square$

As in inverse limit spaces, one can describe a basic (left) shift map on $\mathcal{M}^{\mathbb{Z}}$:

$$\sigma : \mathcal{M}^{\mathbb{Z}} \to \mathcal{M}^{\mathbb{Z}}, \quad \sigma(\{\mathbf{x}_n\}) = \{\sigma(\mathbf{x}_n)\} = \{\mathbf{x}_{n+1}\}.$$

One can show that this map is continuous, so we have a discrete dynamical system on $\mathcal{M}^{\mathbb{Z}}$.

Exercise 269 Use the metric in Equation 7.1.3 to show that the shift map $\sigma : \mathcal{M}^{\mathbb{Z}} \to \mathcal{M}^{\mathbb{Z}}$ is continuous.

Now define a *word* associated with $\mathcal{M}$ as simply a finite string of letters in $\mathcal{M}$. Words help us to distinguish certain sequences by comparing a word with finite "blocks" of letters in the sequence. Indeed, for $i \leq j$, denote a block of length $j - i$ elements in a sequence by $(x_i x_{i+1} \ldots x_j)$. Should $i > j$, consider the block empty. Then we can create subsets of sequences by restricting to those that either include or exclude certain words. Let $\mathcal{F}$ be a set of words. Then

$$\mathcal{X}_{\mathcal{F}} = \left\{ \mathbf{x} \in \mathcal{M}^{\mathbb{Z}} \mid (x_i x_{i+1} \ldots x_j) \notin \mathcal{F} \quad \forall i \in \mathbb{Z} \right\}$$

is the subset of all sequences that do not contain the words of $\mathcal{F}$ anywhere in the sequence.

Exercise 270 Show that for any set of words $\mathcal{F}$, $\mathcal{X}_{\mathcal{F}}$ is a closed subset of $\mathcal{M}^{\mathbb{Z}}$ and invariant under the shift map.

Being a subset of a metric space, then, $\mathcal{X}_{\mathcal{F}}$ is also a metric space. And, by Exercise 270, and the fact that a closed subset of a compact space is compact, it follows that $\mathcal{X}_{\mathcal{F}}$ is compact also. And, since any $\mathcal{X}_{\mathcal{F}}$ is invariant under the shift map, one can create a discrete dynamical system on any space of sequences of symbols described by a set of words $\mathcal{F}$ in those symbols with the shift map. Such an $\mathcal{X}_{\mathcal{F}}$ is called a *shift space*:

- The space $\mathcal{M}^{\mathbb{Z}}$ (along with σ) is often called the *full shift space*, although commonly it is just referred to as the shift. In this case, $\mathcal{F}$ is empty.

- Since each shift space $\mathcal{X}_{\mathcal{F}}$ defined by $\mathcal{F}$ is automatically invariant under the shift map, we often refer to the shift space with the shift map collectively as simply a *shift*.

- A *subshift* is σ restricted to a closed, invariant subset of $\mathcal{M}^{\mathbb{Z}}$. If the closed, invariant subset is $\mathcal{X}_{\mathcal{F}}$, where $\mathcal{F}$ is a finite set of words, then $\mathcal{X}_{\mathcal{F}}$ is said to be a *subshift of finite type (SFT)*. These are also examples of what are sometimes called *sofic systems*, a kind of generalization of this finiteness property, which we will not go into here. The term "sofic" was coined by Benjamin Weiss in 1973 [59].

- A subshift of finite type is a *k-step SFT* if the largest word in $\mathcal{F}$ has $k+1$ letters. A 1-step SFT is also called a *topological Markov chain*.

- When the shift space is an invariant subset of $\mathcal{M}^{\mathbb{Z}}$, we refer to the shift as a *two-sided (sub)shift*. In contrast, the shift map on $\mathcal{M}^{\mathbb{N}}$ is known as *one-sided*.

- It is easy to see that the two-sided full (left) shift is invertible: Simply define the *right two-sided full shift* as the map $\sigma^{-1}(\{x_i\}) = \{x_{i-1}\}$. For one-sided shifts, however, "moving to the right" is not well-defined. That said, it is possible to define a one-sided right shift on $\mathcal{M}_n^{\mathbb{N}} = \{0, 1, \ldots, n-1\}$ as the map σ_i^{-1}, where

$$\sigma_i^{-1}(\{x_0, x_1, x_2, \ldots\}) = \{i, x_0, x_1, x_2, \ldots\}.$$

However, be careful to note that this is only considered a *one-sided inverse*, since $\sigma \circ \sigma_i^{-1}$ does equal the identity map on $\mathcal{M}_n^{\mathbb{N}}$, but $\sigma_i^{-1} \circ \sigma$ does not.

Example 7.35 For $\mathcal{M}_2 = \{0, 1\}$, with $\mathcal{F} = \{(11)\}$, the shift map on $\mathcal{X}_{\mathcal{F}} \subset \mathcal{M}_2^{\mathbb{Z}}$ is a topological Markov chain (a two-sided, 1-step SFT) on the space of all sequences $\{x_i\}_{i \in \mathbb{Z}}$ of 0s and 1s that do not contain two consecutive 1s. This shift space is often referred to as the *Golden Mean shift* or the *Fibonacci shift* because of its properties. For example, the number of distinct allowable words of size $n \in \mathbb{N}$ is the nth Fibonacci number.

Exercise 271 Show that this is true.

Example 7.36 Let $\mathcal{M}_3 = \{0, 1, 2\}$, with $\mathcal{F} = \{(1)\}$. Then the shift map on $\mathcal{X}_{\mathcal{F}} \subset \mathcal{M}_3^{\mathbb{N}}$ is a one-sided subshift of finite type on the space of sequences $\{x_i\}_{i \in \mathbb{N}}$ of 0s and 2s. Now consider the map

$$(7.1.5) \qquad h : \mathcal{M}_3^{\mathbb{N}} \to [0, 1], \quad \{x_i\}_{i \in \mathbb{N}} \mapsto (.x_1 x_2 x_3 \ldots)_3 = \sum_{i=0}^{\infty} \frac{x_i}{3^i},$$

where the image is the ternary expansion of elements in the unit interval, as in Equation 3.4.2. From the discussion in Section 3.4, one can easily see that the image of $\mathcal{X}_{\mathcal{F}}$ this map is the ternary Cantor set C. So, what are the properties of h?

Exercise 272 Show that the map $h : \mathcal{M}_3^{\mathbb{N}} \to [0,1]$ in Equation 7.1.5 is continuous and surjective. But is it injective? Show that it is not by checking on sequences detailed in Equation 3.4.3 in the construction of the ternary Cantor set in Section 3.4. Now restrict h to $\mathcal{X}_{\mathcal{F}}$ and show that it is bijective onto its image and, restricted to its image, has a continuous inverse. It is a homeomorphism onto its image!

Exercise 273 Given $\mathcal{X}_{\mathcal{F}} \subset \mathcal{M}_3^{\mathbb{N}} = \{0,1,2\}^{\mathbb{N}}$, with $\mathcal{F} = \{(1)\}$, as in Example 7.36, show that the map from $\mathcal{X}_{\mathcal{F}}$ to $\mathcal{M}_2^{\mathbb{N}} = \{0,1\}^{\mathbb{N}}$, which replaces every 2 with a 1, is a homeomorphism. Thus $\mathcal{M}_2^{\mathbb{N}}$ is a Cantor set.

Example 7.37 Let $\mathcal{X}_{\mathcal{F}} \subset \mathcal{M}_2^{\mathbb{N}} = \{0,1\}^{\mathbb{N}}$ be the set of all sequences with at most one 1. It is certainly closed and invariant under the shift map. However, it cannot be described via a finite set of forbidden words, and hence $\mathcal{X}_{\mathcal{F}}$ is a subshift, but not of finite type.

The construction of these shift spaces plays a central role in a field of dynamical systems called *symbolic dynamics*, and provides a set of models for common, complicated behavior.

For example, in a shift space, it is easy to determine whether a point is periodic under the shift map, and to determine what is its minimal period. Simply identify the (bi-)infinite sequences that are just (bi-)infinite concatenations of a finite-length word: A point $\mathbf{z} \in \mathcal{M}^{\mathbb{N}}$ is n-periodic under the shift map if there exists a word $w_n = (x_0 x_1 \cdots x_{n-1})$ of length n letters (we will call w_n here an n-word) such that

$$(7.1.6) \qquad \mathbf{z} = \{x_0 x_1 \ldots x_{n-1} x_0 x_1 \ldots x_{n-1} x_0 x_1 \ldots\} = \{w_n w_n w_n \ldots\}.$$

And counting these is only a matter of determining the number of allowable words of length n in the shift space.

Proposition 7.38 *Periodic points are dense for the shift map on $\mathcal{M}_m^{\mathbb{N}}$ and $P_n(\sigma) = m^n$.*

Proof We prove this for $\mathbb{N}$, noting that the case for bi-infinite sequences is entirely the same. Let $\mathbf{x} \in \mathcal{M}_m^{\mathbb{N}}$ and choose $\rho > m$. Then, for $n \in \mathbb{N}$, the open set

$$B_{1/\rho^n}(\mathbf{x}) = \left\{ \mathbf{y} \in \mathcal{M}_m^{\mathbb{N}} \mid d(\mathbf{x}, \mathbf{y}) < \frac{1}{\rho^n} \right\},$$

contains the sequences that agree with $\mathbf{x}$ up to and including the nth place. But the sequence in Equation 7.1.6 given by $\mathbf{z} = \{w_n w_n w_n \ldots\}$, where $w_n = (x_0 x_1 \ldots x_n)$ are the letters of $\mathbf{x}$ up to position n, is a periodic sequence inside $B_{1/2^n}(\mathbf{x})$. As this will be true for any $n \in \mathbb{N}$, it shows that arbitrarily close to any sequence is a periodic point. This establishes the density of periodic points. And, as for counting n-periodic points, one need only count the set of length-n words in m-symbols. □

Exercise 274 Now show that there exists a dense orbit for the shift map on $\mathcal{M}_m^{\mathbb{N}}$. Construct a sequence $\mathbf{s}$ by concatenating all distinct 1-words together, then all 2-words to the end of that, then all 3-words, and so on, inductively. Then show that for any sequence $\mathbf{t} \in \mathcal{M}_m^{\mathbb{N}}$, there will be an iterate $\sigma^n(\mathbf{s})$ arbitrarily close to $\mathbf{t}$. This shows that the shift map on $\mathcal{M}_m^{\mathbb{N}}$ is topologically transitive.

The case of a subshift is a bit more complicated, since if we are disallowing certain words, then using perfectly acceptable words to create periodic sequences may be problematic. For example, in a subshift created for $\mathcal{F} = \{(11)\}$, we cannot concatenate words that begin and end in a 1 to create periodic sequences. However, when a subshift is created by disallowing all words containing a subset of letters, the above result also holds, where m is simply the number of allowed letters. We have the following:

Corollary 7.39 Let $\mathcal{F} = \{(x_{i_1}), (x_{i_2}), \ldots, (x_{i_k})\}$ be a set of k distinct letters in $\mathcal{M}_m$. Then the periodic points are dense for the shift map on $\mathcal{X}_{\mathcal{F}} \subset \mathcal{M}_m^{\mathbb{K}}$, for $\mathbb{K} = \mathbb{N}$ or $\mathbb{Z}$. Further, $P_n(\sigma) = (m-k)^n$.

Example 7.40 In Example 7.36, the subshift $\mathcal{X}_{\mathcal{F}}$ consists of all points whose sequences do not contain the number 1 in $\mathcal{M}_3$. Also, by Exercise 272, there is a homeomorphism between $\mathcal{X}_{\mathcal{F}}$ and the ternary Cantor set $C \subset [0,1]$. So, is there a map on C whose orbits would coincide through h with the shift? Consider the map $f : C \to C, f(x) = 3x \mod 1$. Here, in ternary expansion,

$$f((.x_0 x_1 x_2 \ldots)_3) = (x_0.x_1 x_2 \ldots)_3 \mod 1 = (.x_1 x_2 \ldots)_3.$$

So,

$$f \circ h(\mathbf{x}) = f \circ h(\{x_i\}_{i=0}^\infty) = f((.x_0 x_1 x_2 \ldots)_3) = (.x_1 x_2 \ldots)_3.$$

But also

$$h \circ \sigma(\mathbf{x}) = h \circ \sigma(\{x_i\}_{i=0}^\infty) = h(\{x_i\}_{i=1}^\infty) = (.x_1 x_2 \ldots)_3.$$

Hence orbits go to orbits, and this is a topological conjugacy. So, what does this say about $\mathcal{M}_2^{\mathbb{N}}$, given the homeomorphism from Exercise 273?

For more general $\mathcal{F}$, we can encode allowable and forbidden words via a matrix that captures not just allowable words, but also other essential dynamic information about the shift map on a subshift. For now, let $\mathcal{F}$ be a set of forbidden 2-words, so that $\mathcal{X}_{\mathcal{F}}$ is a topological Markov chain.

Definition 7.41 Let $\mathcal{X}_{\mathcal{F}}$ be a topological Markov chain in $\mathcal{M}_m^{\mathbb{K}} = \{0, 1, \ldots, m-1\}^{\mathbb{K}}$, for $\mathbb{K} = \mathbb{N}$ or $\mathbb{Z}$, where all words in $\mathcal{F}$ are 2-words. The $m \times m$ matrix $T_{\mathcal{F}}$ whose ij-entry is 1 if the block ij is admissible, $(ij) \notin \mathcal{F}$, and 0 if ij is forbidden, $(ij) \in \mathcal{F}$, is called the transition matrix.

Some ready examples include the full shift: Let $\mathcal{F} = \emptyset$. Then all 2-words are admissible, $\mathcal{X}_{\mathcal{F}} = \mathcal{M}_2^{\mathbb{N}}$, and $T_{\mathcal{F}} = \begin{bmatrix} 1 & 1 \\ 1 & 1 \end{bmatrix}$. For the Golden Mean subshift, where $\mathcal{F} = \{(11)\}$, we have $T_{\mathcal{F}} = \begin{bmatrix} 1 & 1 \\ 1 & 0 \end{bmatrix}$. And for $\mathcal{F} = \{(11),(22),(23)\}$ on $\mathcal{M}_3^{\mathbb{N}}$, we have

$$T_{\mathcal{F}} = \begin{bmatrix} 0 & 1 & 1 \\ 1 & 0 & 0 \\ 1 & 1 & 1 \end{bmatrix}.$$

These transition matrices are powerful tools for understanding subshifts. For example, if $t_{ij} = 1$, then the word (ij) is admissible, but we do not yet know if there is a periodic sequence corresponding to the infinite concatenation of (ij) with itself. However, when $t_{ii} = 1$, we know that the letter i can follow the letter i. Thus there will be a fixed point of the shift map, namely, $\mathbf{x} = \{iii\ldots\}$. Hence it is immediate that the trace of the transition matrix $T_{\mathcal{F}}$ counts the number of fixed points of the shift map. In the last example immediately above, the sole fixed point of the subshift is the point $\{333\ldots\}$, since the other two possible ones are not in the subshift.

Similarly, if we wanted to count the number of admissible 3-words in $\mathcal{X}_{\mathcal{F}}$ that start with i and end with j, say, we can simply take any admissible 2-word (ik), and look to see if the word (kj) is also admissible. If so, then (ikj) is admissible. Do this for each letter k. Then the number of such admissible 3-words that start with i and end with j can be calculated from the transition matrix as

$$\sum_{k=1}^{n} t_{ik} \cdot t_{kj}.$$

But this is simply the ij-element of $T_{\mathcal{F}}^2 = T_{\mathcal{F}} \cdot T_{\mathcal{F}}$. Hence $T_{\mathcal{F}}^2$ documents precisely the number of admissible words of length 3. One can then extrapolate so that the ijth element of $T_{\mathcal{F}}^n$ is just the number of admissible words of length $n + 1$ that begin with i and end with j. And then, since an admissible $(n + 1)$-word that begins and ends with the same letter corresponds to a n-periodic sequence (the one whose first n letters repeat infinitely often), we can also conclude immediately that the trace of $T_{\mathcal{F}}^n$ is the number of n-periodic points of the subshift. All of this from the matrix T.

Example 7.42 One can list admissible words for the Golden Mean subshift: (0) corresponds to a fixed point under the shift map, but (1) does not, since (11) is forbidden. (00), (01), and (10) correspond to period-2 points; there are two that begin with 0 and one that begins with 1. Each of (000), (001), (010), and (100) corresponds to a period-3 point, three of which start with 0 and one with 1. And so on ....

Exercise 275 Show that for $a_n = \mathrm{tr}\, T^n$, with T the transition matrix of the Golden Mean subshift, that $\{a_n\}$ is a *Lucas sequence*: a sequence $\{L_n\}$ such that $L_0 = 2$, $L_1 = 1$, and $L_n = L_{n-1} + L_{n-2}$. What is the asymptotic growth rate of $P_n(\sigma)$ as a function of n? Note that the Lucas sequence follows the same recursion as the Fibonacci sequence, but has different starting values.

Example 7.43 For the subshift $\mathcal{X}_{\mathcal{F}} \subset \mathcal{M}_3^{\mathbb{N}}$, where $\mathcal{M} = \{1, 2, 3\}$ and $\mathcal{F} = \{(11), (22), (23)\}$, the early powers of the transition matrix are

$$T_{\mathcal{F}} = \begin{bmatrix} 0 & 1 & 1 \\ 1 & 0 & 0 \\ 1 & 1 & 1 \end{bmatrix}, \quad T_{\mathcal{F}}^2 = \begin{bmatrix} 2 & 1 & 1 \\ 0 & 1 & 1 \\ 2 & 2 & 2 \end{bmatrix}, \quad T_{\mathcal{F}}^3 = \begin{bmatrix} 2 & 3 & 3 \\ 2 & 1 & 1 \\ 4 & 4 & 4 \end{bmatrix}.$$

Here, there is only one fixed point corresponding to (3). For period-2 points, we have the words (12) and (13) that start with the letter 1, only (21) starts with the letter 2, and the two (31) and (33) that start with the letter 3. Note that there is no period-2 point corresponding to (32), since the word (23) is forbidden.

Exercise 276 For the subshift in Example 7.43, list the seven period-3 points, and calculate the number of period-4 and period-5 points. Then show that the number of period-n points satisfies a second-order recurrence. Then use the idea from Exercise 3 to show that the asymptotic growth rate for periodic points in this subshift is 2.

And lastly, the question of the denseness of periodic points in a subshift is a bit tricky. To address this, let $T_{\mathcal{F}}$ be a transition matrix for a subshift. Then each entry is either 0 or 1. We can call such a matrix a *non-negative matrix* (all of its entries are non-negative real numbers). A matrix of 0s and 1s is called *transitive* if there is a natural number $n \in \mathbb{N}$ such that $T_{\mathcal{F}}^n$ has all entries strictly positive. That is, a transition matrix is transitive if it has a natural number power where it is a *positive matrix*. We have the following:

Proposition 7.44 *If T is a transition matrix and there is a natural number n where T^n is positive, then T^m is positive for all natural numbers $m > n$.*

Exercise 277 Prove Proposition 7.44.

We note here that a square, non-negative matrix that has a power that is positive is also called a *primitive matrix*. This will be a very important feature of a transition matrix, as we will see.

Proposition 7.45 *If a subshift $\mathcal{X}_{\mathcal{F}}$ has a transitive transition matrix, then the set of periodic points are dense.*

Proof Given a subshift with a transitive transition matrix, choose $m > 0$ such that T^m is positive. Now, for a sequence $\mathbf{x} \in \mathcal{X}_{\mathcal{F}}$, where $\mathbf{x} = \{x_i\}_{i=0}^{\infty}$, let $w_n = (x_0 x_1 \ldots x_{n-1})$ be a word of length $n > m$ that agrees with $\mathbf{x}$ on the first n letters of the sequence. If the last letter followed by the first letter of w_n form an admissible word $(x_{n-1} x_0)$, then there exists a periodic sequence formed by infinite concatenation of w_n with itself $\mathbf{w}_n = \{w_n w_n w_n \ldots\}$ which is $(1/2^n)$ close to the original sequence. Suppose that this is not the case, and $(x_{n-1} x_0) \in \mathcal{F}$. Then choose an m-word $v_m = (x_n y_0 y_1 \ldots y_{m-2})$ whose first letter is the $(n+1)$th letter in the sequence $\mathbf{x}$, and whose last letter forms the admissible 2-word, $(y_{m-2} x_0)$. Since $T_{\mathcal{F}}^m$ is positive, this is always possible. Then the $(m+n)$-word $w_n v_m$ corresponds to an $(m+n)$-periodic sequence $\mathbf{w}_n = \{w_n v_m w_n v_m \ldots\}$ that is $(1/2^n)$-close to $\mathbf{x}$. And since this is true for all $n \in \mathbb{N}$, the result follows. $\square$

Note that the theory and examples herein refer only to subshifts based on admissible and forbidden words of length 2. This is the "Markov" part of the construction: that the admissability of a word is based only on the current letter and a successor, and not on previous letters in a sequence. To create subshifts based on forbidden words of length greater than 2, the admissability would depend not just on the current letter, but on a string of letters prior to it. This fails the notion of a *memoryless process*.

However, we can transform the process to a construction that is, in effect, memoryless. Indeed, let a forbidden word have length $m > 2$. To untangle the reliance on previous states, create a new alphabet each of whose letters corresponds to admissible words of length $m-1$ in the previous alphabet. Then the concatenation of two new letters will correspond to a word of length $2m-2$ in the old alphabet. Using this, create a transition matrix whose non-zero entries will correspond precisely to the admissible 2-words in the new alphabet, where admissability corresponds to whether the $(2m-2)$-word is admissible in the original alphabet. This new subshift with be memoryless, now a topological Markov chain, and will carry the same information as the original.

Remark 7.46 *This is another form of a "reduction of order" endeavor similar to that of transforming a high-order recursion into a first-order vector recursion from Chapter 4 (See Remark 4.22 in Section 4.3.5). Here, creating new letters based on words in another alphabet (syllables?) not so much removes as encodes the previous states into the new forbidden set $\mathcal{F}$ in the new alphabet. Hence, to analyze the dynamic behavior of finite subshifts, we need only consider forbidden sets of 2-words.*

Example 7.47 Let $\mathcal{F} = \{(111)\}$ in $\mathcal{M}_2^{\mathbb{N}} = \{0,1\}^{\mathbb{N}}$. Here, consider a new alphabet $\mathcal{N} = \{A,B,C,D\}$, where $A=(00)$, $B=(01)$, $C=(10)$, and $D=(11)$ are all admissible. Then the matrix of 2-words in the new alphabet corresponds to

$$\begin{bmatrix} AA & AB & AC & AD \\ BA & BB & BC & BD \\ CA & CB & CC & CD \\ DA & DB & DC & DD \end{bmatrix} = \begin{bmatrix} 0000 & 0001 & 0010 & 0011 \\ 0100 & 0101 & 0110 & 0111 \\ 1000 & 1001 & 1010 & 1011 \\ 1100 & 1101 & 1110 & 1111 \end{bmatrix}.$$

We can readily see which 4-words in the old alphabet are forbidden, and we simply forbid these in the new alphabet, to get

$$T_{\mathcal{F}} = \begin{bmatrix} 1 & 1 & 1 & 1 \\ 1 & 1 & 1 & 0 \\ 1 & 1 & 1 & 1 \\ 1 & 1 & 0 & 0 \end{bmatrix}, \quad \mathcal{F} = \{(BD),(DC),(DD)\}.$$

We note here that $T_{\mathcal{F}}$ here is transitive, since $T_{\mathcal{F}}^2$ is positive (do this calculation), and that $(T_{\mathcal{F}}^2)_{44} = 1$. This means that the only 6-word in the original alphabet that starts and ends with 11 is the word $(11\_\_11) = (110011)$. This corresponds to the word DAD in the new alphabet.

Exercise 278 Given the subshift in Example 7.47, calculate how many admissible 8-words there are that begin with the word 01 and end with 11, and then list them. Equivalently, calculate how many admissible 4-words there are that start with B and end with D.

7.1.6 Markov Partitions

A nice application of a dynamical system involving shifts and shift spaces, along with their shift maps, involves the study of a dynamical system where we record not necessarily where an orbit is at any particular moment in time, but rather what region, among a finite number of regions of the domain, an orbit visits. If these regions are chosen carefully with regard to the map, then the sequences of regions an orbit visits can help to distinguish orbits. This is a coarser study of a map on a space, which can uncover a vast amount of information about the dynamical structure of a map, and will allow us to use these shift spaces and symbolic dynamics as a tool for study.

To begin, let X be a topological space and $\mathcal{A} = \{A_i\}$ an arbitrary collection of subsets of X. We say that $\mathcal{A}$ is a *cover* of X, or that $\mathcal{A}$ *covers* X, if

$$X = \bigcup_i A_i.$$

If all of the A_i are closed sets, then the cover is called closed, and if there are only a finite number of elements in $\mathcal{A}$, we call the cover a *finite cover*.

Definition 7.48 *A (topological) partition of a space X is a finite set $\mathcal{P} = \{P_1, \ldots, P_n\}$ of disjoint open subsets of X, whose closures form a finite closed cover of X. That is, for $i \neq j$, $P_i \cap P_j = \emptyset$, and $X = \bigcup_i \overline{P_i}$.*

Some notes:

- In essence, a partition $\mathcal{P}$ is a finite set of elements of the topology of X where each $x \in X$ is contained in at least one element of the closed cover formed by $\mathcal{P}$. And, here, the elements of $\mathcal{P}$ only overlap on their boundaries. Note that this does mean that there will be points in the closures of a partition that are not in any element of the partition. For our purposes, we will have to deal with these points separately.

- We use open sets as the elements of the partition so that the elements reflect the topological properties of the space. Hence, for spaces like those described in Chapter 3, or Euclidean space, the closed sets of a closed cover of X usually comprise closed sets with non-empty interior, or the closure of elements of the topology on X. To force this to be the case, partition elements are defined as open disjoint sets, and the cover consists of their closures.

- Note that this is different from the idea of a set partition in Definition 3.42 of Section 3.3.2, where the partition elements were mutually exclusive and collectively exhaustive. There, when defining an equivalence relation on a set, any possible topology defined on the set was not considered.

- You have seen topological partitions in your first-semester calculus course: when designing a Riemann sum to approximate a definite integral of a function over an interval $I = [a, b]$, $b > a$, one would choose a finite set of points x_i, $i = 0, \ldots, n$, in I such that

$$a = x_0 < x_1 < x_2 < \ldots < x_{n-1} < x_n = b.$$

There, the set of closed intervals $[x_i, x_i + 1]$, for $i = 1, \ldots, n - 1$ was called a partition of I. In Definition 7.48, we alter the first-semester calculus definition of a partition to say that the open intervals (x_i, x_{i+1}) form the partition, and their closures play the role of the closed intervals used in the Riemann sum.

- When $\mathcal{P}$ partitions a space X, we say that $(X, \mathcal{P})$ is a partitioned space.

For the moment, let X be $I = [0, 1]$, the unit interval, with a partition $\mathcal{P}$ and let $f : I \to I$ be a continuous map.

Definition 7.49 *We say a map $f : I \to I$ respects $\mathcal{P}$ if $\forall i \in \{1, 2, \ldots, n\}$,*

$$f(P_i) = \bigcup_{j=1}^{n} \delta_{ij} P_j, \text{ where } \delta_{ij} = \begin{cases} 1 & \text{if } \exists x \in P_i \text{ such that } f(x) \in P_j, \\ 0 & \text{otherwise.} \end{cases}$$

Here, when $(I, \mathcal{P})$ is a partitioned space and f respects the partition, we say that f has the *Markov condition* and will call $\mathcal{P}$ a *Markov partition* for f on I. Essentially, a map respects a partition if the boundary points of the partition elements are mapped to boundary points. This does imply, though, that if the image of a partition element spans more than one partition element, a point inside the image will be mapped to the boundary points. However, there will only be a countable number of such points in total, and for almost every point inside a partition element, its full (forward) orbit will remain off of the set of boundary points. Thus we can record the orbit by a sequence in the letters assigned to the partition elements. In fact, we can study the map and its effect on the partition via a matrix that catalogs how the images of the partition elements intersect the partition elements:

Definition 7.50 *For $\mathcal{P}$ a Markov partition for f on I, the* transition matrix *is an $n \times n$ matrix T_f where $t_{ij} = \delta_{ij}$.*

In effect, t_{ij} is the cardinality of the set $\{f(P_i) \cap P_j\}$, so that

(7.1.7)
$$T_f = \begin{bmatrix} \#\left(f(P_1) \cap P_1\right) & \#\left(f(P_1) \cap P_2\right) \\ \#\left(f(P_2) \cap P_1\right) & \#\left(f(P_2) \cap P_2\right) \end{bmatrix}.$$

Example 7.51 Consider the map $f : I \to I$ given by

$$f(x) = \begin{cases} -2x + 1, & x \leq \frac{1}{2}, \\ x - \frac{1}{2}, & x > \frac{1}{2}, \end{cases}$$

whose graph is on the left in Figure 146. Let $P_1 = \left(0, \frac{1}{2}\right)$ and $P_2 = \left(\frac{1}{2}, 1\right)$, so that $\mathcal{P} = \{P_1, P_2\}$, satisfies the Markov condition. The transition matrix is then $T_f = \begin{bmatrix} 1 & 1 \\ 1 & 0 \end{bmatrix}$, since P_1 is mapped onto I, while P_2 is only mapped onto P_1. Via the graph of f^2 in the center of Figure 146, we can easily calculate $T_{f^2} = \begin{bmatrix} 2 & 1 \\ 1 & 1 \end{bmatrix}$. Notice that, at least here, $T_{f^2} = \left(T_f\right)^2 = T_f^2$. This is no coincidence. Indeed, for example, one can see readily that the number of times the image of P_1 under f^2 intersects P_1 is equal to the number of pieces of $f(P_1)$ in P_1, counted for each piece of $f(P_1)$ in P_1, plus the number of pieces of $f(P_1)$ in P_2, counted for each piece of $f(P_2)$ in P_1. Thus

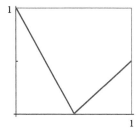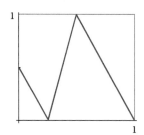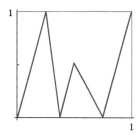

Figure 146 Graphs of f, f^2, and f^3 from Example 7.51.

$$\#\left(f^2(P_1) \cap P_1\right) = \#\left(f(P_1) \cap P_1\right) \cdot \#\left(f(P_1) \cap P_1\right)$$
$$+ \#\left(f(P_1) \cap P_2\right) \cdot \#\left(f(P_2) \cap P_1\right).$$

This will be true in general, so

$$T_{f^n} = T_f^n \quad \forall n \in \mathbb{N}.$$

The effect is that the properties of T_f reflect important aspects of the map f and hence of the dynamics.

Exercise 279 Verify that T_{f^3} is compatible with the graph of f^3 on the right side of Figure 146.

Markov partitions are ways of encoding information about orbits without tracking the precise points in the orbit. In essence, one divides a domain into a finite number of pieces, and then records only the piece an orbit visits at each iterate, a natural number or integer. Now this only works when the map takes each partition element onto a union of other partition elements: partition elements must map to unions of partition elements. And the precise information of this mapping is contained within the $n \times n$ transition matrix, where n is the cardinality of the partition. But this coarse record of orbit behavior can illustrate a lot of dynamical information, as we will see.

Remark 7.52 *The Markov property refers to the "memoryless" property of many stochastic processes in probability theory. The probability of an outcome conditioned on history is equal to conditioning only on the present state. Hence only the current position in the partition determines the future orbit.*

Example 7.53 Let $I = [0, 1]$ and $f : I \to I$ be defined as

$$f(x) = \begin{cases} 3x + \frac{1}{4}, & 0 \le x \le \frac{1}{4}, \\ \frac{4}{3} - \frac{4}{3}x, & \frac{1}{4} \le x \le 1. \end{cases}$$

See Figure 147. Here we create $\mathcal{P} = \{\mathcal{P}_1, \mathcal{P}_2\}$, where $\mathcal{P}_1 = \left[0, \frac{1}{4}\right]$ and $\mathcal{P}_2 = \left[\frac{1}{4}, 1\right]$. Then $\mathcal{P}$ is a Markov partition for f on I, since $f(\mathcal{P}_1) = \mathcal{P}_2$ and $f(\mathcal{P}_2) = \mathcal{P}_1 \bigcup \mathcal{P}_2$. The transition matrix is then $A = \begin{bmatrix} 0 & 1 \\ 1 & 1 \end{bmatrix}$. And f on I is a discrete dynamical system that

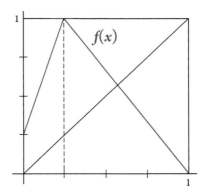

Figure 147 The graph of $f(x)$.

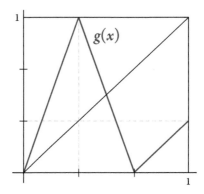

Figure 148 The graph of $g(x)$.

respects $\mathcal{P}$. One can say that, symbolically, f corresponds to the one-sided left subshift on the sequence space $\mathcal{M}^{\mathbb{N}}$, where $\mathcal{M} = \{0, 1\}$, where the invariant subset of allowable sequences consists of those with no consecutive 0's in them.

The number of elements in the partition dictates the "size" of the sequence space, and accordingly the size of the transition matrix. That the map f above is not injective will play a crucial role later, as we will see. For now, contrast this with another example:

Example 7.54 Let $g : I \to I$ be given as in Figure 148. Then a Markov partition for g is $\mathcal{P} = \{\mathcal{P}_1, \mathcal{P}_2, \mathcal{P}_3\}$, where $\mathcal{P}_1 = \left[0, \frac{1}{3}\right]$, $\mathcal{P}_2 = \left[\frac{1}{3}, \frac{2}{3}\right]$, and $\mathcal{P}_3 = \left[\frac{2}{3}, 1\right]$. The transition matrix here is $A = \begin{bmatrix} 1 & 1 & 1 \\ 1 & 1 & 1 \\ 1 & 0 & 0 \end{bmatrix}$. Here, g corresponds to the left shift map on the subshift of finite type on $\{0, 1, 2\}^{\mathbb{N}}$ characterized by the forbidden blocks $\mathcal{F} = \{(23), (33)\}$.

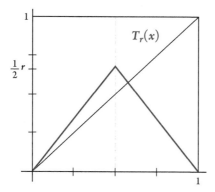

Figure 149 The tent map $T_r(x)$.

Exercise 280 Create a piecewise linear C^0 map on I where $\mathcal{F} = \{(11), (22), (23)\}$, and write out the transition matrix A.

When each element of a partition maps to the union of all of the elements of the partition, there are no forbidden blocks. Here, the subshift is a full shift on $\mathcal{M}$.

Example 7.55 Consider the map on I given by

$$T_r(x) = \begin{cases} rx, & 0 \le x \le \frac{1}{2}, \\ r(1-x), & \frac{1}{2} \le x \le 1, \end{cases}$$

The graph of $T_r(x)$ is called the *tent map* on the unit interval I, as seen in Figure 149. You were introduced to one of this family of maps in Example 7.14 as a unimodal map in Section 7.1.1 when discussing expanding maps, as well as in Exercise 231 in Section 6.3.1 on non-wandering points. Tent maps are also sometimes called *sawtooth functions*. The maximum height, at $x = \frac{1}{2}$, is $\frac{1}{2}r$. The *full tent map* $T_2(x)$, given the obvious partition, seems to share the same transition matrix as the full left shift on the sequence space $\{0, 1\}^\mathbb{N}$; since both partition intervals have closures that are mapped onto I, there are no forbidden blocks here. A slight alteration of notation reveals another interesting fact. Write each sequence as a binary expansion via the map

(7.1.8) $\quad h : \{0, 1\}^\mathbb{N} \to [0, 1], \quad h(\mathbf{x}) = h\left(\{x_i\}_{i \in \mathbb{N}}\right) = (0.x_1x_2x_3\ldots)_2 = \displaystyle\sum_{i=1}^{\infty} \frac{x_i}{2^i}.$

Then the shift map looks like the doubling map $2x$ on unit interval points, at least for $x \in \left[0, \frac{1}{2}\right)$, since the binary expansion for point on this half interval has $x_1 = 0$, so

$$T_2 \circ h(\{x_i\}) = 2 \cdot (.0x_2x_3\ldots)_2 = (0.x_2x_3x_4\ldots)_2$$

and

$$h \circ \sigma(\{x_i\}) = h(\{x_{i+1}\}) = (0.x_2x_3x_4)_2.$$

However, on the other half interval, where the binary expansion of point has $x_1 = 1$, we have

$$T_2 \circ h(\{x_i\}) = T_2 (0.1x_2x_3 \ldots)_2 = 2(1 - (0.1x_2x_3x_4 \ldots)_2) = (0.\tilde{x}_2\tilde{x}_3\tilde{x}_4 \ldots)_2,$$

where the tilde indicates the 2s-complement. Here, h is a continuous surjective map. But it is yet unclear whether it takes orbits to orbits, and can serve as a semiconjugacy.

This last example reveals a close relationship between Markov maps on intervals, like the examples above, and finite subshifts in symbolic dynamics: One can expose this relationship via the *itinerary map*:

Definition 7.56 *Let $\mathcal{P} = \{P_0, \ldots, P_{n-1}\}$ be a finite set partition of a space X with a map $f : X \to X$. Then the map $\iota : X \to \mathcal{M}_n^{\mathbb{N}}$, $\iota(x) = \{\iota_0\iota_1 \ldots\}$, where $\iota_j = k$ iff $f^j(x) \in P_k$, for $j \in \mathbb{N}$, is called the (forward) itinerary map of f on X, given $\mathcal{P}$. If f is invertible, then $\iota : X \to \mathcal{M}_n^{\mathbb{Z}}$ is the (full) itinerary map. And, given a point $x \in X$, its itinerary is its image under ι.*

Some notes:

- It should be easy to see that if $\iota(x) = \{\iota_0\iota_1\iota_2 \ldots\}$, then

$$\iota\big((f(x)\big) = \{\iota_1\iota_2 \ldots\} = \sigma\left(\{\iota_0\iota_1\iota_2 \ldots\}\right),$$

so that $\iota \circ f = \sigma \circ \iota$ and the image $\iota(X)$ is invariant under the shift map σ. Do you recognize the role ι can play in understanding the dynamics of these two dynamical systems? Indeed, depending on the properties of ι, the dynamics of f on X and σ are often very closely linked.

- Evidently, one can show that if f is continuous, then so is ι under certain conditions on the sets on which it is defined—for example, if each partition element is both open and closed, as in the case where X itself consists of a finite number of connected components and each partition element is one such component. More generally, one can choose a finite collection of disjoint subsets of X that do not necessarily cover X (like the open sets of a Markov partition). Then ι is well-defined on all points whose orbits stay in the union of this collection of subsets. We will see this shortly.

- For a topological partition like our Markov partitions above, where we consider the closures of partition elements and allow for overlap on the edges, the map ι is not well-defined on the overlaps. However, this set is small, and one can effectively discount this (countable) set when discussing the general dynamics of the systems.

Example 7.57 For the full tent map T_2 in Example 7.55, with $I_0 = \left[0, \frac{1}{2}\right]$ and $I_1 = \left[\frac{1}{2}, 1\right]$, the itinerary map may have either $\iota\left(\frac{1}{2}\right) = \{01000\ldots\}$ or so $\iota\left(\frac{1}{2}\right) = \{11000\ldots\}$, an indication that ι is not well-defined as a map (it is not a function). One can resolve this issue by either (1) making a choice in which subinterval the critical point resides, effectively creating a set partition, or (2) leaving out the critical point and all of its pre-images from the domain of the itinerary map. In this second option, one considers only the orbits that completely reside within the interiors of I_1 and I_2. On these sets, the itinerary map is well-defined, and, under mild conditions, establishes a bijection between the interval map with the Markov partition, and a subshift of finite type.

There is a third option to resolve the issue of the itinerary map not being well-defined on a subset of the interval: look at the itinerary map in the other direction, where we associate with each infinite sequence of allowable symbols (in a subshift of finite type) a point in an interval under a map f.

Proposition 7.58 *Let $f : I \to I$ be a continuous map on $I = [0,1]$ with a Markov partition $\mathcal{P} = \{P_0, \ldots, P_{n-1}\}$ that is expanding with factor $\beta > 1$ and transition matrix T_f. And let $\mathcal{X}_{\mathcal{F}}$ be the finite subshift with the same transition matrix $T_{\mathcal{F}} = T_f$. Then every allowable n-word $v_n = (s_0 s_1 \ldots s_{n-1})$ corresponds to a nonempty closed interval $I_{v_n} \subset I$ and any infinite allowable sequence $\mathbf{v} \in \mathcal{X}_{\mathcal{F}}$ corresponds to a unique point $\kappa(\mathbf{v}) \in I$, with the map $\kappa : \mathcal{X}_{\mathcal{F}} \to I$ both continuous and onto.*

First, a couple of notes:

- A Markov partition $\mathcal{P}$ on a closed interval I with a map f is *expanding* with expansion factor $\beta > 1$ if f is continuous and surjective on I, f is C^1 on each P_i, and $|f'(x)| \geq \beta$ for all $x \in P_i$. Note that the map f in Example 7.53, as well as the full tent map T^2, have expansive Markov partitions, while the map $g(x)$ in Example 7.54 and the map f in Example 7.51 are not expanding.

- The map κ is the reverse relation to the itinerary map ι. The fact that it is onto, but not necessarily one-to-one, reflects the fact that ι may not be well-defined at certain points, but still establishes a connection between the orbit structures of the two dynamical systems.

- With this proposition, maps with expanding Markov partitions are topologically semi-conjugate to finite subshifts, establishing a close relationship between their dynamics. We will explore the idea of a semiconjugacy in Chapter 8.

Before proving this proposition, we will need a bit more structure. To this end, let $I_i = \overline{P_i}$, $i = 0, \ldots, n-1$, for a piecewise linear map f on the unit interval $I = [0,1]$ with an expanding Markov partition $\mathcal{P} = \{P_0, \ldots, P_{n-1}\}$. Denote the symbols of $\mathcal{X}_{\mathcal{F}}$ by $0, 1, \ldots, n-1$. Now, if (ij) is an admissible word, then $f(I_i) \cap I_j \neq \emptyset$ (so then $(T_f)_{ij} \neq 0$). Recall Definition 7.50). In this case, then, define

$$I_{ij} = I_i \cap f^{-1}(I_j).$$

In essence, I_{ij} is that segment of I_i that gets mapped onto I_j. Recall that even for f non-invertible, the inverse image of a set is still well-defined. We find immediately that $f(I_{ij}) = I_j$, and that (obviously) $I_{ij} \subset I_i$. Going one step further, we can say that

$$I_{ijk} = I_i \cap f^{-1}(I_{jk}) = I_i \cap f^{-1}\left(I_j \cap f^{-1}(I_k)\right) = I_i \cap f^{-1}(I_j) \cap f^{-2}(I_k)$$

and that this closed interval is non-empty iff (ijk) is an admissible word. In general, we can say

$$I_{ij...k\ell} = I_i \cap f^{-1}\left(I_{j...k\ell}\right),$$
$$f(I_{ij...k\ell}) = I_{j...k\ell}, \text{and}$$
$$I_i \supset I_{ij} \supset \cdots \supset I_{ij...k} \supset I_{ij...k\ell} \supset \cdots$$

and each closed interval in this nested sequence is nonempty. For example, below in Figure 150 is the full tent map T_2. One can see the intervals I_0 and I_1, along with $I_{01} = I_0 \cap f^{-1}(I_1)$ and

$$I_{010} = I_0 \cap f^{-1}(I_{10}) = I_0 \cap f^{-1}(I_1 \cap f^{-1}(I_0)) = I_0 \cap f^{-1}(I_1) \cap f^{-2}(I_0).$$

Further, one can see the nested sequence of intervals $I_0 \supset I_{01} \supset I_{010}$.

Proof To start, for any $I_i = [x_1, x_2]$, the length of its image $\ell(f(I_i)) = (x_2 - x_1)f'(x_3)$ for any $x_3 \in (x_1, x_2)$. Hence $\ell(I_i) \leq \beta^{-1}\ell(f(I_i))$ for β the expansion factor of the Markov partition. Denote by L the length of the largest partition element. Then, for $I_{ij} = I_i \cap f^{-1}(I_j)$, we have $f(I_{ij}) = I_j$, so that

$$\ell(I_{ij}) \leq \beta^{-1}\ell(I_j) \leq L\beta^{-1}.$$

By induction, then, we can generalize to say

$$\ell(I_{ijk...l}) \leq \beta^{-1}\ell(I_{jk...l}) \leq \beta^{-2}\ell(I_{k...l}) \leq \cdots \leq \beta^{-m}\ell(I_l) \leq L\beta^{-m},$$

for the m-word $(ijk...l)$. This immediately shows that for $\mathbf{v}$ an infinite admissible sequence, the nested sequence of closed, non-empty intervals has non-empty intersection of a single point.

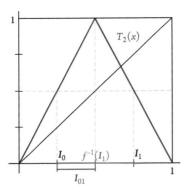

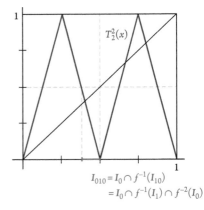

Figure 150 The intervals I_0, I_{01}, and I_{010} of the full tent map.

So, now construct the relation κ that takes infinite admissible sequences, points in $\mathcal{X}_{\mathcal{F}}$, and associates with each a point in I. Since each admissible sequence corresponds to a unique point, the relation is a function. Is the map κ continuous? Let $\epsilon > 0$ and choose $m \in \mathbb{N}$ such that $L\beta^{-m} < \epsilon$ (for natural number choices of m, the sequence $\{L\beta^{-m}\}_{m \in \mathbb{N}} \to 0$). Then, set $\delta < 2^{-(m+1)}$. Then, for two points $\mathbf{x}, \mathbf{y} \in \mathcal{X}_{\mathcal{F}}$ where $d(\mathbf{x}, \mathbf{y}) < \delta$, we know that $x_i = y_i$, for $i = 0, \ldots, m+1$. Then $\kappa(\mathbf{x}), \kappa(\mathbf{y}) \in I_{x_1 \ldots x_m} \le L\beta^{-m}$. But then

$$\left| \kappa(\mathbf{x}) - \kappa(\mathbf{y}) \right| \le L\beta^{-m} < \epsilon.$$

And finally, since $f : I \to I$, every point has at least one itinerary, which is automatically an admissible sequence. Therefore every point in I is the image of a infinite sequence, and the map κ is onto. $\qquad\square$

Some final notes:

- Like the map ι, the map κ satisfies

$$\kappa\left(\sigma\left(\{x_0 x_1 x_2 \ldots\}\right)\right) = \kappa\left(\{x_1 x_2 \ldots\}\right) = f(\mathbf{x} = f\left(\kappa\left(\{x_0 x_1 x_1 \ldots\}\right)\right),$$

  so that $f \circ \kappa = \kappa \circ \sigma$. And since κ is a continuous surjection via Proposition 7.58, κ establishes a topological semiconjugacy between expansive Markov maps of the interval and finite subshifts.

- For Examples 7.51 and 7.54, where $\beta = 1$, one can still establish a semiconjugacy with a finite subshift, although the proof is a bit more subtle. We will not go into any more detail here, though.

- Off of the overlapping edges of the partition elements and their pre-images (interior points of partition elements that are eventually mapped onto a boundary point), the itinerary map is well-defined and a bijection, taking points to sequences representing orbits that never intersect with the partition boundaries. These points represent almost all points in the interval, and thus the corresponding finite subshift can be used to study the dynamics of the interval map, identifying properties of the dynamics like the number of n-periodic points and the growth rate with respect to n, dense orbits, etc.

- Expansiveness as a property of the map with a Markov partition would not work for C^1 maps of the interval, like polynomial maps, for example. Indeed, near the critical points, a C^1 map has a small derivative. One could not construct an expansive partition. However, given a polynomial map, say, with a partition using the critical points, with the resulting map restricted to each partition element a homeomorphism, one can create a similar correspondence to a subshift to relate the resulting dynamics. See Gilmore and Lafranc's book, *The Topology of Chaos*[21], for example, for more information about this general case.

Example 7.59 While continuity was a feature of all of the examples above of piecewise linear maps with a Markov partition, maps with a finite number of discontinuities can also be modelled via symbolic dynamics, as long as the points of discontinuity are endpoints of the partition elements. Consider the map

$$r : [0,1] \to [0,1], \quad r(x) = \tfrac{4}{3}x \mod 1.$$

Evidently, this map has no Markov partition. Indeed, any partition must have 1 as a partition endpoint, and for a partition to be Markov, endpoints must be mapped to endpoints. But the orbit of 1, $\mathcal{O}_1 = \left\{ \left(\tfrac{4}{3}\right)^n \right\}_{n \in \mathbb{N}}$ is not a finite orbit. Hence the number of partition endpoints would have to be infinite, meaning that at least one partition element would have to have no interior. Hence, interval maps with a Markov partition are a bit special.

Exercise 281 Show that the sequence $\left\{ \left(\tfrac{4}{3}\right)^n \right\}_{n \in \mathbb{N}}$ is never an integer, and use this to conclude that $\mathcal{O}_1(r)$ from Example 7.59 has no Markov partition.

Exercise 282 For the following functions, create a Markov partition and construct the transition matrix:

(a) $f(x) = \begin{cases} 1 - \sqrt{2}x, & 0 \le x < \frac{1}{\sqrt{2}}, \\ \sqrt{2}x - 1, & \frac{1}{\sqrt{2}} \le x \le 1; \end{cases}$

(b) $g(x) = \begin{cases} 1 - \sqrt{2}x, & 0 \le x < \frac{1}{\sqrt{2}}, \\ \frac{1}{2 - \sqrt{2}} \left(\sqrt{2}x - 1 \right), & \frac{1}{\sqrt{2}} \le x \le 1. \end{cases}$

You will notice that the maps we used above to study Markov partitions were all surjective but definitely not injective. Hence the symbol spaces we used to compare the dynamics were all based on infinite sequences and not bi-infinite sequences. Bi-infinite sequences and shifts on them also play a role in the study of maps on spaces. However, the spaces are in a sense "bigger" than just an interval. We follow with an interesting example that can serve as a prototype.

7.1.7 Application: The Baker's Transformation

In 1937, Eberhard Hopf [27] devised a construction to showcase some interesting dynamics in a map from the unit square in the plane to itself. His construction, now known as the *baker's transformation* or the *baker's map*, mimics both the folding and kneading a baker may do when working with dough, as well as the stretching and folding that the tent map seems to do on the unit interval. One may say that the baker's map is a 2-dimensional version of the tent map constructed in Example 7.55. Here, we describe this construction.

Let's return to the original construction of $\mathcal{M}_m$ for $m = 2$, using the symbols 0 and 1, so that $\mathcal{M} = \{0, 1\}$. The points of $\mathcal{M}^{\mathbb{Z}}$ are then

$$\mathbf{x} = \{ \cdots x_{-3} x_{-2} x_{-1} x_0 x_1 x_2 \cdots \}, \quad x_i \in \{0, 1\}, \quad \forall i \in \mathbb{Z}.$$

Re-characterize these points as follows:

- create a new variable $y_i = x_{-i}$, for $i \in \mathbb{N}$; and

- create two subsequences as real numbers in binary expansion:

$$\underline{x} = (.x_0x_1x_2\cdots)_2,$$
$$\underline{y} = (.y_1y_2y_3\cdots)_2.$$

As numbers, then, $\underline{x}, \underline{y} \in [0,1]$. One can use this to represent the set of all points in $\mathcal{M}_2$ as elements of the unit square in the plane. In fact, the map

$$h : \mathcal{M}^{\mathbb{Z}} \to [0,1] \times [0,1], \quad h(\{\mathbf{x}\}) = (\underline{x}, \underline{y})$$

is readily seen to be continuous and onto.

Exercise 283 Show that h is both C^0 and surjective.

Note that the surjection h cannot be a homeomorphism, since all of the points in the unit square that have dyadic rationals as coordinates (see Exercise 231 of Section 6.3.1) have more than one unique binary expansion. For example,

$$\tfrac{1}{4} = (0.01\bar{0})_2 = (0.00\bar{1})_2.$$

It is true, though, that there are only a finite number of pre-images of any point in the square under h. In this case, how many can there be?

We now use h to construct a map on the unit square in the plane that corresponds to the shift map on $\mathcal{M}^{\mathbb{Z}}$. Indeed, construct a map as follows:

$$\underline{x} = (0.x_0x_1x_2\cdots)_2 \mapsto (x_0.x_1x_2\cdots)_2 = 2\underline{x} \text{ if } x_0 = 0, \text{ and } 2\underline{x} - 1 \text{ if } x_0 = 1$$

$$\mapsto 2\underline{x} - x_0,$$

$$\underline{y} = (0.y_1y_2y_3\cdots)_2 \mapsto (0.x_0y_1y_2\cdots)_2 = \tfrac{1}{2}\underline{y} \text{ if } x_0 = 0, \text{ and } \tfrac{1}{2}\underline{y} + \tfrac{1}{2} \text{ if } x_0 = 1$$

$$\mapsto \tfrac{1}{2}\left(\underline{y} + x_0\right).$$

Hence we have $\mathbf{b} \circ h = h \circ \sigma$, where

$$\mathbf{b} : [0,1]^2 \to [0,1]^2, \quad \mathbf{b}((x,y)) = \begin{cases} \left(2x, \tfrac{1}{2}y\right) & \text{if } x < \tfrac{1}{2}, \\ \left(2x - 1, \tfrac{1}{2}y + \tfrac{1}{2}\right) & \text{if } x \geq \tfrac{1}{2}. \end{cases}$$

The map $\mathbf{b}$ is also invertible, although we leave the construction of $\mathbf{b}^{-1}$ to the reader. See Figures 151 and 152 for a depiction of the first few iterates of each of $\mathbf{b}$ and $\mathbf{b}^{-1}$. Also, Figure 153 shows a partial overlap of forward and backward images of the unit square inside the unit square. The intersections of these forward and backward images correspond to neighborhoods of sequences in $\mathcal{M}^{\mathbb{Z}}$ that agree on a central block of size 6, in this case. Notice also in this construction that the encoding is not unique on the

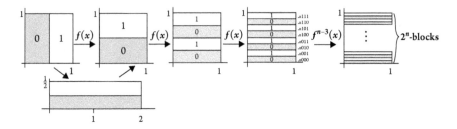

Figure 151 Forward iterates of the baker's transformation.

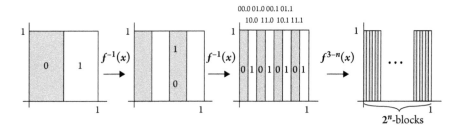

Figure 152 Backward iterates of the baker's transformation.

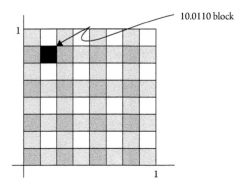

Figure 153 The black box is the set of all sequences that agree on the block stated.

edges of these boxes. But these box edges correspond precisely to the points of the square with dyadic rationals as coordinates. So the full two-sided shift on two letters $\mathcal{M}_2^{\mathbb{Z}}$ is semiconjugate to the baker's transformation via h.

Exercise 284 Construct $\mathcal{X}$, a (not of finite type) subshift of $\mathcal{M}_2^{\mathbb{Z}}$, such that $h : \mathcal{X} \to [0,1]^2$ is also injective.

Exercise 285 Show that the periodic points of the baker's map are dense in the unit square.

7.2 Two-Dimensional Markov Partitions: Arnol'd's Cat Map

In two dimensions, the notion of a partition is easy enough to visualize: think of a family of open, disjoint regions whose closure covers a space, and one may envision the tiling of the plane by triangles in the unfolded triangle billiard of Section 5.3.2 (recall Figure 106 for the fully unfolded table). However, the idea of a map on a 2-dimensional space respecting a partition, satisfying a Markov condition, like that in Definition 7.49 is a little trickier. We cannot simply say that a partition element must be mapped onto a union of partition elements, since, for example, it may not be the case that all directions of the map are expanding. In fact, for a common set of examples where Markov partitions exist, the hyperbolic toral automorphisms, there is always an expanding and a contracting direction. In this case, we can say that in the expanding direction, we would like it to be the case that if the image of a partition element intersects a partition element, then it must span that partition element. But, in the contracting direction, we can only say that if the image of a partition element intersects a partition element, then it must lie completely inside that element. Of course, understanding the expanding and contracting directions involves the stable and unstable manifolds of points in the space, and it will help to use these to create the partition elements. To this end, we will focus on a particular example of a linear transformation of the square torus: the hyperbolic toral automorphism given by Arnol'd's cat map. Recall that this corresponds to the hyperbolic toral automorphism

$f_L : \mathbb{T}^2 \to \mathbb{T}^2$ given by the matrix $L = \begin{bmatrix} 2 & 1 \\ 1 & 1 \end{bmatrix} \in GL(2, \mathbb{Z})$. We will just call the map f. It

has eigendata $\lambda = \frac{3+\sqrt{5}}{2}$ and $\mu = \frac{3-\sqrt{5}}{2}$, with $\mathbf{v}_\lambda = \begin{bmatrix} 1 \\ \frac{1+\sqrt{5}}{2} \end{bmatrix}$ and $\mathbf{v}_\mu = \begin{bmatrix} 1 \\ \frac{1-\sqrt{5}}{2} \end{bmatrix}$. The

vector subspaces of $\mathbb{R}^2$ generated by $\mathbf{v}_\lambda$ and $\mathbf{v}_\mu$ are the invariant subspaces $\mathbb{E}_\lambda$ and $\mathbb{E}_\mu$, corresponding to $\mathbb{E}^u$ and $\mathbb{E}^s$, respectively. And, since the map is linear, $\mathbb{E}^u = \mathcal{W}^u(\mathbf{0})$, and $\mathbb{E}^s = \mathcal{W}^s(\mathbf{0})$.

Definition 7.60 *Given a map $f : M \to M$ on a metric space, a Markov partition of M is a topological partition of M into rectangles $\{A_1, \dots, A_m\}$ such that whenever $x \in A_i$ and $f(x) \in A_j$, then $f\left(\mathcal{W}^u(x) \cap A_i\right) \supset \mathcal{W}^u\left(f(x)\right) \cap A_j$ and $f\left(\mathcal{W}^s(x) \cap A_i\right) \subset \mathcal{W}^s\left(f(x)\right) \cap A_j$.*

Now, given the map $\rho : \mathbb{R}^2 \to \mathbb{T}^2$, the linear eigenspaces $\mathcal{W}^u(\mathbf{0})$ and $\mathcal{W}^s(\mathbf{0})$ of a hyperbolic linear map on the plane have images (via the corresponding hyperbolic toral automorphism) that wrap densely around $\mathbb{T}^2$, as can be gleaned from Exercise 257 in Section 7.1.2. So, construct a rectangular partition of $\mathbb{T}^2$ using $\mathcal{W}^u$ and $\mathcal{W}^s$, as follows:

(1) Create a grid over the square torus in $\mathbb{R}^2$, by extending eigenlines from the origin and lines parallel to these eigenlines at the other three corners of the square torus and some of the nearby integer points in $\mathbb{R}^2$ whose corresponding parallel lines will intersect the square torus. In Figure 154, these eigenlines all have slopes $\lambda_s = \frac{-1+\sqrt{5}}{2}$ and $\mu_s = \frac{-1-\sqrt{5}}{2}$, so some of these eigenlines are

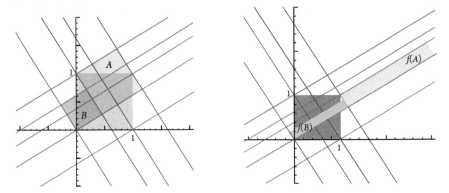

Figure 154 A Markov partition for Arnol'd's cat map.

$$y = \lambda_s x \qquad \text{(bottom edge of partition)},$$
$$y = \mu_s x \qquad \text{(left edge of B)},$$
$$y = \mu_s x + 1 \qquad \text{(right edge of B and left edge of A)},$$
$$y = \mu_s(x - 1) + 1 \quad \text{(right edge of A)},$$
$$y = \lambda_s x + 1 \qquad \text{(top edge of A)},$$
$$y = \lambda_s(x + 1) \qquad \text{(top edge of B)}.$$

(2) Create a set of rectangles using these eigenlines to form a partition of the square torus (so that they are adjacent but not overlapping, and completely cover the torus. The number of rectangles used is not very important, but it will have to be more than one and the edges will have to be parallel to the eigenlines.

> **Remark 7.61** *Kenneth Berg [7], in his PhD thesis, is credited with discovering the existence of a Markov partition for a hyperbolic toral automorphism. And although Roy Adler has been credited with the idea that for any hyperbolic toral automorphism, a partition can always be constructed with only two elements, Ellen Rykken [49] has shown constructively that this is the case. In Figure 154, at left, we have created a partition using two elements {A, B}. Do keep in mind that this covering must be viewed via the wraparound aspects of the torus. In the figure, note how the parts of the rectangles outside of the torus do fill in when considered to wrap around to the other side of the torus. On the right side of Figure 154 are the images of A and B in the plane (before they are wrapped around the square torus. Once wrapped, they also form a partition of the square torus with similar properties to the original.*

Exercise 286 Using Figure 154, at right, draw the square torus (in gray) with the partition elements $f(A)$ and $f(B)$ suitably wrapped around the torus. Then draw the pre-images of A and B and wrap them around the torus also.

Exercise 287 Create such a partition for the hyperbolic toral automorphisms given by
$$\begin{bmatrix} 3 & 2 \\ 1 & 1 \end{bmatrix} \text{ and } \begin{bmatrix} 0 & 1 \\ 1 & 1 \end{bmatrix}.$$

Exercise 288 Create a partition like the above again for Arnol'd's cat map, but using the line $y = \mu_s(x - 2) + 1$ for one of the partition edges.

Now, the question is, does this partition satisfy the definition of a Markov partition? The answer is yes, in that it satisfies the Markov condition: To see this, note that, by Exercise 258, we know that for any $x \in \mathbb{T}^2$, $\mathcal{W}^u(\mathbf{x}) = \{\mathbf{x} + c\mathbf{v}_\lambda \mid c \in \mathbb{R}\}$ and $\mathcal{W}^s(\mathbf{x}) = \{\mathbf{x} + c\mathbf{v}_\mu \mid c \in \mathbb{R}\}$ And, since both $\mathcal{W}^u(\mathbf{0})$ and $\mathcal{W}^s(\mathbf{0})$ are invariant, the images and pre-images of the edges of the partition will remain within $\mathcal{W}^u(\mathbf{0})$ and $\mathcal{W}^s(\mathbf{0})$. We also know that the various images and pre-images of the partition will remain partitions. But we do need more.

Let $\mathbf{x} \in A_i \cap f^{-1}(A_j)$. Then we are assured that $x \in A_i$ and $f(x) \in A_j$. Consider a point $\mathbf{y}$ in the piece of the unstable manifold of x within A_i, so $\mathbf{y} \in \mathcal{W}^u(x) \cap A_i$. Then, there exists a $c \in \mathbb{R}$ such that $\mathbf{y} = \mathbf{x} + c\mathbf{v}_\lambda$. By linearity, $f(\mathbf{y}) = f(\mathbf{x}) + cf(\mathbf{v}_\lambda)$, so $f(\mathbf{y}) \in \mathcal{W}^u(f(\mathbf{x})$. It is also clear that $f(\mathbf{y}) \in f(\mathcal{W}^u(x) \cap A_i)$. But now let c be chosen small enough that

$$y \in \mathcal{W}^u(x) \cap f^{-1}(A_j) \subset \mathcal{W}^u(x) \cap A_i.$$

Then, again, by linearity,

$$f(\mathbf{y}) \in \mathcal{W}^u\left(f(x)\right) \cap f\left(f^{-1}(A_j)\right) = \mathcal{W}^u\left(f(x)\right) \cap A_j.$$

But then $\mathcal{W}^u\left(f(x)\right) \cap A_j \subset f(\mathcal{W}^u(x) \cap A_i)$, as needed. The stable manifold inclusion is entirely similar. Hence the partition is Markov. In Figure 155, $A_i = B$, while $A_j = A$.

Note that this Markov condition is the one we want, in that it implies a "memoryless" process mentioned in Remark 7.52: the image of a point x in a partition is determined only by its current position in the partition and not via the history of the maps iterates.

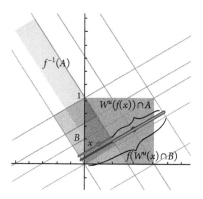

Figure 155 Mapping the unstable manifolds in partition elements.

Now, since we have a map on a space with a Markov partition, we can construct a transition matrix. However, we will need to be careful here. If our goal is to allow for itineraries to help differentiate points, then we will want different itineraries to correspond to different points as much as possible. If we naively define T_f as the 2×2 matrix that records the number of component images of A_i in A_j, so

$$T_{ij} = \# \left\{ f(A_i) \cap A_j \right\},$$

then we would again recover L. But this matrix would not be effective in recording allowable itineraries (as sequences of some finite subshift), since it would not be able to differentiate between the two components of the intersection of $f(A)$ with A (see Figure 156 at top right). In effect, we could not use it to describe a good $\mathcal{F}$ of forbidden words for a finite subshift. We can fix this, though, by refining the Markov partition slightly. There are many ways to do this. One is to break up box A into two pieces by dropping another line $y = \mu_s x + 2$ as the new right edge of A and now the left edge of C, as in

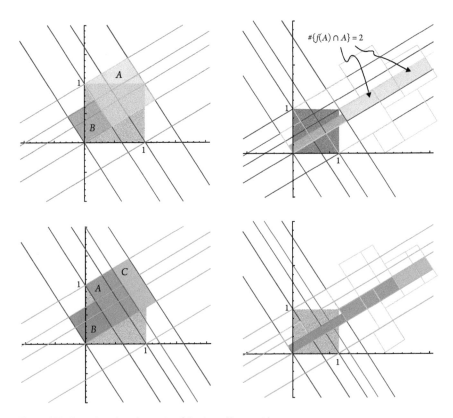

Figure 156 Counting the elements of the transition matrix.

Figure 156 at bottom left. Then, as one can see in the images in Figure 156 at bottom right, one can construct a transition matrix that records the intersections of images of partition elements with partition elements, but also records allowable and nonallowable 2-words.

Definition 7.62 *For a map $f : M \to M$ on a metric space with a Markov partition $\{A_1, \ldots, A_m\}$, the transition matrix T_f is the $m \times m$ matrix of 0's and 1's that records the number of component images of A_i in A_j:*

$$T_{ij} = \# \left\{ f(A_i) \cap A_j \right\}.$$

For the partition $\{A, B, C\}$ we constructed in Figure 156 at bottom left, one can easily show that the transition matrix is precisely the matrix

$$(7.2.1) \qquad\qquad T_f = \begin{bmatrix} 1 & 1 & 1 \\ 1 & 1 & 1 \\ 0 & 1 & 1 \end{bmatrix}.$$

Notice immediately that the spectral radius of T_f is precisely the largest eigenvalue of L.

Exercise 289 Construct a continuous, piecewise linear map of the unit interval that has the transition matrix in Equation 7.2.1.

Exercise 290 Given the Markov partition of Arnol'd's cat map with three elements, as given in Figure 157, calculate a transition matrix.

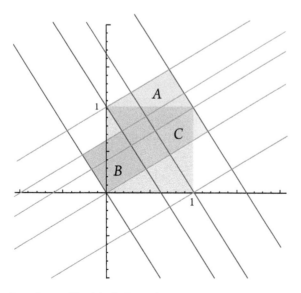

Figure 157 An alternative partition into 3 elements.

7.3 Chaos and Mixing

Recall that a map $f : X \to X$ on a metric space is *topologically transitive* if there exists a dense orbit. The examples that we looked at that had this property included the irrational rotations of S^1 and the irrational linear flows on the two torus $\mathbb{T}^2$. These examples, however, also had the property that every orbit was dense. And there were no periodic points at all. Contrast this with rational rotations of S^1 and rational linear flows on $\mathbb{T}^2$, where there were tons of periodic points, but no dense orbit. And only in the recent examples did we find relatively complicated maps that had lots of periodic points, and these periodic points were varied in their minimal periods. One can ask if these maps are also topologically transitive. Having both of these properties—a dense set of periodic points (and of all periods) and a dense orbit—would seem to be quite complex behavior. How complex?

Definition 7.63 *A continuous map* $f : X \to X$ *of a metric space is said to be* chaotic *if*

- f *is topologically transitive,*
- $\overline{\text{Per}(f)} = X$.

Notes:

- There are many, and sometimes competing, definitions of chaos as a property of a map, since it is a relatively newly defined concept and efforts to finally pin it down continue. But this definition really is one of the better universal definitions. That said, there is still a slight problem even with this definition, which we will expound on in Theorem 7.77 and the adjacent Example 7.78 to follow.
- Either one of these properties without the other means that the dynamics are relatively simple to describe.

Some recent examples of maps with seemingly complicated dynamics:

(1) from Section 7.1, the expanding circle map $E_m : S^1 \to S^1$, for $|m| > 1$;

(2) from Proposition 7.9, the logistic map $f_\lambda : C \to C$, for $\lambda > 4$, restricted to the Cantor set of points whose orbit lies completely within the unit interval;

(3) the hyperbolic toral automorphism $f_L : \mathbb{T}^2 \to \mathbb{T}^2$ from Section 7.1.2, induced from the linear automorphism of the plane determined by the hyperbolic matrix L;

(4) the full shift on n symbols $\sigma : \mathcal{M}_n^{\mathbb{N}} \to \mathcal{M}_n^{\mathbb{N}}$, introduced in Section 7.1.5;

(5) and the function, along with its Markov partition with transition matrix corresponding to the Golden Mean subshift, given in Example 7.51.

In almost all of these cases (except the logistic map), we showed that the periodic points are dense in the respective spaces. Hence each of these dynamical systems is chaotic if we can show that each also has a dense orbit. The same holds for the Cantor map, although we did not actually show that the periodic points are dense. However, showing directly that there exists a dense orbit is not always easy. To do so in these cases, we will instead construct a bit more machinery, and show that these maps possess some stronger properties than transitivity. In this way, we take the opportunity to also study these maps in more detail, and gain some additional insight into their dynamical structure. We start with, in essence, a re-characterization of topological transitivity, attributed to Georg Birkhoff:

Theorem 7.64 (Birkhoff's Transitivity Theorem.) *Let X be a complete separable metric space with no isolated points. For $f : X \to X$ continuous, the following are equivalent:*

(1) f has a dense orbit and is topologically transitive;

(2) f has a dense positive semiorbit;

(3) if $U, V \subset X$ are open and non-empty, $\exists N \in \mathbb{Z}$ such that $f^N(U) \cap V \neq \emptyset$;

(4) if $U, V \subset X$ are open and non-empty, $\exists N \in \mathbb{N}$ such that $f^N(U) \cap V \neq \emptyset$.

Remark 7.65 *Recall from Section 2.2.1 that a metric space X is complete if all Cauchy sequences in X converge in X. And X is separable if there exists a countable dense subset. In essence, one can define a sequence $\{x_i\}_{i \in \mathbb{N}} \subset X$ such that every non-empty open set in X contains at least one element in the sequence. Finally, an isolated point $x \in X$ is one in which there exists an open neighborhood of X containing x and no other points of X. But this means that the point x itself is an open subset of X, which is not something we have often encountered in the spaces we have been constructing. These properties are technical in nature and, while necessary, should not keep you from acquiring a good understanding of how this proposition works on the spaces with which we work. For now, don't worry too much about these technical conditions.*

Obviously, $4 \Rightarrow 3$ and $2 \Rightarrow 1$. If we could show that $3 \Rightarrow 2$ and $1 \Rightarrow 4$, then we would be done. We will not do this, however, since the implication $3 \Rightarrow 2$ does involve some advanced machinery. Indeed, one needs the *Baire Category Theorem*, which we will not explore in this text. The real point of this exposition is to understand the relationship between 1 and 3. To this end, we will prove the statement $1 \Rightarrow 4$.

Proof $(1 \Rightarrow 4)$ Let f be topologically transitive, with a dense orbit given by $\mathcal{O}_x, x \in X$. Then, for any choice of non-empty, open sets $U, V \subset X$, $\exists n \in \mathbb{Z}$ such that $f^n(x) \in U$. Now, without any isolated points, a dense set remains dense with the removal of a finite number of its points. Hence the orbit of the point $f^n(x)$,

$$\mathcal{O}_{f^n(x)} = \left\{f^n(x), f^{n+1}(x), \dots\right\} = \mathcal{O}_x - \left\{x, f(x), \dots, f^{n-1}(x)\right\},$$

is still dense in X. Hence there is an $m > n$ where $f^m(x) \in V$. But then $f^{m-n}(U) \cap V \neq \emptyset$. □

This should make sense, since, by continuity alone, $f^{-n}(U)$, as an open set, would be a neighborhood of x (see Definition 3.4), so that $f^m(x) \in f^m(f^{-n}(U))$. See Figure 158.

Corollary 7.66 *A continuous, open map f of a complete metric space is topologically transitive if and only if there do not exist two disjoint, open f-invariant sets.*

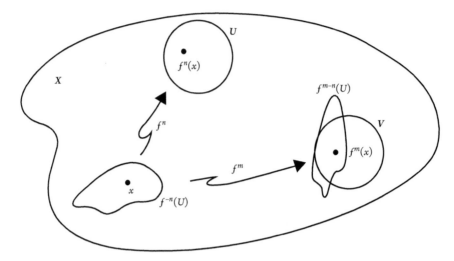

Figure 158 Topological transitivity.

It will help in understanding this last statement to understand the notion of an open map. Roughly speaking, a map is open if it takes open sets to open sets, something that is not generally true for continuous maps (think of a constant map). The idea in the corollary and discussion that finding a dense orbit is equivalent to the notion that the orbit of any open set in X must eventually intersect any other open set in X actually provides a method of discovery for dense orbits. A set $V \subset X$ is f-invariant if $f(V) \subset V$. Now assume that you have such a set V that is open. Now take any other open set U. Whether it is invariant or not, its entire orbit $\mathcal{O}_U$ is a union of all of its images and hence is open in X if the map f is open. The corollary says that an open map is topologically transitive iff we cannot divide the space into two disjoint open sets each of which is invariant under f. Put this way, the two notions look very much alike.

Definition 7.67 *A continuous map $f : X \to Y$ is called* open *if, whenever $U \subset X$ is open, $f(U) \subset Y$ is open also.*

While continuity is common among maps, "openness" is not, and is kind of a special property. When the map f has a continuous inverse, f is open as a map. But this is not that common a property.

Exercise 291 Prove Corollary 7.66 in detail.

Now, we know that expanding maps of S^1 and hyperbolic automorphisms of $\mathbb{T}^2$ look messy dynamically. The question is: How messy are they?

Definition 7.68 *A continuous map $f : X \to X$ is said to be* topologically mixing *if, for any two non-empty, open sets $U, V \subset X$, $\exists N \in \mathbb{N}$, such that $f^n(U) \cap V \neq \emptyset$, $\forall n > N$.*

Notes:

- Do you see how much stronger (more restricting) this is than topological transitivity? For instance,

$$(\text{topologically mixing}) \Rightarrow (\text{topologically transitive}),$$

but not vice versa. To see why, think of the irrational rotations of the circle. The orbit of a small open interval will eventually intersect any other small open interval. But, depending on the rotation, it will most likely leave again for a while before returning. This is not mixing!

Exercise 292 Show that topological mixing implies topological transitivity.

- Actually, the problem with irrational circle rotations is a bit deeper—they are isometries:

Lemma 7.69 *Isometries are not topologically mixing.*

Proof Under an isometry, the diameter of a set $U \subset X$, $\text{diam}(U)$ is preserved. Let $U = B_\delta(x) \subset X$ be a small δ-ball about a point $x \in X$. Here $\text{diam}(U) = 2\delta$ and $\forall n \in \mathbb{N}$, $\text{diam}(f^n(U)) = 2\delta$. Now choose $v_1, v_2 \subset X$, such that the distance between v_1 and v_2 is greater than 4δ. Let $V_1 = B_\delta(v_1)$ and $V_2 = B_\delta(v_2)$ (so that the minimal distance between these two balls is greater than 2δ). If we assume that the isometry $f : X \to X$ is topologically mixing, then there will be a $k \in \mathbb{N}$ such that both $f^n(U) \cap V_1 \neq \emptyset$ and $f^n(U) \cap V_2 \neq \emptyset$. $\forall n > k$. But this is impossible, since V_1 and V_2 are too far apart to both have non-empty intersection with an iterate of U. Hence f cannot be mixing. $\qquad\square$

Proposition 7.70 *Expanding maps on S^1 are topologically mixing.*

Proof For now, suppose that the expanding map is C^1. Differentiable expanding maps have the property that for $f : S^1 \to S^1$, $|f'(x)| \geq \lambda > 1, \forall x \in S^1$. Let $F : \mathbb{R} \to \mathbb{R}$ be a lift. It is an exercise to show that the lift also shares the derivative property, $|F'(x)| \geq \lambda$, $\forall x \in \mathbb{R}$. So, choose a small closed interval $[a, b] \subset \mathbb{R}$, where $b > a$. Then, by the Mean Value Theorem, $\exists c \in (a, b)$ such that

$$|F(b) - F(a)| = |F'(c)||b - a| \geq \lambda(b - a).$$

Hence the length of the iterate of the interval is greater by a factor of λ than the interval. This continues at each iterate of F, so that $\exists n \in \mathbb{N}$ such that $||F^n([a,b])|| > 1$. But then $\pi(F^n([a,b])) = S^1$.

Now simply grab the open interval (a, b), noting that $\pi((a,b))$ will also be open (on small intervals, π is a homeomorphism), and let $U = \pi((a,b))$. With V any other open set in S^1, we are done. $\qquad\square$

Corollary 7.71 *Linear expanding maps of S^1 are chaotic.*

Exercise 293 Without using Proposition 7.70 and topological mixing, show that linear expanding maps of S^1 are chaotic.

Proposition 7.72 $f_L : \mathbb{T}^2 \to \mathbb{T}^2$, the linear hyperbolic automorphism of the 2-torus given by the hyperbolic matrix L is topologically mixing.

Corollary 7.73 f_L is chaotic.

For a brief idea why the previous proposition is true, recall that for f_L given by the matrix $L = \begin{bmatrix} 2 & 1 \\ 1 & 1 \end{bmatrix}$, the eigenvalues were $\lambda = \frac{3 \pm \sqrt{5}}{2}$, and the eigenvalue greater than 1 (the "expanding" eigenvalue) has eigendirection given by the vector $\begin{bmatrix} 1 \\ \frac{-1+\sqrt{5}}{2} \end{bmatrix}$. Choose a small open line segment ℓ along the line $y = \left(\frac{-1+\sqrt{5}}{2} \right) x + c$ within the box representing the torus. As we iterate the map, the orientation (slope) of the line ℓ stays the same, while its length grows by a factor $\lambda = \frac{3+\sqrt{5}}{2}$ at each iterate. For $N >> 1$, we will find that the length of $f_L(\ell)$ will be huge, and wrap around the torus quite densely. In fact, we can choose this N such that $f_L^N(\ell)$ will intersect any ball of radius ϵ in $\mathbb{T}^2$. Hence choose any ϵ-ball V and any other ϵ-ball U, and take as our ℓ the diameter of U in the direction parallel to $y = (-1 + \sqrt{52})x$. Then, with the above choice of N, we will have $f_L^n(U) \cap V \neq \emptyset$, for all $n > N$. Hence f_L is topologically mixing on $\mathbb{T}^2$.

Back to shift maps, using the metric in Equation 7.1.3 that we constructed on $\mathcal{M}_n^N$ (and thus also on finite subshifts $\mathcal{X}_F$), any open set $U \in \mathcal{X}_F$ will contain some small open cylinder around each of the points in U. We can use this to establish also that certain shift maps are also topologically mixing:

Proposition 7.74 For $\mathcal{X}_F$ a finite subshift with transitive transition matrix T_F, the shift map on $\mathcal{X}_F$ is topologically mixing.

Proof Let $U, V \in \mathcal{X}_F$ be two open sets. Choose a point $\mathbf{x} \in U$ and $\mathbf{y} \in V$. Then, for some $k, \ell \in \mathbb{N}$, there exist small open cylinders

$$B_{1/2^k}(\mathbf{x}) = \left\{ \mathbf{w} \mid w_i = x_i, \ \forall i = 0, \dots, k \right\},$$
$$B_{1/2^\ell}(\mathbf{y}) = \left\{ \mathbf{u} \mid u_j = y_j, \ \forall j = 0, \dots, \ell \right\}.$$

If T_F^n is positive, then, for every $m > n$, there is an admissable m-word v_m of length that begins with w_k and ends with u_0. Consider the point $\mathbf{z} \in \mathcal{X}_F$, where

$$\mathbf{z} = \{ w_0 w_1 \dots w_{k-1} v_m u_1 \dots u_\ell z_{k+m+\ell} \dots \}.$$

Here, $\mathbf{z} \in U$ and $\sigma^{m+k}(\mathbf{z}) \in V$. As this is true for all $m > n$, it implies that the shift map is topologically mixing. $\qquad \square$

Exercise 294 Adapt the argument in the proof of Proposition 7.74 to the case for a two-sided finite subshift.

Hence finite subshifts with transitive transition matrices are topologically transitive. Hence the property name is appropriate, even as the objects being named are different. And, since the full shift $\mathcal{M}_n^{\mathbb{N}}$ (as well as $\mathcal{M}_n^{\mathbb{Z}}$) certainly has a transitive transition matrix (T itself is positive), we have the following:

Corollary 7.75 *For $\mathbb{K} = \mathbb{N}$ or $\mathbb{Z}$, a subshift $\mathcal{X} \subset \mathcal{M}_n^{\mathbb{K}}$ with transitive transition matrix is chaotic.*

So, $\sigma : \mathcal{M}_n^{\mathbb{Z}} \to \mathcal{M}_n^{\mathbb{Z}}$ is a chaotic map.

And finally, recall that the quadratic family $f_\lambda : C_\lambda \to C_\lambda$, for $\lambda > 2 + \sqrt{5} > 4$ and C_λ the Cantor set of orbits that lie entirely in $[0, 1]$, are expanding maps (you verified this in Exercise 253). In a manner similar to the above, one can prove that f_λ as above is topologically mixing and hence topologically transitive. Furthermore, it is also true that the set of all periodic points is dense in f_λ, rendering the map chaotic on C_λ. However, we will hold off proving this until Section 8.1 on topological conjugacy (the statement and proof are Corollary 8.7). For now, let's dive a little deeper into the idea of a chaotic map.

7.4 Sensitive Dependence on Initial Conditions

So, the next question is: What information does chaos, as a property, convey about the dynamical system? Flippantly speaking, it tells us that the orbit structure is quite complicated. It tells us that arbitrarily close to a periodic point there are non-periodic points whose orbits are dense in the space. On the other hand, it tells us that arbitrarily close to a point whose orbit is dense in the space there are periodic points of arbitrarily high period. Hence simply being very close to a point of a certain type does not mean that the orbits will be similar. This means that one cannot rely on estimates or precision to help determine orbit behavior. Mathematically, it means the following:

Definition 7.76 *A map $f : X \to X$ of a metric space is said to exhibit a* sensitive dependence on initial conditions *if $\exists \Delta > 0$ (called a sensitivity constant) where $\forall x \in X$ and $\forall \epsilon > 0$, $\exists$ a point $y \in X$ where $d(x, y) < \epsilon$ and $d(f^N(x), f^N(y)) \geq \Delta$ for some $N \in \mathbb{N}$.*

There are lots of comments to make on this topic:

- The idea here is that, for certain constants, no matter how small a neighborhood of a chosen point x you start from, there will always in this neighborhood be a point y whose neighborhood, after a time, will be far away from the orbit of x. For fixed points, this is the definition of unstable. But this applies to every point here.

- The existence of at least one point in each neighborhood of an arbitrary point x whose orbit veers away from the orbit of x is the notion that everywhere there is an expanding direction (think of a differentiable map whose derivative everywhere, as a matrix, has at least one eigenvalue of modulus greater than 1). This should remind you of a hyperbolic toral automorphism.

- This idea is quite profound. Early developers of classical mechanics, like Isaac Newton and Pierre Simon Laplace, tended to believe that eventually we would understand the universe completely as a deterministic mechanical system: eventually, if we can know enough about how the universe works to be able to model it perfectly, then we should be able to completely determine (predict) its future states given its present states and the model. In *A Philosophical Essay on Probabilities*, [28] Laplace wrote "We ought then to regard the present state of the universe as the effect of its anterior state and the cause of the one which is to follow. Given for one instant an intelligence which could comprehend all the forces by which nature is animated and the respective situation of the beings who compose it ..., [then] nothing would be uncertain and the future, as the past, would be present in its eyes."

- Poincaré, in the late nineteenth century [43], saw this phenomenon of a sensitive dependence on initial conditions in certain solutions to the classical three-body problem. He understood immediately that the earlier reasoning was flawed. Indeed, knowing the precise state of all things in the universe was impossible. But, with such a sensitive dependence on initial conditions (even in the simplistic three-body problem), a reasonable approximation to the universe's state in an instant would never be good enough to make good long-term predictions.

- Edward Lorenz, studying early climate models on a computer at MIT in the 1960s, saw his statistical forecasting computer model of weather systems make wildly divergent predictions given extremely small changes to initial data constants. His early attempts were producing regular (read: periodic) patterns, which were hardly realistic. His attempts at more sophisticated modeling led to deviations in the periodic patterns that resulted in his discovery of the initial value sensitivity coming from the nonlinearity of the equations. This discovery, and his further studies into simple systems that exhibit it, give us today the notions that (1) there is no chance of truly using deterministic models to enable long-term predictions involving complicated, nonlinear processes, and (2) even if it were possible, the use of computers to store and process information also inherently creates small but persistent inaccuracies in calculations, some random and some not, which lead to inconsistent model predictions due to small changes in the values of variables upon repeated runs of the code.

- A beautiful metaphor for this sensitivity to initial conditions is the *butterfly effect*, a term coined by Lorenz to express the idea that a small change in conditions at one point can have a huge change later: a prediction of a tornado (its exact time of formation, the exact path taken) is so sensitive to all of the conditions leading up to it that minor perturbations such as the flapping of the wings of a distant butterfly several weeks earlier can influence its formation and fury (or even lead to no tornado at all).

There is an attractor type not covered in this book. The Lorenz Butterfly (see Figure 159) is an example of what is called a "strange attractor" (essentially an attractor with a fractal structure) arising from a simple 3-dimensional system of parameterized ODEs.

For $f(x) = 2x \mod 1$, the distance between nearby points grows exponentially as 2^n. This is a sensitive dependence on initial conditions. Eventually, this distance is larger

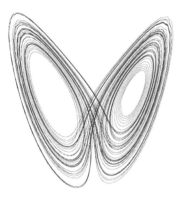

Figure 159 The Lorenz Butterfly.

than 1, and, from this point, future iterates of each orbit tend to look unrelated to each other. For a map exhibiting a sensitive dependence on initial conditions in a compact space (closed and bounded), one can see how the orbit structure can be complicated. If all orbits are moving away from each other, and yet cannot go beyond the boundaries of the space, they just wind up mixing around each other. Think of steam rising from a hot cup of coffee, or smoke from a cigarette, and you can see just how complicated the orbits can be in this case.

Exercise 295 Show that isometries cannot exhibit a sensitive dependence on initial conditions.

Theorem 7.77 *Chaotic maps exhibit a sensitive dependence on initial conditions, except when the entire space consists of one periodic orbit.*

Example 7.78 Let

$$X = \left\{ 0, \tfrac{1}{5}, \tfrac{2}{5}, \tfrac{3}{5}, \tfrac{4}{5} \right\},$$

and $f : X \to X, f(x) = x + \tfrac{1}{5}$ mod 1. Here, f is continuous with respect to the topology that X inherits from $\mathbb{R}$; this is called the subspace topology, defined in Section 3.1. This is really just the trivial topology of X, since it is a finite, discrete subset of $\mathbb{R}$, so each point of X is open. Here, X certainly has a dense orbit (every point of X lives in the orbit of $\tfrac{1}{5}$). And the set of all periodic points of X is dense in X (every point of X is periodic). Hence f is chaotic on X. But there certainly is not a sensitive dependence on initial conditions here.

Example 7.79 The twist map on the cylinder does have a sensitive dependence on initial conditions. To see this, recall that each horizontal circle is invariant, and has a different rotation along it, which is a linear function of height. Now take any point x, and any small neighborhood of x. This small neighborhood will include points on horizontal circles different from that of x. Choose any one of these points. Eventually, x and this other point will wind up pretty much on opposite sides of the cylinder. So what is the sensitivity constant (the largest such Δ)?

Exercise 296 For the twist map and the standard parameterization (and metric) of the circle given by the exponential map $f(x) = e^{2\pi ix}$, show that the sensitivity constant is $\frac{1}{2}$.

Proposition 7.80 *A topological mixing map (on a non-trivial space) exhibits a sensitive dependence on initial conditions.*

Remark 7.81 *Perhaps a better definition of chaos is one that requires a sensitive dependence on initial conditions as a third condition along with the other two. This would discount the "chaotic" map in Example 7.78 (which is hardly chaotic in a non-mathematical sense), while not restricting in any detrimental way the intent of the property. In fact, this is a fairly widely accepted set of conditions for a map to be chaotic. Note also that the condition of sensitive dependence on initial conditions cannot by itself give a chaotic system. The twist map is an example of a system that is hardly in a chaotic state, even though it has a sensitive dependence on initial conditions and the periodic points are dense! It does not have a dense orbit, however. And even the star node, the equilibrium at the origin of the map $\dot{\mathbf{x}} = I_2\mathbf{x}$, exhibits a sensitive dependence on initial conditions. Again, that is hardly chaotic, with neither of the other two conditions satisfied.*

7.5 Quadratic Maps: The Final Interval

Let's return to the quadratic family of maps one last time to study the parameter space interval that we have neglected to this point, namely, the interval $[3,4]$.

Let $I = [0,1]$ and $f_\lambda : I \to I$, $f_\lambda(x) = \lambda x(1-x)$, but this time let $\lambda \in [3,4]$. Recall Definition 2.56 that for $x \in X$ a fixed point of the map $f : X \to X$, the basin of attraction of x is

$$B(x) = \{y \in X \mid \mathcal{O}_y \to x\}.$$

Here, we will explore this notion a bit more:

- Sometimes the basin of attraction is easy to describe:

  Example 7.82 Let $\dot{r} = r(r-1)$, $\dot{\theta} = 1$ be a planar ODE system. It should be obvious now that the only equilibrium solution is at the origin of the plane, and the only other "interesting" behavior is the unstable limit cycle given by the equation $r(t) \equiv 1$. Since solutions are unique on all of $\mathbb{R}^2$ (and hence cannot cross), what starts inside the unit circle stays inside. And since the limit cycle is repelling, and there are no other limit cycles or equilibria inside the unit circle, it must be the case that the origin is attracting (you can also see this directly by noting that $\dot{r} < 0 \; \forall r \in (0,1)$). Hence the basin of attraction of the origin is the open unit disk

  $$B((0,0)) = \left\{ (r,\theta) \in \mathbb{R}^2 \;\middle|\; r < 1 \right\}.$$

- Sometimes, it is not:

Example 7.83 Let $P_c : \mathbb{C} \to \mathbb{C}$, $P_c(z) = z^2 + c$, for $c \in \mathbb{C}$ a constant. For $c = 0$, we get a rather plain model. $\mathcal{O}_z \to 0, \forall |z| < 1$, and $\mathcal{O}_z \to \infty \ \forall |z| > 1$. Do you recognize the map on the unit circle $|z| = 1$? It is the expanding (and chaotic) map $E_2 : S^1 \to S^1$, first introduced in Section 7.1 when we were attempting to count periodic orbits.

For the map E_2 in the circle, recall that the periodic points are dense in S^1 (this was a feature of chaos). And since the map is expanding, you can show that all of these periodic points are actually repelling (simultaneously!). The resulting mess is actually what a "sensitive dependence on initial conditions" is all about. Here again, the origin in $\mathbb{C}$ is an attracting fixed point, and its basin of attraction is everything inside the unit circle.

Now, though, let c be small and non-zero. There will still be two fixed points, right? (Think of solving the equation $z = P_c(z) = z^2 + c$. The solutions will be $z = \frac{1}{2}(1 \pm \sqrt{1 - 4c})$. For $z \in \mathbb{C}$, this always has two solutions!) The one near the origin will still be attracting, while the one near the unit circle will still be a part of a set of repelling periodic points whose closure will form a (typically) fractal structure. This edge of the basin of attraction for the attractive fixed point near the origin can be so complicated that it becomes an interesting object in and of itself for study. We call it the Julia set of P_c for this value of c, and it can be very bizarre looking. We will introduce this set more formally in Section 7.6.1 on complex dynamics.

Definition 7.84 *An m-periodic point p is called* attracting *under a continuous map f if $\exists \epsilon > 0$ such that $\forall x \in X$, where $d(x, p) < \epsilon$, we have $d\left(f^n(x), f^n(p)\right) \xrightarrow{n \to \infty} 0$.*

Exercise 297 Show that for an attracting m-periodic point p, each distinct point in its orbit is also attracting.

Call the basin of attraction for an m-periodic point p the union of the basins of attraction for each point of $\mathcal{O}_p$. That is, for

$$\mathcal{O}_p = \left\{ p, f(p), f^2(p), \ldots, f^{m-1}(p) \right\},$$

the basin of attraction of p is

$$B(p) = \left\{ x \in X \ \middle| \ d(f^{n+k}(x), f^{n+k}(p)) \xrightarrow{n \to \infty} 0 \text{ for some } k \in \mathbb{N} \right\}.$$

Definition 7.85 *The* immediate basin of attraction *of an m-periodic point p is the largest interval $IB(p)$ containing p such that $\forall x \in IB(p)$, $\mathcal{O}_x \to \mathcal{O}_p$. The immediate basin of attraction of a periodic orbit is the union of the immediate basins of attraction of each point in the orbit.*

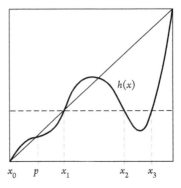

Figure 160 A basin of attraction.

As shown in Figure 160, a basin of attraction of a fixed or periodic point will in general not consist of a single contiguous interval. However, the immediate basin always is. For $h(x)$ in the figure, for example, $B(p) = (x_0, x_1) \cup (x_2, x_3)$, while $IB(p) = (x_0, x_1)$ (draw some mental cobwebs to convince yourself of this). Back to our discussion of the logistic map, we see that the structure of the graph of $f_\lambda(x)$ on $[0,1]$ says a lot about the dynamical structure of the map:

Proposition 7.86 *Let $f : [a,b] \to [a,b]$ be C^2 and concave down, where $f(a) = f(b) = a$. Then f has at most one attracting periodic orbit.*

We will not prove this here, but the idea rests on three important facts:

- The structure of f (twice-differentiable, concave down with images of endpoints equal) implies that it has a unique critical point $x_0 \in (a,b)$.
- The immediate basin of attraction of any attracting periodic orbit must contain x_0 (this is the non-trivial part of the proof).
- Basins of attraction cannot overlap.

This proves very useful in our analysis of the logistic map.

Example 7.87 In all of our examples of $f_\lambda : I \to I$ where $\lambda \in [0,3]$, there was always an attracting fixed point. However, for $\lambda = 3.1$, for example, the fixed point at $x = 0$ is repelling, and there is an attracting period-2 orbit.

Exercise 298 For $\lambda = 3.1$, calculate the points comprising the attractive period-2 orbit directly by solving the quartic $f_\lambda^2(x) = x$ and using derivative information to establish the stability.

Proposition 7.88 *If f_λ has an attracting periodic orbit, then the set outside of the basin of attraction (called the universal repeller) is a nowhere-dense null set.*

Some notes:

- A nowhere dense null set in a metric space is a set that can be covered by balls whose total volume is less than ϵ.

- What can lie within the universal repeller? First, any repelling fixed or periodic points, of course. But since the logistic map is a two-to-one map, the pre-image of a fixed point consists of two points, and includes a point that was not previously fixed.

Example 7.89 Let $\lambda = 3.2 = \frac{32}{10}$. It can be shown that $f_{3.2}$ has an attracting period-2 orbit. And $x_\lambda = 1 - \frac{1}{\lambda} = 1 - \frac{1}{\frac{32}{10}} = \frac{22}{32}$ is fixed under $f_{3.2}$ and repelling (check this!). But the point $1 - x_\lambda = \frac{10}{32}$ also maps to $\frac{22}{32}$. In fact, the point $1 - x_\lambda$ is always the pre-image of the fixed point x_λ owing to where it sits on the graph of f_λ. Neither of these points is in the basic of attraction of any periodic orbit. But, also, $1 - x_\lambda$ is not a periodic point. It is an eventually fixed point, but that is different. Now the point $1 - x_\lambda$ also has two pre-images (find them: cobwebbing them is easy, calculating them?), and these two pre-images also have two pre-images. In fact, there are a countable number of pre-images that eventually get mapped onto x_λ. Cobweb the appropriate points on the right in Figure 162 below to see. All of this set lies outside of the basic of attraction of any attracting periodic orbit, when x_λ is repelling. These points also give a sense of the difference between the basin of attraction and the immediate basin of attraction of an attracting periodic or fixed point. This gives you an idea of what is considered part of the universal repeller. Now think about how this set of pre-images of x_λ sit inside the interval! If you think about it correctly, you start to see just how fractals are born.

Example 7.90 For $\lambda \in \left[3, 1 + \sqrt{6}\right]$, there exists an attracting, period-2 orbit. The basin of attraction is everything except for the points 0, and $x_\lambda = 1 - 1/\lambda$ and all of their pre-images.

Let's work out the situation: For $\lambda \in (1,3]$, 0 is a repelling fixed point, x_λ is an attracting fixed point, and there are no other periodic points. In contrast, for $\lambda \in \left(3, 1 + \sqrt{6}\right)$, both $x = 0$, and x_λ are repelling fixed point, and there now exist an attracting period-2 orbit. This means that we have reached a bifurcation value for λ at $\lambda = 3$. This type of bifurcation is called a *period-doubling bifurcation*, and is visually a "pitchfork" bifurcation for the map f_λ^2.

7.5.1 Period-Doubling Bifurcation

With an immediate comparison with the pitchfork bifurcation from Section 2.4.3, let $f(\alpha, x) = x^3 - (\alpha + 1)x$. Note that this is the negative of the function we used as a prototype for the pitchfork bifurcation. Here, $f(\alpha, x) = x$ precisely when $x(x^2 - (2 + \alpha)) = 0$. Hence we again have two 0-curves, $x = 0$ and $x^2 = 2 + \alpha$ for $\alpha > 0$, as in Figure 161 at left. At the point $(\alpha, x) = (0,0)$, we have the data

$$f(0,0) = 0, \quad \text{and} \quad \frac{\partial f}{\partial x}(0,0) = 3x^2 - (\alpha + 1)\Big|_{(0,0)} = -1.$$

At the origin, we have the following:

- The origin is a non-hyperbolic fixed point, but this time the derivative is -1. What does this mean for the derivative at the origin of the square of f?

- There is only one of the two 0-curves passing through the origin in the αx-plane. The other curve crosses the α-axis at the point $\alpha = -2$.

- A quick analysis by phase lines determines the stability of the fixed-point lines in the plane, rendering $\alpha = -2$ and $\alpha = 0$ as bifurcation points. The bifurcation at $\alpha = -2$ has the look of a pitchfork bifurcation.

Exercise 299 Show that the bifurcation at $(-2, 0)$ is a pitchfork bifurcation for f.

- For $\alpha > 0$, notice that all of the fixed points of f are unstable. This could not happen in a continuous dynamical system, where f would be a vector field (why not?).

To see what is happening here, we examine instead

$$f^2(\alpha, x) = f\left(\alpha, f(\alpha, x)\right) = \left(x^3 - (\alpha + 1)x\right) - (\alpha + 1)(x^3 - (\alpha + 1)x)$$
$$= x - 2x^3 + 3x^5 - 3x^7 + x^9 + 2x\alpha - 4x^3\alpha + 6x^5\alpha$$
$$- 3x^7\alpha + x\alpha^2 - 3x^3\alpha^2 + 3x^5\alpha^2 - x^3\alpha^3.$$

Then, at the point $(0, 0)$, we have

$$f^2(0,0) = 0, \quad \text{and} \quad \frac{\partial (f^2)}{\partial x}(0,0) = 1,$$

where we can calculate the second quantity by the Chain Rule. But we also have here

$$\frac{\partial (f^2)}{\partial \alpha}(0,0) = \frac{\partial^2 (f^2)}{\partial x^2}(0,0) = 0, \quad \frac{\partial^2 (f^2)}{\partial x \alpha}(0,0) = 2, \quad \text{and} \quad \frac{\partial^3 (f^2)}{\partial x^3}(0,0) = 12.$$

By Proposition 2.73, f^2 experiences a pitchfork bifurcation at $(0,0)$. Looking at the bifurcation diagram with this new data added (Figure 161 at right), we have a much better picture of behavior; to fourth order,

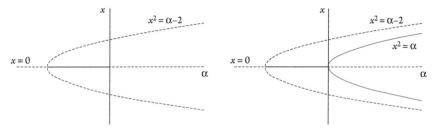

Figure 161 Bifurcation diagram for f (left) and f^2 (right), displaying a period-doubling bifurcation.

$$f^2(\alpha,x) = x - 2x^3 + 2\alpha x + \alpha^2 x + O(4)$$
$$= x + \alpha(2+\alpha)x - 2x^3 + O(4).$$

Exercise 300 Show that the real solutions to $f^2(\alpha,x) - x = 0$, are $x = 0$, $x = \pm\sqrt{\alpha}$, and $x = \pm\sqrt{2+\alpha}$.

Remark 7.91 *Going back to Remark 2.72, this was the complication mentioned; limiting our study to $\alpha \in (-2,1)$ allowed us to neglect to mention another bifurcation, since we (1) did not need it at the time and (2) did not yet have the tools to study it. The function $f(\alpha,x)$ in this section is the negative of the function used in Section 2.4.3, and the change was made to highlight different information.*

We can collect all of this data into the following:

Proposition 7.92 *Consider a family of C^3 maps $x \mapsto f(\alpha,x)$ parameterized by $\alpha \in \mathbb{R}$. Suppose that at (α_0, x_0), we have*

$$f(\alpha_0, x_0) = x_0, \quad \text{and} \quad \frac{\partial f}{\partial x}(\alpha_0, x_0) = -1.$$

Then, if $f^2(\alpha,x) = f(\alpha, f(\alpha,x))$ satisfy all of the conditions of a pitchfork bifurcation at (α_0, x_0), f has a period-doubling bifurcation at $\alpha = \alpha_0$, which is supercritical or subcritical according to whether the pitchfork bifurcation of f^2 is supercritical or subcritical at $\alpha = \alpha_0$.

Back to the logistic map: At $\lambda = 3$, the fixed point is $x_\lambda = 1 - \frac{1}{3} = \frac{2}{3}$, and

$$f_3\left(\tfrac{2}{3}\right) = 3\left(\tfrac{2}{3}\right)\left(1 - \tfrac{2}{3}\right) = \tfrac{2}{3}, \quad \text{and}$$
$$f_3'\left(\tfrac{2}{3}\right) = 3 - 6x\big|_{x=\frac{2}{3}} = -1.$$

Denote by $g_\lambda(x) = f_\lambda^2$ the square of f_λ, so that

$$g_\lambda(x) = \lambda^2 x\left(1 - (1+\lambda)x + 2\lambda x^2 - \lambda x^3\right).$$

It is easy now to verify that

$$\frac{\partial g_\lambda}{\partial \lambda}\left(3, \tfrac{2}{3}\right) = \frac{\partial^2 g_\lambda}{\partial x^2}\left(3, \tfrac{2}{3}\right) = 0$$

and

$$\frac{\partial^2 g_\lambda}{\partial x \partial \lambda}\left(3, \tfrac{2}{3}\right) = 2 > 0 \text{ and } \frac{\partial^3 g_\lambda}{\partial x^3}\left(3, \tfrac{2}{3}\right) = -108 < 0.$$

Hence f_3 has a period-doubling bifurcation, where the stable fixed point x_λ switches from asymptotically stable to unstable and a stable period-2 orbit is born. See Figure 162.

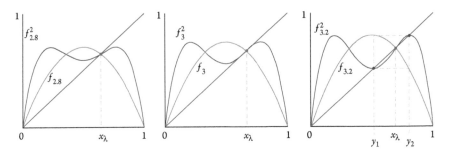

Figure 162 f and f^2, before (left), during (center), and after (right) the period-doubling bifurcation at $\lambda = 3$.

Finally, what happens when $\lambda = 1 + \sqrt{6}$? Basically, the same thing, except that the period-2 orbit becomes a repelling orbit, joining the fixed point as part of the repelling set, and an attracting period-4 orbit is born! This is another period-doubling bifurcation, as can be seen in Figure 163. Note that in the figure on the bottom, we have replaced f^2 with f in order to cobweb the period-4 orbit. One can easily see in the graph of f^4 that both the fixed point and the period-2 orbit are repelling by their slope at the intersection of the graph with the line $y = x$.

Theorem 7.93 *There exists a monotonic sequence of parameter values* $\lambda_1 = 3, \lambda_2 = 1 + \sqrt{6}, \lambda_3 = \ldots$, *such that* $\forall \lambda \in (\lambda_n, \lambda_{n+1})$, *the quadratic map* $f_\lambda(x) = \lambda x (1 - x)$ *has an attracting period-2^n orbit, two repelling fixed points at* $x = 0, x_\lambda$, *and one repelling period-2^k orbit for each* $k = 1, \ldots, n - 1$.

Notes:

- This is called a *period-doubling cascade*.
- At every new λ_n, the previous attracting periodic orbit becomes repelling, and adds (with all of its pre-images) to the universal repeller.
- The lengths of the intervals $(\lambda_n, \lambda_{n+1})$ decrease exponentially as n increases, and go to 0 at a particular value $\lambda = \lambda_c \approx 3.5699456\ldots$.
- In fact, one can calculate the exponential decay of these interval lengths:

$$\delta = \lim_{n \to \infty} \frac{\text{length}\,(\lambda_{n-1}, \lambda_n)}{\text{length}\,(\lambda_n, \lambda_{n+1})} = \lim_{n \to \infty} \frac{\lambda_n - \lambda_{n-1}}{\lambda_{n+1} - \lambda_n} \cong 4.6992016010\ldots$$

This number has a universal quality to it, since it is always the exponential decay rate of the lengths between bifurcation values in period-doubling cascades. It is called the *Feigenbaum number*.

One can follow the cascade of period-doubling bifurcations in a sort of bifurcation diagram, shown here in Figure 164. Note that this version of a bifurcation diagram is a bit different, since it categorizes not only fixed-point behavior, but attractive behavior

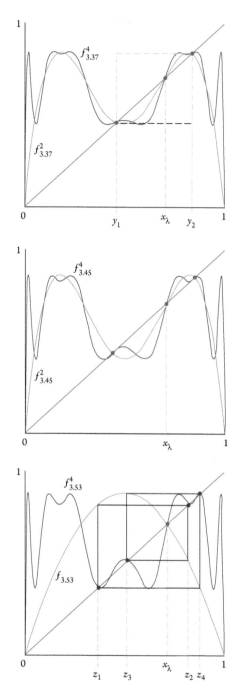

Figure 163 f^2 and f^4, before (top), during (center), and after (bottom) the period-doubling bifurcation at $\lambda = 1 + \sqrt{6}$.

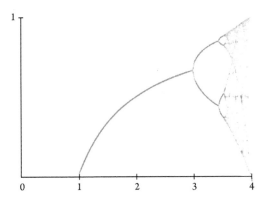

Figure 164 The bifurcation diagram for the logistic map f_λ on the interval $\lambda \in [0, 4]$.

in general. This includes attracting period-n orbits, when they exist. At the end of the period-doubling cascade, where $\lambda = \lambda_c$, is a place called the *transition to chaos*. At this point, there are a countable number of repelling periodic points, namely, all of the order-2^n periodic points, for $n \in \mathbb{N}$. And this repelling collection is dense in a certain subregion of the unit interval bounded below and above by the critical value $f_\lambda\left(\frac{1}{2}\right)$ and its next iterate, respectively. Initially, this subregion is a collection of intervals formed by future iterates of the critical value, arranged in a sort of Cantor-like construction. But, as the parameter increases further, these regions quickly begin to coalesce and, eventually, form a single region. All starting points in $[0, 1]$ have orbits that are attracted to this region, and, once inside, never leave. And, inside the region (the attractor) itself, the dynamics are chaotic (the critical point of f_λ provides an example of a dense orbit). But does this attractor satisfy the definition of what is called a *strange attractor*? A strange attractor is one that has a fractal structure. Many strange attractors are also chaotic, but this is not true of all strange attractors. In the present case, at the onset of chaos, there is definitely a self-similar feature to this subregion: magnify the diagram near any particular arm reaching toward the onset of chaos, and one would see another copy of mostly the entire bifurcation diagram. And, at $\lambda = \lambda_c$, where chaos begins, there are a infinite number of distinct regions comprising the attractor. and magnifying any subinterval would reveal even more of them. P. Grassberger[22], in 1980, estimated the Hausdorff dimension (a notion of fractional dimension similar to what we will define in Section 8.2.3 as box dimension) of the attractor here to be something like 0.538. As was mentioned back in Section 3.4 when we first discussed fractals, fractional dimension is another hallmark of a fractal. And this renders the attractor here as strange! However, as λ continues to increase, these regions coalesce into larger and larger blocks, ultimately taking up the entire interval as λ reaches 4. In Figure 165 is the cobweb of the critical point $x = \frac{1}{2}$ of the first few hundred iterates for $\lambda = 3.58 > \lambda_c$, along with where λ sits in the bifurcation diagram.

Look carefully at the bifurcation diagram within the chaotic region, expanded in Figure 166. Even after the transition to chaos, there seem to be regions of calm (read:

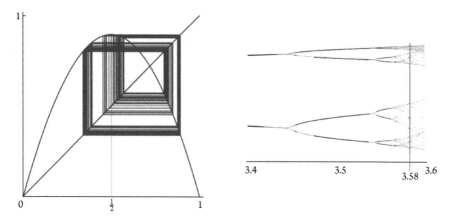

Figure 165 For $\lambda = 3.58$, the orbit of $x = \frac{1}{2}$ and the position where λ sits in the bifurcation diagram.

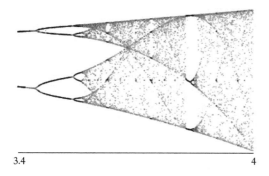

Figure 166 Within the regions of great complexity, there are regions of calm, asymptotically stable periodic behavior and more period-doubling cascades.

asymptotically stable) periodic behavior. These are not artifacts. In fact, there exists an attracting period-3 orbit (can you see it?) for a small band of values of λ.

This attracting period-3 orbit, seen within the map itself in Figure 167, is actually part of a pair of period-3 orbits born simultaneously, as lobes of the graph of f^3 evolve to cross the line $y = x$. The resulting stable period-3 orbit, as the lobe crosses and lengthens, has a slope at the crossing that eventually becomes 0 and then negative and then less than -1, becoming repelling. This bifurcation with a slope of -1 is the period-doubling bifurcation that creates the stable period-6 orbit. And the cascade recommences (period-6 to period-12, etc.). In fact, there exists a period-doubling cascade within this diagram for each prime number n.

Exercise 301 Find two distinct intervals of the parameter $\lambda \in [0,4]$ where the quadratic map $f_\lambda(x)$ has an attracting, periodic point of prime period 6. Approximate, as best as you can, the maximum sizes of these two distinct intervals.

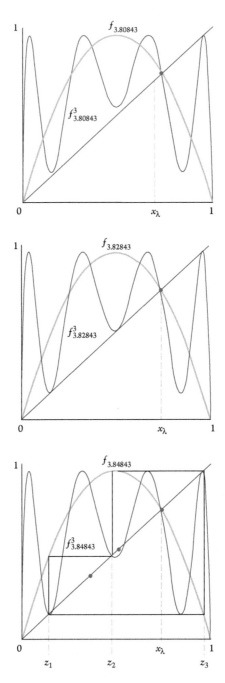

Figure 167 f and f^3, before (top), during (center), and after (bottom) the creation of a stable period-3 orbit.

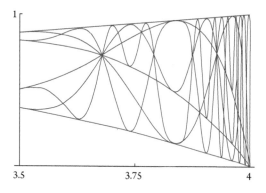

1

3.5   3.75   4

Figure 168 Where do these curves come from? Can you see the stable period-3 interval?

Exercise 302 The curves shown in Figure 168 are reconstructions of the shadows one can see in Figure 166, and are related to the orbit of the critical point $x = \frac{1}{2}$. Recreate this diagram by constructing and graphing these curves.

Now, the fact that these bifurcations occur and the attracting nature of one orbit seems to be handed off to the new orbit at each stage may seem strange at first. However, it is quite common among a class of 1-dimensional maps that have a certain feature; Their *Schwarzian derivative* is negative.

7.5.2 The Schwarzian Derivative

First defined by Joseph Louis de Lagrange in 1781 [16], the Schwarzian derivative was named after Hermann Schwarz due to its use in his study of conformal maps. It was David Singer in 1978 [52] who first applied it to the dynamical study of interval maps. It turns out that functions whose Schwarzian derivatives are negative have quite remarkable properties:

Definition 7.94 *For* f *a* C^3 *function on* $X \subset \mathbb{R}$, *the Schwarzian derivative of* f *at* $x \in X$ *is*

$$Sf(x) = \begin{cases} \dfrac{f'''(x)}{f'(x)} - \dfrac{3}{2}\left(\dfrac{f''(x)}{f'(x)}\right)^2, & f'(x) \neq 0, \\ -\infty, & f'(x) = 0. \end{cases}$$

Note that defining it this way allows us to discuss functions as having a Schwarzian derivative (at least its sign) even at their critical points. And, as a matter of shorthand, if the Schwarzian derivative of a function f is everywhere negative (or everywhere positive), we will write $Sf < 0$ ($Sf > 0$).

Example 7.95 Exponential maps have negative Schwarzian derivative. Let $g(x) = e^{ax}$, $a \in \mathbb{R}$, $a \neq 0$. Then

$$Sg(x) = \frac{a^3 e^{ax}}{ae^{ax}} - \frac{3}{2}\left(\frac{a^2 e^{ax}}{ae^{ax}}\right)^2 = a^2 - \frac{3}{2}a^2 = -\frac{a^2}{2}.$$

Exercise 303 Show that for $a, b, c, d \in \mathbb{R}$, where c and d are not both 0, the Schwarzian derivative of the fractional transformation $(ax + b)/(cx + d)$ is 0.

Exercise 304 Show that $\sin x$ and $2 \arctan x$ have negative Schwarzian derivatives.

In fact, the class of functions that have negative Schwarzian derivative is quite large:

Proposition 7.96 *Let $p(x)$ be a polynomial on $\mathbb{R}$, all of whose critical points are real and distinct (as roots of $p'(x)$). Then $Sp < 0$.*

Proof The proof is simple calculus. Factor $p'(x) = C \cdot \prod_{i=1}^{n}(x - a_i)$ for C a constant and a_i the n roots of the derivative. Then, via logarithmic differentiation,

$$\frac{d}{dx} \ln |p'(x)| = \frac{p''(x)}{p'(x)} = \sum_{i=1}^{n} \frac{1}{x - a_i}.$$

Note that the constant C is gone. And

$$\frac{d}{dx} \left(\frac{p''(x)}{p'(x)} \right) = \frac{p'''(x)p'(x) - \left(p''(x)\right)^2}{\left(p'(x)\right)^2}$$

$$= \frac{p'''(x)}{p'(x)} - \left(\frac{p''(x)}{p'(x)}\right)^2 = -\sum_{i=1}^{n} \frac{1}{(x - a_i)^2}.$$

This derivative is almost the Schwarzian derivative of $p(x)$. In fact,

$$Sp(x) = \frac{p'''(x)}{p'(x)} - \left(\frac{p''(x)}{p'(x)}\right)^2 - \frac{1}{2}\left(\frac{p''(x)}{p'(x)}\right)^2$$

$$= -\sum_{i=1}^{n} \frac{1}{(x - a_i)^2} - \frac{1}{2}\left(\sum_{i=1}^{n} \frac{1}{x - a_i}\right)^2 < 0. \qquad \square$$

Example 7.97 The logistic map $f_\lambda : \mathbb{R} \to \mathbb{R}$ satisfies the supposition of Proposition 7.96 and has negative Schwarzian derivative for all $\lambda \neq 0$. Of course, we can also simply calculate it and obtain

$$Sf_\lambda = -\frac{3}{2}\left(\frac{-2}{1 - 2x}\right)^2.$$

Exercise 305 Show that if $f, g \in C^3$, then on the domain in $\mathbb{R}$ where the composition makes sense, we have

(7.5.1) $$S(f \circ g)(x) = \left(Sf \circ g(x)\right)\left(g'(x)\right)^2 + Sg(x).$$

Call Equation 7.5.1 the Schwarzian Chain Rule. Using it, we can immediately conclude the following: If a function has Schwarzian derivative that is everywhere positive or everywhere negative, then so do all of its iterates.

Proposition 7.98 *Suppose $Sf \leq 0$ (respectively ≥ 0) and $Sg < 0$ (respectively > 0). Then $S(f \circ g) < 0$ (respectively > 0).*

Proof Just use Equation 7.5.1. □

And lastly, a property that sounds innocuous but has remarkable value, as we will see shortly:

Proposition 7.99 *Let $Sf < 0$. Then f' can have neither a positive local minimum nor a negative local maximum.*

Proof At a critical point c of f', we have $f''(c) = 0$. Then

$$0 > Sf(c) = \frac{f'''(c)}{f'(c)}.$$

Hence $f'''(c)$ and $f'(c)$ must have opposite signs. Now use the Second Derivative Test from calculus. □

The property of having a negative Schwarzian derivative intimately ties together the general dynamics of a map f with its critical points. This idea is the central theme of the following result, which we will prove presently:

Theorem 7.100 *Let f be a C^3 map on $\mathbb{R}$ with $Sf < 0$. If f has n critical points, then f has, at most, $n + 2$ attracting periodic orbits.*

Some notes:

- Critical points include the (in)famous turnaround points of polynomial graphs, where a function stops increasing and starts to decrease (a local maximum) or stops decreasing and starts to increase (a local minimum). For a dynamical system based on a map with turn-around points, the dynamics can be quite complicated.
- For the logistic map f_λ, there is only one critical point, at $x = \frac{1}{2}$ (for $\lambda \neq 0$, anyway). Think about the consequences of this.
- Notice the similarities between this theorem and Proposition 7.86. In this case, we will see that the closed, bounded domain affects the possible number of attracting periodic orbits.

Proof First, we will establish that, since f, by supposition, has a finite number of critical points, then so does f^m, for $m \in \mathbb{N}$. Indeed, if the number of critical points is finite, then the cardinality of any inverse image $f^{-1}(x)$ is also finite. This remains true for inverse images of inverse images, so that for any x, $f^{-m}(x)$ is finite for any $m \in \mathbb{N}$. Let c be a critical point of f^m for some $m \in \mathbb{N}$, so $(f^m)'(c) = 0$. By the Chain Rule, then, there is an $i \in \{0, 1, \ldots, m-1\}$ such that $f^i(c)$ is critical for f. Thus the critical point set of f^m comprises points x whose partial orbit $\mathcal{O}_x^{m-1}$ contains a critical point of f.

As this set is then a finite union of the inverse images of the critical set of f (up to m iterates back) and each of these forward partial orbits, each of which is also finite, the union is finite also.

Second, with the suppositions here, we also claim that $P_n(f) < \infty$ for all $n \in \mathbb{N}$ (the number of n-periodic points is also finite). To establish this claim, suppose there is an $m \in \mathbb{N}$ where $P_m(f) = \infty$, and let $g = f^m$, so then $\mathrm{Fix}(g) = \infty$. Then, by the Mean Value Theorem, there are infinitely many points x where $g'(x) = 1$. However, note that these points must be isolated, since if there were an interval on which $g'(x) = 1$, then on this interval $Sg = 0$, but by supposition and Proposition 7.98, $Sg < 0$. So let $x_1, x_2, x_3 \in \mathrm{Fix}(g)$ be adjacent fixed points. Then, there exist points in the interval (x_1, x_3) where $g'(x) < 1$. But $g'(x)$ cannot have a positive local minimum. Hence there must also be points in the interval (x_1, x_3) where $g'(x) < 0$, and hence points where $g'(x) = 0$. But there are an infinity of such intervals, so that there must be an infinite number of critical points of g, contradicting the first claim. Hence the second claim holds.

And finally, let p be an attracting m-periodic point, and let $B_m(p)$ be the immediate basin of attraction for f^m, as a fixed point. Thus $B_m(p)$, as a set, is open, connected, and f^m-invariant. Suppose $\overline{B_m(p)}$ is compact, and denote by $\{y_1, y_2\} = \overline{B_m(p)} - B_m(p)$ its set of endpoints. Then one of the following cases must occur:

- $f^m(y_i) = y_i, i = 1, 2;$
- $f^m(y_1) = y_2$, and $f^m(y_2) = y_1;$ or
- $f^m(y_1) = f^m(y_2).$

Suppose the first case occurs: Then it must be the case that the restricted dynamical system of f^m on $[y_1, y_2]$ has both endpoints repelling, and $(f^m)'(p) < 1$. So there are points $q_1, q_2 \in (y_1, y_2)$ satisfying $q_1 < p < q_2$ and $(f^m)'(q_1) = (f^m)'(q_2) = 1$. But then, as before, $(f^m)'(x)$ can have no positive local minimum on (y_1, y_2), so there must be a place where $(f^m)'(x) < 0$. So there is a point $c \in (y_1, y_2)$ where $(f^m)'(x) = 0$. But then, within $B_m(p)$, there is a critical point. For the second case, where the images are switched on y_1 and y_1, take the map f^{2m}. For the third case, where $f^m(y_1) = f^m(y_2)$, Rolle's Theorem implies that there must be a critical point in (y_1, y_2).

In the case where the immediate basin of attraction of p under f^m is unbounded on one side, it is possible to construct a single additional attracting m-period orbit. Thus there can be at most two additional orbits above the number of critical points. □

Hence, if there is an attracting periodic orbit, then there is a critical point in its basin of attraction. Hence we can always find the attracting periodic orbits by following the orbits of the critical points. Note, however, that it is possible to have no attracting periodic orbit:

Example 7.101 The map $f_4 : [0, 1] \to [0, 1]$ has a single critical point at $x = \frac{1}{2}$. But $\frac{1}{2}$ is a pre-image of the fixed point $x = 0$, which is repelling. Hence there are no attracting periodic orbits of the logistic map at $\lambda = 4$.

Example 7.102 Let $g(x) = 2\arctan x$. By Exercise 304, $Sg < 0$. g also has no critical points, and yet has two attracting fixed points. Can you find them?

And finally, it should be much clearer now just how to construct the bifurcation diagram in Figure 164. Since every attracting orbit must have a critical point in its basin of attraction, and since the logistic map f_λ only ever has one critical point, simply graph the orbit of the critical point to find the attractor. When the attractor is a periodic orbit, after a few iterates, one would see only the periodic orbit. When the attractor is the strange one on a set of subintervals, graphing a large number of iterates will "fill out" the interval, creating the gray regions in the diagram. To construct the diagram, for each value of λ, construct an orbit of a few hundred iterates, then disregard the first few. At that point, the orbits will all be inside (or indistinguishable from) the attractor, and only the attractor will be graphed.

Exercise 306 Using technology (of course), construct the bifurcation diagram for the logistic map on the interval $\lambda \in [0,4]$ (this is Figure 164).

The logistic map has some other very interesting properties. The period-doubling cascades and the strange patterns of calm areas within the chaotic regions all have a particular pattern to them, noticed and documented in the 1960s by Oleksandr Sharkovskii[50].

7.5.3 Sharkovskii's Theorem

One may notice that in the period-doubling cascade in the logistic map's bifurcation diagram, all periodic orbits are powers of 2, and that at the end of the cascade, all orbits are repelling. This marks a transition to a chaotic map where these repelling periodic points are dense in a particular region. Later, once a calm, periodic region occurs, another period-doubling cascade ensues and another transition to chaos. And, on even closer inspection, between the calm regions inside the chaotic region, there seem to be even more calm regions, all of varying periods. Nothing is random here, though, since the occurrence of these periods of calm and the cascades all follow a pattern.

Consider the following reordering of the natural numbers, using ">" as the ordering symbol:

$$3, 5, 7, 9, \ldots, (2n+1) \cdot 2^0, \ldots$$
$$3 \cdot 2, 5 \cdot 2, 7 \cdot 2, \ldots, (2n+1) \cdot 2^1, \ldots$$
$$3 \cdot 2^2, 5 \cdot 2^2, 7 \cdot 2^2, \ldots, (2n+1) \cdot 2^2, \ldots$$
$$\vdots$$
$$3 \cdot 2^r, 5 \cdot 2^r, 7 \cdot 2^r, \ldots, (2n+1) \cdot 2^r, \ldots, \quad \text{for each } r \in \mathbb{N},$$
$$\vdots$$
$$\ldots, 2^s, 2^{s-1}, \ldots, 2^3, 2^2, 2^1, 1 \quad \text{for each } s \in \mathbb{N}.$$

Some notes:

- Here, every natural number is included only once

- This is not a *well-ordering*. Recall that a *total ordering* on a set is a relation $\leq$ that is antisymmetric (if $a \leq b$ and $b \leq a$, then $a = b$), transitive (if $a \leq b$ and $b \leq c$, then $a \leq c$), and total (for all a, b, either $a \leq b$ or $b \leq a$). And a strict total ordering is an ordering relation $<$ where $a < b$ iff $a \leq b$ and $a \neq b$. Then a well-ordered set is one where every non-empty subset has a least element. But in this ordering, the subset

$$\{2^n + 1 \,|\, n \in \mathbb{N}\}$$

has no least element.

- This new ordering of $\mathbb{N}$ is due to Sharkovskii and is called the *Sharkovskii ordering* of the natural numbers.

This leads directly to the following theorem:

Theorem 7.103 (Sharkovskii's Theorem.) *Suppose a continuous $f : \mathbb{R} \to \mathbb{R}$ has a periodic point of prime period m. Then, for every natural number n where $m \succ n$, f also has a periodic point of prime period n.*

We will not prove this theorem here, since we are now simply touring the structure of the quadratic map. A good proof and discussion can be found in, for example, Devaney [18]. This result leads immediately to some rather obvious conclusions:

Corollary 7.104 *For $f : \mathbb{R} \to \mathbb{R}$ continuous, if f has a period-3 point, then f has periodic points of all periods.*

Corollary 7.105 *If a continuous $f : \mathbb{R} \to \mathbb{R}$ has only a finite number of periodic points, then all periods must be powers of 2.*

Corollary 7.106 *If a continuous $f : \mathbb{R} \to \mathbb{R}$ has no period-2 orbit, then it has no period-m orbits for any natural $m > 1$.*

Remark 7.107 *In a 1975 paper entitled "Period three implies chaos" [29], Tien-Yien Li and James Yorke introduced the term* chaos, *showing that any continuous map on $\mathbb{R}$ that has a 3-periodic point has two properties: (1) it has periodic points of all orders and (2) there is an uncountably infinite set S that is scrambled: each pair of points in S has the property that their orbits are at times close together and at other times far apart, and this behavior repeats itself in an irregular pattern. This is a type of mixing feature, and behavior like this is sometimes called "Li–Yorke chaos".*

7.6 Two More Examples of Complicated Dynamical Systems

7.6.1 Complex Dynamics

Consider the polynomial $P_c : \mathbb{C} \to \mathbb{C}$, $P_c(z) = z^2 + c$, first introduced in Example '7.83 in Section 7.5 when discussing basins of attraction. This family of maps of the complex plane is sometimes called the *Douady–Hubbard family of quadratic polynomials*. As maps on the complex plane, they are seen as the simplest one-parameter, nonlinear family of discrete dynamical systems on C. Each has a single critical point at $z = 0$ for all values of $c \in \mathbb{C}$, and each is unimodal and *entire*, an entire function being one that is complex differentiable at all points of $\mathbb{C}$. Of course, the critical value of $P_c(z)$ is always $P_c(0) = c$, and the critical orbit

$$\mathcal{O}_0(P_c) = \left\{ 0, c, c^2 + c, \left(c^2 + c\right)^2 + c, \ldots \right\}$$

is of special interest in the study of $P_c(z)$, as it was for the logistic map family. First, it is obvious that only for $c = 0$ is the origin a fixed point of $P_c(z)$ (along with the real number 1, that is). But, for small (in modulus) values of c, one would expect $\mathcal{O}_0$ to be bounded, and unbounded for large values of c. Studying the dependence of the critical orbit on the value of c becomes an interesting endeavor that, like that of the logistic map, displays highly complicated dynamics. To start, we define a few new concepts useful for this study:

Definition 7.108 *For $f(z)$ a complex rational function (polynomials are, of course, also rational), the Julia set $J(f)$ is the closure of the set of repelling periodic points of f.*

Some notes:

- $J(f)$ is a closed, invariant set.
- If $J(f)$ is a connected set, it will then be a (maybe quite complicated) curve densely filled with repelling periodic points, but with many non-periodic points. Thus $f\big|_{J(f)}$ should display a sensitive dependence on initial conditions (should be chaotic).
- $J(P_0) = S^1$ is a connected set, the unit circle in the complex plane. But, on this unit circle, $P_0\big|_{J(P_0)} = E_2$, the chaotic expanding map on the circle. Recall the map E_2 defined in Equation 7.1.1 in Section 7.1.
- For $c = -2$, the map P_{-2} has Julia set the (real) interval $[-2,2] \subset \mathbb{C}$. And, as you will see soon in Exercise 315 in Chapter 8, the map P_{-2} on $J(P_{-2}) = [-2,2]$ is topologically conjugate to the Ulam map (the logistic map for $\lambda = 4$), $f_4(x) = 4x(1-x)$ on $[0,1]$ (also see Remark 8.5 in Section 8.1). It turns out that these are the only two values of $c \in \mathbb{C}$ where the Julia set is a smooth curve (one with boundary).
- For $c < -2$ and real, $J(P_c)$, as a subset of the real axis in $\mathbb{C}$, is a Cantor set, and $P_c\big|_{J(P_c)}$ is topologically conjugate to the logistic map $f_\lambda : C_\lambda \to C_\lambda$, for some $\lambda > 4$. You will construct the conjugacy in Exercise 315 in Section 8.1.
- $J(P_c)$ is either a connected set (Definition 4.57), a curve (although usually a complicated curve), or totally disconnected (Definition 3.52), like a Cantor set.
- If f is an entire function, as complex polynomials are, then $J(f)$ is the boundary of the set of points whose orbits converge to infinity under f.
- If f is a polynomial, then $J(f)$ is also the boundary of what is called the *filled Julia set*

$$\Lambda = \left\{ z \in \mathbb{C} \mid \mathcal{O}_z(f) \not\to \infty \right\}.$$

For $f = P_c$, we will denote the filled Julia set by Λ_c.

Definition 7.109 *For $f(z)$ a complex rational function on $\mathbb{C}$, a Fatou domain is one of a finite number of open invariant sets such that the union of the domains is dense in the plane, and, on each domain, f acts equally on all of its points. The union of the domains is called the Fatou set, denoted by $F(f)$.*

Again, some notes:

- $F(f)$ is an open set, and each of its domains often corresponds to a basin of attraction for some fixed or periodic point. This is especially true if one thinks about the complex plane as having a single point at infinity, so that f acts on $\mathbb{C} \cup \{\infty\}$. Then one can interpret the orbits that converge to infinity as being in the basin of attraction of infinity.

- As an example,

$$F(P_0) = \left\{ z \in \mathbb{C} \mid |z| < 1 \right\} \cup \left\{ z \in \mathbb{C} \mid |z| > 1 \right\}.$$

- It should be noted that

$$\mathbb{C} = J(f) \cup F(f) \quad \text{and} \quad J(f) \cap F(f) = \emptyset.$$

  For f a polynomial, the complement of the Fatou domain that forms the basin of attraction of infinity is the filled Julia set.

- The Fatou set $F(f)$ is also invariant under f, since it is the complement of $J(f)$.

- The sets $F(f)$ and $J(f)$ are names after the French mathematicians Pierre Fatou and Gaston Julia, owing to their work early in the twentieth century in an area now referred to as *holomorphic dynamics*.

And the last component of our very brief foray into complex dynamics is to document how parameter values affect these sets. With that, let's define a parameter space for $P_c(z) = z^2 + c$, and for each $c \in \mathbb{C}$, let's color it as follows:

- black if the orbit of 0, given c, $\mathcal{O}_{0,c} = \left\{ P_c^n(0) \right\}_{n \in \mathbb{N}}$ is bounded; and
- a choice of spectrum color depending on how fast $\mathcal{O}_{0,c}$ tends to infinity.

Then the set comprising the black points is called the *Mandelbrot set* $\mathcal{M}$, as seen in Figure 169.

Strictly speaking, the Mandelbrot set is the bifurcation diagram of the Douady–Hubbard family of maps $P_c : \mathbb{C} \to \mathbb{C}$, where each point $c \in \mathbb{C}$ is marked by color as to how the critical orbit behaves. This is an intensity plot, a 2-dimensional projection of what otherwise would be a 3-dimensional plot. The "edge" of the black region is highly complicated and fractal in nature, although it is not self-similar. Rather, it is a more general version sometimes called *quasi-self-similar*; At smaller and smaller scales, one can find copies of the original shapes. However, these copies are skewed, or degenerate, and certainly not exact copies.

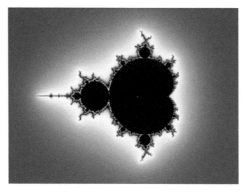

Figure 169 The Mandelbrot set for $P_c(z) = z^2 + c$.

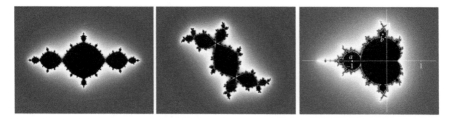

Figure 170 The Julia sets of $P_c(z) = z^2 + c$: the Basilica, for $c = -1$ (left) and the Douady Rabbit, $c = -0.123 + 0.754i$ (center).

As a parameter space, the Mandelbrot set parameterizes a family of Julia sets, one for each value of $c \in \mathbb{C}$, whose shapes and properties seem to be as stunningly beautiful as they are varied. We will not do more than display a few before moving on. Two of the more famous have special names: the Basilica, in Figure 170 at left, for $c = -1$, and the Douady Rabbit, named after the French mathematician Adrien Douady, at center, for $c = -0.123 + 0.754i$.

Notice the following:

- The black regions on these two figures correspond to components of the Fatou set, basins of attraction for either fixed or periodic points.

- In both of these figures, the Julia sets are connected (though complicated) sets. This is true for any value $c \in \mathcal{M}$ (in the black region), indicating that $\mathcal{O}_0$ is bounded. However, for values of c outside of $\mathcal{M}$, the Julia set is totally disconnected. See Figure 171. Because, as sets, they are totally disconnected, it is common to refer to the Julia sets, for values of c outside of the Mandelbrot set, as *Fatou dust*.

Exercise 307 Show that, if $|z_0| \geq \max\{|c|, 2\}$, then $\mathcal{O}_{z_0}(P_c) \to \infty$.

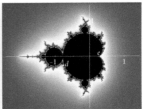

Figure 171 The Julia sets for $c = -\frac{3}{4} - \frac{1}{4}i$ (left) and $c = 0.32$ (center).

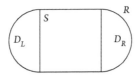

Figure 172 A stadium.

7.6.2 Smale Horseshoe

The linear horseshoe, designed, apparently, by Stephen Smale on the famous Copacabana Beach in Rio de Janeiro, Brazil, in the 1970s, is a highly refined discrete dynamical system in the plane that displays complicated orbit behavior in a fashion much like that of the logistic map f_λ, for $\lambda > 2 + \sqrt{5}$, or the tent map T_r, for $r > 2$. The strength of this map, however, is that it is a linear prototype of behavior easily discoverable in many nonlinear dynamical systems, ensuring at least a level of local complexity in an otherwise hard-to-study system. We present here the basic structure of the map and its orbits as another example of complicated behavior that can be studied symbolically.

To start, let $S \in \mathbb{R}^2$ be the unit square, although we will neglect any coordinates. Cap both sides of S with semicircles D_L and D_R both of diameter 1 to construct a shape commonly referred to as a *stadium*; see Figure 172.

Call this total region R. Now construct a map $f : R \to R$ that does the following:

(a) contracts S vertically by a factor a^{-1} for some $a > 2$, contracting the two caps so that they remain caps of the rectangle $f(S)$;

(b) stretches S horizontally by a, so that $f|_S$ is an area-preserving linear map;

(c) folds the image $f(R)$ into R so that the region D_L is invariant and the region D_R maps into region D_L, all as on the left side of Figure 173.

The map f has some very interesting properties:

1. f can be made a diffeomorphism onto its image. We assume that it is.

2. Restricted to D_L, $f|_{D_L}$ is a contraction.

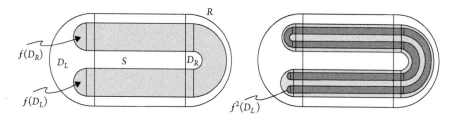

Figure 173 The Smale horseshoe map.

3. $f(D_R) \subset D_L$, so anything that enters either semicircle of the stadium winds up in L in at most the next iterate. Hence everything that lands outside of S is part of the basin of attraction of the fixed point of the contraction.

4. Under iteration by f, most points of S also have orbits that enter D_L at some point.

5. Do any points remain in S indefinitely? That is, is

$$\Lambda = \{x \in S_0 \mid \mathcal{O}_x \subset S_0\}$$

a non-empty set?

To look for points that remain in S throughout their forward orbit, denote the two disjoint regions of $S \cap f(S)$ by S_0 and S_1, so that $S \cap f(S) = S_0 \cup S_1$. Here S_0 and S_1 are both rectangles of unit width and height a^{-1}. Also, we can define the set $S \cap f^{-1}(S) = S^0 \cup S^1$ as the collective pre-images of S_0 and S_1, respectively, so that $f(S^0) = S_0$ and $f(S^1) = S_1$. Here, S^0 and S^1 are disjoint, full-height rectangles of width a^{-1}, as shown in Figure 174.

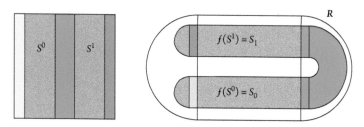

Figure 174 The relationship between S^0 and S_0.

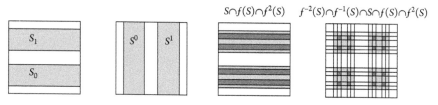

Figure 175 Locating points whose orbits persist in S_0.

Now note that the points whose orbits after two iterates are still in S comprise four disjoint, full-width rectangles of height a^{-2}, which we can encode as

$$S \cap f(S) \cap f^2(s) = S_{00} \cup S_{01} \cup S_{11} \cup S_{10}, \quad S_{ij} = S_i \cap f(S_j),$$

as in Figure 175.

Continuing in this fashion, then, for any n-word $(w_0 w_1 \ldots w_{n-1})$ in the letters 0 and 1 (so an element of $\mathcal{M}_2^n = \{0,1\}^n$), we can define

$$S_{w_0 w_1 \ldots w_{n-1}} = \bigcap_{i=0}^{n-1} f^i(S_{w_i}) \subset \bigcap_{i=0}^{n} f^i(S)$$

as a full-width rectangle of height a^{-n}. There are exactly 2^n of them, one each for each distinct n-word, all mutually disjoint, and each a subset of $S_0 \cup S_1$. In the limit, we see that for any $\mathbf{w} \in \{0,1\}^{\mathbb{N}}$,

$$S_{\mathbf{w}} = \bigcap_{i=0}^{\infty} f^i(S_{w_i}) \subset \bigcap_{i=0}^{\infty} f^i(S)$$

is a horizontal line spanning S, so that the set $\bigcap_{i=0}^{\infty} f^i(S) = [0,1] \times C_v$, a Cantor set of lines in S.

Exercise 308 Show that the set C_v is perfect, compact, and totally discontinuous. (Hint: See the proof of Proposition 7.9 in Section 7.1.1 showing that the points in the unit interval whose orbits remain under the quadratic map f_λ, for $\lambda > 4$, is a Cantor set.)

Here, the set $\bigcap_{i=0}^{\infty} f^i(S) = [0,1] \times C_v$ is the set of all $x \in S$, such that $\mathcal{O}_x^+ \subset S$.

Exercise 309 Show $[0,1] \times C_v$ is not empty via the following: Let $B = S^0 \cap S_0$, and show that $f(B) \subset B$. Then use the Brouwer Fixed-Point Theorem to show that there is a fixed point of f in B. Conclude that this point must be in $[0,1] \times C_v$.

In a similar fashion, notice that the set $S \cap f^{-1}(S) \cap f^{-2}(S)$ can also be encoded so that

$$S \cap f^{-1}(S) \cap f^{-2}(S) = S^{00} \cup S^{01} \cup S^{11} \cup S^{10}, \quad S^{ij} = S^i \cap f^{-1}(S^j).$$

Then, as before, we can extend this so that for any n-word (strangely numbered) $(w_{-(n-1)} \ldots w_{-1} w_0)$, we can associate

$$S^{w_0 w_1 \ldots w_{n-1}} = \bigcap_{i=0}^{n-1} f^{-i}(S^{w_{-i}}) \subset \bigcap_{i=0}^{n} f^{-i}(S)$$

as a full-height rectangle of width a^{-n}. Again, there are exactly 2^n of them, all mutually disjoint, and each a subset of $S^0 \cup S^1$. Note the twist here (and the strange numbering) as these intersections of pre-images will correspond to backwards orbits.

In the limit, we get

$$S^{\mathbf{w}} = \bigcap_{i=0}^{\infty} f^{-i}(S^{w-i}) \subset \bigcap_{i=0}^{\infty} f^{-i}(S)$$

as the Cantor set of vertical, full-height limes $(C_h \times [0,1]) \subset S$. Now let

$$\Lambda = \big([0,1] \times C_v\big) \cap \big(C_h \times [0,1]\big) = C_h \times C_v.$$

Now let $\mathbf{w} \in \mathcal{M}_2^{\mathbb{Z}}$ be a bi-infinite sequence, written as $\mathbf{w} = (\mathbf{v}, \mathbf{u})$, where $v_i = w_{-(i+1)}$, and $u_i = w_i$, for $i \in \mathbb{N}$. Then define a map $h : \mathcal{M}_2^{\mathbb{Z}} \to \Lambda$ such that $h(\mathbf{w}) = S^{\mathbf{v}} \cap S_{\mathbf{u}}$. Then it should be clear that h is both injective and surjective, since, according to this split, every sequence $\mathbf{v}$ and $\mathbf{u}$ is uniquely realized. But also, both h and h^{-1} are continuous, since the image under h of the cylinder

$$B_{1/\rho^n}(\mathbf{w}) = \left\{ \mathbf{x} \in \mathcal{M}_2^{\mathbb{Z}} \;\middle|\; x_i = w_i, i = -n, \ldots, 0, \ldots, n \right\}$$

is precisely the $a^{-n} \times a^{-(n+1)}$ box $S^{w_{-1}\ldots w_{-n}} \cap S_{w_0 \ldots w_n}$.

Exercise 310 Show that $h \circ \sigma = f \circ h$.

With Exercise 310, we have that h is a homeomorphism and establishes a conjugacy between $(\mathcal{M}_2^{\mathbb{Z}}, \sigma)$ and (f, Λ).

Proposition 7.110 *The linear horseshoe $f : \Lambda \to \Lambda$, developed as above, is chaotic.*

Proof Adapt the arguments from Proposition 7.38 and Exercise 274, showing dense periodic points and a dense orbit of the shift map on $\mathcal{M}_2^{\mathbb{N}}$, to show that it is also true for the left shift on $\mathcal{M}_2^{\mathbb{Z}}$. Then, with the above conjugacy, the result is obtained. $\square$

8 Dynamical Invariants

ecall some of our indicators of dynamical complexity from earlier discussions: topological transitivity (Definition 5.15), topological mixing (Definition 7.68), minimality (Definition 5.16), density of periodic orbits, chaos (Definition 7.63), growth rates of periodic orbits, etc. These indicators of complexity all have in common that they are properties of the overall structure of the orbits. Hence, they should persist under a particular homeomorphism that takes orbits to orbits: a topological conjugacy. Indeed, fixed points must be mapped to fixed points under a conjugacy, and a conjugacy would create a one-to-one correspondence between the sets of n-periodic points. The asymptotics of orbits would have to correspond also, so that the stability of an orbit would have to be compatible with the stability of its image under the homeomorphism of the conjugacy. And this remains somewhat true if the map between dynamical systems is only surjective (but still taking orbits to orbits), establishing a semiconjugacy. One may say that if a map f is semiconjugate to a map g, then the dynamics of f are at least as complicated as those of g. For example, one can easily show that topological transitivity persists under a semiconjugacy.

Indeed, recall that, from Birkhoff's Transitivity Theorem 7.64, the definition of topological transitivity (Definition 5.15) as the existence of a dense orbit in a dynamical system is equivalent to the idea that the orbit of any open set must intersect any other open set. In fact, topological transitivity is often defined in the latter way. Presently, we use this fact:

Proposition 8.1 *For $f : X \to X$ semiconjugate to $g : Y \to Y$ via the surjective map $h : X \to Y$, if f is topologically transitive, then so is g.*

Proof Let f be topologically transitive, and $U, V \subset Y$ be non-empty and open. Then, since h is continuous and onto, $h^{-1}(U)$ and $h^{-1}(V)$ are non-empty and open in X. Since f is topologically transitive, there exists an $x \in h^{-1}(U)$ and a $k > 0$ such that $f^k(x) \in h^{-1}(V)$. But then, for $y = h(x), y \in U$ and $g^k(y) = h(f^k(x)) \in V$. $\qquad\qquad\square$

One should be careful not to generalize and assume that under a semiconjugacy, the two dynamical systems are equivalent. For a rather extreme example, consider $R_\alpha : S^1 \to S^1$, a rational rotation of the unit circle. Then the *constant map* $h : S^1 \to \mathbb{R}$, where $h(S^1) = p \in \mathbb{R}$, establishes a semiconjugacy between R_α and the dynamical system given by the map $g : \{p\} \to \{p\}$. h is certainly continuous and onto and takes orbits to orbits. And, even though the dynamics of g are quite simple, g is topologically transitive (in a trivial fashion), but R_α is not. Of course, under a conjugacy, Proposition 8.1 becomes a biconditional (each dynamical system is semiconjugate to the other in a conjugacy, since a homeomorphism and its inverse are both continuous and surjective).

The properties of a dynamical system that persist under a (semi)conjugacy are called *dynamical invariants* of the (semi)conjugacy. Under a conjugacy, we say that two dynamical systems are equivalent in that they are dynamically the same.

- This allows us to study a new dynamical system by establishing a conjugacy with a known one, knowing that the essential properties of the dynamical systems coincide.
- This allows us to classify dynamical systems, since everything conjugate to a known dynamical system has the same dynamical invariants.

And, under a semiconjugacy between the two dynamical systems in Proposition 8.1, we call g a *factor* of f, or say that f is an *extension* of g.

In this chapter, we will discuss the idea of topological conjugacy in more detail and use it to further study some of the features of both new and previously introduced dynamical systems. Then we will define a new dynamical invariant: the *topological entropy* of a system. This will require some new machinery and will give us a better notion of complicated dynamical behavior.

8.1 Topological Conjugacy

8.1.1 Conjugate Maps

Recall Definition 3.12 that two metric space maps $f : X \to X$ and $g : Y \to Y$ are topologically semiconjugate if there exists a surjective map $h : X \to Y$ such that

$$h \circ f = g \circ h,$$

and, if h is a homeomorphism, then h is a conjugacy and f and g are topologically conjugate. In this case, we write $f \sim_h g$.

Remark 8.2 *Recall that homeomorphisms define an equivalence relation (Definition 3.44) on the set of all topological spaces, in the sense that if two spaces are homeomorphic, they are for all intents and purposes equivalent topologically. In a similar fashion, a topological conjugacy defines an equivalence relation on the space of all dynamical systems. In a conjugacy, we can see directly that* $f^m = h^{-1} \circ g^m \circ h$, *so that orbits go to orbits via the*

homeomorphism h. Thus the orbit structures of f and g are the same. This becomes an isomorphism for dynamical systems.

We have already establish a few naive examples of conjugacies, like the coordinate changes in Remark 2.11 and in Exercises 18 and 86. And a more interesting case involved Example 7.40, where an expanding map on the ternary Cantor set is conjugate to a particular subshift on three letters. Here, we showcase a few more.

To motivate a more thoughtful look, recall the full tent map from Example 7.55, T_2 : $[0,1] \to [0,1]$, at left in Figure 176, the continuous, piecewise linear, unimodal interval map given by

$$(8.1.1) \qquad T_2(x) = \begin{cases} 2x & \text{if } 0 \le x \le \frac{1}{2}, \\ 2(1-x) & \text{if } \frac{1}{2} \le x \le 1. \end{cases}$$

In contrast, the linear expanding map E_2 on S^1 has a graph as shown on the right of Figure 176. As a map on S^1, it is certainly continuous. (Here, the point 0 is the same as 1 in both the domain and the range. Hence the map can run off the top of the graph and reappear at the bottom and still be continuous.) As a graph in the unit square, it displays much of the same information as the tent map when the peak is precisely at 1. In fact, we can think of E_2 as an interval map (with a jump discontinuity?) via

$$(8.1.2) \qquad E_2(x) = \begin{cases} 2x & \text{if } 0 \le x < \frac{1}{2}, \\ 2x-1 & \text{if } \frac{1}{2} \le x \le 1. \end{cases}$$

We will make use of this below.

Proposition 8.3 *The linear expanding map $E_2(x) = 2x \mod 1$ on S^1 is topologically semi-conjugate to the logistic map $f_4(x) = 4x(1-x)$ on $[0,1]$ (the Ulam map) via $h_1(x) = \sin^2 \pi x$. The full tent map $T_2(x) = 1 - 2\left|x - \frac{1}{2}\right|$ on $[0,1]$ is topologically conjugate to the Ulam map via the conjugacy $h_2(x) = \sin^2 \frac{\pi}{2} x$.*

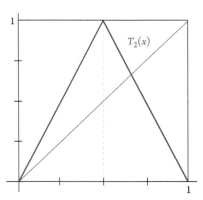
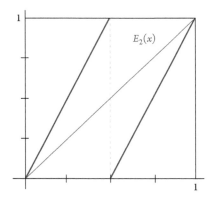

Figure 176 The full tent map $T_2 : [0, 1] \to [0, 1]$, on the left, and $E_2 : S^1 \to S^1$, on the right.

Exercise 311 Show that $h_2(x)$ is a homeomorphism, which is a diffeomorphism on $(0, 1)$, while $h_1(x)$ is surjective but cannot be a homeomorphism.

Proof Here, we will explicitly show the conjugacies. First, we show $h_1 \circ E_2 = f_4 \circ h_1$. This semiconjugacy condition needs to be parsed along the linear pieces of E_2. Hence we want

(8.1.3)
$$h_1(2x) = f_4\left(\sin^2 \pi x\right) \text{ for } 0 \le x \le \tfrac{1}{2}$$

and

(8.1.4)
$$h_1(2x - 1) = f_4\left(\sin^2 \pi x\right) \text{ for } \tfrac{1}{2} \le x \le 1.$$

For the left-hand sides of these two equations, in Equation 8.1.3, we get $h_1(2x) = \sin^2 \pi (2x) = \sin^2 2\pi x$, and in Equation 8.1.4, we have

$$h_1(2x - 1) = \sin^2 \pi (2x - 1) = \sin^2(2\pi x - \pi) = \sin^2 2\pi x,$$

since $\sin(x - \pi) = -\sin x$. On the right-hand side of each, we see

$$f_4\left(\sin^2 \pi x\right) = 4\left(\sin^2 \pi x\right)\left(1 - \sin^2 \pi x\right)$$
$$= 4\left(\sin^2 \pi x\right)\left(\cos^2 \pi x\right)$$
$$= 4\left(\tfrac{1}{2} - \tfrac{1}{2}\cos 2\pi x\right)\left(\tfrac{1}{2} + \tfrac{1}{2}\cos 2\pi x\right)$$
$$= 4\left(\tfrac{1}{4} - \tfrac{1}{4}\cos^2 2\pi x\right)$$
$$= 4\left(\tfrac{1}{4}\sin^2 2\pi x\right) = \sin^2 2\pi x.$$

For the conjugacy $h_2(x)$, we need to show that $h_2 \circ T_2 = f_4 \circ h_2$. Again, we need to parse this condition along the two linear pieces of T_2. The two resulting equations are almost identical to the previous case. In fact, Equation 8.1.3 is precisely the same with all of the factors π replaced by $\tfrac{1}{2}\pi$ (thereby replacing h_1 with h_2). And, for Equation 8.1.4, this time we get

$$h_2(2 - 2x) = \sin^2 \tfrac{1}{2}\pi(2 - 2x) = \sin^2 \pi(1 - x) = \sin^2 \pi - \pi x = \sin^2 2\pi x,$$

since $\sin(\pi - x) = \sin x$. □

Notes:

- The maps h_1 and h_2 are closely related, as are the spaces S^1 and I. Conjugacies, maps between spaces that take orbits to orbits, transfer the dynamics of one system to the other. In this case, both the tent map and the expanding circle map have a certain symmetry about them; $E_2\left(x + \tfrac{1}{2}\right) = E_2(x)$ on I, while $T_2(x) = T_2(1 - x)$. f_4 shares the latter property with T_2, and $T_2(x) = 1 - E_2(x)$ on the interval $\tfrac{1}{2} \le x \le 1$. The sine function has the appropriate property that $\sin \pi x = \sin \pi(1 - x)$. The sine function is also a beautiful way to map S^1 down onto an interval. Indeed,

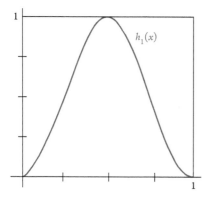

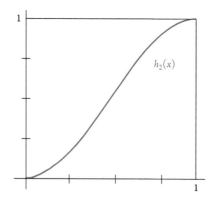

Figure 177 The maps $h_1 : S^1 \to [0, 1]$, and $h_2 : [0, 1] \to [0, 1]$.

view points of S^1 as $e^{2\pi i x}$, for $x \in I$, and the real part of $z = e^{2\pi i x} \in S^1$ is $\cos 2\pi x$. We can scale this as a "tent-like" map on I as the function

$$x \mapsto \frac{1 - \cos 2\pi x}{2} = \tfrac{1}{2} - \tfrac{1}{2} \cos 2\pi x = \sin^2 \pi x.$$

This is precisely h_1 above. For h_2, halving the angle makes h_2 1-1 on I.

- Once a (semi)conjugacy is specified, all of the interesting dynamics of the logistic map for $\lambda = 4$ are present in the tent map for $r = 2$, as well as the linear expanding map E_2 on S^1.

- The semiconjugacy between the logistic map on the unit interval and the expanding map on S^1 is about the best we can do in this situation. There can be no conjugacy between a dynamical system on the closed unit interval and one on S^1 precisely because the two spaces $[0, 1]$ and S^1 are not homeomorphic. There cannot exist a continuous map $h : [0, 1] \to S^1$ that is both onto and also injective with a continuous inverse! Refer back to Example 3.9 and Figure 42 in Section 3.1.

Example 8.4 In contrast to the previous example, the Ulam map $f_4(x) = 4x(1 - x)$ on $[0, 1]$ is not topologically conjugate to $f_\lambda(x) = \lambda x(1 - x)$, for $\lambda \in [0, 1]$, since for each choice of λ, the latter family does not have a non-trivial period-2 point, while $x_* = \frac{5 + \sqrt{5}}{8}$ satisfies $f_4(x_*) \neq x_*$, but $f_4^2(x_*) = x_*$. Check this.

Exercise 312 Show that the map $h : [0, 1] \to [-2, 2]$, $h(x) = 2 \cos \pi x$, establishes a conjugacy between the full tent map T_2 and the map $g(x) = 2 - x^2$. Here, the function $g(x)$, also quadratic, often stands as the "standard" quadratic map, serving as the simplest interval map with complicated dynamics. And its complex cousin is the map that defines the Mandelbrot set.

Exercise 313 Show that the quadratic polynomial $g_{a,b,d}(x) = ax^2 + 2bx + d$ is conjugate to $P_c(x) = x^2 + c$ by the conjugacy $h(x) = ax + b$, where $c = ad + b - b^2$.

Exercise 314 Show that any cubic polynomial is conjugate to $f_{a,b}(x) = x^3 - 3ax + b$.

Exercise 315 Use $h(x) = \frac{1}{2}a - ax$ to conjugate f_λ to $P_c(x) = x^2 + c$ for $\lambda \geq 4$. As a result, write the parameter c as a function of λ.

Remark 8.5 *Note that the result in Exercise 315, originally mentioned back in Section 7.6.1 on complex dynamics, conjugates the Ulam map $f_4 : [0,1] \to [0,1]$ to the Douady–Hubbard map (on $\mathbb{C}$) $P_{-2}(z) = z^2 - 2$, when P_{-2} is restricted to its Julia set $J(P_{-2})$, which is topologically the (real) interval $[-2,2] \subset \mathbb{R} \subset \mathbb{C}$. Further, Proposition 8.3 establishes that the Ulam map is conjugate to the full tent map T_2, and Exercise 312 above establishes that the full tent map is conjugate to the "standard" quadratic map $g(x) = 2 - x^2$ on $[-2,2]$, and $-g(x) = P_{-2}|_{J(P_{-2})}$.*

Now go back to Example 7.55, where we related the full tent map to the full shift $\mathcal{M}_2^\mathbb{N}$ by establishing a surjective map taking sequences in $\{0,1\}^\mathbb{N}$ to binary expansions of numbers (Equation 7.1.8). Now change the codomain to the interval version of the map E_2 given in Equation 8.1.2 to

$$(8.1.5) \qquad h : \{0,1\}^\mathbb{N} \to [0,1], \quad h(\mathbf{x}) = h\left(\{x_i\}_{i \in \mathbb{N}}\right) = (.x_1 x_2 x_3 \ldots)_2 = \sum_{i=1}^{\infty} \frac{x_i}{2^i}.$$

Then, as in Example 7.55, the shift map looks like the doubling map $2x$ modulo 1 on the entire unit interval:

$$E_2 \circ h(\{x_i\}) = 2 \cdot (0.x_1 x_2 x_3 \ldots)_2 \quad \text{mod } 1$$
$$= (x_1.x_2 x_3 x_4 \ldots)_2 \quad \text{mod } 1 = (0.x_2 x_3 x_4)_2, \quad \text{and}$$
$$h \circ \sigma(\{x_i\}) = h(\{x_{i+1}\}) = (0.x_2 x_3 x_4)_2.$$

Hence orbits go to orbits, and the surjective map h establishes the semiconjugacy.

Exercise 316 Show that h here is not a homeomorphism, since it fails to be injective on the dyadic rationals (see Exercise 231 in Section 6.3.1).

Exercise 317 Show that for each integer m, where $|m| > 1$, the expanding map $E_m : S^1 \to S^1$ is topologically semiconjugate to the full shift $\mathcal{M}_{|m|}^\mathbb{N}$. (Hint: Use the n-ary expansion.)

While on the subject of full shifts, we can now return to our logistic map $f_\lambda : C_\lambda \to C_\lambda$, where $\lambda > 2 + \sqrt{5} > 4$. Here C_λ is the Cantor set of orbits that remain in the unit interval, which we constructed in Section 7.1.1. One way to show that this map is, in fact, chaotic is to show that it is dynamically equivalent to a chaotic map.

Recall the construction of C_λ. Letting $C_0 = [0,1]$, we defined $C_1 = f_\lambda([0,1]) \cap [0,1] = I_0 \cup I_1$ as the two disjoint intervals remaining in $[0,1]$ after one iterate of f_λ (refer back to Figure 133 at left for a visualization). Keeping in mind that $C_\lambda \subset C_1$, we can construct the itinerary map $\iota : C_\lambda \to \mathcal{M}_2^\mathbb{N}$ as simply the account of which of the intervals of C_1 the orbit of a point $x \in C_\lambda$ visits at each iterate. Note that since there is no overlap of these intervals, we do not run into the ambiguity that occurred when creating an itinerary map for a Markov map on $[0,1]$ (from Section 7.1.6).

In fact, in the current case, the itinerary map is well-defined and injective. Indeed, suppose $x, y \in C_\lambda$ and $\iota(x) = \iota(y)$, so that x and y have the same itinerary. Recall that for $\lambda > 2 + \sqrt{5}$, f_λ is expanding (see the proof of Proposition 7.9 showing that C_λ was a Cantor set). Thus, if $x \neq y$, then $d(x,y) > 0$, and there would exist an iterate N where $d(f^N(x), f^N(y)) > 1$, so that both x and y could not be on the same side of the critical point at the Nth iterate. Hence their itinerary could not be the same, unless $x = y$. This establishes that ι is 1–1. It is also true that every point whose orbit remains in $[0,1]$ will have an itinerary. Hence the domain for ι is C_λ.

In this case, also, ι is surjective. To see this, we return to the nested interval construction we utilized to show the semiconjugacy of an expanding Markov map and a subshift in Proposition 7.58 of Section 7.1.6 on Markov partitions: Let $\mathbf{v} \in \mathcal{M}_2^\mathbb{N}$. In that construction, for any finite n-word $v = (v_0 \dots v_{n-1})$ that agrees with $\mathbf{v}$ on the first n letters, the closed interval I_v consisted of all points whose initial itinerary was v. The interval I_v was non-empty and closed, and

$$I_v = I_{v_0 \dots v_{n-1}} = I_{v_0} \cap f^{-1}\left(I_{v_1 \dots v_{n-1}}\right),$$
$$I_{v_0 \dots v_{n-1}} \subset I_{v_0 \dots v_{n-2}} \subset \cdots \subset I_{v_0}, \quad \text{and}$$
$$f\left(I_{v_0 \dots v_{n-1}}\right) = I_{v_1 \dots v_{n-1}}.$$

Since for any v, we have the set of nested, closed, non-empty, positive-length intervals, we have that

$$\bigcap_{n \geq 0} I_{v_0 \dots v_{n-1}}$$

is non-empty. And, since ι is also 1–1, this infinite intersection can only consist of one unique point. This point has image $\iota(x) = \mathbf{v}$.

So ι is 1–1 and onto (a bijection). But is ι actually continuous, and does it have a continuous inverse? For the latter point, note that the inverse function is the same κ defined in Proposition 7.58, and the establishment of continuity of κ in this context is precisely the same as that of the proof of the proposition, using $L = \ell(I_0)$, the length of the interval I_0, and

$$\beta = f_\lambda'\left(\frac{1}{2}\left(1 - \sqrt{1 - \frac{4}{\lambda}}\right)\right) > 1, \quad \lambda > 2 + \sqrt{5},$$

following Remark 7.10, just after the proof of Proposition 7.9 showing C_λ is a Cantor set. Hence κ is continuous.

Exercise 318 For $f_\lambda : C_\lambda \to C_\lambda$, with $\lambda > 2 + \sqrt{5}$, show that the itinerary map $\iota : C_\lambda \to \mathcal{M}_2^\mathbb{N}$ is continuous. (Hint: Use the metric on the symbol space given by Equation 7.1.3 and the one on C_λ it inherits as a subset of the unit interval.)

Put all of this together and we get the following:

Proposition 8.6 *The itinerary map $\iota : C_\lambda \to \mathcal{M}_2^{\mathbb{N}}$ establishes a conjugacy between $f_\lambda :$ $C_\lambda \to C_\lambda$, for $\lambda > 2 + \sqrt{5}$, and the full shift on $\mathcal{M}_2^{\mathbb{N}}$.*

Corollary 8.7 *For $\lambda > 2 + \sqrt{5}$, the quadratic map $f_\lambda : C_\lambda \to C_\lambda$ is chaotic.*

Proof f_λ under this condition on λ is transitive by Proposition 8.1, since $\iota^{-1} = \kappa$ establishes a semiconjugacy. And, since periodic points are dense in the full shift on $\mathcal{M}_2^{\mathbb{N}}$ by Proposition 7.38, the conjugacy means the same will be true for f_λ. □

Corollary 8.8 $\mathcal{M}_2^{\mathbb{N}}$ *is a Cantor set.*

Here is another beautiful family of examples:

Example 8.9 For $\alpha > 0$ a real number, let $\varphi_\alpha : \mathbb{R} \to \mathbb{R}$ be defined by

$$\varphi_\alpha(x) = \begin{cases} x^\alpha, & x \geq 0, \\ -|x|^\alpha, & x < 0. \end{cases}$$

Exercise 319 Show φ_α is a homeomorphism on $\mathbb{R}$, and $\varphi_\alpha^{-1} = \varphi_{\frac{1}{\alpha}}$.

This family φ_α, an example of which is graphed in Figure 178, has the following property: Let $g_\lambda : \mathbb{R} \to \mathbb{R}$, $g_\lambda(x) = \lambda x$ be a family of linear maps for $\lambda \in \mathbb{R}$. Then, for $\lambda_* = \varphi_\alpha(\lambda)$, we have $g_\lambda = \varphi_\alpha^{-1} \circ g_{\lambda_*} \circ \varphi_\alpha$, so that g_λ and g_{λ_*} are topologically conjugate.

Exercise 320 Verify that for $\lambda \in \mathbb{R}$, and $\alpha > 0$ real, $g_\lambda = \varphi_\alpha^{-1} \circ g_{\lambda_*} \circ \varphi_\alpha$, when $\lambda_* = \varphi_\alpha(\lambda)$.

This leads to the following: Partition $\mathbb{R}$ into the intervals

$$(-\infty, -1), \{-1\}, (-1, 0), \{0\}, (0, 1), \{1\}, (1, \infty)$$

and define an equivalence relation R on $\mathbb{R}$ with these equivalence classes. We have the following:

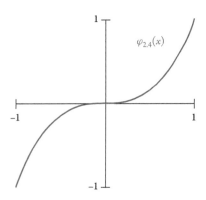

Figure 178 φ_α for $\alpha = 2.4$.

Proposition 8.10 *For the family of linear maps $g_\lambda : \mathbb{R} \to \mathbb{R}$, $g_\lambda(x) = \lambda x$, $\lambda \in \mathbb{R}$,*

$$g_\lambda \sim g_{\lambda'} \quad \textit{iff} \quad \lambda \sim_R \lambda'.$$

Exercise 321 Prove Proposition 8.10.

Some notes about conjugacy:

- The Hartman–Grobman Theorem 4.47 is a statement on the local conjugacy of two flows: In a neighborhood of a hyperbolic equilibrium solution, the flow of $\dot{\mathbf{x}} = f(\mathbf{x})$ is topologically conjugate to a linear flow. See Hartman [24] for details.

- In general, the homeomorphism h establishing the conjugacy (or the surjective map establishing the semiconjugacy) is difficult, if not impossible, to find. And even in the case where both f and g are smooth (C^∞), h need not be differentiable at all! As an example, $f(x) = 2x$ and $g(x) = 4x$ are topologically conjugate, as in Example 8.9. But the conjugacy $\varphi_2(x)$ has an inverse that is not differentiable. However, showing two maps are not conjugate may in fact be quite easy. For example, if one map has a different number of fixed or n-periodic points than another, the two maps are not topologically conjugate.

- In a real sense, and informally speaking, the notion of a topological conjugacy is the same as a coordinate change, since the maps are dynamically equal and the spaces are the same topologically.

Here are a couple more interesting examples:

Example 8.11 [Arrowsmith and Place [3]] Let $f : \mathbb{R} \to \mathbb{R}$ be a diffeomorphism, where $Df(x) > 0$ for some $x \in \mathbb{R}$. Then $f \simeq \varphi^1$, where φ^1 is the time-1 map of the flow $\varphi^t : \mathbb{R}^2 \to \mathbb{R}$ of the differential equation $\dot{x} = f(x) - x$. Indeed, f and φ^1 share many of the same properties as diffeomorphisms:

- Both are strictly increasing functions on $\mathbb{R}$ (do you see why?),
- Both have precisely the same fixed points, which correspond to the equilibria of the ODE.

Exercise 322 Show this.

- The map f can have an arbitrary finite number of fixed points (even zero), a countably infinite number, or even a continuum. However, $\text{Fix}(f)$ is always a closed subset of $\mathbb{R}$.

- If there exist gaps between successive fixed points for f, these gaps form open intervals without fixed points. There can only be a countable number of such fixed point gaps (why?)

  Suppose f is not the identity map on $\mathbb{R}$, so that at least one gap exists between fixed points. And suppose, for the sake of argument, that one of the gaps I_0 is of finite length. Then its endpoints are fixed and $I_0 = (x_0^*, x_1^*)$ (we will leave it to the reader to adapt this argument for I_0 an infinite length interval). Now index the rest of the gap intervals using a subset of $\mathbb{Z}$ compatible with the ordering on $\mathbb{R}$.

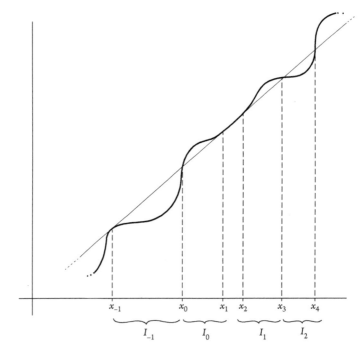

Figure 179 A diffeomorphism $f : \mathbb{R} \to \mathbb{R}$.

- f has no non-trivial periodic points, and hence each gap I_i on the index set is invariant: if $x \in I_i$, then $\mathcal{O}_x \subset I_i$.

Figure 179 is an example of this construction. Here, we explicitly construct $h : \mathbb{R} \to \mathbb{R}$ so that $h \circ \varphi^1 = f \circ h$. Choose I_i and $x_0, y_0 \in I_i$. Construct the orbits

$$(8.1.6) \qquad \mathcal{O}_{x_0,f} = \{x_n\}_{n \in \mathbb{Z}} = \{f^n(x_0)\}_{n \in \mathbb{Z}}$$

and

$$(8.1.7) \qquad \mathcal{O}_{y_0,\varphi^1} = \{y_n\}_{n \in \mathbb{Z}} = \{\varphi^n(y_0)\}_{n \in \mathbb{Z}}.$$

Then $f : [x_n, x_{n+1}] \to [x_{n+1}, x_{n+2}]$ and $\varphi^1 : [y_n, y_{n+1}] \to [y_{n+1}, y_{n+2}]$ are orientation-preserving diffeomorphisms. Construct the homeomorphism (any will do, but we will use the linear one)

$$h_0 : [y_0, y_1] \to [x_0, x_1], \quad h_0(y) = x_0 + (y - y_0)\left(\frac{f(x_0) - x_0}{\varphi^1(y_0) - y_0}\right).$$

Extend this homeomorphism to all of the closed intervals $\bar{I}_i = \left[x_0^*, x_1^*\right]$ via

$$h_n : [y_n, y_{n+1}] \to [x_n, x_{n+1}], \quad h_n(y) = f^n \circ h_0 \circ \varphi^{-n}(y).$$

Here, on the successive interval edges, we have $h_n(y_{n+1}) = h_{n+1}(y_{n+1}) = x_{n+1}$. Then $h_{\bar{I}_i} : \bar{I}_i \to \bar{I}_i$ is defined by

$$h_{\bar{I}_i}(y) = \begin{cases} x_i^*, & y = x_i^*, \\ h_n(y), & y \in [y_n, y_{n+1}], \quad n \in \mathbb{Z}, \\ x_{i+1}^*, & y = x_{x+1}.^* \end{cases}$$

Do this on each interval gap I_i, and extend to all of $\mathbb{R}$ via

$$h(y) = \begin{cases} h_{\bar{I}_i}(y), & y \in \bar{I}_i, \\ y, & \text{otherwise.} \end{cases}$$

Exercise 323 Verify that this h is indeed a homeomorphism.

It is also a conjugacy, since if $y \in \text{Fix}(f)$, then it is obvious that $h \circ \varphi^1(y) = f \circ h(y)$. But also if $y \notin \text{Fix}(f)$, then $y \in [y_n, y_{n+1}] \subset I_i$ for some $n, i \in \mathbb{Z}$. And then

(8.1.8) $\qquad h \circ \varphi^1(y) = h_{n+1} \circ \varphi^1(y) = f^{n+1} \circ h_0 \circ \varphi^{-(n+1)} \circ \varphi^1(y)$

(8.1.9) $\qquad\qquad\qquad\qquad = f\left(f^n \circ h_0 \circ \varphi^{-n}(y)\right) = f \circ h_n(y) = f \circ h(y).$

This example also provides the means to state that all strictly increasing, fixed-point-free maps on $\mathbb{R}$ are conjugate, as the next example details:

Example 8.12 Let $f, g : \mathbb{R} \to \mathbb{R}$ both satisfy $f(x) > x$, $g(x) > x$ $\forall x \in \mathbb{R}$, so that both are strictly increasing functions. Then $f \sim g$. Indeed, let $x_0 \in \mathbb{R}$ and consider $\mathcal{O}_{x_0} = \{f^i(x_0)\}_{i \in \mathbb{Z}} = \{x_i\}_{i \in \mathbb{Z}} \subset \mathbb{R}$. Here, of course, $\mathcal{O}_{x_0}$, for any x_0, is monotonically increasing and defines a partition of $\mathbb{R}$,

$$P_{x_0} = \bigcup_{i \in \mathbb{Z}} [x_i, x_{i+1}].$$

Then, for $y_0 \in (x_0, x_1)$, we have automatically $y_i = f^i(y_0) \in (x_i, x_{i+1})$, $\forall i \in \mathbb{Z}$. Hence, for all $x \in \mathbb{R}$, $\mathcal{O}_x$ has a unique iterate in $[x_0, x_1)$. f defines an equivalence relation on $\mathbb{R}$, and the set of equivalence classes can be represented by the interval $[x_0, x_1)$.

Now choose any $x_0' \in \mathbb{R}$ and, under g, construct another equivalence relation on $\mathbb{R}$ in the same manner. Let h_0 be any orientation-preserving bijection from $[x_0, x_1]$ to $[x_0', x_1']$. Then h_0 defines a unique homeomorphism $h : \mathbb{R} \to \mathbb{R}$, where $h|_{[x_0, x_1]} = h_0$, that takes $\mathcal{O}_{x_0 f}$ to $\mathcal{O}_{x_0' g}$ and indeed takes every orbit of f to a unique orbit of g.

Exercise 324 Formally construct h from h_0 and the two maps f and g, as in Example 8.11.

Now if we vary g, the new map will still be conjugate to the original f as long as the strictly increasing condition holds. To lose this condition would be to lose the 1–1 nature of g, thereby destroying the ability to create the homeomorphism necessary for the conjugacy. And if a perturbed g were to introduce a fixed point, then two maps could not be conjugate, having a different number of fixed points.

There is also a tendency in mathematics to identify maps and/or spaces that seem to do the same thing even though their descriptions are different. Think of the plane as simultaneously the set of all ordered pairs of real numbers and as the set of all 2-vectors. They are not the same, although they are isomorphic, and we often treat them as the same. This is a kind of abuse of logic that we try to get away with when we value understanding over formal accuracy. For another example, recall that S^1 has multiple geometric interpretations.

Example 8.13 Let $X = \mathbb{R}/\mathbb{Z}$ and define the rotation $R_\alpha : X \to X$ by $R_\alpha[x] = [x + \alpha]$, where again $[\cdot]$ denotes the fractional part of any real number (mod 1). Then let $Y = \{z \in \mathbb{C} \mid |z| = 1\}$, and define the map $r_\alpha(z) = z_\alpha z$, where $z_\alpha = e^{2\pi i \alpha}$. Then R_α and r_α are really just topologically conjugate. So, what is the homeomorphism that takes X to Y? It is the exponential map $h : x \mapsto e^{2\pi i x}$. Now show that it is a homeomorphism (on S^1). And what is its inverse?

8.1.2 Conjugate Flows

There is also a notion of topological conjugacy for flows, addressing the fact that orbits are mapped to orbits preserving the time parameter t:

Definition 8.14 *Two flows* $\varphi : \mathbb{R} \times X \to X$ *and* $\psi : \mathbb{R} \times Y \to Y$ *are topologically conjugate if there is a homeomorphism* $h : X \to Y$ *such that*

$$h \circ \varphi^t(x) = \psi^t(h(x)), \quad \forall t \in \mathbb{R}.$$

In essence, this is the requirement that at each time t, the flow's respective time-t maps are topologically conjugate as maps. However, to capture essential dynamical information about a flow, this requirement is arguably too restrictive. As an example, consider the following two flows:

Example 8.15 Let

(8.1.10)
$$\dot{r} = r(r-1), \quad \dot{\theta} = 1,$$

(8.1.11)
$$\dot{r} = 2r(r-1), \quad \dot{\theta} = \tfrac{1}{2}$$

be two first-order systems in the plane, written in the polar coordinate system. Each has a source at the origin and an isolated, asymptotically stable limit cycle given by the solution $r(t) \equiv 1$, with $\theta(t)$ a linear polynomial in t. (You should verify this.) But the cycle in Equation 8.1.10 has period 2π, while the one in Equation 8.1.11 has period 4π. Thus, for example, the time-2π maps cannot be conjugate, since one is the identity map on the plane and the other is not. And hence the flows are also not conjugate.

However, the essential dynamical information in the flows, such as the type, stability, and position of their equilibria, cycles, and other orbits, as well as their directions, can be captured in a slightly less stringent criterion:

Definition 8.16 *Two flows* $\varphi : \mathbb{R} \times X \to X$ *and* $\psi : \mathbb{R} \times Y \to Y$ *are topologically equivalent if there is a homeomorphism* $h : X \to Y$ *that takes orbits to orbits and preserves orientation on each orbit.*

Some notes:

- One can readily see that the two flows in Example 8.15 are topologically equivalent via the homeomorphism that is the identity map on the plane.

  Exercise 325 Explicitly solve the systems in Equations 8.1.10 and 8.1.11, and show that the flows are topologically equivalent.

- In mathematics, we say that the property of topological equivalence is "weaker" than that of topological conjugacy, in that all pairs of flows that are topologically conjugate are also topologically equivalent, but not vice versa.

8.1.3 Conjugacy as Classification

And finally, conjugacy allows us to classify entire families of dynamical systems, as long as they share the appropriate characteristics. For example, recall that the circle homeomorphisms of Section 5.5 have a quite rigid structure in that if they have periodic points at all, these are all of the same period, and are related directly to the rotation number of the map. While simply continuous maps, even homeomorphisms, can be tricky to work with, diffeomorphisms of the circle can be classified to an extent by the rotation number of the map. This is addressed in a theorem by Arnaud Denjoy [17], now called the Denjoy Classification Theorem for degree-1 circle diffeomorphisms, where the (partial) classification is based on conjugacy and only accounts for maps with irrational rotation number. It should be obvious that one cannot do the same with rational rotation number, since in these cases there would be fixed or periodic points, and maps with a different number of fixed points cannot be conjugate.

Theorem 8.17 (Denjoy Classification Theorem) *Suppose* $f : S^1 \to S^1$ *is an orientation-preserving diffeomorphism with irrational rotation number* $\rho(f)$*, where* $f'(x)$ *has bounded variation. Then* f *is topologically conjugate to* $R_{\rho(f)}$*.*

Another classification allows us to neglect the study of the degree-m expanding circle maps (of Section 7.1) that are not $E_m : S^1 \to S^1$, $E_m(\bar{x}) = m\bar{x}$ mod 1. This is because there is no need, dynamically speaking:

Theorem 8.18 *Every expanding, degree-m map of the circle is topologically conjugate to* E_m*.*

We will not prove these theorems here, since they rely on concepts and mathematical structures that are a bit advanced for this text.

Exercise 326 Show that every degree-m expanding map of S^1 is topologically semicon-
jugate to the full shift $\mathcal{M}^N_{|m|}$.

Thus, for example, the only necessary degree-2 expanding map to study dynamically is
the map $E_2 : S^1 \to S^1, E_2(x) = 2x \mod 1$.

8.2 Topological Entropy

Here we define a new dynamical invariant, called the *topological entropy* of a map. Roughly
speaking, it is the exponential growth rate of the number of orbit segments distinguishable
with arbitrary precision. Perhaps a good way to motivate this discussion is to think of this
new concept as a sort of analytical version of Lyapunov exponents.

8.2.1 Lyapunov Exponents

Measuring orbit stability, how nearby orbits behave vis-à-vis a particular chosen orbit, is a
hallmark of a standard differential equations course, at least when the orbit in question is
an equilibrium or periodic orbit. Typically, this is done via a local linearization or tangent
linear map (see Section 4.5 on local linearization), or via a Poincaré section for a periodic
orbit (see Section 4.6). In both cases, it is the eigenvalues of the map that play the pivotal
role. It would take more general tools to study a notion of the stability of an arbitrary
solution of an ODE system or the stability of an arbitrary orbit of a map. But to do so
would also recapture the idea of a sensitive dependence on initial conditions if a typical
solution to an ODE system were unstable, say. To see how this would work, let's first look
a couple of examples.

Example 8.19 Given $f : \mathbb{R} \to \mathbb{R}$, $f(x) = 2x$, and any two points $x_0 \neq y_0 \in \mathbb{R}$, their
images will be twice as far away, since $d\left(f(x_0), f(y_0)\right) = 2d(x_0, y_0)$. Hence orbits
will diverge exponentially and, after n-iterates, will be $2^n |x_0 - y_0|$ apart. Hence the
exponential growth rate of the distance between the orbits is $\ln 2$, since $2^n = e^{n \ln 2}$.
This suggests that all orbits are unstable.

Example 8.20 The same is true for $\mathbf{f} : \mathbb{R}^2 \to \mathbb{R}^2, \mathbf{f}(\mathbf{x}) = \begin{bmatrix} 2 & 0 \\ 0 & \frac{1}{3} \end{bmatrix} \mathbf{x}$, when the two chosen points $\mathbf{x}_0, \mathbf{y}_0$ have second coordinates equal. However, should the two initial points
be chosen so that the first coordinates are equal, the resulting orbits will converge
exponentially, with a growth rate of $-\ln 3$. More generally, orbit distances will grow
or decay with rates in between these extremes. Compare this with the asymptotic
growth rates of the linear map devised as a means to study the Fibonacci sequence in
Section 4.3.5 on saddle points. A good question is, what is the long-term growth rate
of orbit distance for the orbits of two points neither of whose coordinates are equal?
Could you classify all orbits as unstable in this case?

In this last example, we can conclude that the maximum growth rate of orbit distance
for a given orbit is $\ln 2$, and since this quantity is greater than 0, orbits will always have

nearby orbits that diverge. This is the essence of the sensitive dependence on initial conditions from Section 7.4, and, at least when the space is compact, a mark of a chaotic system.

In his doctoral thesis, recorded at Kharkov University in 1892 [30], Alexandr Lyapunov derived what he called a "characteristic number" associated with a time-varying function $x(t)$: λ_0 is the characteristic number of $x(t)$ if for all $\lambda > \lambda_0, x(t)e^{\lambda t} \longrightarrow \infty$ but for all $\lambda < \lambda_0, x(t)e^{\lambda t} \longrightarrow 0$.

Example 8.21 For $x(t)$ a polynomial, it should be clear that $\lambda_0 = 0$. And for $x(t) = e^{at}$, $\lambda_0 = -a$.

Exercise 327 Calculate the characteristic numbers for the three functions $x(t) = e^{t \sin t}$, $y(t)(1+t)^{-1}$, and $z(t) = e^{t^2}$.

In the context of ODE systems and given a reference trajectory $\varphi(t)$, one can study how a nearby trajectory $\mathbf{x}(t)$ tends toward or away from $\varphi(t)$ via the function $\delta_t = \delta(t) = d(\varphi(t), \mathbf{x}(t))$. Then one can recover the characteristic number if $\delta_t \approx e^{\lambda t}\delta_0$. This leads directly to a definition of what we now call the *Lyapunov exponent*:

Definition 8.22 *For $x_0 \in X \subset \mathbb{R}^n$ a reference point and y_0 a point δ_0 away from x_0, the Lyapunov exponent of x_0 in the direction of y_0 is*

$$\lambda = \lim_{t \to 0} \frac{1}{t} \lim_{\delta_0 \to 0} \log \frac{\delta_t}{\delta_0},$$

where δ_t is the separation of $\mathcal{O}_{x_0}$ from $\mathcal{O}_{y_0}$ at time $t > 0$.

Of course, the tendency of trajectories to diverge or converge is dependent on their relative position, as one can glean from Example 8.20 above. But also as in the example, it is the largest one that dominates over time, and determines orbit stability, since for almost all initial nearby points, Definition 8.22 will determine the same number.

Some notes:

- For C^1 dynamical systems, the exponents at a point are related directly to the eigenvalues of the Jacobian matrix evaluated at that point, which is a local linearization of the system. In fact, the idea behind measuring the stability of an orbit involves locally linearizing the flow along that orbit, resulting in a possibly non-autonomous linear system. With some mild extra conditions, the maximal Lyapunov exponent can fully determine the stability of the orbit.

- Calculations of Lyapunov exponents are usually done numerically and only locally. Only rarely can they be calculated analytically or over the entire space.

For a non-autonomous system $\dot{\mathbf{x}} = \mathbf{f}(t, \mathbf{x})$, Lyapunov proposed a method for studying the stability of a general trajectory via the following: Under the substitution $\mathbf{y} = \mathbf{x} - \varphi(t)$ for a given solution $\varphi(t)$, the question of measuring the stability of $\varphi(t)$ reduces to measuring the stability of the equilibrium at the origin of the new system $\dot{\mathbf{y}} = \mathbf{F}(t, \mathbf{y})$. In general, for this transformed system, the local linearization will have time-varying coefficients. Lyapunov observed that when

(8.2.1)
$$\mathbf{F}(t,\mathbf{y}) = P(t)\mathbf{y} + R(t,\mathbf{y}),$$

where $R(t,\mathbf{y})$ is the higher-order part, and the linearized system

$$\dot{\mathbf{y}} = P(t)\mathbf{y}$$

is regular (a rather technical condition, which is satisfied when $P(t)$ has either constant or periodic coefficients), the resulting Lyapunov exponents of this linearized system determine the stability of the solution $\varphi(t)$.

Proposition 8.23 *If the maximum Lyapunov exponent of a solution $\varphi(t)$ of a regular system is negative, then the solution is asymptotically stable. If the maximum Lyapunov exponent is positive, then $\varphi(t)$ is unstable.*

Notes:

- Roughly speaking, a system is *regular* if the sum of the Lyapunov exponents of the linearized system is equal to the lower bound of the exponential growth rate of the fundamental matrix determinant. Otherwise the linearized system is called irregular. Again, if $P(t)$ in Equation 8.2.1 has constant or periodic coefficients, the system is regular.

- in 1930, Oskar Perron [37] showed not only that for a regular n-dimensional system, there are at most n-distinct Lyapunov exponents, but also that this notion of regularity is necessary. Without it, Perron produced an irregular system where all Lyapunov exponents for a solution were negative and yet the solution was unstable, and also a system where all of the Lyapunov exponents were positive and the solution was asymptotically stable.

Example 8.24 Let $\dot{\mathbf{x}} = \begin{bmatrix} 2t(x+1) \\ -y^2 \end{bmatrix}$. The solution passing through the initial value $\mathbf{x}^0 = \begin{bmatrix} x(0) \\ y(0) \end{bmatrix} = \begin{bmatrix} 0 \\ 1 \end{bmatrix}$ is

$$x(t) = e^{t^2} - 1, \quad y(t) = \frac{1}{t+1}.$$

Under the substitution $z_1 = x - x(t)$ and $z_2 = y - y(t)$, we have the new system

$$\dot{\mathbf{z}} = \begin{bmatrix} 2tz_1 \\ -z_2^2 - 2\dfrac{z_2}{t+1} \end{bmatrix} = \begin{bmatrix} 2t & 0 \\ 0 & \dfrac{-2}{t+1} \end{bmatrix} \mathbf{z} + \begin{bmatrix} 0 \\ -z_2^2 \end{bmatrix}.$$

This has the linearized system

$$\dot{\mathbf{z}} = P(t)\mathbf{z} = \begin{bmatrix} 2t & 0 \\ 0 & \dfrac{-2}{t+1} \end{bmatrix} \mathbf{z}$$

with coefficient matrix neither constant nor periodic (in fact unbounded). This linearized system has only one Lyapunov exponent, and hence the sum of the exponents cannot equal the lower bound on the determinant of a fundamental matrix for this linearized system. In fact, a fundamental matrix for this linearized system is

$$Z(t) = \begin{bmatrix} e^{t^2} & 0 \\ 0 & \dfrac{1}{(t+1)^2} \end{bmatrix},$$

whose determinant $e^{t^2}/(t+1)^2$ has lower bound on $I = (-1, \infty)$ of near 0.56.

Example 8.25 Let $\dot{x} = \begin{bmatrix} 2 & 0 \\ 0 & -3 \end{bmatrix} x$. Then $x(t) = \begin{bmatrix} e^{2t} & 0 \\ 0 & e^{-3t} \end{bmatrix} x^0$ is the general solution using the matrix exponential. It is easy to see that for two orbits whose initial y-values are the same, the orbits will separate by e^{2t} over time, resulting in a growth rate of 2. This is a maximum, since

$$\delta_t = d(\mathbf{x}(t), \mathbf{y}(t)) = \sqrt{\left(x_1(t) - y_1(t)\right)^2 + \left(x_2(t) - y_2(t)\right)^2}$$

$$= \sqrt{e^{4t}\left(x_1(0) - y_1(0)\right)^2 + e^{-6t}\left(x_2(0) - y_2(0)\right)^2}$$

$$= e^{2t}\sqrt{\left(x_1(0) - y_1(0)\right)^2 + e^{-10t}\left(x_2(0) - y_2(0)\right)^2},$$

and one can see that growth rates for the separation of almost all orbits (outside of those whose first coordinates are equal, is asymptotically 2.

In the case of a C^1 map on $\mathbb{R}$, the Lyapunov exponent can be defined by the derivative, and

$$\lambda = \lim_{n \to \infty} \frac{1}{n} \ln \left|(f^n)'(x)\right| = \lim_{n \to \infty} \frac{1}{n} \ln \left|\prod_{i=0}^{n-1} f'(x_i)\right| = \lim_{n \to \infty} \frac{1}{n} \sum_{i=0}^{n-1} \ln \left|f'(x_i)\right|$$

by the Chain Rule. In certain cases, this facilitates calculation (and verifies in Example 8.19 above that the Lyapunov exponent is $\ln 2$).

Example 8.26 Given the tent map T_r, defined in Example 7.55 in Section 7.1.6 on Markov partitions, we have that T_r is differentiable except at its peak and endpoints, where the one-sided derivatives do exist, however. And, everywhere else, $\left|T_r'(x)\right| = r$. Thus, for all points $x \in [0, 1]$ that are not pre-images of $\frac{1}{2}$, we have that the Lyapunov exponent is $\ln r$. Notice from this result that as long as $r > 1$, the Lyapunov exponent is positive, indicating that nearby orbits are diverging. Notice in Figure 149 that once r increases past 1, the graph of T_r crosses the line $y = x$ at a unique point, and it is here that the complexity of the dynamics is high.

Another aspect of Lyapunov exponents is that they are independent of the coordinate system as well as the metric used to define distances. In fact, as long as the coordinate map is a diffeomorphism, any conjugacy will render two systems whose Jacobians are similar as matrices. This implies immediately the following:

Proposition 8.27 *The logistic map* $f_4 : [0,1] \to [0,1]$, $f_4(x) = 4x(1-x)$ *has Lyapunov exponent* $\ln 2$ *for almost every orbit.*

Proof The map f_4 is topologically conjugate to the full tent map T_2 via the conjugacy $h_2(x) = \sin^2 \frac{1}{2}\pi x$ from Proposition 8.3. By Exercise 311, h_2 is a diffeomorphism (on a closed set, this is defined as a homeomorphism that is a diffeomorphism on its interior). □

There are inherent problems with the calculation of Lyapunov exponents in general nonlinear systems, however. For systems deemed irregular, the Lyapunov exponents may not determine the stability of a solution. And, even if they did, their calculation is usually difficult and must be done numerically.

To construct a better measure of the tendency of orbits to diverge, and to do so wiihout including derivative information, we will return to a topological description.

8.2.2 Capacity

Let X be a metric space. A set $E \subset X$ is called *r-dense* in X if, using the metric,

$$X \subset \bigcup_{x \in E} B_r(x).$$

That is, X can be covered (recall the definition for a cover from Section 7.1.6 on Markov partitions) by a set of r-balls all of whose centers lie in E. Then, in the case that X is a compact metric space (both closed and bounded), the *r-capacity* of X, with metric d is the minimal cardinality of any r-dense set. Denote the r-capacity of a set X by $S_{X,d}(r)$ (or simply $S_d(r)$ when either the space X is understood or it is not necessary to define it explicitly).

Some notes:

- This is simply a way of denoting the "thickness" of sets that have no actual volume by how they sit inside X (think of how Cantor sets sit inside an interval).
- It does not really matter ultimately, but we will mostly consider closed balls in these calculations.
- Some examples:

  Example 8.28 $\mathbb{Z}$ is r-dense in $\mathbb{R}$ if $r > \frac{1}{2}$ if the balls are open, and $r \geq \frac{1}{2}$ is the balls are closed.

  Example 8.29 $\mathbb{Z}^2$ is r-dense in $\mathbb{R}^2$ if $r > \frac{\sqrt{2}}{2}$ if the balls are open, and $r \geq \frac{\sqrt{2}}{2}$ is the balls are closed. Can you visualize this?

Example 8.30 Let $I = [0, 1]$ be the unit interval. Using open balls here, the $\frac{1}{2}$-capacity of I is 2. The $\frac{1}{4}$-capacity is 3. The $\frac{1}{8}$-capacity is 5, and the $\frac{1}{16}$-capacity is 9. One can show that $S_d\left(\left(\frac{1}{2}\right)^n\right) = 2^{n-1} + 1$.

Exercise 328 Show this.

Exercise 329 Determine a bound on r for which $\mathbb{Z}^3$ is r-dense in $\mathbb{R}^3$.

- These calculations work well with Cantor sets. Studying how $S_d(r)$ changes as r changes (in fact, it is the order of magnitude of $S_d(r)$) leads to a generalized notion of dimension.

A rough notion of dimension for a topological space would be how many coordinates it would take to completely determine a point in the space in relation to the other points. A parameterization often provides this rough notion of dimension. For example, the common description of the 2-sphere S^2 is as the unit sphere in $\mathbb{R}^3$: the set of all unit-length vectors in $\mathbb{R}^3$. However, using spherical coordinates (ρ, θ, ϕ) on $\mathbb{R}^3$ (see the connection?), all of these points have coordinate $\rho = 1$, and hence it only requires two coordinates to differentiate between each point on the sphere. Hence, in a way, S^2 is 2-dimensional as a space. This notion is not mathematically precise, however, since there do exist curves (1-dimensional lines) that can "fill" a 2-dimensional space (Peano curves, some examples are called). Hence is this curve 1-dimensional, or 2-dimensional? Here, we will explore one mathematically precise notion of dimension (there are many), which will be useful in our definition of topological entropy.

8.2.3 Box Dimension

We start with a definition:

Definition 8.31 *A metric space X is called* totally bounded *if $\forall r > 0$, X can be covered by a finite set of r-balls all of whose centers are in X.*

Actually, this definition is technical, and is meant to account both for the general metric space aspect of this discussion and for spaces that may not be compact but for which the notion of a capacity still makes sense. For the notion of entropy to follow, we will always assume that X is compact. That the centers need to be within X is really only a factor when the metric space X is a subspace of another space Y (otherwise there is no "outside" of X). And, in Euclidean space, the notion of totally bounded is just the common notion of bounded with which you are familiar.

Definition 8.32 *For X totally bounded,*

$$bdim(X) := \lim_{r \to 0} \frac{-\log S_{(X,d)}(r)}{\log r}$$

is called the box dimension *of X.*

Notes:

- This concept is also called the *Minkowski–Bouligard dimension,* or the *entropy dimension* or the *Kolmogorov dimension.*
- This is an example of the idea of fractional dimension: some sets may look more than 0-dimensional, yet less than 1-dimensional, for example.
- In the case where this limit may not exist, certainly one can use the limit superior or the limit inferior to gain insight as to the "size" of a set.
- To calculate, find a sequence of r-sizes going to 0, and calculate the r-capacities for this sequence. If the limit exists, then any sequence of r's going to 0, with their associated r-capacities, will determine the same box dimension (why?).

Remark 8.33 *The Minkowski–Bouligard dimension is only one of a number of methods for defining the dimension of a set that (1) agrees with our standard (topological) notion of dimension for easily recognized sets (spaces), but also (2) allows us to discriminate between some of the more notoriously interesting self-similar sets like fractals. Another one that is commonly used is the Hausdorff dimension. Although differences do sometimes exist among these differing definitions of dimension for some fractals, for relatively benign self-similar sets (those that satisfy what is known as the Open Set Condition), many of the definitions agree. We will not explore this topic here, but encourage the reader toward independent study.*

Example 8.34 Calculate $\text{bdim}(I)$, for $I = [0,1]$ with the metric d that I inherits from $\mathbb{R}$. Recall that if we were to use closed balls, then the $(\frac{1}{2})^n$-capacity for I is $S_{(X,d)}\left((\frac{1}{2})^n\right) = 2^{n-1}$. But, for open balls, we have $S_{(X,d)}\left((\frac{1}{2})^n\right) = 2^{n-1} + 1$. The box dimension should be the same for both. Indeed it is: For the harder one,

$$\text{bdim}(I) = \lim_{r\to 0} \frac{-\log S_{(X,d)}(r)}{\log r} = \lim_{n\to\infty} \frac{-\log(2^{n-1}+1)}{\log((\frac{1}{2})^n)} = \lim_{n\to\infty} \frac{\log(2^{n-1}+1)}{\log 2^n}$$

$$\geq \lim_{n\to\infty} \frac{\log 2^{n-1}}{\log 2^n} = \lim_{n\to\infty} \frac{n-1}{n} = 1,$$

and

$$\text{bdim}(I) = \lim_{r\to 0} \frac{-\log S_{(X,d)}(r)}{\log r} = \lim_{n\to\infty} \frac{-\log(2^{n-1}+1)}{\log((\frac{1}{2})^n)} = \lim_{n\to\infty} \frac{\log(2^{n-1}+1)}{\log 2^n}$$

$$\leq \lim_{n\to\infty} \frac{\log 2^{n-1}\cdot n}{\log 2^n} = \lim_{n\to\infty} \frac{\log 2^{n-1}}{\log 2^n} + \lim_{n\to\infty} \frac{\log n}{\log 2^n}$$

$$= \lim_{n\to\infty} \frac{n-1}{n} + \lim_{n\to\infty} \frac{\log n}{n} = 1 + 0 = 1.$$

Hence $\text{bdim}(I) = 1$. Using the closed ball construction is even easier.

Example 8.35 Let C be the ternary Cantor set, constructed in Section 3.4. We show that $\text{bdim}(C) = \log 2/\log 3$. Here, assume that C sits inside I from the previous example,

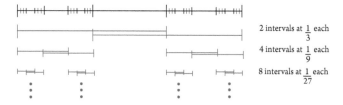

2 intervals at $\frac{1}{3}$ each

4 intervals at $\frac{1}{9}$ each

8 intervals at $\frac{1}{27}$ each

Figure 180 The first few stages of constructing covers for the ternary Cantor set C.

and again inherits its metric d from I. And, since we can choose our sequence of r's going to zero, we will choose $r = (\frac{1}{3})^n$, and consider only closed balls. Then one can show that $S_{(C,d)}((\frac{1}{3})^2) = 2^n$. Indeed, think about this: At each stage C_n of the ternary Cantor set, we have 2^n intervals, each of size $(\frac{1}{3})^n$. We can cover these intervals efficiently using balls of radius $(\frac{1}{3})^{n+1}$ by using only two on each of these intervals, centering them at $\frac{1}{3}$ and $\frac{2}{3}$ of the way across that subinterval. This gives us 2^{n+1} balls of size $(\frac{1}{3})^{n+1}$ covering the set. See Figure 180 for the first three stages.

The calculation is now easy:

$$\mathrm{bdim}\,(C) = \lim_{r \to 0} \frac{-\log S_{(C,d)}(r)}{\log r} = \lim_{n \to \infty} \frac{-\log(2^{n+1})}{\log((\frac{1}{3})^{n+1})}$$

$$= \lim_{n \to \infty} \frac{\log(2^{n+1})}{\log 3^{n+1}} = \lim_{n \to \infty} \frac{n+1}{n+1} \cdot \frac{\log 2}{\log 3} = \frac{\log 2}{\log 3}.$$

Exercise 330 By construction, calculate the r-capacity and hence the box dimension of the Cantor set formed by removing the middle half of each subinterval of the unit interval at each stage.

Exercise 331 Let $B = \{0, 1, \frac{1}{2}, \frac{1}{3}, \frac{1}{4}, \ldots, \frac{1}{n}, \ldots\}$. Calculate $\mathrm{bdim}\,(B)$.

In fact, we have the following:

Proposition 8.36 *Let $C \subset I$ be the Cantor set formed by removing the middle interval of relative length $1 - 2/\alpha$ at each stage. Then*

$$\mathrm{bdim}(C) = \frac{\log 2}{\log \alpha}.$$

Exercise 332 Prove Proposition 8.36.

A special note: All Cantor sets are homeomorphic. Yet, if we change the size of a removed interval at each stage, we effectively change the box dimension. This means that box dimension is NOT a topological invariant (it does not remain the same under topological equivalence). Since a homeomorphism here would also act as a conjugacy between two dynamical systems on Cantor sets, this means that box dimension is also not a dynamical invariant.

Exercise 333 Calculate the box dimension of the Cantor set constructed like the ternary Cantor set but by removing the middle fourth of the remaining intervals at each stage.

Exercise 334 Find the relative length of the removed interval at each stage in the construction of a Cantor set from the unit interval in which the box dimension of the Cantor set is precisely $\frac{2}{3}$.

8.2.4 Bowen–Dinaburg (Metric) Topological Entropy

For the rest of this discussion, we will assume that (X, d) is a compact metric space with a continuous map $f : X \to X$. Consider a sequence of new metrics on X indexed by $n \in \mathbb{N}$:

$$(8.2.2) \qquad d_n^f(x,y) := \max_{0 \le i \le n} d\left(f^i(x), f^i(y)\right).$$

Here, with $d_0^f = d$, the new metrics d_n^f actually measures a "distance" between orbit segments

$$\mathcal{O}_{x,n} = \left\{ x, f(x), \ldots, f^n(x) \right\}$$
$$\mathcal{O}_{y,n} = \left\{ y, f(y), \ldots, f^n(y) \right\}$$

as the farthest that these two sets diverge along the orbit segment, and assigns this distance to the pair x and y.

Exercise 335 Show that for $n \in \mathbb{N}$, d_n^f defines a metric on X.

Now, using the metric d_n^f, we can define an r-ball as the set of all neighboring points y whose nth orbit-segment $\mathcal{O}_{y,n}$ stays within a distance r of $\mathcal{O}_{x,n}$:

$$B_r(x, n, f) = \left\{ y \in X \,\middle|\, d_n^f(x,y) < r \right\}.$$

Convince yourself that as we increase n, the orbit segment is getting longer, and more and more neighbors y will have orbit segments that possibly move away from $\mathcal{O}_{x,n}$. Thus the r-ball will either remain the same size or get smaller as n increases. But, by continuity, the r-balls for any n will always be open sets in X that have x as an interior point. Also, as r goes to 0, the r-balls will also get smaller, right?

Now define the r-capacity of X, using the metric d_n^f and the new r-balls $B_r(x, n, f)$, denoted by $S_{(X,d)}(r, n, f)$. Note that this is the same notion of r-capacity as the one we used for the box dimension! We are only changing the metric on X to d_n^f. But the actual calculations of the r-capacity depend on the choice of metric. As before, as r goes to 0, the r-balls shrink, and hence the r-capacity grows. And also, as n goes to ∞, we use the different d_n^f to measure ultimately the distances between entire positive orbits. This also forces the r-balls to shrink, and hence the r-capacity to grow. What is the exponential growth rate of the r-capacity as $r \to 0$? This is the notion of topological entropy:

Definition 8.37 *Let*

$$h_d(r,f) := \overline{\lim}_{n \to \infty} \frac{\log S_d(r,n,f)}{n}.$$

Then

$$h_d(f) := \lim_{r \to 0} h_d(r,f)$$

is called the topological entropy *of the map f on X.*

There are many things to say about this. To start with:

- Topological entropy is a measure of the tendency of orbits to diverge from each other. It will always be a non-negative number, and the higher it is, the faster orbits are diverging. In Euclidean space, maybe this is not so special (think of the linear map on $\mathbb{R}^2$ given by the matrix λI_2, with $\lambda > 1$: all orbits diverge, but the dynamics is not very interesting), but in a compact space with all orbits diverging, the resulting messy nature of the dynamics can be quite interesting. Thus, topological entropy is a measure of the orbit complexity, and the higher the number, the more interesting (read messy) the dynamical structure.

- If it is important to mention the space as well as the map in the discussion of entropy, one may write $h_d(f,X)$. This can reveal an important distinction between the topological entropy of a map on a space $h_d(f,X)$ and the entropy of a map restricted to a subspace $h_d(f,Y)$, for $Y \subset X$ invariant under f. Another common notation for topological entropy is $h_{\text{top}}(f)$ or $h_T(f)$ or even $h(f)$. These are, in a sense, more accurate since it turns out that the topological entropy of a map does not actually depend on the metric d, at least up to equivalence, chosen for use in its definition. It is possible, however, that inequivalent metrics may lead to either the same or a different entropy. We will use the notation $h(f)$ in our subsequent discussion.

- Contractions and isometries have no entropy:

Proposition 8.38 *Let f be either a contraction or an isometry. Then $h(f) = 0$.*

Proof In the case of f an isometry, for any $n \in \mathbb{N}$, $d_n^f = d$, since distances between iterates of a map are the same as the original distances between the initial points. Hence the r-capacity $S_{(X,d)}(r,n,f) = S_{(X,d)}(r,f)$ does not depend on n, and hence $h(r,f) = 0$. For a contraction, the iterates of two distinct points are always closer together than the original points. Hence also here $d_n^f = d$. This leads to the same conclusion. □

Corollary 8.39 *For $R_\alpha : S^1 \to S^1$ a rotation of the circle, $h_T(R_\alpha) = 0$.*

Proof Rational or not, all rotations of S^1 are isometries. □

Exercise 336 Show that the same is true for any toral translation $R_\alpha : \mathbb{T}^n \to \mathbb{T}^n$, namely, that $h_T(R_\alpha) = 0$.

- Topological entropy measures, in a way, the exponential growth rate of the number of trajectories that are r-separable after n iterations. Suppose this number is proportional to e^{nh}. Then h would be the growth rate for a fixed r, and as $r \to 0$, this h would tend to the entropy.

- Defining the topological entropy for a flow is simply a matter of replacing $n \in \mathbb{N}$ with $t \in \mathbb{R}$ in all of the definitions for the invariant. We can relate the two in the following way: the topological entropy of a flow is equal to the topological entropy of its time-1 map (actually its time-t for any choice of t, since the flow provides the conjugacy of any t-map with any other).

- In practice, topological entropy is quite hard to calculate. However, in many cases, and in response to the last bullet point, the entropy is directly related to the largest Lyapunov exponent of the system, at least for C^1 systems.

Defined in this fashion, it would seem that entropy definitely depends on the metric used to define capacity. However, the quantity $h_d(f)$ is the same for different metrics as long as they are uniformly equivalent:

Proposition 8.40 *If d and d' are uniformly equivalent metrics on X, then for $f : X \to X$ a discrete dynamical system, $h_d(f) = h_{d'}(f)$.*

Proof Since d and d' are equivalent, there exist constants $n, M \in \mathbb{R}$ such that for all $x, y \in X$,

$$md(x,y) \le d'(x,y) \le Md(x,y).$$

But this inequality extends to the construction of d_n and d'_n from Definition 8.2.2. (We intentionally omit the superscript denoting the dependence on f here.) Then, for $B_r(x, d_n)$ an r-ball defined for $r > 0$ around $x \in X$ under the metric d_n, there will be an $\delta(r)$-ball about x in the d'_n metric where

$$B_{\delta(r)}(x, d'_n) \subset B_r(x, d_n).$$

But then the respective capacities will satisfy

$$S_{(X,d')}(\delta(r), n, f) \ge S_{(X,d)}(r, n, f)$$

for each n. This immediately implies that $h_{d'}(f) \ge h_d(f)$. But one can switch the roles of d and d' in this entire argument to generate the conclusion that $h_d(f) \ge h_{d'}(f)$. Hence $h_{d'}(f) = h_d(f)$. $\qquad\square$

Corollary 8.41 *Topological entropy is a dynamical invariant under conjugacy.*

Proof Suppose $f : X \to X$ is topologically conjugate to $g : Y \to Y$ under the homeomorphism $h : X \to Y$, where both X and Y are metric spaces. Then we can define a new metric on Y by

$$d'_Y(x,y) = d_X(h^{-1}(x), h^{-1}(y)).$$

Under this metric on Y, h is a isometry. Hence $h_{d_X}(f) = h_{d'_Y}(g)$. □

Remark 8.42 *There are different notions of equivalence in the context of metrics. One is that of topological equivalence. Two metrics are topologically equivalent if they generate the same topology: an ϵ-ball in one metric is open iff it is also open in the other metric. What we defined back in Section 3.2.2 as uniformly equivalent is also called sometimes strongly equivalent. Strongly equivalent implies topologically equivalent, but not vice versa. In the proof above, the homeomorphism already equates the topologies of the two spaces, so that if there were already a metric defined on Y, the new metric defined via the homeomorphism would be topologically equivalent to the one Y already possessed. And, in cases where X is compact (so that Y is also), the new metric on Y will be strongly equivalent (uniformly equivalent) to the original one.*

So, topologically conjugate maps have the same topological entropy. This will be very useful in calculating the entropy of certain maps on spaces. When two dynamical systems are only semiconjugate, however, we can only say something akin to what we said at the beginning of this chapter: that for a map f semiconjugate to a map g (where g is the factor), the dynamics of f are at least as complicated as those of g. Mathematically speaking, in terms of topological entropy, we can say the following:

Proposition 8.43 *For $f : X \to X$ topologically semiconjugate to $g : Y \to Y$ via the surjective factor map $h : X \to Y$, $h_T(f, X) \geq h_T(g, Y)$.*

Proof In fact, this proof is quite similar to that of Proposition 8.40. Again, under the general supposition that X (and also Y) is compact, the continuous map h is actually uniformly continuous (Definition 3.17). Thus, there will exist a $\delta = \delta(\epsilon)$ such that

$$\text{if } d_X(x,y) < \delta(\epsilon), \text{ then } d_Y(f(x), f(y)) < \epsilon.$$

But, again, this automatically means that

$$S_{(X,d_X)}(\delta(\epsilon), n, f) \geq S_{(Y,d_Y)}(\epsilon, n, g)$$

for each n. Thus $h_T(f, X) \geq h_T(g, Y)$. □

Example 8.44 Given Proposition 8.43, we can now say that the entropy of $E_2 : S^1 \to S^1$, the linear expanding map on the circle, cannot be bigger than the topological entropy of the full shift on two letters $\sigma : \mathcal{M}_2^{\mathbb{N}} \to \mathcal{M}_2^{\mathbb{N}}$, since they are semiconjugate via the surjective map in Equation 8.1.5. And, by Exercise 317, the same is true for all natural numbers, namely, $h_T(\sigma, \mathcal{M}_{|m|}^{\mathbb{N}}) \geq h^T(E_m, S^1)$, for $|m| > 1$.

Example 8.45 Proposition 8.43, paired with Proposition 8.3, also stipulates now that $h_T(f_4, [0,1]) = h_T(T_2, [0,1])$ owing to the conjugacy between the logistic map f_λ for $\lambda = 4$ and the full tent map T_2 on the unit interval, and that the topological entropy of either cannot be greater than $h_T(E_2, S^1)$, the linear expanding map of the unit circle.

Exercise 337 Given $f : X \to X$ a discrete dynamical system on a compact space X, show that if $Y \subset X$ is a closed, f-invariant subspace, then $h_T(X,f) \geq h_T(Y,f_Y)$.

And here are some other properties of topological entropy: For $f : X \to X$ a continuous map on a compact metric space, we have the following:

- Claim: $h_T(f^m) = m h_T(f)$, for $m \in \mathbb{N}$. Indeed, one can easily see that $d_n(f^m(x), f^m(y)) \leq d_{mn}(x,y)$. Thus the corresponding capacities will dictate that $h_T(f^m) \leq m h_T(f)$. However, owing to uniform continuity, for any $\epsilon > 0$, we can also find a $\delta(\epsilon)$ such that we know $d(x,y) < \delta(\epsilon)$ implies $d(f^i(x), f^i(y)) < \epsilon$ for all $i = 0, \ldots, m$. Hence, as in the above arguments, we then know that

$$S_{(X,d)}(\delta(\epsilon), n, f^m) \geq S_{(X,d)}(\epsilon, mn, f).$$

  Thus $h_T(f^m) \geq m h_T(f)$.
- Claim: If f^{-1} exists, then $h_T(f^{-1}) = h_T(f)$. We will leave this as Exercise 338.

Exercise 338 If $f : X \to X$ is invertible on a compact metric space X, then show that $h_T(f^{-1}) = h_T(f)$.

For some of the maps that we have been highlighting over the course of this text, we now compute their topological entropy:

Exercise 339 Let $T_r : [0,1] \to [0,1]$ be the tent map. For $r \in [0,1]$, show that $h_T(T_r) = 0$.

Exercise 340 Let $f_\lambda : [0,1] \to [0,1]$ be the logistic map for $\lambda \in [0,2]$. Then show that $h_T(f_\lambda) = 0$.

Proposition 8.46 *For the expanding map* $E_2 : S^1 \to S^1$, *where* $E_2(x) = 2x \mod 1$, $h(E_2) = \log |2|$.

Proof To start, note that, by definition, the r-ball around a point $x \in S^1$ under the arc-length metric (with total length of S^1 set to 1) $B_r(x,n,E_2)$ is the set of points whose distance away from x is less than r after n iterates of E_2. As the map is locally expanding by a factor of 2, distances double after each iterate. Hence we will have to get closer to x when we start iterating to remain within r as we iterate. Hence $B_r(x,n,E_2)$ will shrink in size as n increases.

Suppose that $r = \frac{1}{4}$. Choose an $x \in S^1$, and recall that

$$B_{\frac{1}{4}}(x,0,E_2) = \left\{ y \in S^1 \,\middle|\, d_0^{E_2}(x,y) = d(x,y) = |x - y| < \tfrac{1}{4} \right\}.$$

The radius of $B_r(x,n,E_2)$ is $\frac{1}{4}$ here. After one iterate, however,

$$B_{\frac{1}{4}}(x,1,E_2) = \left\{ y \in S^1 \,\middle|\, d_1^{E_2}(x,y) = \max\left\{ |x-y|, |2x - 2y| \right\} < \tfrac{1}{4} \right\}.$$

It is obvious here that the condition $d_1^{E_2}(x,y) = |2x - 2y| = 2|x - y| < \frac{1}{4}$ means that the actual distance between x and y will have to be $|x - y| < \frac{1}{4} \cdot \frac{1}{2} = \frac{1}{8}$. Hence the radius of $B_{\frac{1}{4}}(x, 1, E_2)$ is only $\frac{1}{8}$. Similarly, the radius of $B_{\frac{1}{4}}(x, 2, E_2)$ is only $\frac{1}{16}$, and in general we have that

$$\text{radius}\left(B_{\frac{1}{4}}(x, n, E_2)\right) = \frac{1}{4} \cdot \frac{1}{2^n}.$$

But, really, the initial size of r does not determine the relative sizes of the r-balls with respect to each other. Hence we can say that, for any choice of $r > 0$, we have

$$\text{radius}\left(B_r(x, n, E_2)\right) = r \cdot \frac{1}{2^n}.$$

Recall that the r-capacity, $S_{(S^1,d)}(r, n, E_2)$ is the minimum number of r-balls $B_r(x, n, E_2)$ it takes to cover S^1. Think of S^1 as being parameterized by the unit interval $[0, 1]$ with the identification of 0 and 1. Then we really only need to find out how many r-balls are required for a given iterate n to cover an interval of length 1. Call this number K_n. Hence, we solve the equation (actually it is an inequality, but since adding one more ball to each quantity will not change the limit, this is an acceptable simplification)

$$\#\left(B_r(x, n, E_2)\right) \cdot 2 \cdot \text{radius}\left(B_r(x, n, E_2)\right) = K_n \cdot 2 \cdot r \cdot \frac{1}{2^n} = 1.$$

Which is solved by $K_n = (1/r) \cdot 2^{n-1}$. This is $S_{(S^1,d)}(r, n, E_2)$.
We now calculate

$$h(E_2, r) = \varlimsup_{n \to \infty} \frac{\log S_{(S^1,d)}(r, n, E_2)}{n}$$

$$= \lim_{n \to \infty} \frac{\log(1/r) \cdot 2^{n-1}}{n}$$

$$= \lim_{n \to \infty} \left(\frac{\log(1/r)}{n} + \frac{\log 2^{n-1}}{n}\right)$$

$$= 0 + \log 2 \cdot \left(\lim_{n \to \infty} \frac{n-1}{n}\right) = \log 2.$$

Here again, the r-topological entropy does not depend on r at all, so that

$$h(E_2) = \lim_{r \to 0} h(E_2, r) = \lim_{r \to 0} \log 2 = \log 2. \qquad \square$$

Exercise 341 Adapt the argument in the proof of Proposition 8.46 to show that the topological entropy of E_m, for $|m| > 1$, is $\log |m|$.

Exercise 342 Adapt the argument in the proof of Proposition 8.46 to show that the topological entropy of the full tent map $T_2 : [0, 1] \to [0, 1]$, $T_2(x) = 1 - 2|x - \frac{1}{2}|$ is $\log 2$.

394 | Dynamical Invariants

Note: Exercise 342 now establishes, along with Proposition 8.3, that $h_T(f_4, [0,1]) = \log 2$.

Exercise 343 Now adapt the argument from Exercise 342 to show that for the tent map $T_r : [0,1] \to [0,1]$, $T_2(x) = \frac{1}{2}r\left(1 - 2\left|x - \frac{1}{2}\right|\right)$ is $\log r$, for $r \in [1,2]$.

Proposition 8.47 For $f_L : \mathbb{T}^2 \to \mathbb{T}^2$, a hyperbolic toral automorphism, where $L \in GL(2,\mathbb{Z})$ has λ and λ^{-1} as eigenvalues, with $|\lambda| > 1$, $h_T(f_L) = \log|\lambda|$.

Proof Consider the metric on $\mathbb{R}^2$ given by the following: Choose $\mathbf{v}_1$ and $\mathbf{v}_2$ unit-length eigenvectors of $\lambda_1 = \lambda$ and $\lambda_2 = \lambda^{-1}$, respectively. Then denote

$$\mathbf{x} - \mathbf{y} = a_1\mathbf{v}_1 + a_2\mathbf{v}_2$$

and define $d(\mathbf{x},\mathbf{y}) = \max\{|a_1|,|a_2|\}$. Note that this is just the *max metric* in the plane, first defined in Example 25 in Section 2.1.3 on set metrics, but using a coordinate system defined by the eigenvectors instead of the standard basis in the plane. Being a translation-invariant metric, d induces a metric on $\mathbb{T}^2$ via the map $\rho : \mathbb{R}^2 \to \mathbb{T}^2$ we defined as the vectorized exponential map in Section 5.4 on toral translations. A good way to think of this induced metric is that it is the planar metric restricted to the square torus for small distances. This new metric is also uniformly equivalent to the Euclidean metric in the plane (see Exercise 99), and both induced metrics on $\mathbb{T}^2$ will also remain equivalent.

Using this new metric d on $\mathbb{T}^2$, define

$$B_r(\mathbf{x}) = B_r(\mathbf{x},0,f_L) = \left\{\mathbf{y} \in \mathbb{T}^2 \mid d(\mathbf{x},\mathbf{y}) \leq r\right\}.$$

Here $B_r(\mathbf{x})$ is a parallelogram with side lengths $2r$ each and total area $4r^2$. Note that we will use closed balls here for simplicity. Note also that the Euclidean area of $B_r(\mathbf{x})$ will be $C4r^2$, where the constant is related to the Euclidean angle between the eigendirections.

Now consider three squares in the plane: the usual square torus, the box $\left[\frac{1}{4},\frac{3}{4}\right]^2$, and the box $\left[-\frac{1}{2},\frac{3}{2}\right]^2$, with respective Euclidean areas 1, $\frac{1}{4}$, and 4. If we tile the plane using these r-balls, and choose an r small enough, then no tile that intersects the inner box will extend outside the square torus, and no tile that extends beyond the outer box will intersect the square torus. Then $S_{(\mathbb{T}^2,d)}(r,0,f_L)$, the r-capacity of the square torus, will be strictly greater than the area of the inner box divided by the area of each r-ball, and strictly less than the area of the outer box divided by the area of each ball, so that

$$\frac{1}{4}\left(\frac{1}{C4r^2}\right) = \frac{1}{C16r^2} < S_{(\mathbb{T}^2,d)}(r,0,f_L) < 4\left(\frac{1}{C4r^2}\right) = \frac{1}{Cr^2}.$$

Now, using the orbit segment metric d_1, we can see that in the expanding direction, distances will be stretched by $|\lambda|$, and contracted in the other eigendirection. Thus, to keep both the original points and their first iterates within an r-ball, we see that $B_r(\mathbf{x},1,f_L)$ will have side lengths $2r$ and $2r|\lambda|^{-1}$, so that the total area is $4r^2|\lambda|^{-1}$. And for d_n, in general, we get

$$\frac{|\lambda|^n}{C16r^2} < S_{(\mathbb{T}^2,d)}(r,0,f_L) < \frac{|\lambda|^n}{Cr^2}.$$

Thus the relative growth rates of each of these quantities are

$$\lim_{n\to\infty} \frac{1}{n}\log\left(\frac{|\lambda|^n}{C16r^2}\right) < \lim_{n\to\infty} \frac{1}{n}\log S_{(\mathbb{T}^2,d)}(r,0,f_L) < \lim_{n\to\infty} \frac{1}{n}\log\left(\frac{|\lambda|^n}{Cr^2}\right).$$

Each of the outer quantities limits to $\log|\lambda|$, and hence we have

$$h_T(f_L, r) = \log|\lambda|,$$

so we can conclude that $h_T(f_L) = \log|\lambda|$. $\qquad\qquad$ □

Example 8.48 From Proposition 8.47, we now know that the topological entropy of Arnol'd's cat map is $h_T(f_L) = \frac{3+\sqrt{5}}{2}$, for $L = \begin{bmatrix} 2 & 1 \\ 1 & 1 \end{bmatrix}$.

Note that in both of the above cases, for expanding maps on S^1 and these hyperbolic toral automorphisms, the topological entropy of the map is equal to the maximum positive Lyapunov exponent of the system.

Going back to shift maps, it is now straightforward to create small neighborhoods about points in a subshift space, extend the notion of distance to finite orbits, and then count how many are needed to cover the space. Recall that we can create neighborhoods of points in a symbol space by collecting all sequences that match a particular sequence on its first number of letters. These cylinders provide a way for the small neighborhoods to be generated by admissible blocks of a certain size, and counting the admissible block of a certain size is tied directly to the entries of the transition matrix. We just need to be able to count correctly!

Indeed, let $(x_0 x_1 \cdots x_n)$ be an admissible n-word on the full shift on m letters $\mathcal{M}_m^{\mathbb{N}}$. Also, for the metric in $\mathcal{M}_m^{\mathbb{N}}$, given in Equation 7.1.3, choose an appropriate $\rho > m$, as stipulated in the discussion following Equation 7.1.3. Then the neighborhood is the cylinder

$$B_{1/2^n}(x) = \left\{y \in M_m^{\mathbb{N}} \mid x_i = y_i, \ i = 0,1,\ldots,n\right\}.$$

To construct r-neighborhoods in the metric d_n^σ, we first see that

$$B_r(\mathbf{x},0,\sigma) = \left\{\mathbf{y} \in \mathcal{M}_m^{\mathbb{N}} \mid d(\mathbf{x},\mathbf{y}) < r\right\} \subset \left\{\mathbf{y} \in \mathcal{M}_m^{\mathbb{N}} \mid x_i = y_i, \ i = 0,1,\ldots,k\right\}$$

for some natural number k such that $1/\rho^{k+1} \leq r \leq 1/\rho^k$.

Now, for points to stay nearby after one iterate and remain inside this r-neighborhood, they will need to be close enough so that the $(k+1)$th sequence elements also agree. Hence $B_r(\mathbf{x},1,\sigma)$, the ball of radius r in the metric d_1^σ, will have radius $r \cdot (1/\rho)$ in the original metric. And so, $B_r(\mathbf{x},n,\sigma)$ will have radius $r \cdot (1/\rho^n)$ in the original metric. So, choose $r = 1/\rho^s$ for some natural number s. Then, as $s \to \infty$, we have $r \to 0$, and

$$B_{1/\rho^s}(\mathbf{x}, n, \sigma) \text{ has radius } \frac{1}{\rho^{s+n}}.$$

So, how many of these are needed to cover $\mathcal{M}_m^{\mathbb{N}}$?

Well, for balls of radius 1, we have

$$B_1(\mathbf{x}, 0, \sigma) = \left\{ \mathbf{y} \in \mathcal{M}_m^{\mathbb{N}} \mid x_0 = y_0 \right\},$$

and we would need exactly m of these to cover $\mathcal{M}_m^{\mathbb{N}}$. And, for balls of radius $1/\rho$, we would need exactly m^2 of them, since

$$B_{1/\rho}(\mathbf{x}, 0, \sigma) = \left\{ \mathbf{y} \in \mathcal{M}_m^{\mathbb{N}} \mid x_i = y_i \ \ i = 0, 1 \right\}.$$

Hence for balls of radius $1/\rho^{s+n}$, we will need exactly m^{s+n} of them. Hence we have

$$S_{(\mathcal{M}_m^{\mathbb{N}}, d)}\left(\frac{1}{\rho^s}, n, \sigma\right) = m^{s+n}.$$

Then

$$
\begin{aligned}
h_T\left(\sigma, \frac{1}{\rho^s}\right) &= \varlimsup_{n\to\infty} \frac{\log S_{(\mathcal{M}_m^{\mathbb{N}}, d)}\left(1/\rho^s, n, \sigma\right)}{n} \\
&= \lim_{n\to\infty} \frac{\log m^{s+n}}{n} \\
&= \lim_{n\to\infty} \left(\frac{\log m^s}{n} + \frac{\log m^n}{n}\right) \\
&= \log m,
\end{aligned}
$$

a constant. Thus, again, the r-topological entropy does not depend on r at all, so that

$$h_T(\sigma) = \lim_{s\to 0} h_T(\sigma, r) = \lim_{s\to 0} \log m = \log m.$$

Notice the correspondence between the radius of the balls and the number of allowable blocks of words in $\mathcal{M}_m^{\mathbb{N}}$. In the full shift, every initial n-word is admissible and there are precisely m^n such words. This gives us a sort of physical interpretation of topological entropy: it is simply the relative growth rate of allowable n blocks. And, counting them, for a choice of n, is quite easy. It is simply the number of choices per position multiplied by the number of positions in the n-block, or m^n. Then, one can call $S_{(\mathcal{M}_m^{\mathbb{N}}, d)}(n, \sigma) = m^n$ the n-capacity of $\mathcal{M}_m^{\mathbb{N}}$, and

$$h_T(\sigma) = \lim_{n\to\infty} \frac{\log S_{(\mathcal{M}_m^{\mathbb{N}}, d)}(n, \sigma)}{n} = \lim_{n\to\infty} \frac{\log m^n}{n} = \log m.$$

We have established the following:

Proposition 8.49 $h_T\left(\sigma, \mathcal{M}_m^{\mathbb{N}}\right) = \log m.$

Two points to make here:

- Adapting the argument establishing Proposition 8.49 to the full 2-sided shift on $\mathcal{M}^{\mathbb{Z}}$ reveals the same result, that $h_T\left(\sigma, \mathcal{M}_m^{\mathbb{Z}}\right) = \log m$.

- And, with topological conjugacy, we can then establish the topological entropy of other discrete dynamical systems:

  Corollary 8.50 For $f : \Lambda \rightarrow \Lambda$ the horseshoe map (from Section 7.6.2), $h_T\left(f, \Lambda\right) = \log 2$.

  Corollary 8.51 For $f_\lambda : C_\lambda \rightarrow C_\lambda$ the logistic map, with $\lambda > 2 + \sqrt{5}$ (from Section 7.1.1), $h_T(f_\lambda) = \log 2$.

Now let $\mathcal{X}_{\mathcal{F}} \subset \mathcal{M}_m^{\mathbb{N}}$ be a finite subshift corresponding to a finite set of forbidden words, and let $T_{\mathcal{F}}$ be the transition matrix (we will assume that all forbidden words are of length 2, owing to the discussion in Section 7.1.5. Then we can understand the idea of topological entropy in precisely the same fashion, as the relative growth rate of allowable n-blocks. The problem is properly counting n-blocks as a function of n. But the transition matrix contains enough information to do precisely that. First, some linear algebra:

Definition 8.52 *Given a square matrix $A_{m \times m}$, the* grand sum *of A is the sum of all of the entries of A, denoted by*

$$gs(A) = \sum_{i,j=1}^{m} a_{ij}.$$

Also, denote by $\mathbf{1}_m$ the m-vector all of whose entries are 1, which we will refer to as the *one-vector*; it is simply the sum of all of the basis vectors in the standard basis of $\mathbb{R}^m$. One can also say that the 1-norm for non-negative vectors of $\mathbb{R}^m$, which is really just the sum of the magnitudes of the elements of the vector, can be expressed using the one-vector. Let $\mathbf{v} \in \mathbb{R}^m$ have all non-negative elements. Then

$$||\mathbf{v}||_1 = \sum_i |v_i| = \sum_i v_i = \mathbf{v} \cdot \mathbf{1}_m.$$

We will find this useful presently.

Exercise 344 Show that for $A_{m \times m}$, $gs(A) = \mathbf{1}_m^T \cdot A \cdot \mathbf{1}_m$.

Exercise 345 Let $A_{m \times m}$ have all of its entries be 1. Then show that $gs(A^n) = m^{n+1}$, $n \in \mathbb{N}$.

Then, with these straightforward linear algebra (indeed combinatorial) exercises, we can use this new method of counting to establish the capacity and hence the topological entropy on the full shift m letters. The capacity is

$$S_{(\mathcal{M}_{m,d}^{\mathbb{N}})}(n,\sigma) = m^{n+1}.$$

Thus

$$h_T\left(\sigma, \mathcal{M}_m^{\mathbb{N}}\right) = \lim_{n\to\infty}\frac{\log m^{n+1}}{n} = (\log m)\lim_{n\to\infty}\frac{n+1}{n} = \log m.$$

And for a subshift? Well, if the transition matrix is transitive, then we already know that the dynamics are complicated by Propositions 7.45, 7.74, and 7.75. So entropy should reflect the complexity of the dynamics. But a transitive transition matrix also means that, eventually, all (beginning with a certain letter and ending with a certain other letter) sufficiently large n-blocks are admissible. But counting these, in effect finding $gs(T_F^n)$ as a function of n, is not straightforward. To do this, we will need some more linear algebra:

Definition 8.53 *A square matrix $A_{m\times m}$ is called* irreducible *if for every $i,j = 1,\ldots,m$, there is a natural number N such that $(A^N)_{ij} \neq 0$.*

Some notes:

- The matrix $A = \begin{bmatrix} 1 & 0 \\ 1 & 1 \end{bmatrix}$ fails to be irreducible, since $A^n = \begin{bmatrix} 1 & 0 \\ n & 1 \end{bmatrix}$. In this case, A is called *reducible*.

- All transitive matrices are irreducible.

- The matrix $A = \begin{bmatrix} 0 & 1 \\ 1 & 0 \end{bmatrix}$ is irreducible but not transitive. Indeed,

$$A^2 = \begin{bmatrix} 0 & 1 \\ 1 & 0 \end{bmatrix}\begin{bmatrix} 0 & 1 \\ 1 & 0 \end{bmatrix} = \begin{bmatrix} 1 & 0 \\ 0 & 1 \end{bmatrix}, \quad A^3 = \begin{bmatrix} 0 & 1 \\ 1 & 0 \end{bmatrix}, \ldots$$

To utilize the transition matrix to "count" correctly, we now present a rather famous result attributed to Oskar Perron [38] (for transitive matrices) and Georg Frobenius [20] (for more general irreducible matrices), called the Perron–Frobenius Theorem:

Theorem 8.54 (Perron–Frobenius Theorem) *For $A_{m\times m}$ an irreducible non-negative matrix matrix with $\rho(A) = r$, we have the following:*

(i) *there is a simple, real, largest (in modulus) eigenvalue $r > 0$, called the Perron–Frobenius eigenvalue;*

(ii) *the eigenspace corresponding to the Perron–Frobenius eigenvalue is generated by a vector of all positive entries; and*

(iii) *the value r satisfies*

$$\min_i \sum_j a_{ij} \leq r \leq \max_i \sum_j a_{ij}.$$

The proof of this theorem, at least for positive matrices, is rather straightforward. However, we will omit it here for expediency. But, in the context of shift maps, it facilitates our counting mechanism.

Proposition 8.55 Let $\mathcal{X}_{\mathcal{F}}$ be a finite subshift on m letters, with transitive transition matrix $T_{\mathcal{F}}$. Then $h_t(\sigma, \mathcal{X}_{\mathcal{F}}) = \rho(T_{\mathcal{F}})$.

Proof Let $\mathcal{X}_{\mathcal{F}}$ be a finite subshift as above, noting that the transitive $T_{\mathcal{F}}$ is irreducible and non-negative. Then the Perron–Frobenius eigenvalue $\lambda = \rho(T_{\mathcal{F}})$, and we cannot, by the third assertion of the Perron–Frobenius Theorem 8.54, have $\lambda \leq 1$. Also, denote by β the strictly positive eigenvector of λ. Then λ^n is also an eigenvalue of $T_{\mathcal{F}}^n$, with eigenvector β.

If we denote $c = \min_i \beta_i$ and $C = \max_i \beta_i$, then it is obvious that, for $i = 1, \ldots, m$,

$$\frac{\beta_i}{C} \leq 1 \quad \text{and} \quad \frac{\beta_i}{c} \geq 1.$$

And, since A is non-negative,

$$\frac{\beta_i}{C}\lambda^n = \frac{(\beta \cdot A^n)_j}{C} \leq \left(\mathbf{1}_m^T A^n\right)_j \leq \frac{(\beta \cdot A^n)_j}{c} = \frac{\beta_j}{c}\lambda^n.$$

Dot each of these with the one-vector (effectively adding the vector components), and we get

$$\frac{\lambda^n}{C}\beta \cdot \mathbf{1}_m \leq \mathbf{1}_m^T A^n \mathbf{1}_m \leq \frac{\lambda^n}{c}\beta \cdot \mathbf{1}_m,$$

$$\frac{\lambda^n}{C}\|\beta\|_1 \leq \mathrm{gs}(A^n) \leq \frac{\lambda^n}{c}\|\beta\|_1.$$

Then the relative growth rate of the grand sum of $T_{\mathcal{F}}^n$ with respect to n can be found from this inequality via squeezing:

$$\lim_{n\to\infty} \frac{\log\left(\frac{\lambda^n}{C}\|\beta\|_1\right)}{n} \leq \lim_{n\to\infty} \frac{\log \mathrm{gs}(A^n)}{n} \leq \lim_{n\to\infty} \frac{\log\left(\frac{\lambda^n}{c}\|\beta\|_1\right)}{n},$$

$$\lim_{n\to\infty}\left(\frac{\log\left(\frac{\|\beta\|_1}{C}\right)}{n} + \frac{\log \lambda^n}{n}\right) \leq h_T(\sigma, \mathcal{X}_{\mathcal{F}}) \leq \lim_{n\to\infty}\left(\frac{\log\left(\frac{\|\beta\|_1}{c}\right)}{n} + \frac{\log \lambda^n}{n}\right).$$

Hence $h_T(\sigma, \mathcal{X}_{\mathcal{F}}) = \log \lambda$. □

We will end this text with a couple of notes: First, there is so much more to notions like topological entropy than we can possibly do justice to here, partly because of the insufficient development of the machinery necessary for a full exploration of the concept. And second, we have sidestepped a vital concept central to much of topological dynamics: the notion of a measure, a topological generalization of the idea of geometric volume. But we will end with a theorem by Rufus Bowen [11] that presents a powerful way to characterize the entropy of a large class of dynamical systems by simply understanding the few orbits we can more or less identify easily. We offer this theorem without proof, but hope that the interested reader will continue to explore this area by using a theorem like this as a platform for support.

Theorem 8.56 *If* $f : X \to X$ *is a map on a compact metric space, then* $h_T(f) = h_T(f_{\mathcal{NW}(f)})$. *That is, the topological entropy of* f *on* X *is equal to the topological entropy of* f *restricted to the non-wandering set of* f.

Thus, for example, if the non-wandering set of f on a compact space is finite, then it consists of periodic orbits, and thus $h_T(f) = 0$.

Bibliography

[1] A. A. ANDRONOV, E. A. LEONTOVICH, AND A. G. MAÄ-ER, *Theory of Bifurcations of Dynamic Systems on a Plane* (translated from the Russian), Halsted Press, New York–Toronto, 1971.

[2] V. ARNOL'D AND A. AVEZ, *Ergodic Problems in Classical Mechanics*, Benjamin, New York, 1968.

[3] D. K. ARROWSMITH AND C. M. PLACE, *An Introduction to Dynamical Systems*, Cambridge University Press, Cambridge, 1990.

[4] Z. ARTSTEIN, *Sensitivity estimates via Lyapunov functions and Lyapunov metrics*, Contemporary Mathematics, 568 (2012), pp. 1–17.

[5] S. BANACH, *Sur les opérations dans les ensembles abstraits et leur application aux équations intégrales*, Fundamenta Mathematicae, 3 (1922), pp. 133–81.

[6] I. BENDIXSON, *Sur les courbes définies par les équations différentielles*, Acta Mathematica, 24 (1901), pp. 1–88.

[7] K. BERG, *On the conjugacy problem for k-systems*, PhD Thesis, University of Minnesota (1967).

[8] C. BESSAGA, *On the converse of the Banach fixed-point principle*, Colloquium Mathematicum, 7 (1958), pp. 41–3.

[9] G. D. BIRKHOFF, *Proof of Poincaré's last geometric theorem*, Transactions of the AMS, 14 (1913), pp. 14–22.

[10] G. D. BIRKHOFF, *On the periodic motions of dynamical systems*, Acta Mathematica, 50 (1927), pp. 359–79.

[11] R. BOWEN, *Topological entropy and axiom A*, Proceedings of Symposia on Pure Mathematics, (1970), pp. 23–41.

[12] L. E. J. BROUWER, *Über Abbildungen von Mannigfaltigkeiten*, Mathematische Annalen, 71 (1910), pp. 97–113.

[13] E. M. BRUINS, ed., *Codex Constantinopolitanus: Palatii Veteris no. 1*, vol. II, Janus; revue internationale de l'histoire des sciences, de la médecine, de la pharmacie et de la technique, 1964.

[14] G. CANTOR, *Über unendliche, lineare Punktmannigfaltigkeiten V*, Mathematische Annalen, 21 (1883), pp. 545–91.

[15] C. CARATHÉODORY, *Über den Wiederkehrsatz von Poincaré*, Sitzungsberichte der Preussischen Akademie der Wissenschafte, 34 (1919), pp. 580–4.

[16] J. L. DE LAGRANGE, *Sur la construction des cartes géographiques*, Nouveaux mémoires de l'Académie royale des sciences et belles-lettres de Berlin, 4 (1779), pp. 637–92.

[17] A. DENJOY, *Sur les courbes définies par les équations différentielles á la surface du tore*, Journal de mathématiques pures et appliquées 9e série, 11 (1932), pp. 333–75.

[18] R. L. DEVANEY, *An Introduction to Chaotic Dynamical Systems*, 2nd ed., Addison-Wesley, Reading, Ma, 2003.

[19] F. Dyson and H. Falk, *Period of a discrete cat mapping*, The American Mathematical Monthly, 99 (1992), pp. 603–14.

[20] G. Frobenius, *Üeber matrizen aus nicht negativen elementen*, Sitzungsberichte der Königlich Preussischen Akademie der Wissenschaften, 26 (1912), pp. 456–77.

[21] R. Gilmore and M. Lefranc, *The Topology of Chaos*, 2nd ed. Wiley-VCH, Weinheim, 2012.

[22] P. Grassberger, *On the Hausdorff dimension of fractal attractors*, Journal of Statistical Physics, 26 (1981), pp. 173–9.

[23] V. Gutenmacher and N. Vasilyev, *Lines and Curves*, Birkhäuser, Boston, 2004.

[24] P. Hartman, *A lemma in the theory of structural stability of differential equations*, Proceedings of the AMS, 11 (1960), pp. 610–20.

[25] P. Hartman, *Ordinary Differential Equations*, John Wiley & Sons, New York, 1964.

[26] B. Hasselblatt and A. Katok, *A First Course in Dynamics*, Cambridge University Press, Cambridge, 2003.

[27] E. Hopf, *Ergodentheorie*, Springer-Verlag, Berlin, 1937.

[28] P.-S. Laplace, *Essai philosophique sur les probabilités*, Mme. Ve. Courcier, Paris, 1814.

[29] T.-Y. Li and J. A. Yorke, *Period three implies chaos*, The American Mathematical Monthly, 82 (1975), pp. 985–92.

[30] A. M. Lyapunov, *The general problem of stability of motion* (in Russian), Doctoral dissertation, University of Kharkov, Kharkov Mathematical Society, (1892).

[31] B. B. Mandelbrot, *Fractals: Form, Chance, and Dimension*, Freeman, San Francisco, 1977.

[32] B. B. Mandelbrot, *The Fractal Geometry of Nature*, Freeman, San Francisco, 1982.

[33] B. B. Mandelbrot, *24/7 Lecture on Fractals*, Ig Nobel Awards. Improbable Research, 2006.

[34] R. M. May, *Simple mathematical models with very complicated dynamics*, Nature, 261 (1976), pp. 459–67.

[35] K. Menger, *Allgemeine räume und cartesische räume, I*, Proceedings of the Koninklijke Akademie van Wetenschappen, Amsterdam, 29 (1926), pp. 476–82.

[36] O. B. Isaeva, S. P. Kuznetsov, I. R. Sataev, D. V. Savin, and E. P. Seleznev, *Hyperbolic chaos and other phenomena of complex dynamics depending on parameters in a nonautonomous system of two alternately activated oscillators*, International Journal of Bifurcation and Chaos, 25 (2015), pp. 1530033-1–15.

[37] O. Perron, *Die Stabilitäutsfrage bei Differentialgleichungen*, Mathematische Zeitschrift, 32 (1885), pp. 703–28.

[38] O. Perron, *Zur Theorie der Matrices*, Mathematische Annalen, 64 (1907), pp. 248–63.

[39] C. E. Picard, *Sur l'application des méthodes d'approximations successives à l'etude de certains équations différentielles ordinaires*, Journal de mathématiques pures et appliquées, (1893), p. 217.

[40] G. Pick, *Geometrisches zur Zahlenlehre*, Sitzungsberichte des deutschen naturwissenschaftlich-medicinischen Vereines für Böhmen "Lotos" in Prag, 19 (1899), pp. 311–19.

[41] H. Poincaré, *Sur l'equilibre d'une masse fluide animée d'un mouvement de rotation*, Acta Mathematica, 7 (1885), pp. 259–380.

[42] H. Poincaré, *Sur les courbes définies par les équations différentielles*, Journal de mathématiques pures et appliquées, 1 (1885), pp. 167–244.

[43] H. Poincaré, *Sur le probléme des trois corps et les équations de la dynamique*, Acta Mathematica, 13 (1890), pp. 1–270.

[44] H. Poincaré, *Les méthodes nouvelles de la mécanique céleste, tomes I-III*, Gauthiers-Villars, Paris, 1899.

[45] H. POINCARÉ, *Sur un théorème en géométrie*, Rendiconti del Circolo Matematico di Palermo, 33 (1912), pp. 375–407.

[46] I. PRIGOGINE AND R. LEFEVER, *Symmetry breaking instabilities in dissipative systems. II*, Journal of Chemical Physics, 48 (1968), pp. 1695–700.

[47] H. RICARDO, *A Modern Introduction to Differential Equations*, Houghton Mifflin, Boston, 2003.

[48] S. ROBERTS, *Historical note on Dr. Graves's theorem on conical sections*, Proceedings of the London Mathematical Society, s1-12 (1880), pp. 120–2.

[49] E. RYKKEN, *Markov partitions for hyperbolic toral automorphisms of T^2*, Rocky Mountain Journal of Mathematics, 28 (1998), pp. 1103–24.

[50] O. M. SHARKOVSKII, *Co-existence of cycles of a continuous mapping of the line into itself*, Ukrainian Mathematics Journal, 16 (1964), pp. 61–71.

[51] W. SIERPIŃSKI, *Sur une courbe cantorienne qui contient une image biunivoque et continue de toute courbe donnée*, Comptes rendus hebdomadaires des séances de l'Académie des sciences, Paris, 162 (1916), pp. 629–32.

[52] D. SINGER, *Stable orbits and bifurcation of maps of the interval*, SIAM Journal on Applied Mathematics, 35 (1978), pp. 260–7.

[53] A. N. TYCHONOFF, *Über die topologische Erweiterung von Räumen*, Mathematische Annalen, 102 (1930), pp. 544–61.

[54] D. VAN DANTZIG, *Ueber topologisch homogene Kontinua*, Fundamenta Mathematicae, 15 (1930), pp. 102–25.

[55] B. VAN DER POL, *A theory of the amplitude of free and forced triode vibrations*, Radio Review (later Wireless World), 1 (1920), pp. 701–10.

[56] P.-F. VERHULST, *Notice sur la loi que la population suit dans son accroissement*, Correspondance mathématique et physique, 10 (1838), pp. 113–21.

[57] L. VIETORIS, *Über den höheren Zusammenhang kompakter Räume und eine Klasse von zusammenhangstreuen Abbildungen*, Mathematische Annalen, 97 (1927), pp. 454–72.

[58] H. VON KOCH, *Sur une courbe continue sans tangente, obtenue par une construction géométrique élémentaire*, Arkiv för Matematik, Astronomi och Physik, 1 (1904), pp. 681–704.

[59] B. WEISS, *Subshifts of finite type and sofic systems*, Monatshefte für Mathematik, 77 (1973), pp. 462–74.

[60] H. WEYL, *Über die Gleichverteilung von Zahlen mod. Eins*, Mathematische Annalen, 77 (1916), pp. 313–52.

Index